Two Manuals Combined into One Book

Mine Countermine Operations

And

Military Police Operations

Department of Defense

Pentagon
Publishing

*FM 20-32

Field Manual
No. 20-32

Headquarters
Department of the Army
Washington, DC, 29 May 1998

MINE/COUNTERMINE OPERATIONS

Table of Contents

Page

LIST OF ILLUSTRATIONS x
Figures x
Tables xv

PREFACE xvii

CHAPTER 1. INTRODUCTION 1-1
MECHANICS OF MINES 1-1
Characteristics and Functioning 1-1
Components and Initiating Actions 1-2
ANTITANK MINES 1-4
Types of Kills 1-4
Types of Sensing 1-5
Types of Warheads 1-5
ANTIPERSONNEL MINES 1-5
Types of Kills 1-5
Types of Sensing 1-6
Types of Effects 1-6
ANTIHANDLING DEVICES 1-6

Part One. Mine Operations

CHAPTER 2. MINE-WARFARE PRINCIPLES 2-1
MINE-WARFARE CONCEPTS 2-1
TYPES OF MINEFIELDS 2-1
Protective Minefields 2-2
Tactical Minefields 2-3
Nuisance Minefields 2-4
Phony Minefields 2-4
PROTECTIVE VERSUS TACTICAL MINEFIELDS 2-5
TACTICAL MINEFIELDS 2-5
Minefield Variables 2-7
Design 2-10

DISTRIBUTION RESTRICTION: Approved for public release; distribution is unlimited.

*This manual supersedes FM 20-32, 30 September 1992.

Page

TACTICAL-OBSTACLE INTEGRATION PRINCIPLES ... 2-14
Obstacle Emplacement Authority ... 2-14
Obstacle Control ... 2-14
Obstacle Control Measures ... 2-15
Fratricide Prevention ... 2-19
Maneuver-Plan Support ... 2-19
SITING AND EMPLACING TACTICAL MINEFIELDS ... 2-32
Coordinating with the Maneuver Commander ... 2-32
Siting the Minefield ... 2-37
Emplacing Minefields ... 2-39
Determining Resource Requirements ... 2-39
MINEFIELD SUPPLY OPERATIONS ... 2-39
Resupply Nodes ... 2-41
Resupply Rules ... 2-43
Supply Location ... 2-44
Resupply Methods ... 2-44
MINEFIELD MARKING ... 2-49
Criteria ... 2-49
Perimeter ... 2-50
Techniques ... 2-50
MINEFIELD TURNOVER ... 2-52
MINEFIELD INSPECTION AND MAINTENANCE ... 2-55

CHAPTER 3. SCATTERABLE MINES AND MINE DELIVERY SYSTEMS ... 3-1
GENERAL CHARACTERISTICS ... 3-1
Antipersonnel Mines ... 3-1
Antitank Mines ... 3-3
CAPABILITIES ... 3-5
Faster Response ... 3-5
Remote Placement ... 3-5
Increased Tactical Flexibility ... 3-5
Efficiency ... 3-5
Increased Lethality ... 3-5
LIMITATIONS ... 3-5
Extensive Coordination ... 3-5
Proliferation of Targets ... 3-6
Visibility ... 3-6
Accuracy ... 3-6
Orientation ... 3-6
LIFE CYCLE ... 3-6
LETHALITY AND DENSITY ... 3-7
Lethality and Tactical-Obstacle Effect ... 3-7
Density ... 3-8
COMMAND AND CONTROL ... 3-9
AUTHORITY ... 3-9
COORDINATION ... 3-10
EMPLOYMENT AND EMPLACEMENT ... 3-10
Area-Denial Artillery Munitions and Remote Antiarmor Mines ... 3-11
Gator ... 3-14
Volcano ... 3-17
Modular Pack Mine System ... 3-27

Page

MARKING ... 3-26
Safety Zones ... 3-27
Fragment Hazard Zones ... 3-27
Fencing ... 3-28

CHAPTER 4. SPECIAL-PURPOSE MUNITIONS ... 4-1
M18A1 CLAYMORE ... 4-2
SELECTABLE LIGHTWEIGHT ATTACK MUNITION ... 4-3
Operating Modes ... 4-3
Antitamper Feature ... 4-6
M93 HORNET ... 4-6
Employment Considerations ... 4-7
Employment Roles ... 4-7
Tactical Emplacement ... 4-8
Recording and Marking ... 4-15

CHAPTER 5. CONVENTIONAL MINES ... 5-1
ANTITANK MINES ... 5-1
M15 ... 5-1
M19 ... 5-2
M21 ... 5-2
ANTIPERSONNEL MINES ... 5-3
M14 ... 5-3
M16 ... 5-4
EMPLACING MINES ... 5-4
Mines With Prongs ... 5-4
Mines With Pressure Plates ... 5-4
Mines With Tilt Rods ... 5-6
Bearing Boards ... 5-6
Concealment ... 5-6
Maneuver Assistance ... 5-8

CHAPTER 6. ROW MINING ... 6-1
USE ... 6-1
RULES ... 6-1
LOGISTICS ... 6-3
Calculations ... 6-3
Task Organization ... 6-14
Site Layout ... 6-16
Mine-Laying Vehicles ... 6-18
Laying a Row Minefield ... 6-18
Immediate-Action Drill ... 6-24
Squad Drill ... 6-24
Marking, Recording, and Reporting Row Minefields ... 6-25
STANDARDIZED TACTICAL ROW MINEFIELDS ... 6-25
Disrupt and Fix ... 6-28
Turn ... 6-29
Block ... 6-31
HASTY PROTECTIVE ROW MINEFIELDS ... 6-33
Rules ... 6-34
Site Layout ... 6-34

Page

CHAPTER 7. STANDARD-PATTERN MINEFIELDS **7-1**
COMPONENTS 7-1
Mine Strips 7-1
Mine Clusters 7-1
Rules for Positioning Clusters Within a Strip 7-2
Standard-Pattern Minefield Rules 7-4
LOGISTICAL CALCULATIONS 7-9
Cluster Calculation 7-9
Platoon Organization 7-10
Mine-Emplacement Procedures 7-11
Mine Emplacement 7-13
NUISANCE MINEFIELDS 7-17
Siting 7-17
Location 7-17
Laying 7-18
Inspection and Maintenance 7-18
Handover 7-19

CHAPTER 8. REPORTING AND RECORDING **8-1**
MINEFIELD/MUNITION FIELD REPORTS 8-1
Report of Intention 8-1
Report of Initiation 8-1
Report of Completion 8-2
Report of Transfer 8-2
Report of Change 8-3
Progress Reports 8-3
MINEFIELD/MUNITION FIELD RECORDS 8-3
Minefield Record 8-4
Hasty Protective Row Minefield Record 8-17
Nuisance Minefield 8-20
SCATTERABLE MINEFIELD/MUNITION FIELD REPORTING AND RECORDING 8-20
MINEFIELD/MUNITION FIELD OVERLAY SYMBOLS 8-25

Part Two. Counteroperations

CHAPTER 9. COUNTERMINE OPERATIONS **9-1**
DEFINITIONS 9-1
Obstacle 9-1
Reduction 9-1
Breaching 9-1
Area Clearance 9-1
Route Clearance 9-1
Mine Neutralization 9-1
Proofing 9-2
Demining 9-2
BREACHING OPERATIONS 9-2
Intelligence 9-2
Fundamentals 9-4
Organization 9-4
Mass 9-5
Synchronization 9-5

Page

CLEARING OPERATIONS ... 9-6
Upgrade of Breach Lanes ... 9-6
Area Clearance ... 9-7
Demining ... 9-7

CHAPTER 10. MINEFIELD REDUCTION ... 10-1
DETECTING ... 10-1
Visual ... 10-1
Physical ... 10-2
Electronic ... 10-3
Mechanical ... 10-6
REPORTING ... 10-7
REDUCING ... 10-7
Explosive ... 10-7
Mechanical ... 10-14
Electronic ... 10-22
Manual ... 10-22
PROOFING ... 10-24
MARKING ... 10-24
Lane-Marking Terms ... 10-25
Levels of Lane Marking and Patterns ... 10-27
Commander's Guidance for Lane Marking ... 10-33
Lane-Marking Devices ... 10-34
Marking Requirements of the North Atlantic Treaty Organization ... 10-36

CHAPTER 11. ROUTE AND AREA CLEARANCE ... 11-1
ROUTE CLEARANCE ... 11-1
Planning ... 11-1
Planning Considerations ... 11-3
Task Organization ... 11-7
Methods and Types ... 11-11
AREA CLEARANCE ... 11-15
Planning ... 11-16
Planning Considerations ... 11-17
Task Organization ... 11-18
Methods and Types ... 11-18
IMPROVISED MINE THREAT ... 11-19
MINE LOCATIONS ... 11-20
DISPOSITION OF MINES ... 11-20
Mine-Removal Techniques ... 11-21
Hand Neutralization ... 11-21
SAFETY ... 11-22
REPORTS ... 11-22
Situation Report ... 11-23
Progress Report ... 11-23
Mine Incident Report ... 11-26

Page

Part Three. Special Mining Operations

CHAPTER 12. MINING OPERATIONS IN SPECIAL ENVIRONMENTS 12-1
STREAMBED AND RIVER MINING 12-1
Employment 12-1
Emplacement 12-2
Recovery 12-3
Recording 12-3
Safety 12-3
URBAN-TERRAIN MINING 12-6
Antipersonnel Mines 12-7
Conventional Antitank Mines 12-12
Scatterable Mines 12-13
Deception Measures 12-15
SPECIAL ENVIRONMENTS 12-16
Cold Regions 12-16
Jungles 12-17
Deserts 12-17

CHAPTER 13. BOOBY TRAPS AND EXPEDIENT DEVICES 13-1
Section I. Setting Booby Traps 13-1
TACTICS 13-2
SITING 13-4
TYPES OF TRAPS 13-4
COMPONENTS AND PRINCIPLES 13-5
ACTUATION METHODS 13-5
METHODS OF CONNECTION 13-5
Remote 13-5
Direct 13-8
PLANNING, SETTING, AND RECORDING 13-8
Timeliness 13-8
Orders and Briefing 13-8
Rehearsal 13-9
Organization and Procedure 13-9
Reporting and Recording 13-10
SITES 13-14
SAFETY 13-14
Section II. Clearing Booby Traps 13-14
INDICATIONS 13-15
DETECTION 13-15
CLEARING METHODS 13-15
COMBAT CLEARANCE 13-16
CLEARANCE IN SECURE AREAS 13-17
Policy and Planning 13-17
Control Point 13-18
Control and Size of Parties 13-18
Marking 13-18
Clearing of Open Areas 13-18
Clearing of Buildings 13-19
Exterior Reconnaissance and Entry 13-19
Search Techniques 13-19

Page

Clearing Installations and Facilities 13-21
Clearing Obstacles 13-21
Clearing Secure Areas 13-21
CLEARANCE METHODS 13-22
IMPROVISED TRAPS 13-23
NONEXPLOSIVE TRAPS 13-23
Punji 13-23
Closing Trap 13-23
Spike Board 13-28
Venus Flytrap 13-28
Section III. Expedient Devices 13-29
AUTHORIZATION 13-29
EMPLOYMENT AND CONSTRUCTION TECHNIQUES 13-29
High-Explosive, Artillery-Shell Antitank Device 13-30
Platter Charge 13-31
Improvised Claymore 13-31
Grapeshot Antipersonnel Device 13-32
Barbwire Antipersonnel Device 13-32

APPENDIX A. INSTALLATION AND REMOVAL OF US MINES AND FIRING DEVICES A-1
Section I. Antipersonnel Mines A-2
M14 .... A-2
Characteristics A-2
Installation A-3
Removal A-5
M16 .... A-6
Characteristics A-6
Installation A-7
Removal A-10
Section II. Antitank Mines A-11
M15 .... A-11
Characteristics A-12
Installation Using the M624 Fuse A-13
Removal Using the M624 Fuse A-17
Installation Using the M603 Fuse A-17
Removal Using the M603 Fuse A-20
M19 .... A-21
Characteristics A-22
Installation A-22
Removal A-24
M21 .... A-24
Characteristics A-25
Installation A-26
Removal A-29
Section III. Firing Devices and Activators A-29
M5 PRESSURE-RELEASE FIRING DEVICE (MOUSETRAP) A-30
Characteristics A-30
Installation A-31
Removal A-31

Page

M142 MULTIPURPOSE FIRING DEVICE A-32
Characteristics A-33
Arming and Disarming A-33
M1 AND M2 ACTIVATORS A-33

APPENDIX B. CONTROLS AND COMPONENTS OF SPECIAL-PURPOSE MUNITIONS B-1
SELECTABLE LIGHTWEIGHT ATTACK MUNITION B-1
M93 HORNET B-1

APPENDIX C. THREAT MINE/COUNTERMINE OPERATIONS C-1
MINE OPERATIONS C-1
CHEMICAL MINES C-6
COUNTERMINE OPERATIONS C-7
Organization C-7
Equipment C-11

APPENDIX D. AIR VOLCANO D-1
COMPONENTS D-1
M87-Series Mine Canister D-1
M139 Dispenser D-2
LIMITATIONS D-2
EMPLOYMENT D-2
Deep Operations D-3
Close Operations D-4
Rear Operations D-5
Minefield Effects D-6
Planning D-8
EMPLACEMENT D-18
Outside Friendly Territory D-20
Within Friendly Territory D-20
REPORTING D-22
Scatterable Minefield Warning D-22
Scatterable Minefield Report and Record D-22

APPENDIX E. SAFETY AND TRAINING E-1
STORAGE E-1
LIVE-MINE TRAINING E-3
LIVE-MINE DEMONSTRATIONS E-5
M16 Antipersonnel Mine E-5
M18A1 Antipersonnel Munition E-6
M15, M19, and M21 Antitank Mines E-7
RISK ASSESSMENT FOR LIVE-MINE DEMONSTRATIONS E-8
RISK ASSESSMENT FOR LIVE-MINE TRAINING E-10

APPENDIX F. MINE AWARENESS F-1
SOLDIER F-1
Visual Indicators F-1
Probing F-2
AN/PSS-12 Metallic Mine Detector F-3
Evacuation Drills F-13

Page

LEADER F-17
Risk Management F-17
Recording and Mine-Data Tracking F-20
Mine-Incident Report F-21
TRAINING F-21
Individual Training F-21
Leader Training F-22
Unit Training F-23

APPENDIX G. COUNTERMINE DATA G-1
BREACHING ASSETS VERSUS THREAT OBSTACLES G-1
FOREIGN MINE DATA G-1
FOREIGN MINEFIELD EMPLACEMENT DATA G-1
FOREIGN MINE DELIVERY SYSTEMS G-1

APPENDIX H. METRIC CONVERSION CHART H-1

GLOSSARY Glossary-1

REFERENCES References-1

INDEX Index-1

LIST OF ILLUSTRATIONS

Figures

Page

Figure 1-1. Mine components 1-2
Figure 1-2. Methods of actuating mines 1-3
Figure 1-3. Types of fuses 1-4
Figure 1-4. AHD incorporating a release mechanism 1-7
Figure 1-5. AHD not attached to the mine 1-7
Figure 1-6. Hand-emplaced US AHDs 1-8
Figure 2-1. Tactical versus protective obstacles 2-6
Figure 2-2. Tactical-obstacle effects 2-6
Figure 2-3. Minefield variables 2-7
Figure 2-4. Vehicle mine encounter probability versus minefield density 2-9
Figure 2-5. Disrupt-effect group 2-10
Figure 2-6. Fix-effect group 2-11
Figure 2-7. Turn-effect group 2-12
Figure 2-8. Block-effect group 2-13
Figure 2-9. Obstacle zones 2-16
Figure 2-10. Obstacle belts 2-17
Figure 2-11. Obstacle groups 2-18
Figure 2-12. TF defense COA 2-24
Figure 2-13. TF direct-fire analysis 2-25
Figure 2-14. TF obstacle-intent integration and priorities 2-26
Figure 2-15. Obstacle-plan refinement 2-28
Figure 2-16. Scheme-of-obstacle overlay 2-30
Figure 2-17. Sample obstacle-execution matrix 2-31
Figure 2-18. Minefield siting 2-38
Figure 2-19. Example of minefield resourcing 2-40
Figure 2-20. Mine resupply 2-41
Figure 2-21. Supply-point resupply method 2-46
Figure 2-22. Service-station resupply method 2-47
Figure 2-23. Tailgate resupply method 2-48
Figure 2-24. Minefield marking 2-51
Figure 2-25. Marking of minefields and obstacle groups 2-52
Figure 2-26. Sample obstacle-turnover work sheet 2-54
Figure 3-1. AP SCATMINEs 3-2
Figure 3-2. AT SCATMINE 3-3
Figure 3-3. Emplacement of ADAMs and RAAMs 3-11
Figure 3-4. Gator SCATMINE system 3-15
Figure 3-5. Gator minefield 3-17
Figure 3-6. Volcano mine system 3-18
Figure 3-7. Volcano components 3-18
Figure 3-8. Volcano disrupt and fix minefields 3-21
Figure 3-9. Volcano turn and block minefields 3-22
Figure 3-10. MOPMS 3-22

Page

Figure 3-11. MOPMS emplacement and safety zone 3-23
Figure 3-12. MOPMS in a disrupt minefield 3-25
Figure 3-13. MOPMS in a fix minefield 3-26
Figure 3-14. Ground Volcano minefield 3-27
Figure 4-2. M18A1 claymore 4-2
Figure 4-3. SLAM 4-3
Figure 4-4. SLAM in bottom-attack mode 4-4
Figure 4-5. SLAM in side-attack mode 4-5
Figure 4-6. SLAM in timed-demolition mode 4-5
Figure 4-7. SLAM in command-detonation mode 4-5
Figure 4-8. M93 Hornet 4-6
Figure 4-9. Hornet reinforcing a conventional minefield 4-9
Figure 4-10. Hornet reinforcing a Volcano minefield 4-10
Figure 4-11. Hornet area-disruption obstacle 4-11
Figure 4-12. Hornet gauntlet obstacle (one cluster) 4-12
Figure 4-13. Hornet gauntlet obstacle (platoon) 4-13
Figure 4-14. Hornet-enhanced turn-and fix-obstacle groups 4-14
Figure 5-1. AT mines 5-1
Figure 5-2. AP mines 5-3
Figure 5-3. Prong-activated AP mine 5-5
Figure 5-4. Trip-wire-activated AP mine 5-5
Figure 5-5. Buried mine with pressure plate 5-6
Figure 5-6. Buried mine with tilt rod 5-7
Figure 5-7. Buried and concealed mines 5-7
Figure 6-1. Minefield requirements computation work sheet 6-5
Figure 6-2. Step-by-step procedures for completing the minefield requirements computation work sheet 6-9
Figure 6-3. Site layout 6-19
Figure 6-4a. Laying a minefield 6-20
Figure 6-4b. Laying a minefield (continued) 6-21
Figure 6-5. Laying an IOE short row 6-21
Figure 6-6. Sample strip feeder report 6-22
Figure 6-7. Laying a row minefield 6-23
Figure 6-8. Measuring distances between mines with sandbags 6-25
Figure 6-9a. Sample DA Form 1355 for a row minefield (front) 6-26
Figure 6-9b. Sample DA Form 1355 for a row minefield (back) 6-27
Figure 6-10. Standardized disrupt and fix row minefields 6-28
Figure 6-11. Standardized turn row minefield 6-30
Figure 6-12. Standardized block row minefield 6-32
Figure 6-13. Site layout 6-35
Figure 7-1. Minefield layout 7-2
Figure 7-2. Cluster compositions 7-3
Figure 7-3. Arrangement of clusters in a mine strip 7-3
Figure 7-4. IOE baseline with short strips 7-4
Figure 7-5. Clusters on an IOE short strip 7-7
Figure 7-6. Minefield lanes and gaps 7-8
Figure 7-7. Mine-emplacement procedures 7-11
Figure 7-8. Laying and fusing mines 7-14
Figure 7-9. Lane closure 7-16

Page

Figure 8-1. Conventional minefield/munition field reporting chain 8-2
Figure 8-2a. Sample DA Form 1355 (front side) for a standard-pattern minefield/munition field 8-5
Figure 8-2b. Sample DA Form 1355 (inside) for a standard-pattern minefield/munition field 8-6
Figure 8-2c. Sample DA Form 1355 (back side) for a standard-pattern minefield/munition field 8-7
Figure 8-3a. Sample DA Form 1355 (front side) for a Hornet minefield/munition field 8-12
Figure 8-3b. Sample DA Form 1355 (back side) for a Hornet minefield/munition field 8-13
Figure 8-4. Sample DA Form 1355-1-R 8-18
Figure 8-5. Hasty protective row minefield/munition field record 8-19
Figure 8-6a. Sample DA Form 1355 (front side) for a nuisance minefield/munition field 8-21
Figure 8-6b. Sample DA Form 1355 (inside) for a nuisance minefield/munition field 8-22
Figure 8-7. Scatterable minefield/munition field report and record work sheet 8-23
Figure 8-8. Sample SCATMINWARN 8-24
Figure 8-9. Scatterable minefield/munition field report and record for an ADAM/RAAM artillery mission 8-24
Figure 8-10. Sample SCATMINWARN for an artillery mission 8-25
Figure 8-11. Minefield/munition field overlay symbols 8-26
Figure 9-1. Sample OBSTINTEL report 9-3
Figure 10-1. AN/PSS-12 mine detector 10-3
Figure 10-2. ASTAMIDS 10-4
Figure 10-3. IVMMD components 10-5
Figure 10-4. MICLIC 10-8
Figure 10-5. AVLM 10-8
Figure 10-6. MICLIC employment in a minefield less than 100 meters deep 10-10
Figure 10-7. MICLIC employment in a minefield of uncertain depth or greater than 100 meters 10-10
Figure 10-8. Skip zone 10-11
Figure 10-10. APOBS 10-13
Figure 10-11. Bangalore torpedo 10-13
Figure 10-12. Skim technique 10-15
Figure 10-13. MCB 10-15
Figure 10-14. Mine-blade width compared to track-vehicle widths 10-16
Figure 10-15. MCR 10-17
Figure 10-16. Mine-roller width compared to track-vehicle widths 10-17
Figure 10-17. Panther 10-19
Figure 10-18. MiniFlail 10-20
Figure 10-19. Grizzly 10-20
Figure 10-20. CEV with mine rake 10-21
Figure 10-21. Tripod 10-23
Figure 10-22. Initial lane marking 10-28
Figure 10-23. Intermediate lane marking 10-30
Figure 10-24. Full lane marking 10-32
Figure 10-25. Marking devices 10-35
Figure 10-26. NATO standard marker 10-37
Figure 10-27. NATO lane-marking conversion 10-37
Figure 10-28. NATO standard marking for limited visibility 10-38
Figure 11-1. IBASIC 11-6

Page

Figure 11-2. Platoon-size sweep team 11-9
Figure 11-3. Squad-size sweep team 11-10
Figure 11-4. Sweep teams in echelon 11-11
Figure 11-5. Linear clearance method 11-12
Figure 11-6. Combat clearance method 11-13
Figure 11-7. Deliberate route clearance 11-14
Figure 11-8. Hasty route clearance 11-15
Figure 11-9. Area clearance site layout 11-19
Figure 11-10. Sample enemy obstacle report 11-24
Figure 11-11. Sample route status report 11-25
Figure 11-12. Sample mine incident report 11-26
Figure 12-1. Outrigger techniques 12-3
Figure 12-2a. Sample DA Form 1355 (front side) for river mining 12-4
Figure 12-2b. Sample DA Form 1355 (inside) for river mining 12-5
Figure 12-3. Building sketch and mine plan (DA Form 1355) 12-7
Figure 12-4. Underground passageway 12-8
Figure 12-5. Open spaces 12-8
Figure 12-6. Street obstacles 12-9
Figure 12-7. Roof obstacles 12-9
Figure 12-8. Building obstacles 12-10
Figure 12-9. Probable M14 AP mine emplacement 12-11
Figure 12-10. Probable M16 AP mine emplacement 12-11
Figure 12-11. Probable M18A1 AP mine emplacement 12-12
Figure 12-12. AT mine emplacement in urban areas 12-13
Figure 12-13. AT mine emplacement in industrial and transportation areas 12-13
Figure 12-14. ADAM/RAAM employment 12-14
Figure 12-15. MOPMS employment 12-16
Figure 13-1. Typical electric and nonelectric booby traps 13-6
Figure 13-2. Methods of actuation 13-7
Figure 13-3. Remotely connected traps 13-7
Figure 13-4. Standard booby-trap sign 13-9
Figure 13-5a. Sample DA Form 1355 (front side) for a booby-trapped area 13-12
Figure 13-5b. Sample DA Form 1355 (inside) for a booby-trapped area 13-13
Figure 13-6. Improvised electrical FDs 13-24
Figure 13-7. Improvised nonelectric FDs (shear-pin operated) 13-25
Figure 13-8. Improvised nonelectric FDs (spring-operated) 13-25
Figure 13-9. Improvised, electric delay devices 13-26
Figure 13-10. Improvised, nonelectric delay devices 13-26
Figure 13-11. Typical punjis 13-27
Figure 13-12. Side-closing trap 13-27
Figure 13-13. Spike board 13-28
Figure 13-14. Venus fly trap 13-28
Figure 13-15. HE, artillery-shell AT device 13-30
Figure 13-16. Platter charge 13-31
Figure 13-17. Improvised claymore device 13-32
Figure 13-18. Grapeshot AP device 13-33
Figure 13-19. Barbwire AP device 13-33
Figure A-1. M14 AP mine A-2
Figure A-2. M22 wrench A-3
Figure A-3. M14 mine in ARMED position A-4
Figure A-4. Removal of safety clip A-4

Page

Figure A-5. Bottom view of M14 mine A-5
Figure A-6. M16A1 AP mine A-6
Figure A-7. M16A1 mine and M25 wrench A-7
Figure A-8. M605 fuse A-8
Figure A-9. Safety pins A-9
Figure A-10. Buried mine with a trip wire A-9
Figure A-11. Metal collar on an M605 fuse A-10
Figure A-12. M15 AT mine A-12
Figure A-13. M20 wrench A-13
Figure A-14. Correct safety-pin configuration A-14
Figure A-15. Greasing the M624 fuse A-14
Figure A-16. Tightening the fuse with the extension rod A-15
Figure A-17. M15 mine in the hole A-15
Figure A-18. Extension-rod assembly .... A 16
Figure A-19. Assembly of the extension rod into the fuse ring A-16
Figure A-20. Removal of safety pin A-17
Figure A-21. ARMED position A-18
Figure A-22. SAFE position A-18
Figure A-23. Safety fork A-19
Figure A-24. Clearance test A-20
Figure A-25. M15 mine in the hole A-20
Figure A-26. M19 AT mine A-21
Figure A-27. Removal of the pressure plate A-22
Figure A-28. Firing pin A-23
Figure A-29. M21 AT mine A-25
Figure A-30. M607 fuse A-26
Figure A-31. M26 wrench A-26
Figure A-32. Buried M21 mine A-27
Figure A-33. Removing the band and the stop A-28
Figure A-34. M5 FD A-30
Figure A-35. Arming the M15 .... A-31
Figure A-36. M142 FD A-32
Figure A-37. M1 activator A-34
Figure B-1. SLAM components B-1
Figure B-2. Hornet components B-3
Figure B-3. Hornet controls and indicators B-4
Figure C-1. GMZ armored tracked mine layer C-2
Figure C-2. Threat-style rapidly emplaced minefield C-3
Figure C-3. Threat-style antitrack minefield C-3
Figure C-4. Threat-style antihull minefield C-4
Figure C-5. Threat-style AP minefield C-4
Figure C-6. UMZ SCATMINE system C-6
Figure C-7. Chemical-mine employment C-7
Figure C-8. BAT-M with BTU bulldozer blade C-8
Figure C-9. KMT-4 plow C-8
Figure C-10. IMP portable mine detector C-9
Figure C-11. DIM mine detector C-9
Figure C-12. KMT-5 plow-roller combination C-10
Figure C-13. IMR armored engineer tractor C-10
Figure C-14. M1979 armored mine clearer C-11

Page

Figure D-1. Air Volcano system ... D-1
Figure D-2. Turn obstacle ... D-6
Figure D-3. Block obstacle ... D-7
Figure D-4. Disrupt obstacle ... D-7
Figure D-5. Fix obstacle ... D-8
Figure D-6. Site layout ... D-15
Figure D-7. Sample Volcano card ... D-17
Figure D-8. Fencing for an air Volcano minefield ... D-21
Figure E-1. M16 AP mine ... E-6
Figure E-2. M18A1 AP mine ... E-7
Figure E-3. M15 and M19 AT mines ... E-8
Figure E-4. M21 AT mine ... E-8
Figure E-5. Excerpt from Risk-Assessment Techniques Manual, prepared by the Department of Transportation's Transportation Safety Institute, August 1986 ... E-9
Figure E-6. Preliminary hazard-analysis work sheet (arming M15) ... E-11
Figure E-7. Preliminary hazard-analysis work sheet (disarming M15) ... E-12
Figure E-8. Preliminary hazard-analysis work sheet (arming M16) ... E-13
Figure E-9. Preliminary hazard-analysis work sheet (disarming M16) ... E-14
Figure E-10. Preliminary hazard-analysis work sheet (arming M19) ... E-15
Figure E-11. Preliminary hazard-analysis work sheet (disarming M19) ... E-16
Figure E-12. Preliminary hazard-analysis work sheet (arming M21) ... E-17
Figure E-13. Preliminary hazard-analysis work sheet (disarming M21) ... E-18
Figure E-14. Preliminary hazard-analysis work sheet (command detonation) ... E-19
Figure E-15. Preliminary hazard-analysis work sheet (peripheral factors) ... E-20
Figure F-1. AN/PSS-12 metallic mine detector ... F-4
Figure F-2. AN/PSS-12 packed components ... F-4
Figure F-3. Electronic unit ... F-5
Figure F-4. Battery installation ... F-5

Tables

Table 2-1. Echelons of obstacle control and effect ... 2-15
Table 2-2. Planning factors for the mine dump ... 2-21
Table 2-3. Planning factors for work rates ... 2-21
Table 2-4. Planning factors for standardized row minefields ... 2-22
Table 2-5. Planning factors for scatterable minefields ... 2-22
Table 2-6. Ranges of common weapons ... 2-23
Table 2-7. Personnel requirements for a Class IV/V supply point ... 2-42
Table 2-8. Class IV/V haul capacity ... 2-45
Table 3-1. Characteristics of AP SCATMINEs ... 3-2
Table 3-2. Characteristics of AT SCATMINEs ... 3-4
Table 3-3. SD windows ... 3-7
Table 3-4. Emplacement authority ... 3-9
Table 3-5. Coordination responsibilities ... 3-10
Table 3-6. RAAM and ADAM minefield density and size ... 3-14
Table 3-7. Marking scatterable minefields ... 3-26
Table 3-8. Safety and fragment hazard zones ... 3-28
Table 4-1. Hornet minimum emplacement distances ... 4-15

Page

Table 5-1. Characteristics of AT mines 5-2
Table 5-2. Characteristics of AP mines 5-3
Table 5-3. Sympathetic detonation chart 5-8
Table 7-1. Platoon organization and equipment 7-10
Table 7-2. Sample mines tally sheet 7-15
Table 8-1. Minefield/munition field obstacle numbering system 8-8
Table 8-2. Abbreviations for obstacle types 8-9
Table 9-1. Lane widths 9-4
Table 10-1. Lane-marking levels, unit responsibilities, and trigger events 10-33
Table 10-2. Guidelines for lane-marking devices 10-34
Table 11-1. Sample task organization for a route clearance 11-2
Table 11-2. Personnel and equipment requirements for a sweep team 11-8
Table 11-3. Sample task organization for an area clearance 11-17
Table 13-1. Tactical reports 13-11
Table 13-2. Clearing equipment 13-17
Table C-1. Normal parameters for threat-style minefields C-2
Table D-1. Air Volcano capabilities and limitations D-4
Table D-2. Air Volcano minefield data D-6
Table D-3. Planning process (H-hour sequence) D-11
Table D-4. Air Volcano dispensing times based on air speed D-19
Table E-1. Mine color-coding system E-2
Table F-1. Risk-assessment criteria F-18
Table F-2. Sample risk assessment F-19
Table G-1. Mounted breaching assets versus threat obstacles G-2
Table G-2. Dismounted breaching assets versus threat obstacles G-5
Table G-3. Foreign track-width AT mines G-9
Table G-4. Foreign full-width AT mines G-10
Table G-5. Foreign side-attack AT mines G-11
Table G-6. Foreign pressure-fused AP mines G-11
Table G-7. Foreign trip-wire/break-wire-fused AP mines G-12
Table G-8. Foreign emplaced minefields G-13
Table G-9. Foreign mine delivery systems G-14
Table H-1. Metric conversion chart H-1

Preface

Field Manual (FM) 20-32 provides United States (US) armed forces with tactical, technical, and procedural guidance for conducting mine and countermine operations. It applies to all elements of the combined arms team for maneuver and engineer staff planning and coordination. The manual is presented in three parts—mine operations, counteroperations, and special-mining operations.

The guidance provided focuses on individual skills of emplacing and removing mines, team and squad tasks, platoon and company organization and planning, and battalion/task force (TF) organization and coordination for successful obstacle reduction and breaching operations.

The provisions of this publication support existing doctrine established by FMs 5-34, 5-100, 90-7, and 90-13-1. It also contains new and improved techniques for emplacing row mines; marking, reporting, and recording minefields; reducing simple and complex obstacles; and emplacing a standard-pattern minefield. This manual reflects new doctrine from FMs 5-10, 5-71-2, and 5-71-3.

This publication implements the following International Standardization Agreements (STANAGs) between North Atlantic Treaty Organization (NATO) forces:

- STANAG 2036. *Land Minefield Laying, Marking, Recording, and Reporting Procedures.* Edition 5.
- STANAG 2889. *Marking of Hazardous Areas and Routes Through Them.* Edition 3.
- STANAG 2990. *Principles and Procedures for the Employment in Land Warfare of Scatterable Mines with a Limited Laid Life.* Edition 1.

NOTE: US policy regarding the use and employment of antipersonnel land mines (APLs) outlined in this FM is subject to the Convention on Certain Conventional Weapons and Executive Orders. Current US policy limits the use of non-self-destructing APLs to (1) defending the US and its allies from armed aggression across the Korean demilitarized zone and (2) training personnel engaged in demining and countermine operations. The use of the M18A1 claymore in the command-detonation mode is not restricted under international law or Executive Order.

All references to US employment of non-self-destructing APLs (such as row mining) in this manual are intended to provide doctrine for use in Korea only. This information is provided in bold lettering throughout the manual. Detailed doctrine on APLs is also provided to ensure that US forces recognize how the enemy can employ these weapons.

As the US military seeks to end its reliance on APLs, commanders must consider the increased use of other systems such as the M18A1 claymore, nonlethal barriers (such as wire obstacles), sensors and surveillance platforms, and direct and indirect fires.

This publication includes the following appendixes:

- Appendix A. Installation and Removal of US Mines and Firing Devices.
- Appendix B. Controls and Components of Special-Purpose Munitions.
- Appendix C. Threat Mine/Countermine Operations.
- Appendix D. Air Volcano.
- Appendix E. Safety and Training.
- Appendix F. Mine Awareness.
- Appendix G. Countermine Data.
- Appendix H. Metric Conversion Chart.

The proponent for this publication is Headquarters, US Army Training and Doctrine Command (TRADOC). Forward comments and recommendations on Department of the Army (DA) Form 2028 to Commandant, US Army Engineer School, ATTN: ATSE-DME-MWF, Fort Leonard Wood, Missouri 65473-5000.

Unless this publication states otherwise, nouns and pronouns do not refer exclusively to men.

Chapter 1

Introduction

This chapter provides the mechanics and characteristics of antitank (AT) mines and munitions, antipersonnel (AP) mines and munitions, and antihandling devices (AHDs). The information contained in this chapter also provides a foundation for the rest of the manual.

Land-based mines and munitions are hand-emplaced, remote-delivered, ground-delivered, or air-delivered:

- Hand-emplaced mines and munitions require manual arming and are labor-, resource-, and transport-intensive.
- Remote- and air-delivered mines and munitions require less time and labor; however, they are not as precisely placed as hand-emplaced mines and munitions.
- Ground-delivered mines are less resource-intensive than hand-emplaced mines. They are not precisely placed; however, the minefield boundaries are.

Soldiers can surface lay or bury mines and munitions and can place AHDs on hand-emplaced AT mines.

NOTE: Some countries employ AHDs on AP mines, but US forces are not authorized to employ AHDs on any type of AP mine.

MECHANICS OF MINES

CHARACTERISTICS AND FUNCTIONING

A land mine is an explosive device that is designed to destroy or damage equipment or personnel. Equipment targets include ground vehicles, boats, and aircraft. A mine is detonated by the action of its target, the passage of time, or controlled means. There are two types of land-based mines—AT and AP. Mines generally consist of the following parts (Figure 1-1, page 1-2):

- Firing mechanism or other device (sets off the detonator or igniter charge).
- Detonator or igniter (sets off the booster charge).
- Booster charge (may be attached to the fuse or the igniter or be part of the main charge).
- Main charge (in a container; usually forms the body of the mine).
- Casing (contains all the above parts).

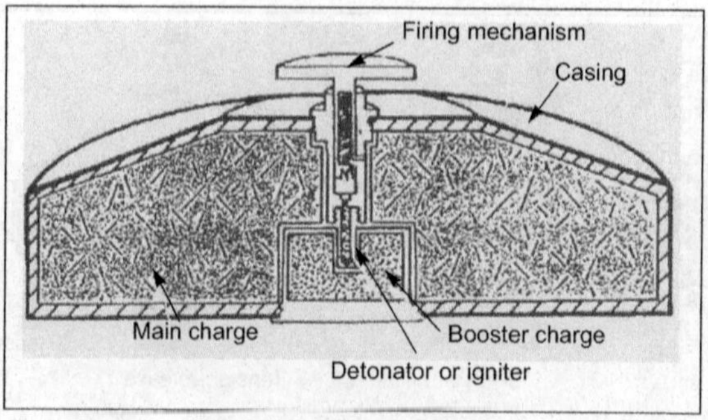

Figure 1-1. Mine components

Components and Initiating Actions

A firing mechanism prevents the mine from exploding until it makes contact with, or is influenced by, its target. Once a mine has been armed, the firing mechanism may be actuated by the following methods (Figure 1-2):

- Applying pressure (including tilt rod).
- Pulling a trip wire.
- Releasing tension or breaking a trip wire.
- Releasing pressure.
- Passage of time (time-delay mechanism).
- Impulses.
  - Electrical.
  - Vibration.
  - Magnetic-influence.
  - Electromagnetic-frequency.
  - Infrared-sensored.
  - Acoustic.

To arm some mines, you must position the igniter, set the mechanism properly, and disengage the safety device (usually by removing a safety pin). The fuse is the initial component in the firing chain; it has a low-explosive (LE) powder but is highly sensitive. The fuse is actuated by an initiating action. Although mines are issued with a standard fuse, alternate fuses are issued separately for some mines.

The four main fuse types are shown in Figure 1-3, page 1-4.

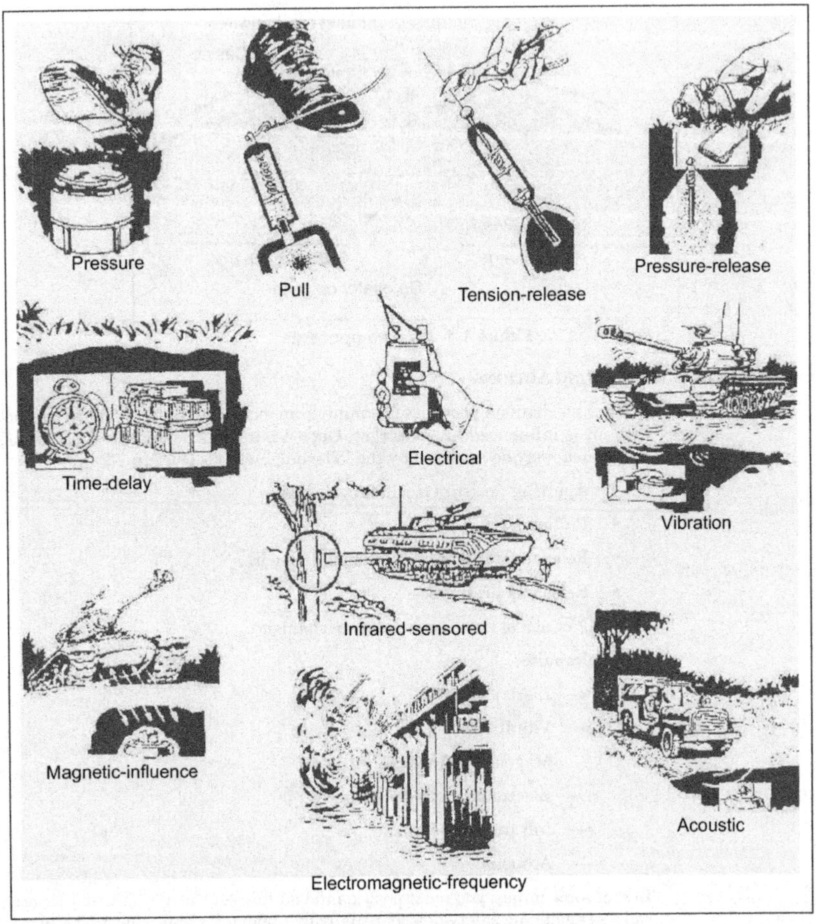

Figure 1-2. Methods of actuating mines

Mechanical. A spring drives a striker against a percussion cap, which fires the detonator.

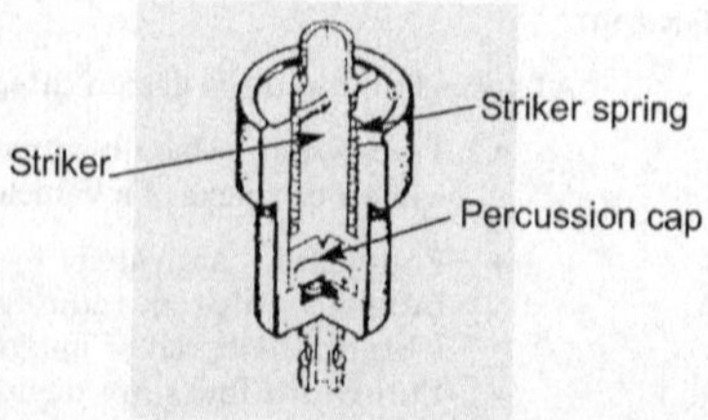

Chemical. A small container of a chemical compound is broken by the initiating action. The chemical compound reacts with another substance to generate heat, which ignites the detonator.

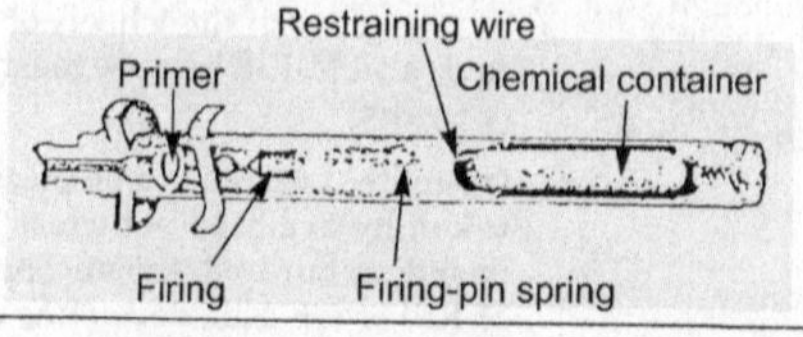

Friction. The initiating action ignites substances inside the fuse by friction. The flame fires the detonator.

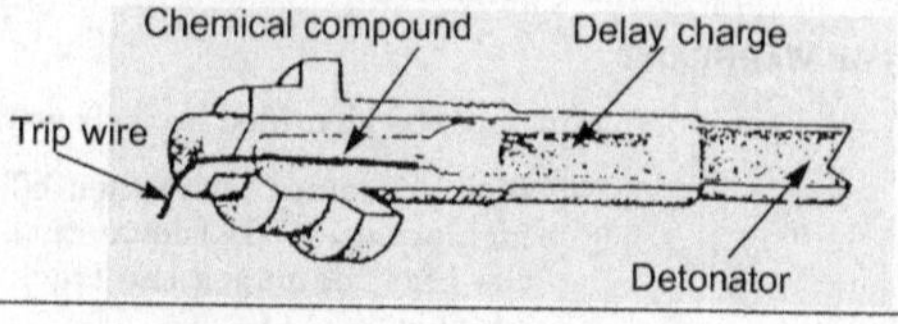

Electrical. The initiating action closes an electrical circuit, which functions an electrical detonator.

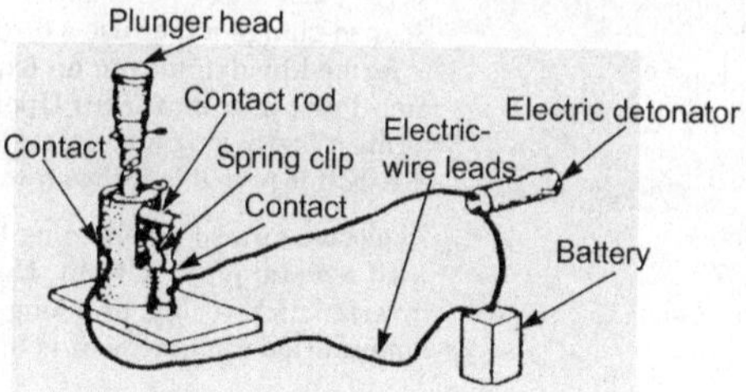

Figure 1-3. Types of fuses

ANTITANK MINES

AT mines are designed to immobilize or destroy vehicles and their occupants.

TYPES OF KILLS

An AT mine produces a mobility kill (M-Kill) or a catastrophic kill (K-Kill). An M-Kill destroys one or more of the vehicle's vital drive components (for example, breaks a track on a tank) and immobilizes the target. An M-Kill does not always destroy the weapon system and the crew; they may continue to function. In a K-Kill, the weapon system and/or the crew is destroyed.

TYPES OF SENSING

AT fuses fall into three design categories:

- Track-width. Usually pressure-actuated, requiring contact with the wheels or tracks of a vehicle.
- Full-width. Activated by several methods—acoustics, magnetic-influence, tilt-rod, radio-frequency, infrared-sensored, command, or vibration. Tilt-rod or magnetic-influence fuses are the most common. Full-width fuses are designed to be effective over the entire target width and can cause a K-Kill from penetration and spalling metal or from secondary explosions. When a full-width fuse is activated solely by contact with the wheels or tracks of the target vehicle, it usually causes an M-Kill because most of the energy is absorbed by the wheels or tracks.
- Off-route. Designed to be placed along the side of a route likely to be taken by armored vehicles. It has numerous fuzing possibilities, including infrared, seismic, break wire, and magnetic. It produces an M-Kill or a K-Kill, depending on the location of the target at the time of mine detonation.

TYPES OF WARHEADS

AT mines can be identified by their warheads:

- Blast AT mines derive their effectiveness from the force generated by high-explosive (HE) detonation. They usually produce an M-Kill when the blast damages the track or the vehicle, but a K-Kill is also possible.
- Shaped-charge mines use a directed-energy warhead. A shaped charge is formed by detonating an explosive charge behind a cone of dense metal or other material. Upon detonation, the cone collapses and forms a metal slug and a gaseous metal jet that penetrate the target. A K-Kill is probable if the crew or ammunition compartment is hit.
- Explosive-formed penetrating (EFP) mines have an explosive charge with a metal plate in front. Upon detonation, the plate forms into an inverted disk, a slug, or a long rod. A K-Kill is probable if the crew or ammunition compartment is hit.

ANTIPERSONNEL MINES

TYPES OF KILLS

AP mines can kill or incapacitate their victims. The injuries and deaths they cause commit medical resources, degrade unit morale, and damage nonarmored vehicles. Some types of AP mines may break or damage the track on armored vehicles.

TYPES OF SENSING

AP mines can be fused in many ways, to include pressure, seismic, wire, or command detonation:

- Pressure fuses usually activate an AP mine when a load is placed on the fuse.
- Seismic fuses activate an AP mine when the sensor detects vibrations.
- Trip wires or break wires activate an AP mine when something disturbs barely visible wires.
- Command-detonated mines are activated by a soldier when he detects the enemy in the mines' blast area.

TYPES OF EFFECTS

AP mines contain five types of effects:

- Blast. Cripples the foot or leg of a soldier who steps on it; can also burst the tires of a wheeled vehicle that passes over it.
- Bounding-fragmentation. Throws a canister into the air; the canister bursts and scatters shrapnel throughout the immediate area.
- Direct-fragmentation. Propels fragments in the general direction of enemy soldiers.
- Stake-fragmentation. Bursts and scatters shrapnel in all general directions.
- Chemical. Disperses a chemical agent to whoever activates it; contaminates the surrounding area.

ANTIHANDLING DEVICES

AHDs perform the function of a mine fuse if someone attempts to tamper with the mine. They are intended to prevent moving or removing the mine, not to prevent reduction of the minefield by enemy dismounts. An AHD usually consists of an explosive charge that is connected to, placed next to, or manufactured in the mine. The device can be attached to the mine body and activated by a wire that is attached to a firing mechanism. US forces can employ AHDs on conventional AT mines only. Other countries employ AHDs on AT and AP mines.

Some mines have extra fuse wells that make it easier to install AHDs (Figure 1-4). An AHD does not have to be attached to the mine; it can be placed underneath the mine (Figure 1-5). Mines with AHDs are sometimes incorrectly called *booby-trapped* mines.

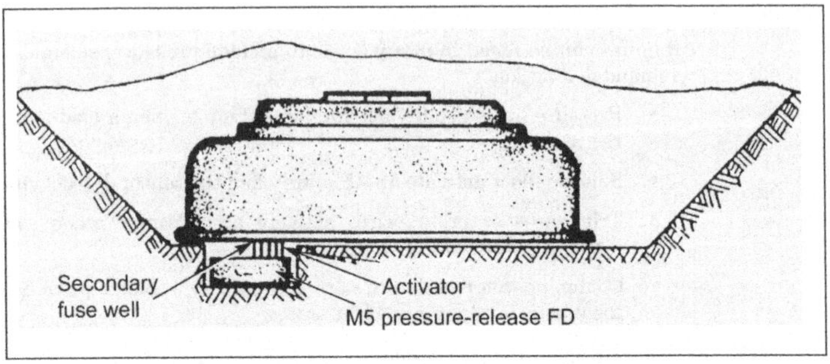

Figure 1-4. AHD incorporating a release mechanism

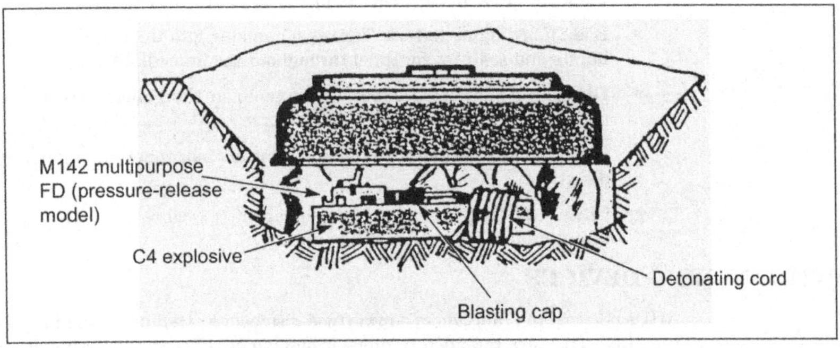

Figure 1-5. AHD not attached to the mine

The following hand-emplaced AHDs are used by US forces (Figure 1-6, page 1-8):

- M5 pressure-release firing device (FD).
- M142 multipurpose FD.

These devices use a spring-loaded striker with a standard base, and they function in one or more modes—pressure, pressure-release, tension, and/or tension-release. When an FD is employed as an AHD on certain AT mines, it requires the use of an M1 or M2 activator. FDs and activators are described in Appendix A.

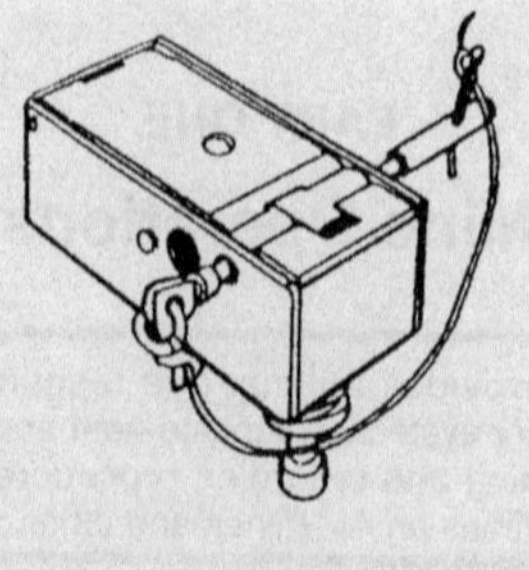

M5 Pressure-Release FD

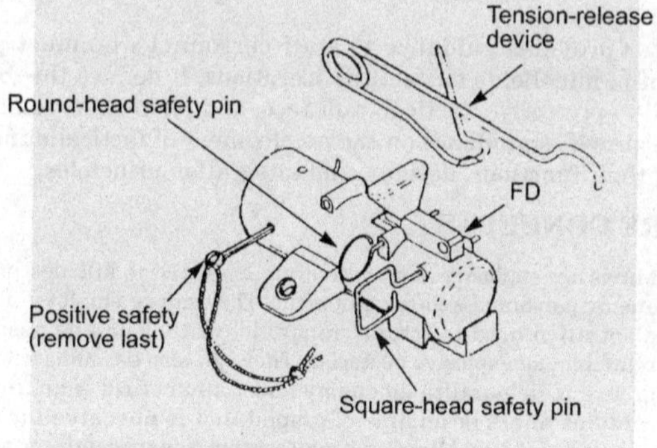

M142 Multipurpose FD

Figure 1-6. Hand-emplaced US AHDs

PART ONE

Mine Operations

This part of the manual provides tactical and technical information on mines, minefields, and mine-delivery systems; emplacement and employment methods and responsibilities; and reporting and recording procedures for US mine operations. Current US policy for restrictions on AP mines and other devices is also discussed.

Chapter 2

Mine-Warfare Principles

This chapter provides guidance to staff personnel who must plan the employment of minefields for tactical operations. It defines the four types of minefields—protective, tactical, nuisance, and phony. The remainder of the chapter provides guidance on the employment of tactical minefields—specifically their functions, designs, and integration principles.

MINE-WARFARE CONCEPTS

Mines are explosive devices that are emplaced to kill, destroy, or incapacitate enemy personnel and/or equipment. They can be employed in quantity within a specified area to form a minefield, or they can be used individually to reinforce nonexplosive obstacles. They can also be emplaced individually or in groups to demoralize an enemy force. A minefield is an area of ground that contains mines or an area of ground that is **perceived** to contain mines (a phony minefield). Minefields may contain any type, mix, or number of AT and/or AP mines. Minefields are used to—

- Produce a vulnerability on enemy maneuver that can be exploited by friendly forces.
- Cause the enemy to piecemeal his forces.
- Interfere with enemy command and control (C^2).
- Inflict damage to enemy personnel and equipment.
- Exploit the capabilities of other weapon systems by delaying enemy forces in an engagement area (EA).
- Protect friendly forces from enemy maneuver and infiltration.

TYPES OF MINEFIELDS

There are four general types of minefields—protective, tactical, nuisance, and phony. Each type is determined by its distinct battlefield purpose. Therefore, minefields are employed differently, and they target the enemy in unique ways that support the overall concept of the operation.

- Protective minefields are employed to protect soldiers, equipment, supplies, and facilities from enemy attacks or other threats.
- Tactical minefields directly effect the enemy's maneuver in a way that gives the defending force a positional advantage.
- Nuisance minefields impose caution on enemy forces and disrupt, delay, and sometimes weaken or destroy follow-on echelons.
- Phony minefields deceive the enemy about the exact location of real minefields. They cause the attacker to question his decision to breach and may cause him to expend his reduction assets wastefully. Phony minefields may be employed in conjunction with other minefields, but should be used only after the enemy has become mine-sensitive.

It is important to distinguish the difference between the types of minefield and the means of emplacement. Volcano, Modular Pack Mine System (MOPMS), standard-pattern, and row mining are not types of minefields; they are just some of the means used to emplace tactical, nuisance, and protective minefields. They may also be the method of emplacement that is replicated by a phony minefield.

PROTECTIVE MINEFIELDS

Protective minefields are employed to protect soldiers, equipment, supplies, and facilities from enemy attacks or other threats. Other threats range from enemy surveillance to theft of supplies and equipment. In tactical operations, protective minefields provide friendly forces with close-in protection and defeat the enemy's ability to maneuver or utilize the tenants of offense. They deny mechanized penetration and dismounted infiltration. In military operations other than war (MOOTW), protective minefields may focus on preventing unauthorized access to facilities and installations, rather than assisting in the destruction of an enemy force.

Protective minefields are usually employed and emplaced at the small-unit level (platoon or company/team). The authority to emplace protective minefields is normally delegated to the company/team commander. In some cases, such as a hasty defense, protective minefields are laid on short notice by units that use mines from their basic load or local stock. More commonly, protective minefields are used as part of a unit's deliberate defense. The mines are laid so that they are easy to detect and recover by the laying unit.

An important aspect of protective minefields is the requirement to recover them before leaving the area. This is often overlooked and is difficult to control because they are emplaced at the small-unit level. When a unit is being relieved in place by an adjacent unit, protective minefields are turned over to the relieving unit (minefield turnover is further defined later in this chapter). The decentralized emplacement of protective obstacles makes consolidating reports and records difficult and requires command involvement.

Much like final protective fires (FPF), protective minefields provide the defender with close-in protection during the enemy's final assault. Protective minefields serve two purposes. First, they impose a delay on an attacker that allows the defender time to break contact as the unit displaces to another battle position. Secondly, they break up the enemy's assault to complete its destruction. The composition of a protective minefield is driven by the vulnerability of the defender:

- Dismounted infantry is the greatest close-combat threat to a defending tank company/team. Protective minefields encountered in this case consist predominantly of AP mines that limit enemy dismounts from closing with the armor defender.
- A tank force is the greatest threat to an infantry defense. Protective minefields in this case consist predominantly of AT mines that reduce the enemy's ability to close quickly onto the infantry's position.

Neither AP nor AT mines are used in isolation. The preponderance of mine composition is designed against the most severe close-combat threat and the likelihood of that threat.

A protective minefield may take many forms. It may be only a few mines in front of a platoon, or it may be a standard-pattern minefield around an airfield. Protective minefields are used in both close and rear operations, and they are classified as either hasty or deliberate:

- Hasty protective minefields are temporary in nature and are used as part of a unit's defense perimeter. They are usually laid by units using mines from their basic load. If time permits, mines should be buried to increase their effectiveness; but they can be laid on top of the ground. AHDs and low-metallic mines are not used so that the minefield can be easily recovered. Mines are employed outside the hand-grenade range but within the range of small-caliber weapons. All mines are picked up by the emplacing unit upon leaving the area, unless enemy pressure prevents mine retrieval or the minefield is being transferred to a relieving commander. The brigade commander has the initial authority to employ hasty protective minefields. This authority may be delegated to a battalion or company commander on a mission basis. Procedures for emplacing a hasty protective row minefield are contained in Chapter 6.
- Deliberate protective minefields are more permanent, require more detailed planning, and usually require more resources. They are commonly used to protect static assets (vital sites)—logistical sites, communication nodes, depots, airfields, missile sites, air-defense artillery (ADA) sites, and permanent-unit locations. A typical deliberate protective minefield is the standard-pattern minefield; however, a row minefield can also be used. Deliberate protective minefields are usually emplaced for extended periods of time and can be transferred to another unit. Techniques for emplacing deliberate protective minefields are discussed in Chapters 6 and 7.

Tactical Minefields

Tactical minefields are employed to directly attack enemy maneuver and to give the defender a positional advantage over the attacker. Tactical minefields may be employed by themselves or in conjunction with other types of tactical obstacles. They attack the enemy's maneuver by disrupting its combat formations, interfering with its C^2, reducing its ability to mass fires, causing him to prematurely commit limited breaching resources, and reducing his ability to reinforce. The defender masses fires and maneuver to exploit the positional advantage created in part by tactical obstacles.

Tactical minefields add an offensive dimension to the defense. They are a commander's tool for recapturing and maintaining the initiative that is normally afforded to an attacker. Combined with fires, tactical obstacles force the attacker to conform to the defender's plan.

Tactical minefields may be emplaced during offensive operations to protect exposed flanks, isolate the objective area, deny enemy counterattack routes, and disrupt enemy retrograde. This chapter further discusses the principles behind designing, integrating, siting, and emplacing tactical minefields.

Nuisance Minefields

Nuisance minefields are a form of tactical minefields. They are mainly used to impose caution on enemy forces and to disrupt, delay, and sometimes destroy follow-on echelons. Once nuisance minefields are emplaced, they do not require cover by observation or direct fire. Nuisance minefields are usually irregular in size and shape; they can be a single group of mines or a series of mined areas. They can be used to reinforce existing obstacles and can also be rapidly emplaced on main avenues of approach (AAs). Conventional mines and scatterable mines (SCATMINEs) may be used in nuisance minefields.

Phony Minefields

Phony minefields are areas of ground that are altered to give the same signature as a real minefield and thereby deceive the enemy. Phony minefields serve two primary functions. First, they confuse an attacker's breach cycle and cause him to question his breach decision. Secondly, they may cause an attacker to wastefully expend reduction assets to reduce mines that are not really there.

The success of phony minefields depends on the enemy's state of mind. The bluff succeeds best when the enemy is mine-conscious and has already suffered the consequences of a mine encounter. A fear of mines can quickly evolve into paranoia and break the momentum of the enemy's attack. Therefore, phony minefields are normally employed in conjunction with real minefields and are seldom employed alone. Once the enemy has become mine-conscious, phony minefields may produce considerable tactical effects with very little investment in time, labor, and material. Phony minefields may also be used to extend the front and depth of live minefields when mines or labor are in short supply or when time is restricted. They may be used to conceal minefield gaps through live minefields. There is **no** guarantee that phony minefields will achieve their purpose.

There are two mission-essential tasks inherent in the employment of a phony minefield:

- The phony minefield must completely replicate a live minefield in every detail, using a specific method of emplacement as a model. This becomes the deception story, and every aspect of the phony minefield must support the deception story. For example, if the deception story is a buried row minefield, the depth, front, and marking must be similar to that of a live buried row minefield. The ground should be disturbed, and tracks should be made on the ground in the same pattern as other minefields to give the ground the same signature. Occasional empty mine crates, discarded fuses, or other mine-laying supplies add to the deception.

- The deception story must never be compromised. Once emplaced, the phony minefield must be regarded by friendly forces as live until the tactical situation no longer warrants maintaining the deception. This can be extremely painful for the friendly unit. There is great temptation to drive through, rather than around, a known phony minefield—particularly if it is intended to be a gap between live minefields. However, one vehicle driving through a phony minefield and observed by enemy reconnaissance compromises the minefield's effectiveness.

Live mines are never laid in a phony minefield. A minefield designated as phony implies that the area contains no live mines. Emplacing even a single live mine within a phony minefield makes it a live minefield. Empty tins and such may be laid in a phony minefield but is seldom worthwhile. Minefield marking and covering fire should be the same as for a live minefield. Employment authority and reporting requirements are the same as for the minefield being simulated.

PROTECTIVE VERSUS TACTICAL MINEFIELDS

As discussed, minefields can be tactical or protective obstacles (Figure 2-1, page 2-6). Tactical and protective obstacles have different purposes with regard to the enemy's maneuver. This difference causes them to have a particular relative place on the battlefield. Tactical obstacles attack enemy maneuver and are placed on the battlefield where the enemy maneuvers from march, prebattle, and attack formations. Protective obstacles are used to protect the force from the enemy's final assault onto the force's position. Protective obstacles are close to defensive positions and are tied in with the FPF of the defending unit. Additional information on obstacles can be found in FM 90-7.

TACTICAL MINEFIELDS

Tactical minefields are designed, sited, emplaced, and integrated with fires to produce four specific tactical-obstacle effects—disrupt, turn, fix, and block (Figure 2-2, page 2-6). Each obstacle effect has a specific impact on an enemy's ability to maneuver, mass, and reinforce. Obstacle effects also increase the enemy's vulnerability to friendly fires. They support the friendly scheme of maneuver by manipulating the enemy in a way that is critical to the commander's intent. Minefield design is the means by which an emplacing unit varies minefield width, minefield depth, mine density, mine composition, the use of AHDs, and the irregular outer edge (IOE) to best achieve one of the four tactical-obstacle effects. Modifying these variables is at the heart of tactical minefield employment principles.

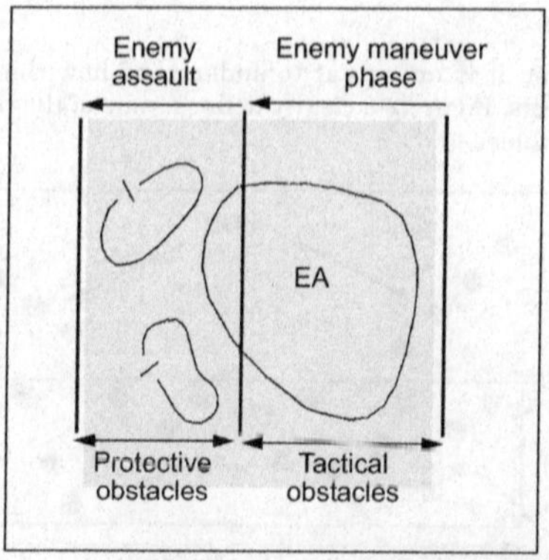

Figure 2-1. Tactical versus protective obstacles

Obstacle Effect	Application	Examples Conveying Intent	Description
Disrupt 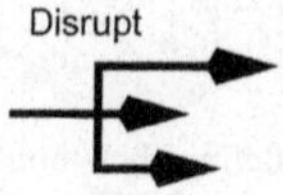	The short arrow indicates where an enemy is attacked by obstacles. The longer arrows indicate where the bypass is allowed and attacked by fires.	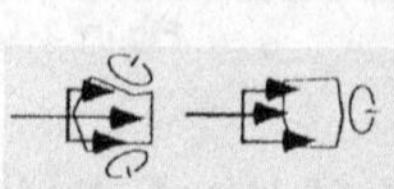	Causes an enemy to break up its formation and tempo, interrupt its timetable, commit its reduction assets prematurely, and piecemeal the attack.
Turn 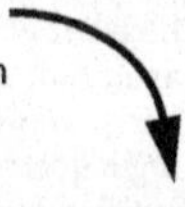	The heel of the arrow is the anchor point. The direction of the arrow indicates the desired direction of turn.		Manipulates an enemy's maneuver in a desired direction.
Fix 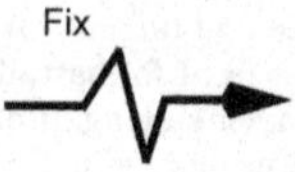	The irregular part of the arrow indicates where an enemy advance is slowed by obstacles.	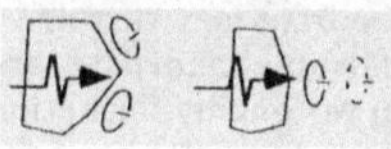	Slows an enemy within a specific area, normally an EA. Gives the defender time to acquire, target, and destroy the attacking enemy throughout the depth of an EA.
Block 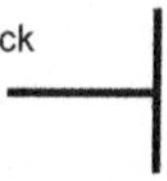	The ends of the vertical line indicate the limit of an enemy's advance and where obstacles tie in to no-go terrain.	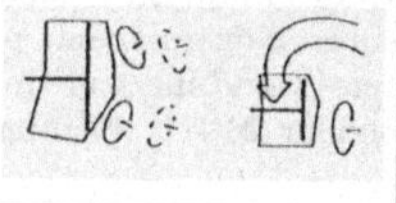	Stops an enemy along a specific AA or prevents it from passing through an EA.

NOTES:
1. Arrows indicate the direction of enemy attack.
2. Minefields must be integrated with fires to achieve the desired effect.

Figure 2-2. Tactical-obstacle effects

Minefield Variables

First, it is important to understand how the variables relate to minefield effects. Figure 2-3 clearly defines some of the terms used to discuss minefield variables.

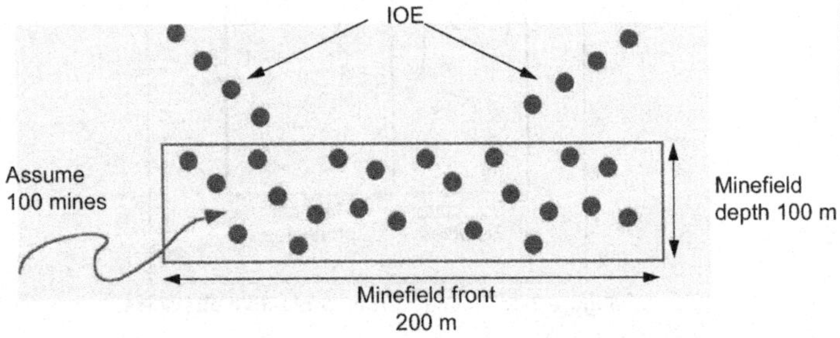

Linear density: 100 mines ÷ 200 meters = 0.5 mine per meter of front
Area density: 100 mines ÷ (100 x 200) = 0.005 mine per square meter

Figure 2-3. Minefield variables

Front

Minefield front is the dimension of a minefield that defines how much of the attacking enemy formation is affected by the minefield. The front of a minefield is based on the desired obstacle effect (disrupt, turn, fix, or block) and the attack front of a company-size enemy force. The front of an attacking enemy depends largely on the type of enemy force (armored, motorized, or dismounted infantry) and norms by which the enemy army fights. For armored warfare, the minefield front is based on effecting a doctrinal company attack front of 500 meters (13 to 18 combat vehicles). For dismounted warfare, the minefield front is based on effecting a company attack front of 150 meters. The front may vary and require a study of enemy force and terrain. Groups of individual minefields are employed to achieve a larger front for battalion and larger enemy formations. For example, a battalion consisting of 52 to 72 combat vehicles has a front of 1,500 meters and requires more minefields.

Depth

Minefield depth is based on the amount of reduction assets that will be required by an enemy to reduce a lane. The standard should start with 100 meters and increase in depth if denying the enemy the use of a mobility corridor (MC) is the intent (turn or block).

Density

Minefield density depicts the number of mines in a minefield. It is expressed in linear or area density:

- Linear density is the average number of mines within a 1-meter-wide path through the minefield's depth, anywhere along the front. In Figure 2-3, page 2-7, the minefield contains 100 mines, with a minefield front of 200 meters. The linear density is 0.5 mine per meter of front (100 mines/200 meters of front).
- Area density is the average number of mines within a square meter, anywhere in the minefield. In Figure 2-3, the minefield contains 100 mines within a 20,000-square-meter area. The area density is 0.005 mine per square meter (100 mines/[200 meters x 100 meters]). Area density is normally used to express the density of scatterable minefields.

Mine Composition

This variable includes the effective use of different types of mines. By using full-width mines, the probability of kill increases for the minefield. AT mines with AHDs **(Korea Only: as well as AP mines)** are used where the enemy is expected to use dismounted reduction techniques.

Antitank Mines

If the enemy is an armored force, tactical obstacles are predominantly AT mines. Track-width mines (M15s with the M603 fuse) have a lower probability of kill (M-Kill or K-Kill) than full-width mines (M21s and M15s with the M624 fuse). The ratio of full-width versus track-width mines in a minefield depends on the kill required. In general, a track-width minefield does not adequately affect the enemy's maneuver.

Antipersonnel Mines

AP mines target dismounted soldiers. Their composition in tactical minefields depends on the threat and the enemy's reduction assets. Based on current technology, most breaching operations are accomplished by mechanical or explosive means. **(Korea Only: If the minefield group's intent is to exhaust the enemy's breaching assets, AP mines should be integrated to attack its dismounted reduction ability.)**

Probabilities of Encounter and Kill

The probabilities of encounter and kill measure a minefield's lethality.

Probability of Encounter

The probability of encounter is measured by the chance (in percent) that a vehicle, blindly moving through a minefield, will detonate a mine. The probability of encounter is based on mine density, the type of mine, and the type of enemy vehicle. In short, the more dense a minefield, the higher the probability is of encountering a mine. Probability of encounter also depends on the fuse capability of the mines. Tilt-rod and magnetic-influence mines will detonate if they are encountered anywhere along the width of the enemy vehicle. Pressure-fused mines detonate if a vehicle's track or wheel actually runs over them. The probability of encounter is also affected by the type of enemy vehicle. The smaller the width or track signature of the vehicle, the less likely it will encounter and detonate a mine.

Figure 2-4 illustrates the relationship between mine density and the probability of encounter for light versus heavy tracked vehicles and for track-width versus full-width mines. Figure 2-4 also provides general guidance for varying the mine density to yield the necessary probability of encounter when developing disrupt, fix, turn, and block minefields. Varying mine density is further discussed later in this chapter.

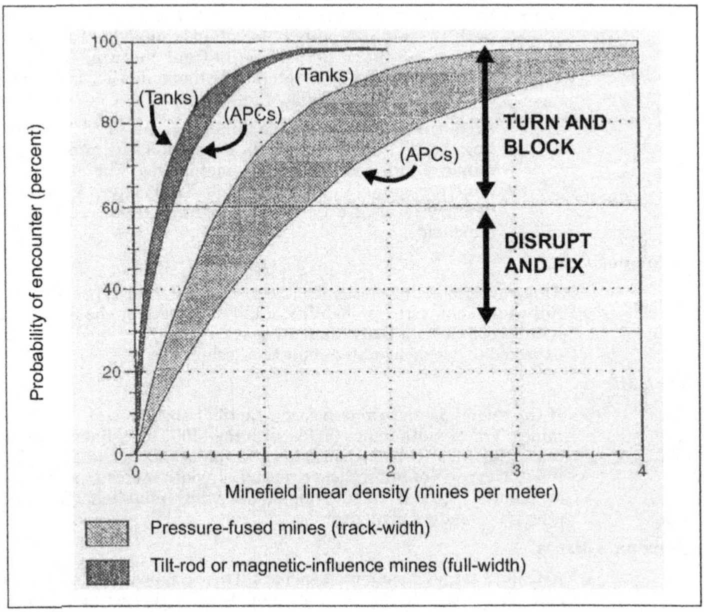

Figure 2-4. Vehicle mine encounter probability versus minefield density

Probability of Kill

The probability of kill is measured by the chance (in percent) that a vehicle will no longer be mission-capable (M-Kill or K-Kill) because of mine effects. It is a function of the combined probability that a vehicle will encounter a mine and the probability that the mine effect will produce an M-Kill or a K-Kill.

Antihandling Devices

Emplacing AHDs on mines is time-intensive. AHDs are added to a minefield to discourage manual removal and reuse of mines by the enemy and to demoralize the enemy who is attempting to reduce the minefield. AHDs do not prevent an enemy from reducing the minefield; they only discourage manual reduction methods.

Irregular Outer Edge

An IOE is a strip/row or multiple strips/rows of mines that normally extend toward the enemy from the first (enemy side) row of mines. An IOE is employed to break up the otherwise regular pattern of a minefield. It is used to confuse the enemy about the exact limits of the minefield, particularly its leading edge. An IOE adds an unknown quality to a minefield that makes the enemy's decision of whether to breach or bypass more difficult. The effect an IOE has on enemy actions may increase the overall lethality of a minefield.

Design

Modifying minefield variables to achieve the desired obstacle effect is a challenge for the engineer, both technically (resourcing and designing) and tactically (supporting the maneuver scheme). Experience will provide the best basis for designing minefields. Figures 2-5 through 2-8, pages 2-10 through 2-13, provide guidelines for varying minefield depth, front, density, and composition to best achieve disrupt, fix, turn, and block effects.

These are guidelines, not fixed rules. Minefield designs must be based on a threat analysis. The designs are simply considerations or parameters to use when designing tactical minefields, regardless of the emplacement method. They apply to conventional mine-laying techniques as well as the employment of SCATMINE dispensers. These parameters give the engineer the flexibility to design and emplace tactical minefields based on mission, enemy, terrain, troops, time available, and civilian considerations (METT-TC) (particularly resources and terrain) and still achieve the required effect. These norms are also the basis for developing minefield packages and emplacement procedures outlined throughout this manual. Chapter 3 discusses the characteristics and emplacement procedures for each of the SCATMINE systems, Chapter 6 outlines procedures for row mining using conventional mines, and Chapter 7 is dedicated to the standard-pattern minefield. Each chapter describes standard disrupt, fix, turn, and block minefield packages particular to that method of emplacement or dispensing system.

Each tactical-obstacle effect has a specific resourcing factor. In short, this numeric value helps determine the amount of linear obstacle effort that is needed to achieve the desired effect. The resource factor is multiplied by the width of the AA or MC to get the total amount of linear obstacle effort required. The linear obstacle effort is then divided by the minefield front norm for the specific effect (rounded up) to yield the number of individual minefields required in the obstacle group.

Disrupt

A disrupt effect (Figure 2-5) focuses fire planning and obstacle effort to cause the enemy to break up its formation and tempo, interrupt its timetable, commit reduction assets prematurely, and piecemeal the attack. It also deceives the enemy about the location of friendly defensive positions, separates combat echelons, or separates combat forces from their logistical support. A disrupt effect should not be time-, manpower-, or resource-intensive. It should not be visible at long range but easily detected as the enemy nears it. Commanders normally use the disrupt effect forward of EAs.

Resource factor	0.5 (3 point obstacles) x AA
Group dimensions	W = 0.5 x AA
Probability of kill	50%
Minefield front	250 m
Minefield depth	100 m
AT mines	Yes (pressure/tilt)
AP mines	No **(Korea Only: optional, based on threat analysis)**
AHD	Optional, based on threat analysis
IOE	No

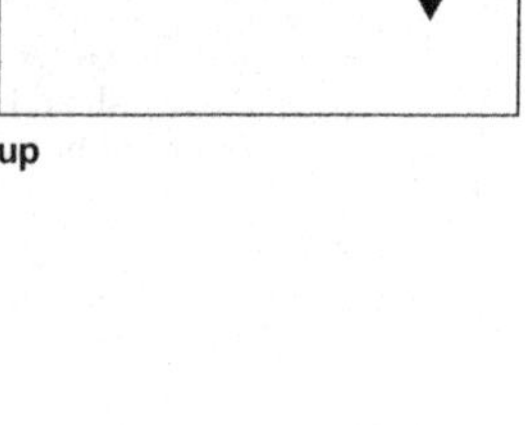

Figure 2-5. Disrupt-effect group

Normally, only half of the enemy's AA is attacked with minefields or other tactical obstacles to achieve a disrupt effect. For a minefield to disrupt an enemy company, half of the formation must react to the minefield. The typical width of a disrupt-effect minefield is 250 meters with a standard minefield depth of 100 meters. When designing a disrupt effect to attack an enemy battalion, three disrupt-effect minefields are arrayed in a group to achieve a width that is about half the size of the battalion's attack front (750 meters of minefield). Multiply the resource factor of 0.5 by the width of the AA to provide the amount of linear obstacle effort required for the disrupt effect. When the AA is narrow, an alternative disrupt group is three point obstacles along the AA.

Disrupt-effect minefields should be designed with approximately 50 percent probability of mine encounter to achieve the desired disrupt effect (see Figure 2-4, page 2-9). They should contain predominantly track-width AT mines and include full-width AT mines at the leading edge of the minefield to increase the probability of mine encounter. This should cause the enemy to commit its reduction assets.

AHDs can be added to disrupt-effect minefields to frustrate the enemy's breaching and clearing operations. However, adding AHDs may be too resource-intensive for the return in effect. An IOE is not required.

Fix

A fix effect (Figure 2-6) focuses fire planning and obstacle effort to slow an attacker within a specific area, normally an EA. The fix effect is primarily used to give the defender time to acquire, target, and destroy the attacking enemy throughout the depth of an EA or AA. A fix effect may be used to generate the time necessary for the force to break contact and disengage as the enemy maneuvers into the area (typically used for delays). Fix-effect minefields in the obstacle group must be employed in depth, causing the enemy formation to react and breach repeatedly. Fix-obstacle groups must span the entire width of the AA. Commanders normally use the fix effect inside the EA.

Resource factor	1.0 x AA
Group dimensions	W = 1.0 x AA
Probability of kill	50%
Minefield front	250 m
Minefield depth	120 m
AT mines	Yes (pressure/tilt)
AP mines	No **(Korea Only: optional, based on threat analysis)**
AHD	Optional, based on threat analysis
IOE	Yes

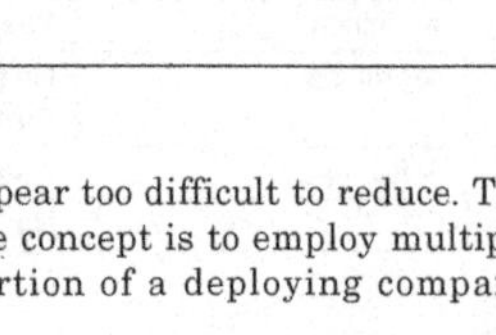

Figure 2-6. Fix-effect group

Individual fix-effect minefields must not appear too difficult to reduce. The enemy should be enticed into the area. The concept is to employ multiple minefields that individually attack a portion of a deploying company formation. Therefore, the fix-effect minefield front is 250 meters. It takes on the characteristics of a disrupt-effect minefield with a similar density, composition, and probability of encounter (Figure 2-4, page 2-9), with two exceptions. First, AHDs are not used because the application of massed direct and indirect fires complicate the enemy's breaching effort. Secondly, an IOE is added to further delay the enemy and confuse the attacker on the exact orientation of individual minefields. This also serves to increase the effective lethality of the minefield. The majority of mines are track-width AT, but full-width AT mines are used in the IOE and the leading edge of the minefield because they are the most lethal.

While individual minefields are designed to attack only portions of an enemy company formation, the fix-obstacle group is resourced, arranged, and sited to attack the entire front of an enemy battalion. Figure 2-6, page 2-11, depicts a fix group effect on an attacking enemy battalion. In this case, six fix-effect minefields are arrayed in an area the full width of the battalion AA (1,500 meters) by 1,500 meters deep. Accordingly, the resource factor for a fix-effect minefield group is one; the amount of linear minefield that should be resourced equals the width of the AA.

Turn

A turn effect (Figure 2-7) manipulates the enemy's maneuver in a desired direction. One technique or a combination of techniques aids in achieving the turn effect. First, in order to entice the enemy to maneuver in the desired direction rather than reduce the obstacle, the obstacle must have a subtle orientation relative to the enemy's approach. Secondly, the obstacle and fires must allow bypass in the direction desired by the friendly scheme of maneuver. Obstacles in the start of the turn are visible and look more complex than those in the direction of the turn. Finally, the obstacle is tied into severely restricted terrain (restricted terrain as a minimum) at the initial point of the turn. The point where the severely restricted terrain feature and the minefield meet is known as the *anchor point*. Commanders normally use the turn effect on the flanks of an EA.

Resource factor	1.2 x AA
Group dimensions	W = 1.0 x AA
Probability of kill	75%
Minefield front	500 m
Minefield depth	300 m
AT mines	Yes (pressure/tilt)
AP mines	No **(Korea Only: optional, based on threat analysis)**
AHD	Optional, based on threat analysis
IOE	No

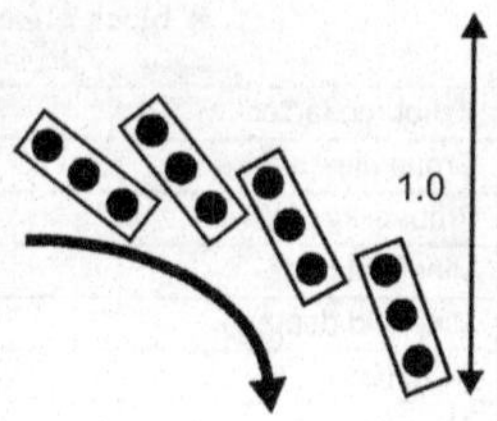

Figure 2-7. Turn-effect group

The standard turn-effect minefield has a width of 500 meters and a depth of 300 meters. One turn-effect minefield affects the entire width of an enemy company's front. It must be deep enough to cause multiple applications of line-charge reduction assets. In short, the minefield must discourage any attempts to breach and must entice the enemy to bypass rather than reduce. Figure 2-7 depicts a turn effect on an attacking battalion, with turn-effect minefields arrayed in a group across a 1,500-meter AA. The angle of the minefields should be subtle, encouraging the enemy to bypass individual minefields. Each minefield overlaps another one to tie the minefields together and prevent gaps. This is considered in the resource factor (1.2) for a turn-obstacle group. This factor, multiplied by the width of the AA, equals the amount of linear obstacle effort required for this turn effect.

As shown in Figure 2-4, page 2-9, turn-effect minefields must be extremely lethal and achieve approximately 80 percent probability of encounter. In other words, an enemy vehicle attempting to reduce or pass through the minefield will likely encounter a mine. This forces the small-unit commander to make an immediate decision—breach or bypass. A lethal minefield that is covered by intense fires and has an easily detectable bypass reduces breach decisions to instinct and causes the enemy to choose the bypass (turn). To produce this lethality, the majority of mines should be full-width AT. Full-width mines in the first rows the enemy encounters and in the depth of the minefield either exhaust the enemy's breaching assets or convince him to bypass early. AHDs are not required because the enemy force will seldom commit to dismounted breach when faced with intense direct and indirect fires. An IOE should not be used because the enemy must be able to determine the orientation of the minefield and the bypass.

Block

A block effect (Figure 2-8) integrates fire planning and obstacle effort to stop an attacker along a specific AA or prevent him from passing through an EA. Block obstacles are complex and are integrated with intense fires; they do not stop an attacker by themselves. Individual block obstacles are employed successively in a relatively shallow area. When the enemy reduces one block obstacle, it is critical that he encounters another, thus denying him to project combat power and maintain momentum. Block obstacles must defeat the enemy's breaching effort (mounted and dismounted) as well as his maneuver. A block effect must span the entire width of an AA and prevent a bypass.

Resource factor	2.4 x AA
Group dimensions	W = 1.0 x AA
Probability of kill	80%
Minefield front	500 m
Minefield depth	300+ m
AT mines	Yes (pressure/tilt)
AP mines	No **(Korea Only: Yes)**
AHD	Yes
IOE	Yes

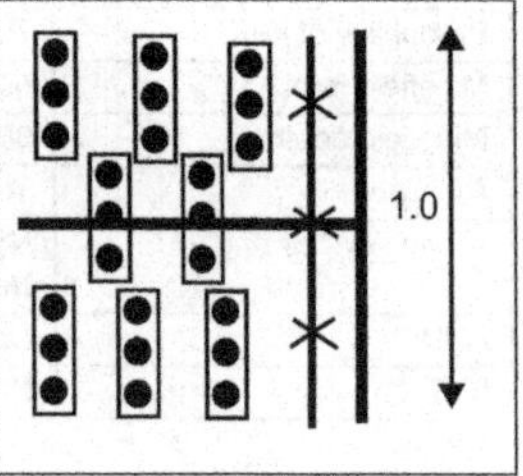

Figure 2-8. Block-effect group

The typical block-effect minefield is 500 meters wide and 300+ meters deep (includes an IOE). Figure 2-8 depicts a block effect on an attacking battalion. Note how individual minefields are arrayed to affect the entire width of the

AA but in a relatively shallow depth. Eight block-effect minefields are required in this example to achieve the necessary depth and width. The block group is the most resource-intensive. A resource factor of 2.4 is multiplied by the width of the AA to determine the linear obstacle effort required.

The lethality of a block-effect minefield (80 percent or higher) is similar to that of a turn-effect minefield (see Figure 2-4, page 2-9). The lethality of the group is considerably higher, since there are enough minefields in the group to cover more than twice the width of the AA. This lethality is produced by a density slightly greater than one mine per meter of front and the use of predominantly full-width AT mines.

A block-effect minefield must be capable of defeating mechanical and dismounted breach efforts. Therefore, AHDs **(Korea Only: and AP mines)** are used to target dismounted breaching. An IOE confuses the attacker about the exact minefield limits and complicates his employment of mechanical reduction assets. The depth of the block-effect minefield requires employing multiple line charges.

The above minefields are not standard solutions to every situation. The terrain could dictate a decrease or an increase in the effort required. Incorporating other reinforcing obstacles (AT ditches, road craters, wire, SCATMINEs) aid in attacking the different reduction assets.

TACTICAL-OBSTACLE INTEGRATION PRINCIPLES

Tactical minefields are considered tactical obstacles and follow the same basic integration principles. Tactical obstacle C^2 focuses on obstacle emplacement authority and obstacle control.

Obstacle Emplacement Authority

Obstacle emplacement authority is the jurisdiction that a unit commander has to emplace tactical obstacles. In a theater of operations (TO), theater commanders have the authority to emplace obstacles. In most cases, they delegate that authority to corps commanders who further delegate it to division commanders. Division commanders then have obstacle emplacement authority in their area of operations (AO), unless the authority is subsequently withheld or restricted by a higher commander. Commanders subordinate to corp and division do not have the authority to emplace obstacles unless the higher commander delegates authority for a current operation. Commanders use control measures and other specific guidance or orders to grant obstacle emplacement authority to subordinate commanders. Emplacement authority for SCATMINEs is covered in Chapter 3.

Obstacle Control

Commanders exercise obstacle control to ensure that obstacles support current and future operations. Obstacle control ensures that subordinate commanders emplace obstacles to best support the higher commander's scheme of maneuver. It also ensures that subordinate commanders do not interfere with future operations. Commanders maintain obstacle control by focusing or withholding emplacement authority or by restricting the types or locations of obstacles. Commanders use control measures, specific guidance, and orders to maintain obstacle control.

Obstacle Control Measures

Obstacle control measures are specific control measures that simplify granting obstacle emplacement authority and providing obstacle control for all levels of command (Table 2-1). Obstacle control measures are classified as—

- Zone.
- Belt.
- Group.
- Restriction.

Table 2-1. Echelons of obstacle control and effect

Obstacle Control Measure	Echelon	Specific Obstacle Effects Assigned	Size of Enemy AA/MC		Planning Guidance
			Armored	Light vs Armored	
Zone	Corps or division	Optional	Division/ brigade	Brigade/ battalion	Requires anticipating belts and intents
Belt	Brigade	Optional but normal	Brigade/ battalion	Battalion/ company	Requires anticipating groups and intents
Group	Corps, brigade, division, or battalion/TF	Mandatory	Battalion/TF	Company/ platoon	Based on individual obstacle norms
Restriction	All	NA	NA	NA	Used only when necessary to support the scheme of maneuver

A protective obstacle is the only obstacle that can be employed outside a designated obstacle zone, belt, or group.

A specific obstacle effect (disrupt, turn, fix, or block) may be assigned to obstacle control measures. This enables the commander to direct the overall effect of obstacles within a designated zone, belt, or group to support his plan. At corps and division levels, assigning specific effects to obstacle zones is optional. At brigade level, the commander will normally assign a specific effect to the obstacle belts. At TF and battalion levels, obstacle effects for obstacle groups are required. This ensures that subordinate commanders emplace tactical obstacles that support the maneuver and fire plans. Assigning a specific obstacle effect to a control measure becomes obstacle intent, giving the obstacle effect, target, and location. Obstacle intent provides a direct link between the obstacle plan, the scheme of maneuver, the commander's intent, and the fire plan (direct and indirect). Obstacle intent is critical at brigade level and below, and it becomes the foundation for obstacle group development and design at the TF level.

Obstacle Zones

Obstacle zones (Figure 2-9, page 2-16) are graphic control measures that corps and division commanders use to grant obstacle emplacement authority to brigades (including armored cavalry regiments [ACRs] and other major subordinate units). Corps and division commanders use zones to ensure that subordinates emplace obstacles that support the higher commander's scheme of maneuver, and to ensure that the obstacles do not interfere with future operations.

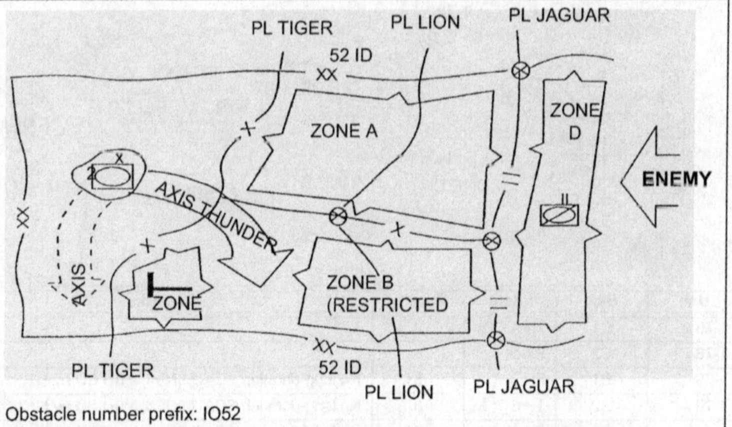

Unit	Zone	Priority	Remarks
1st Brigade	A	3	
3d Brigade	B	2	No SCATMINEs with SD after H+2
3d Brigade	C	1	Block intent
Cavalry Squadron	D	4	

Figure 2-9. Obstacle zones

Obstacle Belts

Obstacle belts (Figure 2-10) are graphic control measures that brigade commanders use to constrain tactical-obstacle employment, and the belts should not cross unit boundaries. Commanders plan obstacle belts within assigned obstacle zones to grant obstacle emplacement authority to their major subordinate units. This is normally the first level in which the commander assigns an intent to the obstacle plan. It gives TF commanders the necessary guidance on the overall effect of obstacles within a belt. It does not designate that all obstacle groups within the belt must be the same effect. It simply means that the sum effect of groups within the belt must achieve the assigned belt effect. This serves to synchronize the obstacle effort within the brigade, particularly between adjacent TFs. Obstacle belts also focus obstacles in support of the brigade scheme of maneuver and ensure that obstacles do not interfere with the maneuver of any higher headquarters.

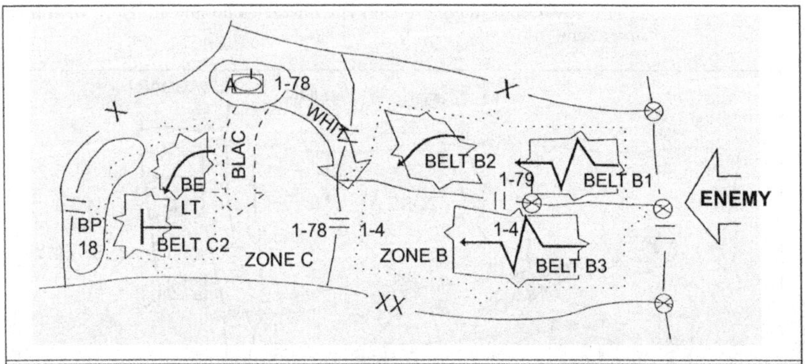

Unit	Belt	Intent	Priority	Remarks
1-78	C1	Turn	2	
1-78	C2	Block	1	
1-79	B1	Fix	5	No long-duration SCATMINEs; must SD by H+2
1-79	B2	Turn	3	No long-duration SCATMINEs; must SD by H+2
1-4	B3	Fix	4	No long-duration SCATMINEs; must SD by H+2

Figure 2-10. Obstacle belts

Belts are planned to attack enemy regiments based on an analysis of enemy battalion MCs. Brigades allocate maneuver companies based on the motorized rifle battalion (MRB) AAs and organize TFs to defeat the motorized rifle regiment (MRR). Obstacle belts and their intents are directed against MRR AAs. This provides the appropriate level of guidance while preserving the TF's need to refine the obstacle intent, based on how the TF will fight its allocated companies.

Obstacle Groups

Obstacle groups (Figure 2-11, page 2-18) contain one or more individual obstacles that are grouped together to provide a specific obstacle effect. TFs use obstacle groups to ensure that company teams emplace individual obstacles that support the TF scheme of maneuver. In rare cases, brigades, divisions, or even corps may use obstacle groups for specific tactical obstacles

Also, units perform detailed integration of obstacle groups with direct- and indirect-fire plans.

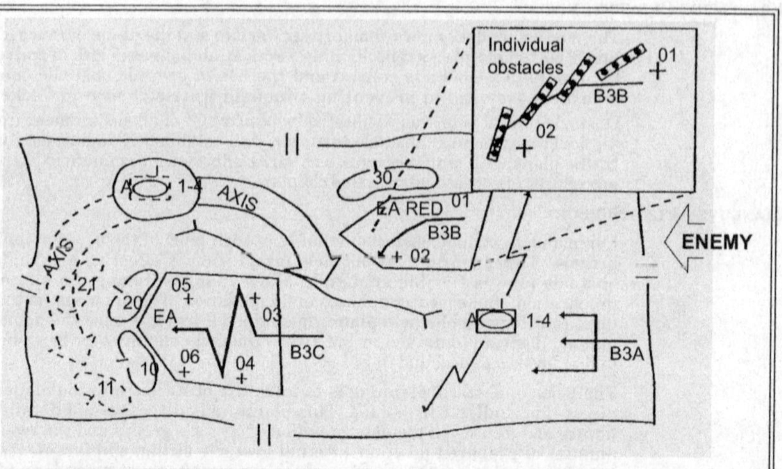

Unit/BP	Group	Intent	Priority	Remarks
Company A/Screen	B3A	Disrupt	3	All groups restricted: • Long-duration SCATMINEs are not allowed. • Short-duration SCATMINEs must be emplaced by H-3.
Team B/BP 30	B3B	Turn	1	
Team D/BP 10	B3C	Fix	2	

Figure 2-11. Obstacle groups

A TF allocates platoons against motorized rifle company (MRC) MCs and task-organize them into companies/teams to defeat MRBs. Likewise, direct-fire plans are designed based on the maneuver of MRBs and independent MRCs. Therefore, obstacle groups are used to attack the maneuver of MRB-size forces. Groups are designed specifically to support the direct-fire plan of the TF. The TF designates groups rather than obstacles because the location of individual obstacles hinges on siting at the company/team level. The group effect or obstacle intent drives obstacle siting and is therefore more important to convey to commanders. There can be more than one type of obstacle group to support the overall intent of an obstacle belt. This is because the belt design is based on the brigade's scheme of maneuver, without knowing the TF's direct-fire plan and scheme of maneuver. Groups are developed once the fire plan is established.

Obstacle Restrictions

Commanders at all levels may use obstacle restrictions to provide additional obstacle control. They may use obstacle restrictions to limit the specific types of obstacles used (for example, no buried mines and no SCATMINEs that do not self-destruct within 48 hours). These restrictions ensure that subordinates do not use obstacles with characteristics that impair future operations. It also allows commanders to focus the use of limited resources for the main effort by restricting their use elsewhere. Commanders may also use restrictions to prevent subordinates from emplacing obstacles in a certain area. This type of restriction may be shown graphically as an obstacle-restricted area.

Fratricide Prevention

The modern tendency toward maneuver warfare and the disappearance of the linear battlefield places repositioning forces at an increased risk of fratricide by minefields. Obstacle control and the use of graphic obstacle control measures are vital in preventing minefield fratricide at every echelon.

Obstacle control is further facilitated by positive C^2 of all sustainment traffic, tactical repositioning, obstacle turnover, well-established and disseminated traffic plans with traffic control, and strict adherence to minefield marking procedures (discussed later in this chapter).

Maneuver-Plan Support

Commanders include obstacle planning in each level of the decision-making process. This ensures that obstacle integration is effective and that the obstacle plan is flexible enough to allow changes during the planning, preparation, and execution phases of an operation. The following method is used to integrate obstacle planning at the TF level; it uses the decision-making doctrine contained in FM 101-5. Obstacle planning for brigade and higher levels can be found in FM 90-7.

The focus of obstacle planning is to integrate obstacles into the maneuver direct- and indirect-fire plans. This planning is directive and detailed in nature and focuses on the determination of obstacle groups and the type and amount of prepared positions. Actual obstacle siting, emplacement, and position location are the purview of the company/team commander and are normally supported by an engineer platoon.

TF-level defensive planning is part of the military decision-making process. The engineer battlefield assessment (EBA) process provides the basis for integrating engineer issues into the decision-making process.

Mission Analysis

The key activities during mission analysis are to—

- Determine facts and assumptions.
- Analyze relative combat power.
- Analyze the engineer battalion's/brigade's mission and the commander's intent.
- Issue the commander's guidance.

Determine Facts and Assumptions

Defensive planning normally starts with the receipt of a warning order (WO) from the higher headquarters to defend. The company executive officer (XO) and the battlefield information control center (BICC) (or the TF engineer and the Intelligence Officer [US Army] [S2]) begin by developing a situation template (SITEMP) that includes a modified combined obstacle overlay (MCOO). The MCOO is a product developed during the intelligence preparation of the battlefield (IPB) process. The MCOO development is a joint effort of the engineer and the intelligence section of the TF's tactical operations center (TOC).

The MCOO should define the AAs and the MCs within the TF's AO. This information is vital to obstacle planning. Obstacles are placed on AAs to attack enemy maneuver. The AA analysis details potential EAs and indicates where forces can defend with limited survivability construction, because a reverse slope or undulating terrain provides natural concealment and cover.

The threat evaluation and the enemy course-of-action (COA) development detail how the enemy will potentially attack. They also provide an insight as to what and where the enemy's objective and routes might be. The SITEMP helps the engineer to understand how the enemy will traverse through the TF's sector and allows the engineer to gain an understanding of how and where he can best attack the enemy's maneuver.

The SITEMP also depicts how the enemy's reconnaissance forces will enter the sector. This is especially important when countering the enemy's ability to reconnoiter obstacle efforts.

The engineer must articulate the current capabilities of the engineer forces, its current combat power, and its ability to support the TF. Assumptions of future capability or potential reinforcement by other engineers should be analyzed. Specific characteristics of special engineer equipment and SCATMINE systems are detailed for the staff. An initial Class IV/V supply-point location and an operation plan should be developed with the TF staff (note that the TF has responsibility for Class IV/V supply-point operation). The XO/first sergeant (1SG) works with the TF and the engineer battalion Supply Officer (US Army) (S4) to ensure that delivery of Class IV/V barrier material supports the TF countermobility plan.

Analyze Relative Combat Power

The engineer compares friendly and enemy combat power and identifies possible obstacle requirements that offset potential enemy breaching and direct- and indirect-fire capabilities. The actual inclusion of the obstacles normally occurs after COA development. During this phase, the engineer finishes his EBA to gain an understanding for the engineer company's ability to support the TF.

Analyze the Engineer Battalion's/Brigade's Mission and the Commander's Intent

The staff analyzes and identifies information from the maneuver brigade order and the commander's intent that will potentially impact defensive planning. The engineer analyzes the maneuver brigade/engineer battalion commander's intent to determine potential obstacle placement, obstacle intent, and construction priority based on his concept of the operation.

The TF must identify tasks and limitations imposed from the brigade operation order (OPORD). These might include obstacle belts with or without specific intents, obstacle-restricted areas, or restrictions on the type of obstacles. Also, the brigade OPORD might specify reserve, situational, or directed obstacle groups.

The engineer must identify the TF's total obstacle capabilities. Available assets include engineer units, SCATMINE systems, and other units that can provide additional manpower for obstacle construction. Engineer-equipment status, work rates, and the time available must be evaluated in detail.

NOTE: Work rates should only be used in the absence of unit-developed planning factors.

Table 2-2 provides planning factors for the mine dump. Tables 2-3 through 2-6, pages 2-21 through 2-23, provide planning factors for obstacles. Special focus to limited visibility work rates and engineer squad strengths must be made when making total-capability estimates. Plan to accomplish obstacle siting and Class IV/V supply-point setup during daylight hours, and plan to emplace mines during limited visibility hours as much as possible.

Table 2-2. Planning factors for the mine dump

Number of Personnel	Quantity of Mines	Required Equipment
2-man team (2 minutes per mine)	25 mines per hour	Shears, metal cutting Grease, automotive and artillery Rags Work gloves Flashlight Night-vision goggles Pliers
Squad (7 soldiers and an NCO)	100 mines per hour	
Platoon (with leadership)	300 mines per hour; 3,600 mines per day	
Company	10,800 mines per day	
NOTE: Soldiers work 50 minutes per hour, 12 hours per day.		

Table 2-3. Planning factors for work rates

Survivability	Time Required to Construct		
	With D7F Dozer	With ACE	With SEE
Hull-defilade position	1 BTH	1.5 BTH	NA
Turret-defilade position	2.5 BTH	3.5 BTH	NA
HMMWV TOW position	1.5 BTH	2 BTH	NA
Vehicle-protective position	0.75 BTH	1 BTH	NA
Dismount-crew position	NA	NA	1 SEEH
Individual-fighting position	NA	NA	0.5 SEEH
Countermobility	**With D7F Dozer**	**With ACE**	**In Man-Hours**
Antitank ditch	1 BTH/70 m	1 BTH/50 m	NA
Standardized disrupt minefield	NA	NA	1.5 PH
Standardized fix minefield	NA	NA	1.5 PH
Standardized turn minefield	NA	NA	3.5 PH
Standardized block minefield	NA	NA	5 PH
Triple-standard concertina	NA	NA	1 PH/300 m
Road crater	NA	NA	1.5 SH
Point minefield	NA	NA	1 SH
Concertina roadblock	NA	NA	1 SH
Bridge demolition (massive)	NA	NA	2 SH
Bridge demolition (steel)	NA	NA	1 SH
Mine preparation at the TF Class IV/V supply point	NA	NA	1 SH/100 mines
LEGEND: BTH (blade team hour). One blade team working for one hour. A blade team consists of two engineer blades (two dozers, two ACEs, or one ACE and one dozer). One vehicle digs (cutter) while the other spreads the spoil (striker). A dozer-ACE blade team uses the dozer BTH. SEEH (SEE hour). One SEE working for one hour. PH (platoon hour). One platoon (3 squads) working for one hour. SH (squad hour). One squad working for one hour.			

Table 2-4. Planning factors for standardized row minefields

Effect	Resource Factor	Front	Depth	Full-Width AT Mines	Track-Width AT Mines	Frag AP Mines
Disrupt	0.5	250 m	100 m	42	84	NA
Fix	1.0	250 m	120 m	63	84	NA
Turn	1.2	500 m	300 m	336	168	NA
Block	2.4	500 m	320 m	378	168	**84 (Korea Only)**

Table 2-5. Planning factors for scatterable minefields

System	Minefield Size	SD Time	Arming Time
ADAM	400 x 400 m 200 x 800 m	4 hr 48 hr	Within 1 min after ground impact
RAAM	400 x 400 m 200 x 800 m	4 hr 48 hr	2 min 45 sec
Volcano (one load = 160 canisters or 960 mines [800 AT and 160 AP])	Turn or block (1 per load): • Ground: 555 x 320 m • Air: 557 x 320 m Fix or disrupt (4 per load): • Ground: 277 x 120 m • Air: 278 x 120 m	4 hr 5 days 15 days	2 min
MOPMS	70 x 35 m	4 hr*	89 sec

*Can be recycled 3 times for a total of 13 hr

Table 2-6. Ranges of common weapons

Weapon		Maximum Effective Range	Planning Range*
FRIENDLY WEAPON SYSTEMS			
M16A2		580 m	400 m
M249 SAW		1,000 m	800 m
M60		1,100 m	1,100 m
M203	Area	350 m	350 m
	Point	160 m	160 m
M2, .50 Cal	Area	1,830 m	1,830 m
	Point	1,200 m	1,200 m
MK19	Area	2,200 m	2,200 m
	Point	1,600 m	1,600 m
AT4		300 m	300 m
M47 Dragon		1,000 m	800 m
Javelin		2,000 m	2,000 m
M1 Abrams tank	105 mm	2,500 m	2,000 m
	120 mm	3,000 m	2,500 m
M2 Bradley ITV	25 mm (APDS)	3,000 m	1,700 m
	25 mm (HEI-T)	3,000 m	1,700 m
	TOW2	3,750 m	3,750 m
60-mm mortar	HE	3,400 m	50 m (min)
	WP	4,800 m	50 m (min)
	ILLUM	931 m	50 m (min)
81-mm mortar	HE	4,595 m	75 m (min)
	WP	4,595 m	75 m (min)
	ILLUM	3,150 m	75 m (min)
4.2-in mortar	HE	6,840 m	770 m (min)
	WP	5,650 m	920 m (min)
	ILLUM	5,490 m	400 m (min)
SOVIET-STYLE WEAPON SYSTEMS			
BMP, 73 mm		800 m	800 m
AT3 missile		3,000 m	3,000 m
AT5 missile		4,000 m	4,000 m
BMP-2		2,000 m	2,000 m
BTR, 14.5 mm		2,000 m	1,000 m
T-72 tank, 125 mm		2,100 m	2,000 m
T-80 tank, 125 mm		2,400 m	2,000 m
T-80 AT8		4,000 m	4,000 m

*The planning range is based on ideal weather conditions during daylight.

Issue the Commander's Guidance

The commander should be as specific as possible with his initial obstacle guidance. If the commander narrows the COA focus, he should also provide obstacle guidance. His guidance is a key factor to an early start and must be solicited if not offered.

Course-of-Action Development

Detailed planning begins following the COA development (Figure 2-12). The engineer focuses on four specifics of obstacle planning in the scheme of engineer operations (SOEO) for the defensive plan:

- Direct-/indirect-fire analysis.
- Obstacle-intent integration.
- Obstacle priority.
- Mobility requirements.

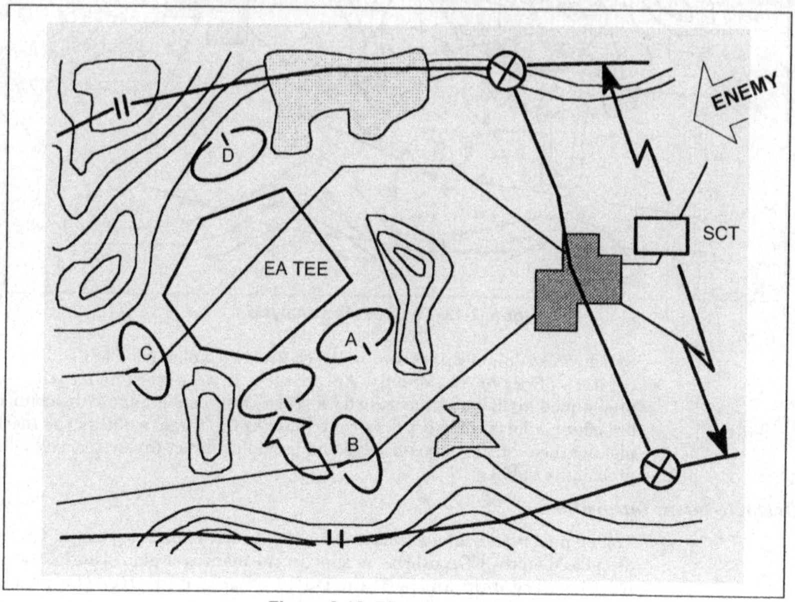

Figure 2-12. TF defense COA

Direct-/Indirect-Fire Analysis

The direct-/indirect-fire analysis examines how engineers can best use obstacles (within the commander's intent) to enhance the direct-/indirect-fire plan. Figure 2-13 shows a sample direct-fire analysis. This analysis can be used to formulate obstacle locations with the direct-fire plan. The engineer must have a fundamental understanding of the direct-/indirect-fire and maneuver plans and the TF's organization of the EA to effectively integrate obstacles with the direct-/indirect-fire plan.

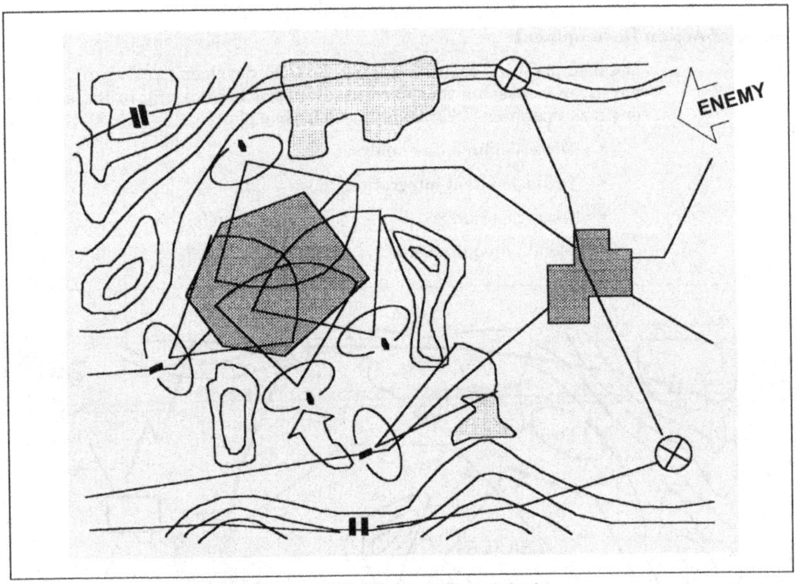

Figure 2-13. TF direct-fire analysis

Synchronization of direct and indirect fires with obstacles multiplies the relative effect on the enemy. An obstacle is an excellent location for preplanned artillery and mortar fires. These fires can eliminate dismounted breaching efforts. Indirect fires contribute to the threat's ability to breach, making the obstacle more effective and providing direct-fire systems a higher probability to kill.

Obstacle-Intent Integration

The engineer determines locations for the directed obstacle groups. Groups are placed on the COA overlay to support the maneuver plan. This location is for planning only and will normally be adjusted after the ground reconnaissance.

Obstacle groups target specific enemy elements based on the SITEMP. The engineer generally allocates an obstacle group against a battalion-size AA. This approach mirrors the staff's placement of a company/team against the same enemy force. The company's/team's fire responsibility drives the placement of the obstacle groups. The engineer advises the commander on which specific effect each directed obstacle group must achieve. He plans obstacle groups to—

- Disrupt the enemy.
- Turn the enemy into an area where friendly units can mass fires.
- Fix the enemy in the EA and enhance his direct-fire destruction.
- Block the enemy from using an AA.

The engineer integrates directed obstacle groups with the COA. The obstacle effects are shown on the COA overlay using obstacle-effect graphics (Figure 2-14). The engineer draws the obstacle-group graphic to reflect the location, the target, and the specific intent of the group as accurately as possible. The engineer should visualize how the terrain naturally effects maneuver. Terrain visualization is vital to proper obstacle-group design.

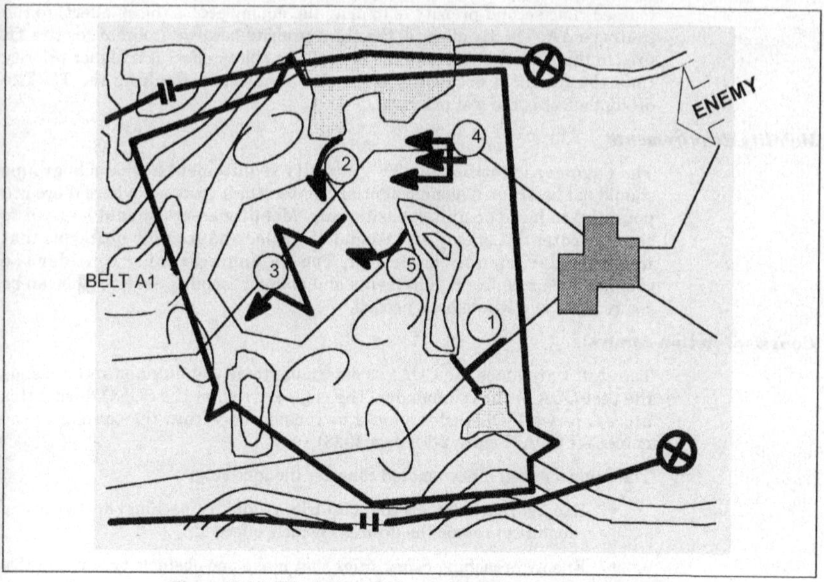

Figure 2-14. TF obstacle-intent integration and priorities

Note the placement and the effect of obstacle groups in Figure 2-14. First, the engineer must manipulate the MRB into the EA. The turn groups (2 and 5), combined with a heavy volume of AT fires from a company team at the turning point, achieves this. In the EA, particularly where the TF fires are massed, a fix group (3) slows the enemy and increases the effects of the fires. A block group (1) in the south, along with direct fires from a company team, will stop the advance of any element along the southern AA. A disrupt group (4), with indirect fires, will break up the C^2 and the tempo of the attacking force.

Protective obstacles in front of all team positions protect the teams from the enemy's final assault.

Obstacle Priority

The staff determines the priority of each group depicted on the overlay. Priority is established by the commander's intent and the most likely enemy COA. The obstacle priority should reflect the TF's greatest obstacle requirement. The primary obstacle effort can be with an economy of force where the commander needs more obstacles to overcome a shortage of direct-fire systems. The TF engineer should be cognizant of flank protection, weapon types and ranges, and the overall commander's intent for the entire force before placing obstacle priority on the main EA. Priorities assist the engineer in allocating resources and ensuring that the most critical obstacle groups are emplaced first.

In Figure 2-14, the first priority is to turn the enemy where the fires are massed. The second priority is to deny the enemy access (block effect) to the southern AA. The fix effect is the third priority because it enhances the TF fires in the EA but only slows the enemy). The block effect is a higher priority than the fix effect because it stops the enemy from flanking the TF. The disrupt effect is the last priority.

Mobility Requirements

The engineer identifies the TF's mobility requirements. Obstacle groups should not be arrayed along potential counterattack routes or where there is a potential to hamper unit repositioning. Mobility assets should be used to counter potential enemy situational obstacles and friendly obstacles that might hinder friendly maneuver. The TF engineer must consider the commander's mobility requirements and plan for mobility assets so he can be ready when and where he is needed.

Course-of-Action Analysis

The staff war-games the COAs to determine their viability and recommends the best COA to the commander. The engineer refines the SOEO during this process as well. Obstacles should be considered within the context of the maneuver COA (Figure 2-15, page 2-28).

The engineer staff officer should consider the following:

- Enemy reactions at the obstacle groups (breaching or bypassing capability) versus the desired obstacle effect.
- Enemy breaching capabilities that make one obstacle type preferable to another (such as an AT ditch versus a minefield).
- Obstacle locations that hinder friendly maneuver.
- The compatibility of obstacle effects and weapon-system capabilities.
- Adequate direct-/indirect-fire control measures and targeting that support the obstacle effect. The effects of artillery and obstacles must be synchronized to gain the desired effect on the enemy's maneuver.
- Locations and types of enemy situational obstacles that make one type of breaching asset preferable to another.

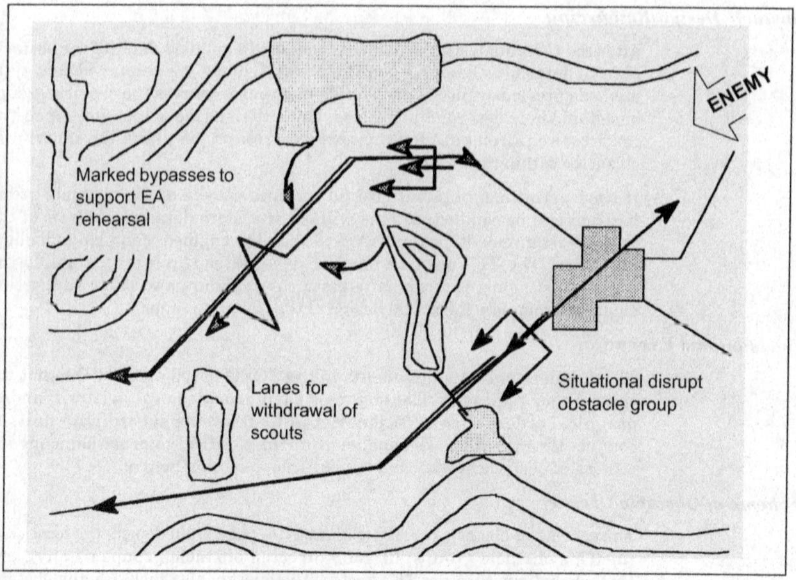

Figure 2-15. Obstacle-plan refinement

After war gaming, the staff adjusts the COA (including the obstacle plan) by—

- Changing the location of obstacle groups.
- Changing the obstacle effects.
- Adding more situational obstacles.
- Adding more reserve obstacles.
- Identifying other mobility requirements.
- Refining artillery targets based on obstacle-group changes.

Mobility Requirements

The staff determines which obstacles require lanes and determines the closure criteria for the lanes. It also determines obstacle-restricted areas that support the TF's maneuver. Lanes and bypasses are determined using tactical repositioning requirements developed during the COA analysis. Requirements for rehearsal movement, placement of the target reference point (TRP), and logistical support of forward TF elements are also considered in lane development. Mobility requirements identified during COA development are synchronized and refined during COA analysis. Additional mobility requirements identified during war gaming are resourced and planned. (Lane marking is discussed later in this chapter.)

Obstacle Design/Resourcing

After the COA analysis, the engineer conducts a detailed study of the obstacle plan to determine resource requirements. Groups are resourced using the methods previously discussed. The TF engineer resources the obstacle groups based on their assigned priorities. Once the engineer has developed the resource requirements for the obstacle groups, he plans the individual obstacles within the group.

If time permits, a detailed ground reconnaissance of the obstacle-group location can be conducted. This will allow a more detailed analysis of the obstacle requirement for that AA, and then the engineer can plan individual obstacles. The TF engineer usually designates the intent to guide the companies/teams; the company/team commanders and their supporting engineers complete the actual design of the obstacle groups.

Decision and Execution

The engineer makes adjustments to the SOEO based on the COA that the commander approves. The engineer then provides oral, written, and/or graphical orders with sufficient detail to allow the subordinate units to conduct the operation. The engineer provides critical information using the scheme-of-obstacle overlay and the obstacle-execution matrix.

Scheme-of-Obstacle Overlay

The scheme-of-obstacle overlay (Figure 2-16, page 2-30) depicts the location of the TF's obstacle groups, brigade-directed obstacle groups (if any), and obstacle belts within the TF's sector. The overlay also includes any obstacle restrictions dictated from a higher headquarters. The overlay depicts the obstacle groups using the standardized obstacle-effect symbols. The overlay does not generally show individual obstacles unless the engineer has had sufficient time to conduct a thorough ground reconnaissance where exact obstacle locations have been identified. The engineer must exercise extreme caution if he uses individual obstacles on the overlay. He must ensure that inexperienced leaders do not attempt to emplace obstacles exactly as shown on the overlay, but instead, properly site the obstacle with the company/team commander. The scheme-of-obstacle overlay graphically depicts how the commander seeks to influence enemy maneuver through obstacles.

Obstacle-Execution Matrix

The obstacle-execution matrix includes specific instructions and detailed information concerning the obstacle groups shown on the scheme-of-obstacle overlay. Develop an obstacle-execution matrix for all situational, reserve, and direct obstacles within the TF area. As a minimum, the matrix should include the information shown in Figure 2-17, page 2-31.

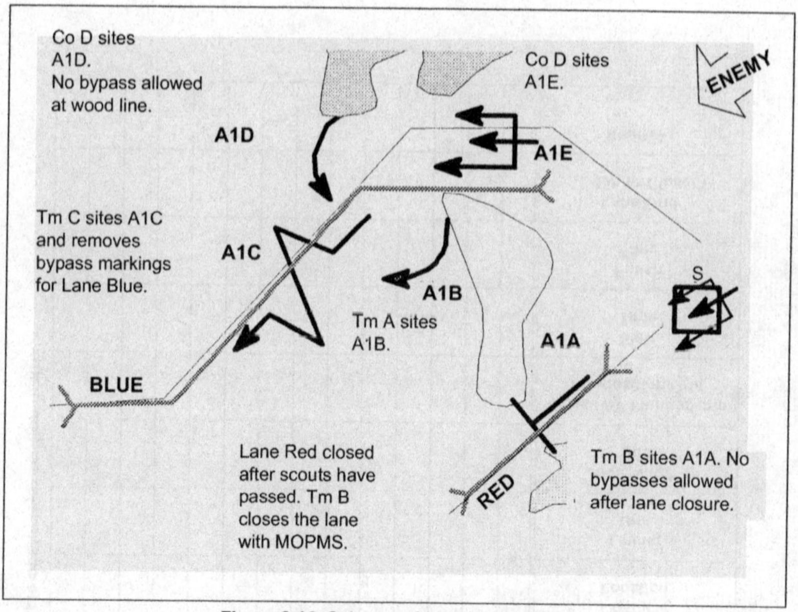

Figure 2-16. Scheme-of-obstacle overlay

Percent Complete	Priority	Standard Obstacle Number				Obstacle Location	Laying Unit	Overwatch Unit	Class IV/V Planned and Pre-position Grid	Start Time	Finish Time	Lane Grid (Open or Closed)	Remarks
		Obstacle Authority	Zone, Belt, or Group	Obstacle Type and Number	Status								

Figure 2-17. Sample obstacle-execution matrix

SITING AND EMPLACING TACTICAL MINEFIELDS

This section outlines the principles for siting tactical minefields to support the company/team. These principles apply to all methods of emplacement—standard pattern, row mining, and SCATMINE systems. The focal point of the discussion on siting is the coordination that must occur between the emplacing engineer (normally the engineer platoon leader) and the maneuver company commander. Coordination between the engineer platoon and the maneuver company is perhaps the most vital component of effective obstacle integration, and it is also a vital component in EA development. Obstacles are directly integrated with weapon effects, capability, and the fire plan at this level. The two subcomponents of tactical-obstacle siting are coordinating with the maneuver commander and siting the minefield.

Coordinating with the Maneuver Commander

Effective coordination with the maneuver company/team commander who will fight the obstacle(s) is essential to realizing the full potential of minefields as a combat multiplier. In short, the emplacing engineer becomes the maneuver company/team commander's team engineer for the mission. The engineer, the fire-support team (FIST), and the maneuver commander must work closely to ensure complete integration of the minefield into all aspects of the company plan. The engineer must be integrated into the maneuver company/team EA development process. Throughout each step of the process, the engineer must provide the maneuver commander and the FIST with the engineer expertise necessary to ensure complete and effective obstacle integration.

Before the emplacing engineer can conduct effective coordination, he must have tools and information from the TF order that serves as common ground between the emplacing engineer, the FIST, and the maneuver commander. The order drives the integration of tactical obstacles into the fire plan and ensures that the obstacles affect their intended enemy target in a way that supports the scheme of maneuver.

Modified Combined Obstacle Overlay

The MCOO is a product from the IPB process that graphically depicts the maneuverability of the terrain. It depicts slow-go and no-go terrain relative to the type of enemy force. It also defines AAs and MCs that the enemy may use for its attack. Since tactical obstacles attack the enemy's maneuver and must complement the existing terrain, the MCOO is vital to obstacle siting. It helps ensure that obstacles correctly address the enemy AAs and MCs. It also helps select how and what part of the enemy formation will be directly attacked by obstacles, and it shows the effect the obstacles will have on the enemy's maneuver.

Situation Template

The SITEMP is developed by the maneuver battalion S2 and the TF engineer during the IPB. It estimates how the enemy will attack, in terms of size and type of units, and the formations it will use. Tactical obstacles are employed to produce specific effects on specific enemy targets. Therefore, the SITEMP helps the engineer and the maneuver commander site and emplace obstacles in a way that attacks the intended target. The SITEMP may also depict the likely routes for enemy reconnaissance elements. This helps the engineer and the maneuver commander analyze requirements for reconnaissance and surveillance (R&S) patrols that defeat enemy attempts to reconnoiter the obstacles and reduce enemy effectiveness before the attack. The type of formations the S2 expects the enemy to use during the entire course of the attack is also vital information. The SITEMP should identify when the enemy is in march, prebattle, and attack formations. The enemy formation may impact on the necessary front of obstacle groups and the obstacle groups' effectiveness in achieving the intended effect on enemy maneuver.

Commander's Intent

The emplacing engineer, the FIST, and the maneuver commander must have a common understanding of the battalion commander's intent. The battalion commander's intent is his vision of the battle and normally outlines what actions the unit must do to accomplish the mission. The commander's intent may include key aspects of the plan that he wants to emphasize to subordinates to synchronize the actions of subordinates toward a single purpose. The engineer must understand the commander's intent and how it relates to integrating obstacles. The engineer should always ensure that the obstacles he is emplacing support the commander's overall intent.

Maneuver Graphics and the Fire Plan

In order to fully support the scheme of maneuver, the engineer must have and understand the maneuver graphics on the battalion's operational overlay. The maneuver graphics use symbols to depict the missions of each subunit within the battalion. Maneuver control measures such as battle positions, sectors, phase lines, passage lanes/points, and counterattack axis are vitally important to understanding the plan and integrating tactical obstacles. The maneuver graphics may include direct-fire control measures that direct how and where combat forces will mass, shift, and lift fires to destroy the enemy. Direct-fire control measures include EAs, trigger lines, and TRP and unit boundaries. In short, they dictate the direct-fire responsibilities of each subordinate. Understanding the direct-fire plan and the organization of the engagement is fundamental to integrating obstacles with fires. The maneuver graphics also give the engineer an appreciation of how tactical obstacles supporting one unit must complement the adjacent units. This is particularly true of adjacent EAs or plans requiring any tactical repositioning of forces.

Obstacle-Execution Matrix

The obstacle-execution matrix includes specific instructions and detailed information concerning the obstacle groups shown on the scheme-of-obstacle overlay. This matrix gives the engineer critical information on minefield groups that will be emplaced within the company's/team's AO. As a minimum, the obstacle-execution matrix should include the information shown in Figure 2-17, page 2-31.

Scheme-of-Obstacle Overlay

At the maneuver battalion level, the scheme-of-obstacle overlay depicts the location of brigade-directed belts, TF obstacle groups, and any directed obstacles within the battalion sector. Any obstacle restrictions attached to an obstacle control measure (belt or group) that preclude the employment of certain types of obstacles are annotated on the overlay. A scheme-of-obstacle overlay is a graphic control measure that defines the general location of the obstacle groups and the effect to be achieved by them. The scheme-of-obstacle overlay does not normally depict individual obstacle locations. The location of individual obstacles within a group is determined during the siting process between the emplacing engineer and the maneuver company commander. When overlaid on the maneuver graphics and the SITEMP, the scheme-of-obstacle overlay should depict the essential elements of obstacle integration:

- Enemy targeted by the obstacle.
- Location of the obstacle on the battlefield.
- Unit covering the obstacle.
- Directed link between the obstacle effects and the fire plan.

Fire-Support Plan

The emplacing engineer should be familiar with the key elements of the fire-support plan. He must understand the general scheme of fires and how the fires support the scheme of maneuver, the commander's intent, and the obstacle plan. Normally, the emplacing engineer does not need the entire fire-support overlay depicting the location of all targets. However, he should know the location of fire-support targets directed by the battalion to cover obstacles. The emplacing engineer should know who has the priority of fires for each phase of the battle. The emplacing engineer should also know the location and the type of priority targets or FPF allocated to the maneuver company he is supporting. During coordination with the maneuver company, the emplacing engineer should discuss the fire-support plan with the company FIST and get updates on changes to the plan as well as any company-level fire-support plans that may impact on the integration of obstacles.

Combat Service Support

The engineer must be familiar with the plan for combat service support (CSS). In particular, the engineer must know the location of major supply routes into and through the battalion area, the location of the battalion logistics release point (LRP), what routes the maneuver company will take from its position to the LRP, and the location of key battalion logistics nodes. The emplacing engineer must always be cognizant of the sustainment traffic flow and the impact obstacle emplacement has on sustainment operations. Of particular concern to the engineer is the location of the Class IV/V supply point and the routes to it.

Battlefield Operating System

During coordination with the maneuver company commander, a checklist or a framework is a useful tool for organizing thoughts and formulating questions. Below is a list of considerations or points of coordination that should drive the integration of the emplacing engineer and the maneuver company commander. The list is organized using the Battlefield Operating System (BOS) because it provides a logical sequence and a framework that is easily remembered.

- Intelligence.

  - Enemy AAs and MCs (mounted and dismounted).
  - Likely enemy COA and possible reactions to the obstacles.
  - Enemy breach capability.
  - Enemy reconnaissance routes and friendly counterreconnaissance or R&S plans, particularly company-level patrols.
  - Likely enemy formations and transitions between formations.
- Maneuver.
  - Specified, implied, and essential tasks of the maneuver company.
  - Higher commander's intent.
  - Organization of the defensive position, including—
    - Task organization (type of weapons).
    - Decisive point or defeat mechanism.
    - Organization of direct fires in the EA.
    - Location and marking of direct-fire control measures.
    - Position of weapons to cover assigned direct-fire responsibilities.
  - Tactical mobility requirements of the maneuver company and any adjacent units, including—
    - Counterattack axis.
    - Repositioning of forces and their routes to alternate, supplementary, or subsequent battle positions.
    - Employment of reserves.
    - Passage of lines.
  - Obstacle protection measures.
- Mobility/survivability.
  - Intent of tactical obstacles covered by the maneuver company (enemy target, obstacle location, and obstacle effect).
  - Obstacle control measures and restrictions imposed by higher headquarters.
  - Mobility requirements (lanes/gaps), as identified above in maneuver.
  - Mutual support between the obstacle location, the fire plan, obstacle effects, and survivability positions.
  - Security for engineers provided by the maneuver unit supported.
- Fire support.
  - Location of the company FIST and frequency of fire support.
  - Updates on the tentative fire-support plan.

- Allocation of fires to the company, including—
  - > Artillery or mortar targets.
  - > Priority targets, types of targets, and the FPF.
- Covering obstacles and their effects with indirect fires.
- Indirect-fire control measures to synchronize direct fires, indirect fires, and obstacles.
- Area-denial artillery munition (ADAM)/remote antiarmor mine (RAAM) use (lane closure and breached obstacle repair).
- Registering fires. (Deconflict with obstacle emplacement; registration should occur after obstacles are sited but before emplacement.)
- Company fire-support execution matrix.
- Means for obtaining fire support, if enemy contact is made during emplacement.

• Air defense.

- Enemy air AAs during emplacement.
- Update on changes to air-defense warning and weapons status.
- Location of air-defense systems that can cover engineers emplacing obstacles.
- Method of obtaining early air-defense warning.

• CSS.

- Tentative location of the mine dump, if used, within the company position and routes from the mine dump to obstacles.
- Routes the company plans on using to conduct logistics package (LOGPAC) operations that must remain open.
- Manpower assistance for operations at the Class IV/V supply point and the mine dump.
- Casualty evacuation routes for scouts, observation posts (OPs), and ADA systems.

• C^2.

- Location of the commander during defensive preparation.
- Frequency-modulated (FM) net of the supported company and the means of communication.
- Unit boundaries affecting obstacle emplacement.
- Time and place of the company/team order.
- Coordination that must occur with adjacent units.
- Obstacle reporting and recording requirements.

- Control and execution of situational and reserve obstacles.
- Lane-closure responsibilities and procedures.
- Time and method of obstacle turnover, including lanes.
- Company/team understanding of the obstacle intent.

SITING THE MINEFIELD

The emplacing engineer and the company/team commander site individual obstacles to achieve synchronization between the obstacle effect and fires. Siting is a key component to the EA development process, and it represents the final adjustments to the obstacle location and the fire control measures before emplacement.

Certain preconditions are necessary to site individual obstacles. First, the company/team commander decides where he plans to mass fires and marks the necessary fire control measures on the ground. The location of these control measures must be clear, since they are the basis for obstacle siting. The commander then identifies tentative locations for key weapons within the position or the sector. Finally, the commander and the engineer must both understand the intent of the obstacle group.

Obstacle siting concentrates on marking the obstacle group as a whole instead of marking each individual obstacle. In broken terrain, however, it may be easier to site individual obstacles. The company/team commander and the emplacing engineer use vehicles or soldiers from the company/team, the engineer platoon, or both to simulate the enemy force and do the physical marking. The simulated enemy force moves into the EA to the enemy side of the obstacle group. The engineer platoon leader and the company/team commander collocate near the weapons covering the obstacle. As a technique, one or all of the tanks, Bradleys, and other crew-served weapons may occupy their position and contribute to the siting process. All participants in the siting process use a common FM net to communicate during siting.

The simulated enemy force moves into the EA, simulating the enemy's attack. It deploys into a formation of front that is similar to the expected enemy formation. Once it is near the marked fire control measures, it places markers at intervals as it drives the trace of the obstacle-group effect (or individual obstacles in broken terrain). It remains oriented on key fire control measures to ensure that obstacle location and effect are synchronized with fires. During the process, each participant verifies that he can cover the obstacle, notes the location of fire control measures and obstacles, and records the appropriate data on range cards. As the simulated force drives the obstacle trace, siting participants also identify dead space and requirements to refine the location of the obstacle group and fire control measures. Figure 2-18, page 2-38, illustrates how the engineer and the company/team commander work together to site turn- and fix-obstacle groups.

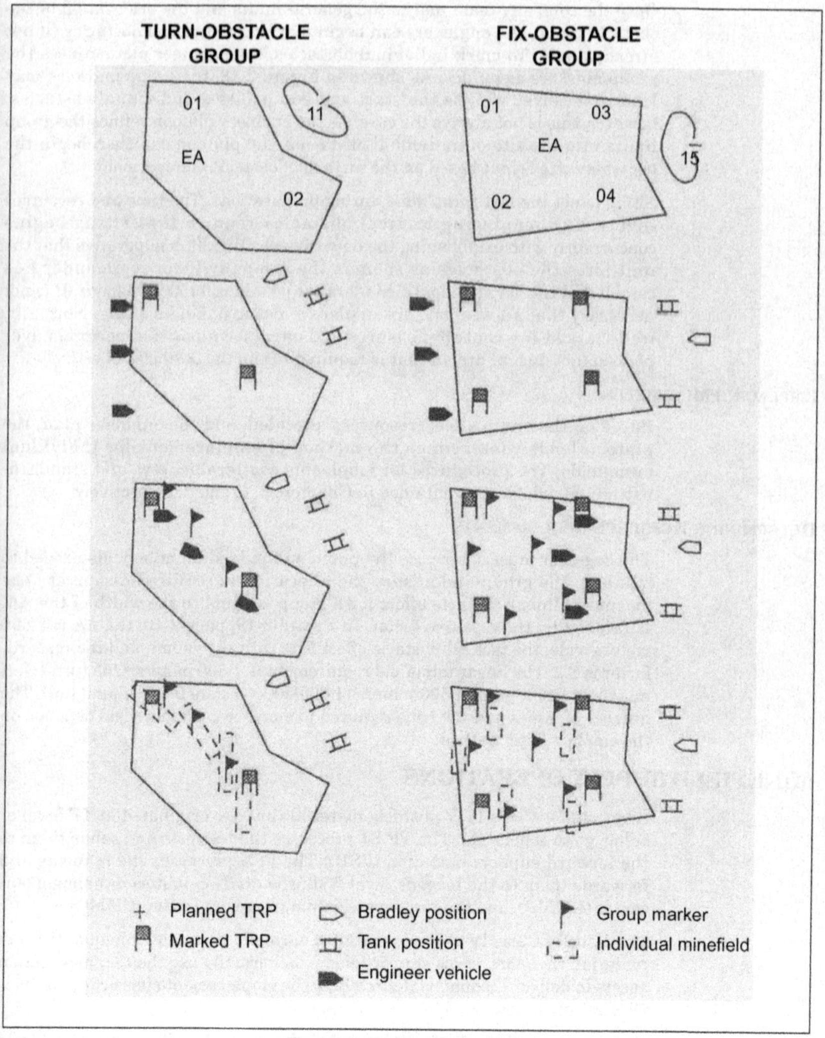

Figure 2-18. Minefield siting

Once the company/team marks the general limits and the orientation of the obstacle group, the engineers can begin marking individual obstacles (if not already done). To mark individual obstacles, the engineer platoon uses the group markers as guides. As shown in Figure 2-18, the group markers may lend themselves well as the start and end points of individual obstacles; however, this is not always the case. As the engineer platoon refines the group limits into the site of individual obstacles, the platoon can then begin the necessary site layout based on the method of obstacle emplacement.

Siting is not the last thing done during preparations. The time and resources involved in emplacing tactical obstacles require that siting begins concurrently with establishing the defensive position. It is imperative that the unit sites the obstacles as soon as the company/team commander has established the EA and identified tentative positions for key weapons. It is not necessary that all weapons are in place and dug in before siting. Normally, well-marked fire control measures and one known position per maneuver platoon (not dug in) are all that is required to site the obstacles effectively.

Emplacing Minefields

Based on the group effect, resources allocated, and the engineer plan, the platoon leader determines the method of emplacement for individual minefields. The procedures for emplacing scatterable, row, and standard-pattern minefields are contained in Chapters 3, 6, and 7, respectively.

Determining Resource Requirements

The engineer must determine the number of individual minefields needed to make up the group and ensure the allocation of required resources. The amount of linear obstacle effort for a group is equal to the width of the AA, multiplied by the resource factor. In Figure 2-19, page 2-40, the AA is 1,500 meters wide, the tactical-obstacle effect is to turn the enemy, and the resource factor is 1.2. The linear minefield requirement is 1,800 meters. One turn-effect minefield has a front of 500 meters (1,800/500 = 4 minefields [round up]). The number of mines and the time required to emplace each minefield depends on the emplacement method.

MINEFIELD SUPPLY OPERATIONS

Requests for Class IV/V obstacle material that are originated at TF level or below go to the TF S4. The TF S4 processes the requests and sends them to the forward support battalion (FSB). The FSB processes the requests and forwards them to the brigade-level FSB, the division material-management center (DMMC), and the corps material-management center (CMMC).

The issue of Class IV obstacle material normally involves large quantities of material; therefore, corps support elements normally use their transportation assets to deliver the material directly to the emplacement sites or to the Class IV/V supply point.

Units request Class V obstacle material somewhat differently. The TF S4 notifies the brigade S4 of Class V requirements. The brigade S4 notifies the division ammunition officer (DAO) in the DMMC, who authorizes Class V issue by the ammunition transfer point (ATP). Class V obstacle material,

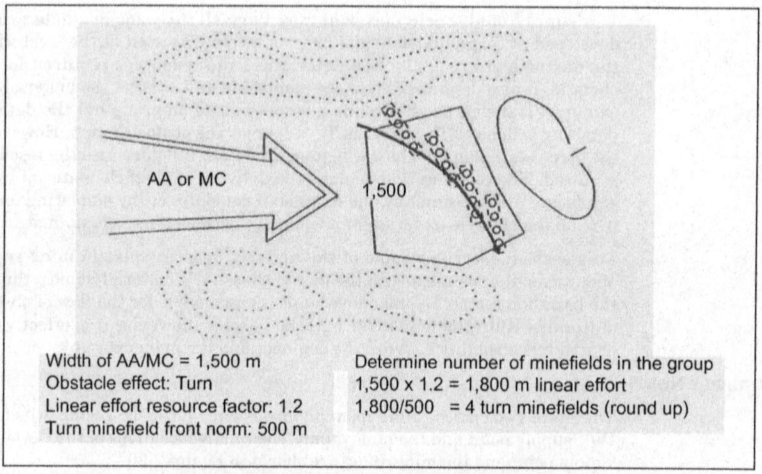

Figure 2-19. Example of minefield resourcing

unlike most ammunition, is delivered to the user at the obstacle emplacement site or the Class IV/V supply point.

A supply request includes the quantity, the required delivery time, the transportation responsibilities, and the desired delivery location. The quantity includes the total for each type of obstacle. There may be several Department of Defense identification codes (DODICs) and national stock numbers (NSNs) involved, depending on the types of obstacles required. The required delivery time is very important for ensuring an early start on the preparation of the battlefield because the lack of material could adversely affect the mission. Transportation responsibilities must be clearly understood. Material handling equipment (MHE) is required to ensure a rapid turnaround of haul assets.

In addition, the brigade staff coordinates with the TF staff to identify the location of Class IV/V supply points in the TF sectors. Prompt identification of TF Class IV/V supply points is required if the obstacle material is forwarded from the corps into the TF sector. If the material is not forwarded into the TF sector, the brigade is responsible for delivering it to the TF.

At the TF level, sustaining obstacle operations is an extremely difficult task. Centralized throughput operations by the corps or the division stop at the TF level. Mass quantities of obstacle material, especially mines, are centrally received, broken down into minefield packages, and then distributed throughout the sector based on the obstacle plan. At some point in the distribution plan, the TF turns over control of the obstacle material to engineers who then emplace the obstacles. Obstacle logistics (especially for mine warfare) at the TF level can be complex, require prudent use of scarce haul assets and MHE, and demand positive C^2.

The supply of obstacle material may vary slightly for obstacle group developed at corps, division, and brigade levels. The staff at the level wher the obstacle group is planned determines the resources required for th obstacle. It also determines how the emplacing unit will get the material. Fo example, if the corps staff plans a reserve obstacle group but the detaile planning is done at TF level, the TF resources the obstacle group. However, i the corps staff plans the obstacle group in detail, it determines the resource required. The corps staff also plans the delivery of obstacle material to th emplacing unit. Alternately, the corps staff could direct the emplacing unit t pick up the obstacle material from a location such as the corps storage area.

This section describes some of the underlying principles in mine suppl operations. It concentrates on the flow of Class IV/V material (mines) throug the battalion sector. The maneuver unit is responsible for the flow of obstac material within the maneuver battalion sector; however, it is effectively shared responsibility between the engineer and the maneuver unit.

Resupply Nodes

There are two critical mine resupply nodes within the TF sector—the Clas IV/V supply point and the mine dump. The relative locations of the Class IV/ supply point and the mine dump are shown in Figure 2-20.

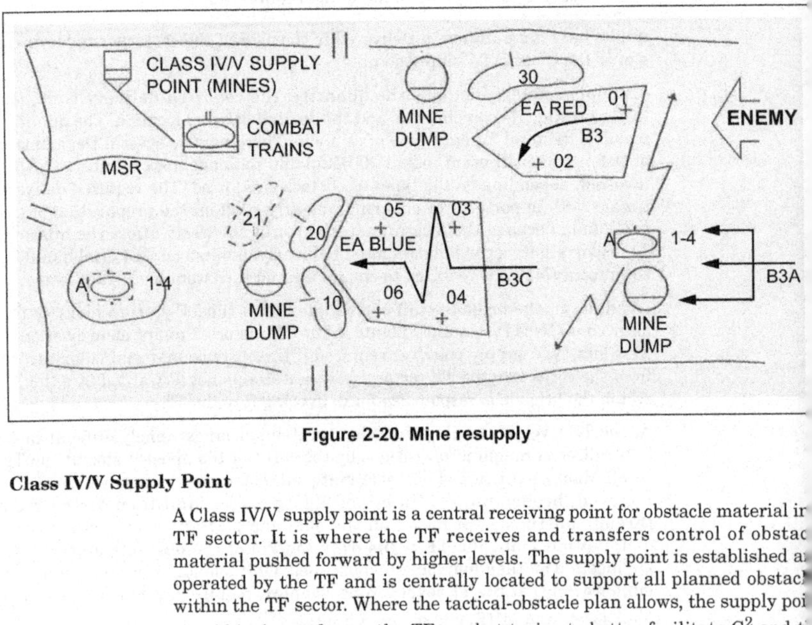

Figure 2-20. Mine resupply

Class IV/V Supply Point

A Class IV/V supply point is a central receiving point for obstacle material ir TF sector. It is where the TF receives and transfers control of obstac material pushed forward by higher levels. The supply point is established a operated by the TF and is centrally located to support all planned obstac within the TF sector. Where the tactical-obstacle plan allows, the supply poi should be located near the TF combat trains to better facilitate C^2 and t availability of equipment. The TF combat trains command post (CTCP) shou provide C^2 of operations at the Class IV/V supply point. The TF engineer will normally furnish a representative that provides technical assistance to the CTCP.

The main purpose of a Class IV/V supply point operation is to receive obstacle material and reconfigure it based on the requirements for each obstacle group. Each supply point must have a dedicated S4 representative to track the flow of obstacle material in and out of the supply point. The supply point should have dedicated MHE to off-load the bulk quantities of obstacle material and reconfigure them into minefield packages as required. Obstacle material is normally broken down into minefield packages if the material is not already delivered in combat-configured loads. This may require a dedicated engineer representative to ensure that obstacle material is configured properly. Table 2-7 shows personnel requirements for a Class IV/V supply point.

Table 2-7. Personnel requirements for a Class IV/V supply point

Personnel	Responsibilities
TF S4/S4 NCOIC	Provides overall C^2 and Class IV/V accountability Coordinates for MHE Is assisted by the RTO
TF engineer representative	Supervises organization of Class IV/V material into packages to support obstacle groups Provides technical assistance on Class IV/V supply-point setup and mine uncrating and inspection
Infantry squad*	Downloads incoming Class V trucks
Infantry squad*	Downloads incoming Class IV trucks
Infantry squad*	Uploads outgoing Class IV/V trucks
Two infantry squads*	Uncrates and inspects mines
Support PL/PSG	Provides C^2 for cross loading of Class IV/V material
Infantry squad*	Provides local security for the Class IV/V supply point
* or equivalent-sized element from a supported unit	

The most labor-intensive task at the Class IV/V supply point is the uncrating of mines. This requires dedicated manpower from the supported maneuver force and the tools needed to break shipping bands and uncrate the mines from their containers. Another important aspect of uncrating mines is tracking fuses and booster charges. As the mines are uncrated, fuses and booster charges are separated. However, the same number and type of fuses and booster charges must be task-organized with minefield packages. This requires strict supervision; mistakes can quickly lead to confusion and a waste of emplacement time.

Because of the assets involved at a Class IV/V supply point, a TF is normally capable of operating only one supply point at any given time. If the TF sector is extremely wide or deep, several supply points may be planned; however, only one can be operated at a time, based on the commander's priorities for obstacle emplacement.

Mine Dump

A mine dump is the most forward mine resupply node. It is where mines are task-organized into mine-strip packages and then inspected, prepared, and loaded onto emplacing vehicles. A mine dump is not a permanent supply point, and it is not always used. Its use depends on the method of minefield resupply, and these techniques are discussed in more detail later in this chapter. When used, one mine dump supports a single obstacle group. It is activated or deactivated upon initiation and completion of the obstacle group's emplacement. Mine-dump operations are primarily an engineer company or platoon responsibility. However, it is a good technique to augment mine-dump operations with personnel from the company/team overwatching the obstacle group being emplaced. A mine dump may be located in the vicinity of the company/team position or closer to the obstacle group.

There are three critical tasks that must be accomplished at the mine dump:

- The minefield packages are further task-organized into mine-strip packages (complete with the right number, type, and mix of fuses and boosters) as they are transported to the mine dump. For example, if a platoon is emplacing a standardized disrupt row minefield, mines are task-organized into three mine-strip packages. As the engineer platoon moves to the mine dump to resupply, each emplacing vehicle loads a designated mine-strip package.
- The mines are prepared for emplacement; but they are not fused at the mine dump. Preparation includes loosening and greasing fuse and booster wells and ensuring proper functioning of the mine.
- The mines are transloaded onto emplacing vehicles or a mine-delivery system.

Transportation of mines from the Class IV/V supply point to the mine dump is a supported TF responsibility. However, it is normally shared between the engineer company and the TF, since neither one has the haul capability to simultaneously service all active mine dumps.

RESUPPLY RULES

The following rules govern mine resupply:

- Mines should be uncrated at the Class IV/V supply point to preserve transportation assets going forward.
- Mines are task-organized into minefield packages at the Class IV/V supply point.
- Transportation from the Class IV/V supply point to the mine dump is a shared engineer and maneuver unit responsibility when a mine dump is used.
- Mines are inspected and prepared at the last supply node (Class IV/V supply point or mine dump) before they are loaded onto an emplacing vehicle or a mine-delivery system.

- Class IV/V supply points are set up using authorized ammunition procedures and distance requirements.

SUPPLY LOCATION

Consider the following when selecting a location for the Class IV/V supply point or the mine dump:

- Carrying capacity. The location of key supply nodes and the type of resupply method used depends in part on the type, amount, and availability of haul assets. The carrying capacity plays a large role. In short, the more material a vehicle can carry, the more turnaround time you can afford. Table 2-8 provides the mine haul capacity for various types of vehicles.
- Traffic circuit. Vehicles must be able to enter, load, unload, and exit without interfering with the loading and unloading of other vehicles.
- Camouflage and cover. Protection from observation and thermal imaging is desired. Protection from artillery and air attack should be considered. Residue must be removed.
- Defense. The site must be organized for defense against enemy patrols and saboteurs.
- Time. Time factors for handling the obstacle material, to include all unloading, uncrating, inspecting, and loading, must be considered.
- Operators. Leaders and soldiers must be specifically allocated to operate Class IV/V supply points and mine dumps. They will probably remain there until the task is complete. The supply node may have to be collocated with or be near a source of manpower. Table 2-7, page 2-42, provides general guidance on how much manpower is required to sustain mine resupply operations.

RESUPPLY METHODS

The methods for obstacle material resupply are—

- Supply point.
- Service station.
- Tailgate.

In each method, corps or division transport delivers Class IV/V supplies forward to a designated Class IV/V supply point in each TF sector. The primary differences in the methods are how the material is delivered from the Class IV/V supply point to the obstacle location and whether or not mine dumps are established.

Supply Point

The supply-point method (Figure 2-21, page 2-46) requires that the emplacing engineer platoon return to the Class IV/V supply point each time it resupplies. This method does not require a separate mine dump. In effect, it moves the normal tasks associated with a mine dump to the supply point. Mines are prepared and inspected at the supply point as they are being loaded onto an emplacing vehicle or a mine-delivery system.

Table 2-8. Class IV/V haul capacity

Vehicle	Concertina Wire[1]	M15 AT Mine	M19 AT Mine	M21 AT Mine	M16 AP Mine	M14 AP Mine	MOPMS Mine	Volcano Mine	MICLIC Reload[2]	Hornet
HMMWV 1,124 kg, 6 cu m	2	51	34	27	55	56	15	1	NA	1
M35 2½-ton truck 2,250 kg, 12.5 cu m	4	102	69	55	111	113	30	2	2	2
M1078 2½-ton truck 2,250 kg, 13.4 cu m	4	102	69	55	111	113	30	2	2	2
M54 5-ton truck 4,500 kg, 13.6 cu m	7	204	138	109	222	227	61	5	3	5
M1083 5-ton truck 4,500 kg, 15.6 cu m	8	204	138	109	222	227	61	5	3	5
M930 5-ton dump truck (without sideboards) 4,500 kg, 3.8 cu m	2	112	64	32	168	71	23	3	2	2
M930 5-ton dump truck (with sideboards) 4,500 kg, 8.2 cu m	4	204	138	70	222	153	51	5	3	4
M1090 5-ton dump truck 4,500 kg, 3.8 cu m	2	112	64	32	168	71	23	3	2	2
HEMTT truck 9,000 kg, 15 cu m	8	408	277	128	444	317	94	10	7	8
12-ton S&T 10,800 kg, 24.5 cu m	13	489	333	208	533	514	148	12	9	13
40-ton lowboy 36,000 kg, 49.3 cu m	27	1,466	1,035	419	1,777	1,035	308	30	27	27
M548 cargo 5,400 kg, 14.9 cu m	8	244	166	125	266	272	74	6	4	6
M1077 PLS flat rack 14,900 kg, 17.6 cu m	9	440	352	164	586	293	110	11	9	9
No of mines per box	NA	1	2	4	4	90	21	240	NA	30
Weight per box (kg)	531	22	33	41	21	20	73	833	1,195	810
Size of box (cu m)	1.8	0.04	0.05	0.12	0.03	0.06	0.16	1.6	1.8	1.8

[1]The number of concertina = bundles; 1 bundle = 40 rolls
[2]Line charge + rocket

Several considerations may drive the use of supply-point resupply. First, if there are no additional haul assets to transport obstacle material forward from the Class IV/V supply point, the supply-point method may be the only viable technique. Secondly, the minefield group may be close enough to the supply point that any other method is less efficient.

- Advantages.
  - Minimizes unloading and loading of material.
  - Requires minimal augmentation of haul assets.

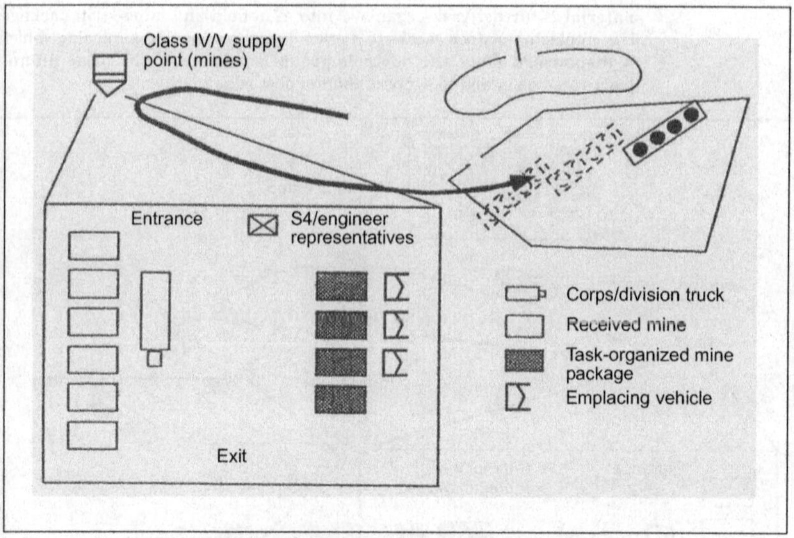

Figure 2-21. Supply-point resupply method

- Allows manpower and equipment to be massed at a single supply point.
- Streamlines C^2 of material.

• Disadvantages.

- Requires more movement of the platoon, which may take away from emplacement time.
- Requires that the platoon move in and out of the area where the minefields are being emplaced, increasing the risk of fratricide.
- May disrupt the emplacement of individual obstacles when emplacing vehicles cannot carry enough material to start and complete the obstacle. This causes emplacing vehicles to stop work, reload, and pick up where they left off.
- Requires a larger Class IV/V supply point that is capable of receiving mass quantities of obstacle material and multiple loading platoons simultaneously.

Service Station

The service-station method (Figure 2-22) centers on the activation of a mine dump forward of the Class IV/V supply point. The mines are transported to a mine dump using a combination of engineer and TF haul assets that are normally under the control of the emplacing engineer. At the mine dump, material is stockpiled and prepared by the mine-dump party. Obstacle material is further task-organized into minefield and mine-strip packages. The emplacing platoon moves to a mine dump to resupply emplacing vehicles or dispensers. Once the obstacle group is emplaced, the mine dump is deactivated or moved to support another obstacle group.

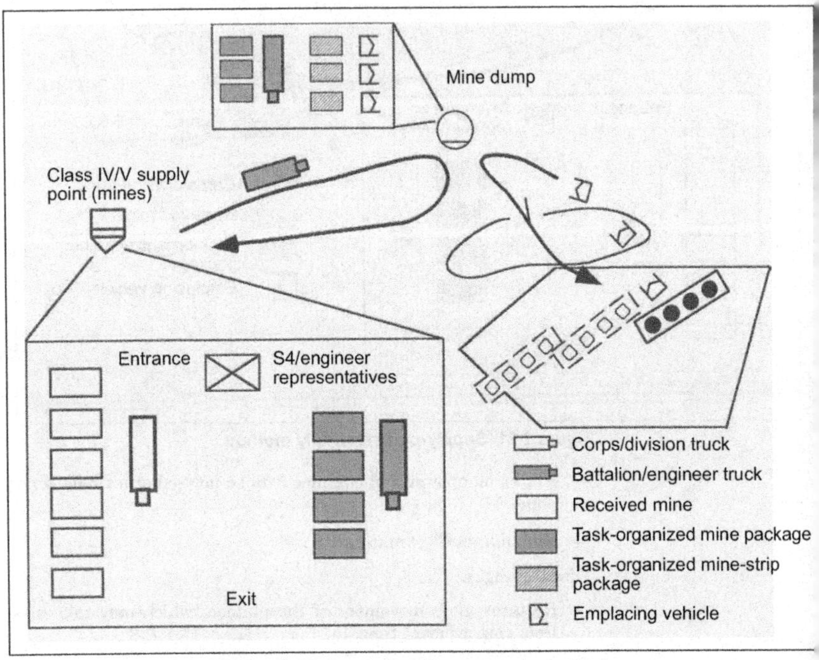

Figure 2-22. Service-station resupply method

There are several considerations for using the service-station method. First, it is used when the obstacle group is located too far from the Class IV/V supply point to allow efficient turnaround. Since this method provides for prestocking obstacle material forward, it is used when available haul assets have a relatively small capacity. This requires making frequent, short-duration resupply trips and stocking mines to keep pace with emplacement. It also streamlines emplacement since there is an opportunity to task-organize the mines into mine-strip packages, based on the emplacement method and the type of minefield. Finally, while it still requires the emplacing platoon to stop laying and resupply, it minimizes the distance and the amount of time the platoon must travel to reload. This requires that a small party be left at the minefield to help pick up where emplacement stopped.

- Advantages.

— Allows prestockage of obstacle material to keep pace with emplacement.

— Minimizes the distance and the amount of time the emplacing platoon must travel to reload.

— May provide additional manpower and security if it is located near a company/team.

- Disadvantages.

— Requires additional loading and unloading of obstacle material.

— May require augmentation with haul assets.

— Disrupts emplacement by requiring the emplacing platoon to stop obstacle emplacement, move to the supply point, reload, and return to the minefield.

Tailgate

The tailgate resupply method (Figure 2-23) transports obstacle material directly from the Class IV/V supply point to the emplacing platoon at the obstacle site. Obstacle material is transported with TF and engineer haul assets. Obstacle material is transloaded by emplacing engineers to emplacing vehicles or dispensers at the obstacle site.

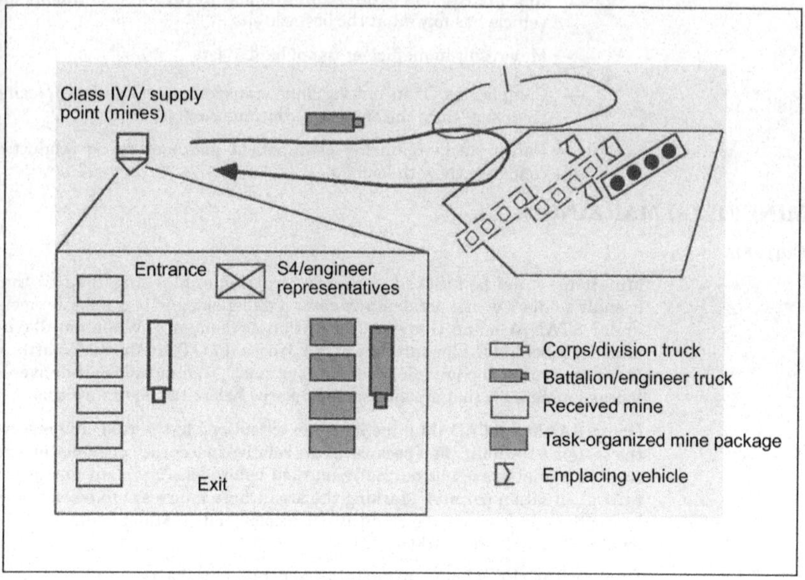

Figure 2-23. Tailgate resupply method

Two overriding considerations drive the decision to use the tailgate resupply method. First, if obstacle emplacement is being conducted during limited visibility, the tailgate method is the primary resupply method. It minimizes disruption of emplacement and the chance of fratricide as engineers move back into the work area after reloading. Secondly, tailgate resupply is the primary method used when establishing a hasty defense or when the situation is unclear and an attack can happen at anytime. Since obstacle material remains loaded until transferred to an emplacing vehicle, the tailgate method enables engineers to quickly break contact without risking a loss of obstacle material to the enemy. The tailgate resupply method is the preferred method for light forces.

- Advantages.
  - Minimizes loading and unloading of obstacle material.
  - Allows engineers to break contact rapidly in the event of enemy attack without losing obstacle material to the enemy.
  - Minimizes the movement of platoons in and out of the minefield (suitable for limited visibility).
- Disadvantages.
  - Requires augmentation by high-capacity transportation assets that are capable of offsetting the loss in turnaround time if the vehicle has to wait at the obstacle site.
  - May result in inefficient use of haul assets.
  - Complicates C^2 in linking mine transport assets with emplacing engineers since the engineers continue emplacement.
  - Causes task-organizing of minefield packages to be conducted concurrently with loading.

MINEFIELD MARKING

CRITERIA

Minefields must be marked to prevent fratricide. Marking ensures that friendly soldiers do not accidentally enter a minefield, and it is a requirement under STANAGs and Geneva Convention agreements. When emplacing minefields behind the forward line of own troops (FLOT) (in the main battle or rear area), mark the minefields on all four sides. This includes air-delivered Volcano minefields that are sited and emplaced before the enemy attacks.

Gator, RAAM, and ADAM minefields are exceptions to the rule. To preserve the system's flexibility and because of the relative inaccuracy of emplacement, these minefields are not normally marked before emplacement unless the tactical situation permits. Marking the area where mines are to be emplaced by artillery or fixed-wing aircraft is not recommended. Mines could likely be emplaced outside the marked area.

Forward of the FLOT, minefields are not generally marked before emplacement. However, commanders must make every attempt to mark these minefields as soon as the tactical situation allows. For scatterable minefields, a commander may choose to remove markings once the self-destruct (SD) time of the mines has expired; but the location of the minefield must still be recorded and forwarded to higher and adjacent units in case some of the mines did not self-destruct.

PERIMETER

Construct a perimeter fence (Figure 2-24) to mark a minefield. Start emplacing the perimeter fence before emplacing mines, preferably during site layout if the tactical situation permits. For conventional minefields, ensure that the perimeter fence is at least 15 meters outside the nearest mine or cluster. For scatterable minefields, the area inside the perimeter fence must include the safety zone. (See Chapter 3 for more details.)

Place warning signs for areas containing emplaced mines. Space the warning signs 10 to 50 meters apart, depending on the terrain. If using pickets and barbwire to mark the minefield, ensure that the wire is waist-high. If using concertina wire, use a one-roll height. Place additional strands of barbwire or rolls of concertina at the discretion of the commander.

TECHNIQUES

A commander may decide to mark individual minefields in an obstacle group or to mark the group as a whole (Figure 2-25, page 2-52). Depending on the size and the location of minefields, either technique may have the advantage of using fewer resources or labor. Normally, marking individual minefields in a fix-obstacle group requires less resources than marking the entire obstacle group. The opposite is usually true for disrupt-, turn-, and block-obstacle groups. The decision to mark individual minefields or obstacle groups should not be based solely on logistical considerations. A commander must consider the amount of tactical and/or sustainment movement required in and around the obstacle groups as well as the capability of the unit's C^2 forces.

The following advantages and disadvantages to marking individual minefields versus marking the obstacle group are provided to help commanders make the best decision.

Marking Individual Minefields

- Advantages.
  - Returning units forward of the minefields have more routes (tactical lanes or bypasses) through the obstacle group.
  - Tactical lanes need only pass through individual minefields.
  - Mine recovery is easier.
- Disadvantages.
  - Obstacle may not provide the desired effect.
  - Enemy units can more easily bypass individually marked minefields in a fix- or block-obstacle group.

Marking Obstacle Groups

- Advantages.

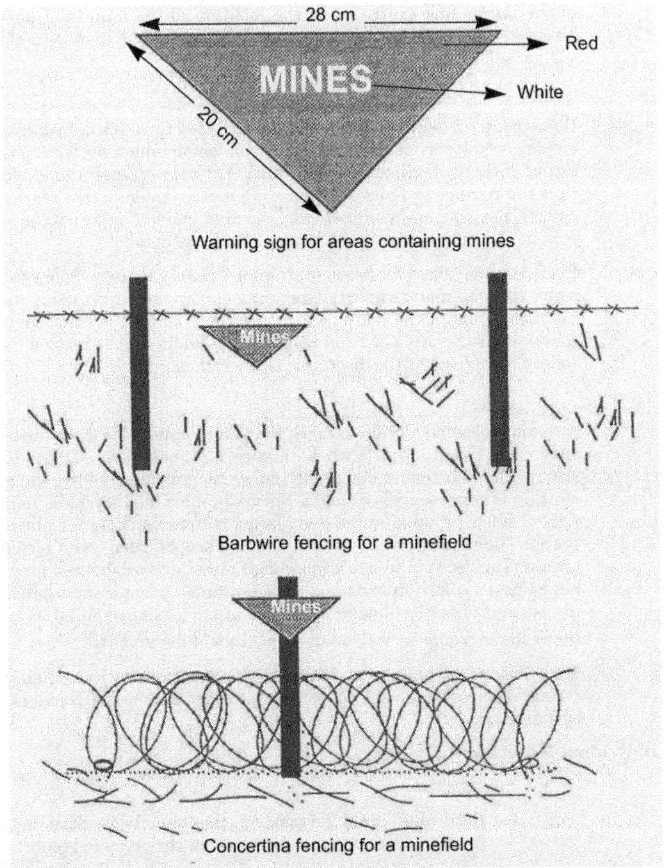

Figure 2-24. Minefield marking

- Obstacle is more likely to provide the desired effect.
- The enemy cannot easily discern individual minefields and decide when to employ breach assets in a fix-or block-effect minefield.

• Disadvantages.
  - Friendly patrols cannot clearly see if the minefield is tampered with unless they are within the perimeter fence.

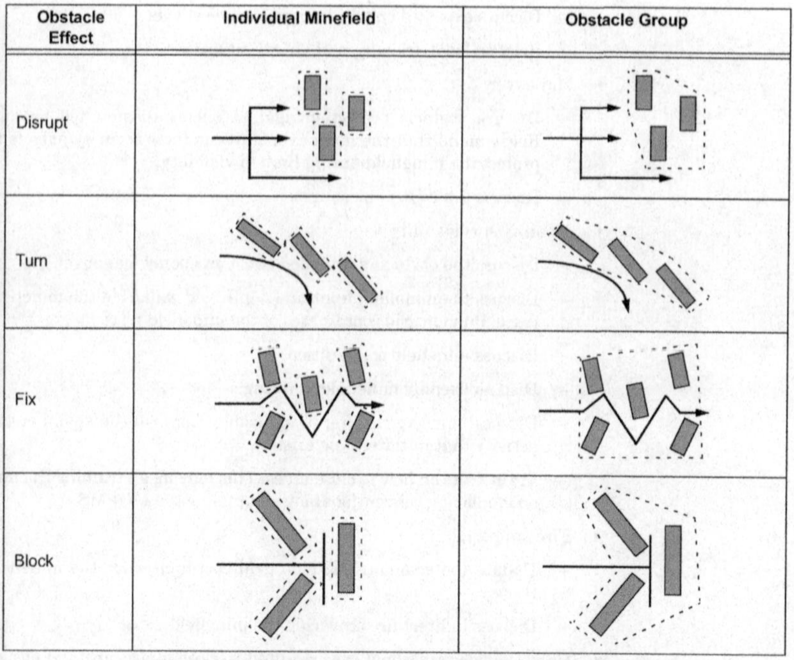

Figure 2-25. Marking of minefields and obstacle groups

— Tactical lanes need to pass through the entire obstacle group. Friendly units passing through the lanes will be slowed considerably.

MINEFIELD TURNOVER

Once an obstacle group is completed, the emplacing unit conducts minefield or obstacle turnover with the owning unit. Occasionally, the owning unit will transfer responsibility for a minefield to another unit. Minefield or obstacle turnover ensures that the commander of the owning unit is familiar with the minefield and understands his responsibilities concerning it. Turnover is conducted whether or not there are lanes/gaps to be closed. Minefield turnover is a must; the time and the location for the turnover is established during the initial siting coordination.

The engineer must verbally address the following items with the overwatching company/team:

- Intelligence.
  — Provide an update on enemy activity forward of the minefield.

- — Discuss expected enemy reconnaissance efforts.
- — Brief on local, friendly, and enemy situations.

- Maneuver.
  - — Discuss obstacle protection against enemy dismounted patrols. Recommend that the maneuver unit conducts security/patrols to protect the minefield during limited visibility.
  - — Discuss fire control measures.
- Mobility/survivability.
  - — Discuss the obstacle's intended effect on enemy maneuver.
  - — Discuss the minefield front and depth and walk/ride the minefield trace. Provide grid coordinates of the minefield trace.
  - — Discuss minefield composition.
  - — Discuss friendly minefield marking.
  - — Discuss lane/gap closure, if applicable. Confirm the signal or the activity that initiates lane closure.
  - — Train units on how to close lanes. This may mean training the unit on emplacing conventional mines or using the MOPMS.
- Fire support.
  - — Update the company FIST on grid coordinates for the minefield trace.
  - — Discuss indirect fires covering the minefield.
- CSS. Provide mines/material required to close lanes/gaps and ensure that all necessary material is available and prepared.
- C^2.
  - — Transfer graphics and documentation (minefield records demolition-target folders, or other written records).
  - — Report completion of the turnover to the higher engineer and supported unit headquarters.
  - — Complete an obstacle-turnover report (a sample work sheet is shown in Figure 2-26, page 2-54) and submit it to higher headquarters.
  - — Forward the written minefield report and record (DA Form 1355 or 1355-1-R) to the next higher commander common to both units.

OBSTACLE-TURNOVER/TRANSFER REPORT		
OBSTACLE DATA		
Obstacle Number	Obstacle Type	Emplacing Authority
Obstacle Location *(8-digit UTM grid to center of mass)*	Obstacle Effect	Emplacing Unit
TURNOVER DATA		
Transferring Unit	Receiving Unit	Date/Time of Transfer
Next Higher Command Common to Both		Date Obstacle Emplaced

TURNOVER CHECKLIST

☐ Intelligence.
- Enemy activity forward of the obstacle *(discuss enemy recon and breaching efforts)*.
- Friendly activity in the vicinity of the obstacle *(discuss LOGPAC, recon, counterrecon maneuver)*.

☐ Maneuver.
- Obstacle protection against enemy recon/breaching *(discuss overwatch, counterrecon patrols, sunrise sweep)*.
- Location of TRPs and indirect-fire targets *(point out location on graphics and terrain)*.
- Rearward/forward passage of lines *(discuss planned and routine activity around the obstacle)*.

☐ Mobility/survivability.
- Obstacle trace *(discuss corner grids on map and point them out on terrain)*.
- Obstacle marking *(rear side, all four sides, fence, signs, HEMMS poles, or no marking)*.
- Obstacle composition.
- Location of gaps/lanes.
- Method of closing gaps/lane *(MOPMS, ADAM, RAAM, conventional mines)*.
- Location of material to close gaps/lanes.

☐ CSS. Planned casualty evacuation and supply routes in the vicinity of the obstacle.

☐ C^2.
- DA Form 1355 or 1355-1-R, dated ________, transferred to receiving unit.
- Obstacle overlay, dated ________, transferred to receiving unit.
- Target folder, dated ________, preparing headquarters ____________, transferred to receiving unit.
- Higher headquarters (transferring, receiving, and emplacing units) notified of transfer.
- DA Form 1355 or 1355-1-R forwarded. (**NOTE: Three copies are required for transfer—one to transferring unit, one to receiving unit, and one to next higher headquarters common to both.)**

CERTIFICATION

The undersigned certify that the receiving unit has assumed full responsibility for the obstacle indicated above. The unit commander understands all the information related to the obstacle, the obstacle intent, and his responsibility to transfer or recover the obstacle upon his departure.

____________________	____________________
Signature of Transferring Unit Commander	Signature of Receiving Unit Commander
____________________	____________________
Printed Name and Rank	Printed Name and Rank

Figure 2-26. Sample obstacle-turnover work sheet

MINEFIELD INSPECTION AND MAINTENANCE

Mines left in the ground for a long time may deteriorate and malfunction for one or more of the following reasons:

- Moisture may have entered the igniter or the body of the mine and either neutralized the explosive or corroded the metal parts. Such actions may be aggravated by local factors (soil acidity or wide temperature swings).
- Frost or heat may have subjected the mine to mechanical strain and caused distortion.
- Insects or vegetation may have caused obstructions.
- Animals may have turned mines over or detonated them.

Technical inspections should only be made by experienced engineers or explosive ordnance disposal (EOD) personnel. When a minefield deteriorates below the operating level, additional mine strips/rows are added to restore its effectiveness. They are sited to the front or the rear of the existing minefield to increase its depth. New mine strips/rows are treated as separate minefields.

Technical inspections of minefields are normally done at three-month intervals. They are done more frequently during extreme weather conditions. Detailed information on mine inspections can be found in Technical Manual (TM) 9-1345-203-12.

This chapter implements STANAG 2990.

Chapter 3

Scatterable Mines and Mine Delivery Systems

SCATMINEs are laid without regard to a classical pattern. They are designed to be delivered or dispensed remotely by aircraft, artillery, missile, or a ground dispenser. All US SCATMINEs have a limited active life and self-destruct after that life has expired. The duration of the active life varies with the type of mine and the delivery system.

SCATMINE systems enable a tactical commander to emplace minefields rapidly in enemy-held territories, contaminated territories, and in most other areas where it is impossible for engineers to emplace conventional minefields. Some systems allow for rapid emplacement of minefields in friendly areas. As with all minefields and obstacles, scatterable minefields are an engineer responsibility.

Based on the tactical plan, the maneuver commander's staff engineer determines the minefield location, size, density, and emplacement and SD times. With this information and a thorough understanding of the available systems, he can then recommend the type of minefield (conventional or scatterable) to be emplaced. If a scatterable minefield is selected, he recommends the delivery system and coordinates the minefield with appropriate staff officers.

GENERAL CHARACTERISTICS

Most US SCATMINEs have similar characteristics. SCATMINEs are much smaller in size and weight than conventional mines. For example, a standard AT SCATMINE weighs approximately 1.8 kilograms and has 600 grams of explosive; an M15 conventional mine weighs 13.5 kilograms and has 10 kilograms of explosive. Arming mechanisms, arming times, and SD times of SCATMINEs differ based on the dispensing system.

ANTIPERSONNEL MINES

There are two general categories of AP SCATMINEs—wedge-shaped and cylindrical (Figure 3-1, page 3-2). Table 3-1, page 3-2, summarizes the characteristics of each AP SCATMINE.

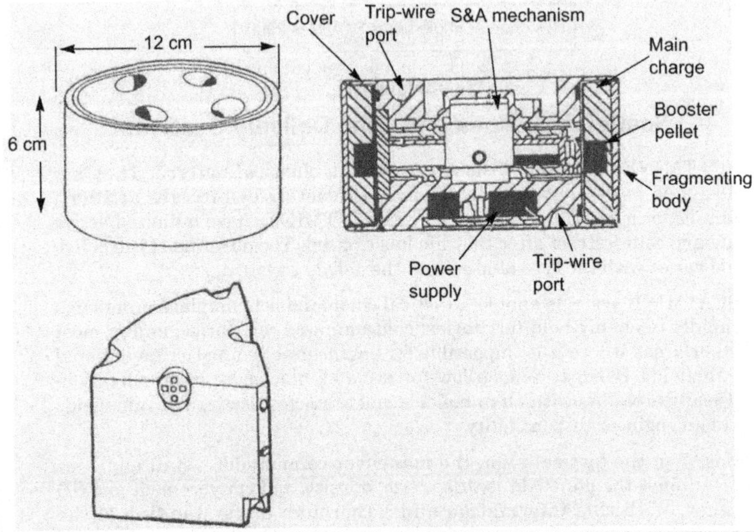

Figure 3-1. AP SCATMINEs

Table 3-1. Characteristics of AP SCATMINEs

Mine	Delivery System	DODIC	Arming Time	Fuse	Warhead	AHD	SD Time	Explosive Weight	Mine Weight	Number of Mines
M67	155-mm artillery (ADAM)	D502	Within 1 min after ground impact	Trip wire	Bounding frag	20%	4 hr	21 g Comp A5	540 g	36 per M731 projectile
M72	155-mm artillery (ADAM)	D501	Within 1 min after ground impact	Trip wire	Bounding frag	20%	48 hr	21 g Comp A5	540 g	36 per M692 projectile
BLU 92/B	USAF (Gator)	K291 K292 K293	2 min	Trip wire	Blast frag	100%	4 hr 48 hr 15 days	540 g Comp B4	1.44 kg	22 per CBU 89/B dispenser
M77	MOPMS	K022	2 min	Trip wire	Blast frag	0%	4 hr (recycle up to 3 times)	540 g Comp B4	1.44 kg	4 per M131 dispenser
Volcano	Ground/ air	K045	2 min	Trip wire	Blast frag	0%	4 hr 48 hr 15 days	540 g Comp B4	1.44 kg	1 per M87 canister

The M67 and M72 AP SCATMINEs are wedge-shaped and dispensed from an ADAM projectile, which is a special 155-millimeter artillery munition. Each mine weighs 540 grams and is 7 centimeters high.

The M74, BLU 92/B, M77, and Volcano AP SCATMINEs are all cylindrical in shape. They are 6 centimeters high and 12 centimeters in diameter. Cylindrical AP SCATMINEs kill enemy soldiers through the combined effects of blast and fragmentation. Each mine contains 540 grams of composition B4 as its main charge. The charge detonates upon actuation and shatters the mine's metal casing to produce shrapnel. Shrapnel is propelled upward and outward from the mine and produces fatal casualties to a distance of 15 meters. Each mine has eight trip wires (four on the top and four on the bottom) that deploy after ground impact up to 12 meters from the mine. Trip wires are similar in appearance to very fine thread; they are olive-drab green in color and weighted at the free end. A tension of 405 grams applied to one trip wire is enough to create a break in the electrical circuit and cause the mine to detonate.

ANTITANK MINES

All AT SCATMINEs (Figure 3-2) have similar functional characteristics. They are cylindrical in shape, weigh approximately 1.8 kilograms, contain 585 grams of cyclonite (RDX) explosive as the main charge, and have a magnetically induced fuse. The characteristics of each AT SCATMINE are summarized in Table 3-2, page 3-4.

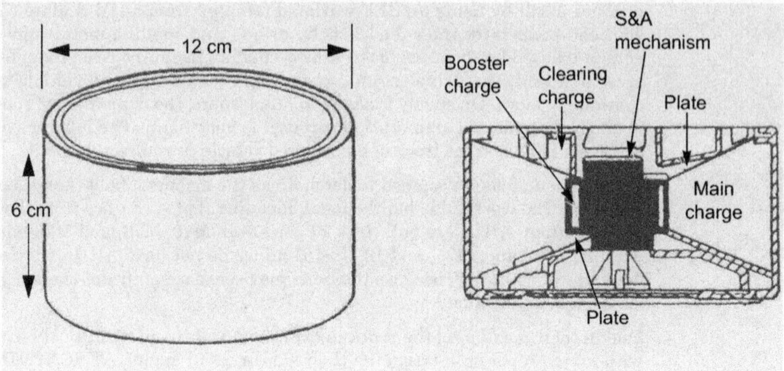

Figure 3-2. AT SCATMINE

Table 3-2. Characteristics of AT SCATMINEs

Mine	Delivery System	DODIC	Arming Time	Fuse	Warhead	AHD	SD Time	Explosive Weight	Mine Weight	Number of Mines
M73	155-mm artillery (RAAM)	D503	Within 1 min after ground impact	Magnetic	M-S plate	20%	48 hr	585 g RDX	1.7 kg	9 per M718 projectile
M70	155-mm artillery (RAAM)	D509	Within 1 min after ground impact	Magnetic	M-S plate	20%	4 hr	585 g RDX	1.7 kg	9 per M741 projectile
BLU 91/B	USAF (Gator)	K291 K292 K293	2 min	Magnetic	M-S plate	NA	4 hr 48 hr 15 days	585 g RDX	1.7 kg	72 per CBU 89/B dispenser
M76	MOPMS	K022	2 min	Magnetic	M-S plate	NA	4 hr (recycle up to 3 times)	585 g RDX	1.7 kg	17 per M131 dispenser
Volcano	Ground/ air	K045	2 min 30 sec	Magnetic	M-S plate	NA	4 hr 48 hr 15 days	585 g RDX	1.7 kg	5 per M87 canister; 6 per M87A1 canister

AT SCATMINEs are designed to produce a K-Kill instead of an M-Kill. They produce a kill by using an SFF warhead (created from an M-S plate). The warhead penetrates the vehicle's belly armor, and spalling metal from the vehicle (caused by the mine blast) kills occupants instantly. Even though the crew is killed, the drive train may be undamaged and the vehicle may continue to move. On enemy tanks with autoloaders, the detonation of rounds in the belly-mounted ammunition carousel is very likely. The mine may not achieve a kill when the track of an armored vehicle runs directly over it.

The magnetic fuse is designed to detonate as the magnetic field changes over the mine. The warhead is bidirectional, meaning that it can fire from the top or the bottom. AHDs are built into 20 percent of M70, M73, and M75 mines. Although Volcano, M76, and BLU 91/B mines do not have AHDs, they may detonate when moved, because the mine may sense a significant change from its original orientation.

Due to their small size, the reduced explosive, and the possibility of landing with an improper orientation (on their side or at an angle), AT SCATMINEs have less chance of destroying a vehicle than a conventional full-width AT mine. An armored vehicle will not always be destroyed after an encounter with an AT SCATMINE. Further, the effectiveness of SCATMINEs in water obstacles is reduced even more, because 5 centimeters of water prevents the formation of the M-S slug. Although the blast wave is accentuated by underwater placement (attacking hatches and covers), mining of banks and approaches is recommended instead.

CAPABILITIES

Faster Response

SCATMINEs can be emplaced more rapidly than conventional mines, so they provide a commander with greater flexibility and more time to react to changes in situations. The commander can use SCATMINEs to maintain or regain the initiative by acting faster than the enemy. Using SCATMINEs also helps preserve scarce mine resources.

Remote Placement

All SCATMINEs are remotely emplaced. This enhances battlefield agility and allows the maneuver commander to emplace mines rapidly to best exploit enemy weaknesses. SCATMINEs can be used as situational obstacles or to attack enemy formations directly through disrupt, fix, turn, and block obstacles. Modern fusing, sensing, and AHDs allow SCATMINEs to better defeat enemy attempts to reduce the minefield.

Increased Tactical Flexibility

Upon expiration of the SD time, the minefield is cleared and the commander can move through an area that was previously denied to enemy or friendly forces. In many cases, the SD period may be set at only a few hours. This feature allows for effective counterattacks to the enemy's flank and rear areas.

Efficiency

SCATMINEs can be emplaced by a variety of delivery methods. They can be deployed by fixed-wing aircraft, helicopters, artillery, manpack, or ground vehicles. They satisfy the high mobility requirements of modern warfare. Manpower, equipment, and tonnage are reduced for their emplacement.

Increased Lethality

AT SCATMINEs utilize an SFF that is created from two M-S plate charges to produce a full-width kill. In simple terms, a metal plate is formed into a high-velocity slug that punches a hole in the belly of a tank. The effect produces an M-Kill against the vehicle's engine, track, or drive train; or it produces a K-Kill when the on-board ammunition is set off and the crew is killed or incapacitated or the vehicle's weapon system is destroyed. AT SCATMINEs are designed to destroy any tank in the world. In order to form an SFF, the mine requires a certain standoff between the vehicle and the target. Mines must also be nearly perpendicular to the target (laying on either side). The M-S plate is actually two plates—one facing the top of the mine and one facing the bottom. This ensures that it will successfully attack the target while lying on either side.

AP SCATMINEs are actuated by a trip wire and utilize a blast-fragmentation warhead.

LIMITATIONS

Extensive Coordination

Because SCATMINEs are a very dynamic weapon system, great care must be taken to ensure that proper coordination is made with higher, adjacent, and subordinate units. To prevent friendly casualties, all affected units must be notified of the location and the duration of scatterable minefields. Recording and reporting procedures for SCATMINEs are discussed in detail in Chapter 8, and they were specifically designed to minimize friendly casualties.

PROLIFERATION OF TARGETS

SCATMINEs may be regarded by some commanders as easy solutions to tactical problems. Target requests must be carefully evaluated, and a priority system must be established because indiscriminate use of weapon systems will result in rapid depletion of a unit's basic load. Controlled supply rates (CSRs) will probably be a constraint in all theaters.

VISIBILITY

SCATMINEs are highly effective, especially when fires and obscurants strain the enemy's C^2. SCATMINEs lay on the surface of the ground, but they are relatively small and have natural coloring.

ACCURACY

SCATMINEs cannot be laid with the same accuracy as conventional mines. Remotely delivered SCATMINE systems are as accurate as conventional artillery-delivered or tactical aircraft-delivered munitions.

ORIENTATION

Between 5 and 15 percent of SCATMINEs will come to rest on their edges; mines with spring fingers will be in the lower percentile. If there is mud or snow more than 10 centimeters deep, the number will be in the higher percentile. When employing ADAMs or RAAMs in more than 10 centimeters of snow or mud, high-angle fire should be used and the number of mines increased. AP mines may be less effective in snow, because the deployment of trip wires is hindered. Melting of the snow may also cause the mines to change positions and activate AHDs.

LIFE CYCLE

All SCATMINEs have a similar life cycle, although specific times vary based on the SD time and the dispensing system.

For safety reasons, SCATMINEs must receive two arming signals at launch. One signal is usually physical (spin, acceleration, or unstacking), and the other is electronic. This same electronic signal activates the mine's SD time.

Mines start their safe-separation countdown (arming time) when they receive arming signals. This allows the mines to come to rest after dispensing and allows the mine dispenser to exit the area safely. Table 3-1, page 3-2, and Table 3-2, page 3-4, show arming times for individual SCATMINEs.

Mines are armed after the arming time expires. The first step in arming is a self-test to ensure proper circuitry. Approximately 0.5 percent of mines fail the self-test and self-destruct immediately.

After the self-test, mines remain active until their SD time expires or until they are encountered. Mines actually self-destruct at 80 to 100 percent of their SD time. The time period from when the mines begin to self-destruct and when they finish is called the SD window (Table 3-3). No mines should remain active after the SD time has been reached. Two to five percent of US SCATMINEs fail to self-destruct as intended. Any mines found after the SD time must be treated as unexploded ordnance (UXO). For example, mines with a 4-hour SD time will actually start self-destructing at 3 hours and 12 minutes. When the 4-hour SD time is reached, no unexploded mines should exist.

Table 3-3. SD windows

SD Time	SD Window Begins
4 hours	3 hours 12 minutes
48 hours	38 hours 24 minutes
5 days	4 days
15 days	12 days

LETHALITY AND DENSITY

LETHALITY AND TACTICAL-OBSTACLE EFFECT

Scatterable minefields are employed to reduce the enemy's ability to maneuver, mass, and reinforce against friendly forces. They increase the enemy's vulnerability to fires by producing specific obstacle effects (disrupt, fix, turn, and block) on the enemy's maneuver. To achieve this aim, individual minefields must be emplaced with varying degrees of lethality. During emplacement, lethality is varied primarily by changing the minefield density. Therefore, there is a direct correlation between the obstacle effect and the minefield density. In order to achieve the tactical-obstacle effect, use the following guidance when selecting minefield density:

- Disrupt.
  - Low density.
  - Probability of encounter: 40 to 50 percent.
  - Linear density: 0.4 to 0.5 mine per meter.
- Fix.
  - Medium density.
  - Probability of encounter: 50 to 60 percent.
  - Linear density: 0.5 to 0.6 mine per meter.
- Turn.
  - High density.
  - Probability of encounter: 75 to 85 percent.

— Linear density: 0.9 to 1.1 mines per meter.

- Block.
  - High density.
  - Probability of encounter: 85+ percent.
  - Linear density: More than 1.1 mines per meter.

DENSITY

Density is normally expressed as linear or area. For conventional mines, linear density is normally used and is expressed in the average number of mines per meter of minefield front. For SCATMINE systems, area density is normally used and is expressed as the average number of mines per square meter. Since SCATMINE systems normally employ a preset combination of AT and AP mines, the area density includes both. For example, a scatterable minefield with an area density of 0.006 mine per square meter may have an AT density of 0.004 AT mine per square meter and an AP density of 0.002 AP mine per square meter. Due to the varying dimensions of scatterable minefields that can be created by the different types of employment devices, the exact density of a scatterable minefield cannot be determined. However, an estimate of the average density can be determined by using the following formulas:

- Linear density equals the number of mines divided by the minefield front.

$$\frac{\textit{number of mines}}{\textit{minefield front}} = \textit{mines per meter}$$

- Area density equals the number of mines divided by the minefield area.

$$\frac{\textit{number of mines}}{\textit{front} \times \textit{depth}} = \textit{mines per square meter}$$

- Area density can be converted to linear density by multiplying the area density by the minefield depth. **(NOTE: Converting area density to linear density is not always accurate due to the space between minefield strips.)**

$$\textit{area density} \times \textit{minefield depth} = \textit{linear density}$$

EXAMPLE: A 650- by 200-meter Gator minefield contains 564 mines (432 AT and 132 AP).

- Area density: 564 ÷ (200 x 650) = 0.004 mine per square meter.
  - AT area density: 432 ÷ (200 x 650) = 0.003 mine per square meter.

  - AP area density: 132 ÷ (200 x 650) = 0.001 mine per square meter.
- Linear density: 564 ÷ 650 = 0.87 mine per meter.
  - AT linear density: 432 ÷ 650 = 0.67 mine per meter.
  - AP linear density: 132 ÷ 650 = 0.2 mine per meter.

COMMAND AND CONTROL

Due to the delivery means, C^2 of SCATMINEs is more complex than conventional mines. SCATMINEs are very dynamic weapon systems because they can be rapidly emplaced and then cleared via their SD capability. Also, the physical boundary of a scatterable minefield is not clearly defined. These characteristics require impeccable communications and coordination to ensure that all friendly units know where mines are located, when they will be effective, and when they will self-destruct.

AUTHORITY

The corps commander has emplacement authority for all scatterable minefields within the corps AO. He may delegate this authority to lower echelons according to the guidelines contained in Table 3-4.

Table 3-4. Emplacement authority

System Characteristics	Emplacement Authority
Ground- or artillery-delivered, with SD time greater than 48 hours (long duration)	The corps commander may delegate emplacement authority to division level, who may further delegate it to brigade level.
Ground- or artillery-delivered, with SD time of 48 hours or less (short duration)	The corps commander may delegate emplacement authority to division level, who may further delegate it to brigade level, who may further delegate it to TF level.
Aircraft-delivered (Gator), regardless of SD time	Emplacement authority is normally at corps, theater, or army command level, depending on who has air-tasking authority.
Helicopter-delivered (Volcano), regardless of SD time	Emplacement authority is normally delegated no lower than the commander who has command authority over the emplacing aircraft.
MOPMS, when used strictly for a protective minefield	Emplacement authority is usually granted to the company, team, or base commander. Commanders at higher levels restrict MOPMS use only as necessary to support their operations.

Based on how the commander wants to shape the battlefield, he must specifically delegate or withhold the authority to employ SCATMINE systems. The commander's guidance concerning SCATMINEs is found in the unit's OPORD/operation plan (OPLAN). Additional information is included in their engineer and fire-support annexes, if used.

Due to the complete control a commander has over the MOPMS, emplacement authority guidelines do not apply to the MOPMS. It is impractical for the corps or brigade commander to authorize every MOPMS protective minefield. Therefore, authority to emplace MOPMS minefields is specifically delegated. In general, units can emplace MOPMS protective minefields as required for their own self-defense and report them as they do any protective obstacle. Any MOPMS minefield used as part of an obstacle plan must be reported as a scatterable minefield.

COORDINATION

Basic responsibilities of key commands, staff elements, and units are outlined in Table 3-5. The fire-support coordinator (FSCOORD) is involved in planning artillery-delivered (ADAM and RAAM) SCATMINEs, and the air liaison officer (ALO) is involved in planning air-delivered (Gator and Volcano) SCATMINEs. The engineer has primary responsibility for planning and employing SCATMINE systems. It is vital that coordination be conducted with all units and subunits that will be effected by the employment of SCATMINEs. A scatterable minefield warning (SCATMINWARN) will be sent to all effected units before the emplacement of the minefield (see Chapter 8 for more details).

Table 3-5. Coordination responsibilities

Element	Responsibilities
G3/S3 with Engineer FSCOORD/ALO	Plan and coordinate the minefield location, size, composition, density, SD time, safety zone, and emplacement time
	Designate and brief the emplacing unit
	Incorporate the minefield and the safety zone into the obstacle plan
	Receive and forward the scatterable minefield report and record
	Disseminate information concerning the minefield in the SCATMINWARN to adjacent and subordinate units prior to laying
	Post operation maps with the minefield location, safety zone, and DTG of the SD time; and disseminate the SCATMINWARN 1 hour prior to initiation of the SD sequence
Emplacing Unit	Calculate the logistical requirements
	Calculate the safety zone
	Emplace the minefield
	Report the amount of ammunition expended
	Prepare and forward the scatterable minefield report and record to the authorizing commander via appropriate channels
Maneuver Units	Be aware of the calculated safety-zone boundary and advise subunits of its location

EMPLOYMENT AND EMPLACEMENT

Employment considerations and emplacement techniques and procedures differ for each type of SCATMINE and delivery system. This section discusses the characteristics of each delivery system and provides tactical considerations for the employment of each system on the battlefield. Techniques and procedures for emplacing minefields intended to disrupt, fix, turn, and block are also discussed; and they build on tactical-obstacle design principles discussed in Chapter 2.

AREA-DENIAL ARTILLERY MUNITIONS AND REMOTE ANTIARMOR MINES

ADAMs and RAAMs are delivered by a 155-millimeter howitzer (Figure 3-3). There are no special modifications or adaptations necessary for the firing system. Mines are contained within a projectile and are dispensed while the projectile is in the air. The effective range for the M109 howitzer is 17,500 meters, and for the M198 howitzer, 17,740 meters.

Figure 3-3. Emplacement of ADAMs and RAAMs

The M692 (long-duration) and the M731 (short-duration) ADAM projectiles deliver AP mines with different SD times. Each ADAM round contains 36 mines. The M731/M731A1 round contains M72 AP mines with 4-hour SD times; the M692/M692A1 round contains M67 AP mines with 48-hour SD times. SD times are preset during the manufacturing process and cannot be changed.

The wedge-shaped ADAM is a bounding-fragmentation mine that deploys up to seven tension-activated trip wires 6 meters away from the mine. After ground impact, trip wires are released and the mine is fully armed. The ADAM contains a metal-jacketed sphere that is filled with 21 grams of composition A5 as its main charge. A liquid-explosive propelling charge positions itself at the bottom of the sphere after impact with the ground. When the mine is jarred or tilted, or when one of its trip wires receives a tension of at least 405 grams, the sphere propels upward 0.6 to 2.4 meters and detonates. The lethal casualty radius is between 6 and 10 meters.

The M741 (short-duration) and the M718 (long-duration) RAAMs are artillery-delivered AT mines. Each RAAM round contains nine mines. The M741/M741A1 round contains M70 AT mines with 4-hour SD times; the M718/M718A1 round contains M73 AT mines with 48-hour SD times. The SD times are preset during the manufacturing process and cannot be changed. The RAAM mine utilizes an SFF warhead, has a magnetic-influence fuse, weighs 1.7 kilograms, and has a small (12 centimeters in diameter by 6 centimeters in height) cylindrical shape.

The new model ADAM and RAAM mines (designated by an A1 suffix) have a 45-second arming time; the older models have a 2-minute arming time. The new model RAAM has a built-in feature that defeats magnetic, signature-duplicating breaching devices.

Employment

The ADAM and RAAM systems were designed to provide a flexible, rapid-response mining capability. These systems provide the maneuver commander with the capability to emplace mines directly on top of, in front of, or behind enemy forces. This is one of their greatest advantages. Their responsiveness allows the mission to be executed quickly and allows the commander to effectively influence a rapidly changing battlefield. They also allow the commander to emplace minefields while maintaining maximum standoff from the target. In short, their emplacement does not require committing any force (ground or air) forward. ADAM and RAAM systems may be used for the following purposes:

- Defense.
  - Develop targets for long-range AT weapons.
  - Close gaps and lanes in other obstacles.
  - Delay or disrupt attacking forces.
  - Deny the enemy unrestricted use of selected areas.
  - Disrupt movement and commitment of second-echelon forces.
  - Disrupt and harass enemy C^2, logistics (excluding medical), and staging areas.
  - Reinforce existing obstacles.
  - Disrupt or delay river crossings.
- Offense.
  - Supplement flank reconnaissance and security forces to protect flanks along AAs.
  - Suppress and disrupt enemy security elements once contact has been made.
  - Hinder the withdrawal of enemy forces.

— Hinder the ability of the enemy to reinforce the objective area.

The time and the number of rounds required to install effective ADAMs and RAAMs limit their use. Their range is limited to 17,500 or 17,740 meters, depending on which howitzer (M109 or M198, respectively) is used. Many of the deep-interdiction missions that support force-projection doctrine require a greater distance. Due to the large footprint created when the minefield is fired, many mines will scatter outside the planned minefield area. It is therefore necessary to plot the safety zone in order to prevent fratricide. The fire-support element (FSE) is responsible for plotting the safety zone, and the staff engineer should be familiar with the process and the expected results. The staff engineer ensures that the safety zone is plotted on the tactical command post (TCP)/TOC operation overlay.

Emplacement

ADAM and RAAM mining missions are requested through normal artillery-support channels. Although the actual numbers vary based on the unit and the mission, a representative basic load for an artillery battalion consists of approximately 32 ADAM and 24 RAAM (short SD time) rounds per artillery piece. **NOTE: The rounds with long SD times are normally used for preplanned targets and are issued from an ammunition supply point (ASP) on a mission-by-mission basis.**

Once the proper authorization has been received to employ the mines, requests for ADAMs and RAAMs are processed in the same way as other requests for fire support, including targets of opportunity. Allocate enough time for processing the request and completing firing procedures. This ensures that the enemy has not moved out of the target area before execution. (FM 90-7 contains more information on this process.) The use of ADAMs and RAAMs for preplanned fires requires close coordination among the Assistant Chief of Staff, G3 (Operations and Plans) (G3)/Operations and Training Officer (US Army) (S3), the staff engineer, and FSE sections. Coordination should also be made with the S2 and the S3 during the development of the decision support template (DST) to identify the proper named areas of interest (NAIs), target areas of interest (TAIs), trigger points, and decision points.

There are two critical aspects when emplacing ADAM and RAAM minefields:

- Designing the minefield to achieve the required effect.
- Ensuring the technical correctness of resourcing and delivering the minefield.

The following discussion provides general guidance for designing the minefield to achieve the desired effect and for determining the safety zone to assess the impact on maneuver. Appendix H of FM 6-20-40 serves as the primary source for technically resourcing and delivering artillery-delivered minefields.

ADAM and RAAM minefields can be emplaced to achieve disrupt, fix, turn, and block effects based on the principles outlined in Chapter 2. The engineer is responsible for deciding the required location, the density, the size, the composition, and the duration of the minefield based on the tactical-obstacle plan and the obstacle restrictions of the higher unit. The engineer provides this information to the FSE. Table 3-6 provides guidance on the minefield density and size necessary to achieve the desired obstacle effect.

Table 3-6. RAAM and ADAM minefield density and size

Obstacle Effect	RAAM		ADAM		Width (meters)	Depth (meters)
	Area[1]	Linear[2]	Area[1]	Linear[2]		
Disrupt	0.001	0.2	0.0005	0.1	200	200
Turn	0.002	0.8	0.001	0.4	400	400
Fix	0.002	0.4	0.0005	0.1	200	200
Block	0.004	0.6	0.002	0.8	400	400

[1]Area density = mines per square meter
[2]Linear density = mines per meter

The FSE determines all the technical aspects for delivering the minefield, such as the number of rounds required per aim point, the number of aim points required, the size of the safety zone, and the time required to emplace mines. There is a wide variety of factors involved in determining the number of rounds, the size of the safety zone, and the emplacement time. These factors are the range-to-target time, the battery-to-minefield angle, the high- or low-angle trajectory, and the method of firing (observer adjust or meteorological data plus velocity error [Met+VE] transfer). The FSE must tell the engineer whether the minefield mission is feasible. Feasibility is based on the number of rounds available, the scheme of indirect fires, and the availability of artillery tubes.

The engineer is primarily concerned with two technical aspects of delivery provided by the FSE—the safety zone and the emplacement time. The engineer uses the safety zone and the minefield duration to assess the impact of the minefield on the mobility requirements of the scheme of maneuver. The engineer depicts the safety zone on the obstacle overlay. He also uses the safety zone to identify requirements for minefield marking if the unit leaves or turns over the area before the SD time. The engineer and the FSE use the emplacement time to synchronize the delivery of the minefield with the tactical plan.

GATOR

The Gator (Figure 3-4) has a longer range than any other SCATMINE system. It provides a means to rapidly emplace minefields anywhere that can be reached by tactical aircraft. The Gator is produced in two versions—the United States Air Force (USAF) CBU-89/B system that contains 94 mines (72 AT and 22 AP) per dispenser and the United States Navy (USN) CBU-78/B system that contains 60 mines (45 AT and 15 AP) per dispenser.

The mines used with the Gator are the BLU-91/B AT mine and the BLU-92/B AP mine. They are similar to the mines used with the Volcano system. The mines are capable of three field-selectable SD times (4 hours, 48 hours, and 15 days). Both types of mines are encased in a plastic, square-shaped protective

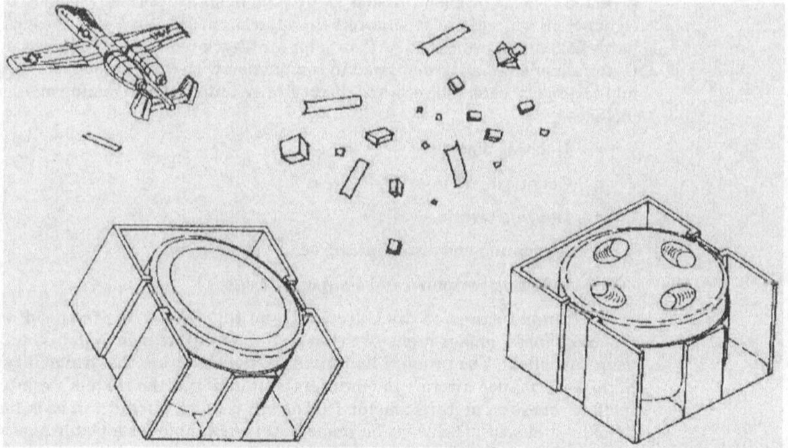

Figure 3-4. Gator SCATMINE system

casing that is designed to aid dispersion and lessen ground impact upon delivery.

The mines are contained inside tactical munition dispensers (TMDs) that are attached under the wings of high-performance, fixed-wing aircraft. The TMD is a USAF dispenser that was designed for common use with cluster munitions. The Gator is compatible with the USAF A-10, F-4, F-15, F-16, B-1, and B-52 aircraft and with the USN A-6, A-7, F-4, FA-18, and AV-8B aircraft.

The TMD is released in the air and allowed to fall free. Four linear charges along the edge of the TMD cut the outer casing, and the mines are aerodynamically dispersed. The maximum delivery speed is 800 knots at altitudes of 75 to 1,500 meters. The area of minefield coverage depends on the number of munitions carried, the aircraft speed and altitude, and the altitude where the fuse functions and opens the dispenser. The average area covered is approximately 200 by 650 meters.

Employment

Gator missions are primarily used at long range to disrupt, fix, turn, or block enemy troop movement beyond the fire-support coordination line (FSCL). For use in interdiction missions beyond the FSCL, submit requests for Gator missions as early as possible to nominate targets for the theater air-tasking order. Gator munitions are well-suited for placing minefields on specific concentrations of forces (artillery, logistic, and C^2) that are out of range of conventional artillery.

While the Gator can provide close combat support, deep-interdiction mining is expected to be its primary mission. Gator minefields are normally employed in conjunction with other deep indirect-fire attacks, such as area of interest (AI), battlefield air interdiction (BAI), or joint air-attack team (JAAT). However, a Gator minefield may be employed in conjunction with close air support (CAS) and covered by close indirect- and direct-fire systems. Typical mining missions include—

- Isolating objectives.
- Countering ADA/artillery fires.
- Denying terrain.
- Disrupting and disorganizing support activities.
- Inflicting personnel and equipment losses.

The extended range of the Gator system, together with its speed and responsiveness, makes it one of the most influential weapon systems on the deep battlefield. The primary limitations of the Gator are the availability of high-performance aircraft to emplace the mines and the system's relative ineffectiveness on units in column. During any conflict, aircraft will be in high demand and will not always be immediately available for a Gator mission when required. Communications may also pose a problem because mission execution is a joint US Army-USAF operation.

The Gator is well suited to support contingency operations and amphibious landing operations in an immature theater when there is no danger to friendly forces or host-nation assets. Gator minefields are one of the light-force commander's few durable, long-range antiarmor weapons.

Emplacement

As an aircraft-delivered munition, the Gator is a corps asset. The Gator is a BAI mission and is controlled by the tactical air control center (TACC). Missions should be requested as early as possible (no later than 36 hours in advance) through fire-support channels to the corps FSE. As a mine system, Gator missions must be approved by corps. The corps FSE passes the mission to the theater or army air headquarters to be included on the theater air-tasking order for execution. In support of BAI or CAS, Gator sorties may be allocated down to battalion level, with final control exercised by the battalion ALO. Immediate Gator missions can also be requested directly from the maneuver unit's TACC. The same records and reports applicable to other SCATMINE systems are used with the Gator mine system. Close cooperation and coordination among the G3/S3, the staff engineer, and the ALO are required for planning and executing Gator missions.

As with artillery-delivered minefields, the engineer is primarily responsible for identifying the minefield location, size, duration, and density. Minefield density is varied by changing the orientation of the minefield with respect to the target AA. Figure 3-5 illustrates how minefield orientation is changed to achieve a fix or block effect. Normally, Gator is employed as a fix obstacle with a front of 650 meters. Emplacing a fix-obstacle group along a battalion AA (1,500 meters) requires two Gator sorties, each delivering one minefield. Each Gator minefield would have a front of 650 meters and a depth of 200 meters.

The minefields would be delivered at different locations so that the group covers the entire AA and affects the entire enemy battalion.

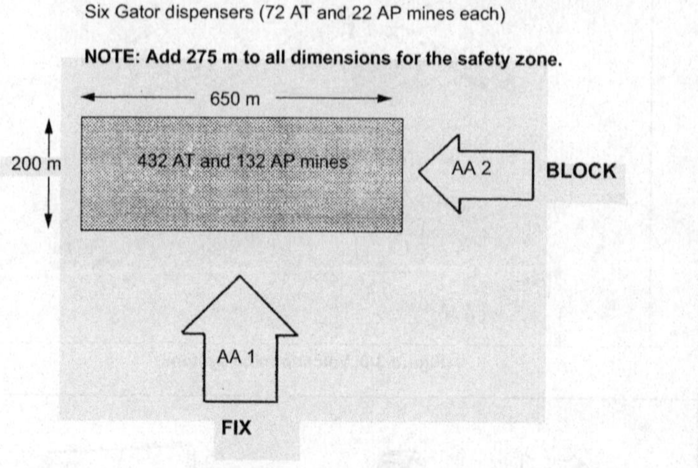

Figure 3-5. Gator minefield

VOLCANO

The Volcano multiple-delivery mine system (Figure 3-6, page 3-18) can be dispensed from the air or on the ground. It can be mounted on any 5-ton truck, an M548 tracked cargo carrier, a heavy expanded mobility tactical truck (HEMTT), a palletized load system (PLS) flat rack, or a UH-60A Blackhawk helicopter. The Volcano uses modified Gator mines and consists of four components (Figure 3-7, page 3-18)—the mine canister, the dispenser, the dispenser control unit (DCU), and the mounting hardware (aircraft also require a jettison kit). The Volcano uses M87 and M87A1 mine canisters. The M87 mine canister is prepackaged with five AT mines, one AP mine, and a propulsion device inside a tube housing. The M87A1 mine canister is prepackaged with six AT mines and a propulsion device. The mixture of mines is fixed and cannot be altered. Mines are electrically connected with a web that functions as a lateral dispersion device as the mines exit the canister. Spring fingers mounted on each mine prevent it from coming to rest on its edge. All canisters are capable of dispensing mines with 4-hour, 48-hour, and 15-day SD times. The SD times are field-selectable prior to dispensing and do not require a change or modification to the mine canister. The arming time is 2 minutes 15 seconds for AT and AP mines. The reload time (not including movement time to the reload site) for an experienced four-man crew is approximately 20 minutes.

Figure 3-6. Volcano mine system

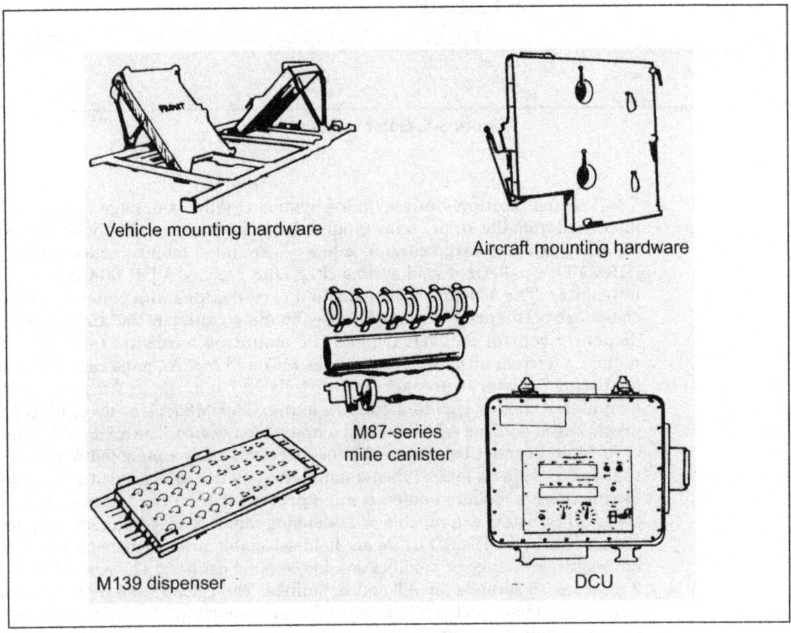

Figure 3-7. Volcano components

The dispenser consists of an electronic DCU and four launcher racks. Four racks can be mounted on a vehicle, and each rack can hold 40 M87-series mine canisters. The racks provide the structural strength and the mechanical support required for launch and provide the electrical interface between the mine canisters and the DCU. Mounting hardware secures the racks to the vehicle or the aircraft. Mounting hardware for the Blackhawk includes a jettison subassembly to propel the Volcano racks and canisters away from the aircraft in the event of an emergency.

The operator uses the DCU to control the dispensing operation electrically from within the carrier vehicle. The DCU provides controls for the arming sequence and the delivery speed and sets mine SD times. The DCU allows the operator to start and stop mine dispensing at anytime. A counter on the DCU indicates the number of remaining loaded canisters on each side of the carrier.

Mines are dispensed from their canisters by an explosive propelling charge. For ground vehicles, the mines are dispensed 25 to 60 meters from the vehicle at ground speeds of 8 to 90 kph. The average time to emplace one ground Volcano load (160 canisters) is 10 minutes.

Employment

The primary mission of the Volcano is to provide US forces with the capability to emplace large minefields rapidly under varied conditions. The Volcano can be rapidly attached to air or ground vehicles. It is used to emplace tactical minefields; reinforce existing obstacles; close lanes, gaps, and defiles; protect flanks; and deny probable enemy air-defense sites. Volcano minefields are ideal for providing flank protection of advancing forces and for operating in concert with air and ground cavalry units on flank guard or screen missions.

The air Volcano is the fastest method for emplacing large tactical minefields. When employed by combat aviation elements in support of maneuver units, close coordination between aviation and ground units assures that Volcano-dispensed mines are emplaced accurately and quickly. Although mine placement is not as precise as it is with ground systems, air Volcano minefields can be placed accurately enough to avoid the danger inherent in minefields delivered by artillery or jet aircraft. Air Volcano minefields can be emplaced in friendly and enemy territory. They should not be planned in areas of enemy observation and fire because the helicopter is extremely vulnerable while flying at the steady altitude, the speed, and the path required to emplace the minefield. The air Volcano is the best form of an obstacle reserve because a minefield can be emplaced in minutes.

The ground Volcano is designed to emplace large minefields in depth. It is normally employed by combat engineer units. These mounted dispensers are primarily used to emplace tactical minefields oriented on enemy forces in support of maneuver operations and friendly AT fires. The system is vulnerable to direct and indirect fires, so it must be protected when close to the FLOT. It is ideal for use as an obstacle reserve, employed when the enemy reaches a decision point that indicates future movement. Obstacles can then be emplaced in depth on the avenues the enemy is using, leaving other avenues open for friendly movement.

Emplacement

The principles and procedures of Volcano emplacement are significantly different for air- and ground-delivery systems. This section outlines the use of the ground Volcano system to emplace disrupt, fix, turn, and block minefields. The air Volcano system is discussed in detail in Appendix D. Both air and ground Volcano systems are capable of emplacing nonstandard minefields. However, the emplacement norms below streamline identifying resource requirements and conducting emplacement drills.

Air and ground Volcano systems emplace a minefield with an average AT linear density of 0.72 mine per meter and an AP linear density of 0.14 mine per meter. These densities may vary slightly since some mines will fail the arming sequence and self-destruct 2 to 4 minutes after dispensing. Additionally, some mines may not orient correctly, will not deliver their full mine effect, and will not produce a K-Kill. The probability of failing the arming sequence and misorienting is relatively small and does not appreciably degrade the minefield's lethality. For tracked vehicles, the AT density yields more than 80 percent probability of encounter. Volcano AT mines do not have AHDs but are highly sensitive to any movement once they are armed. Any attempt to remove the mines will likely result in detonation.

The basic site layout is extremely important, and it is the same for air and ground Volcano minefields. The limits of Volcano minefields are marked before emplacement when the situation (planned targets within the main battle area [MBA] of a defensive operation) allows it. The minefield is not premarked when the situation (offensive operations or situational obstacles) does not allow it. If the mines have not self-destructed, the minefield is marked before the unit leaves the area or turns it over to an adjacent unit. Minefield marking must include the safety zone, which is 40 meters from the start and end points and 80 meters to the left and right of the centerline. The start and end points of the strip centerline are marked based on the minefield front and the number of strips. For a ground Volcano minefield, guide markers are emplaced along the path of the centerline but are offset left to allow the host vehicle to remain on the centerline. When using a ground-delivery system, minefield marking must leave a gap along each centerline for vehicle entrance and exit. The number of guide markers used depends on the terrain and the visibility. Guide markers are not required for an air Volcano minefield because the pilot will use the start and end points of the centerline as reference points.

Figure 3-8 illustrates the emplacement pattern for standard disrupt and fix minefields using the ground or air Volcano. Disrupt and fix minefields use only one centerline to give a minefield depth of 120 meters (ground) or 140 meters (air), not including the safety zone. The strip centerline is 277 meters (ground) or 278 meters (air) long. The host vehicle moves toward the start point, achieving and maintaining the ground or air speed selected on the DCU. The operator depresses the launch switch on the DCU when the vehicle passes the start marker, and he stops dispensing mines when the vehicle passes the end marker. The operator dispenses 40 canisters (20 on each side) along the centerline. One full load of ground or air Volcano emplaces four disrupt or fix minefields. For ground emplacement, the vehicle moves out of the minefield, marks the exit, and waits a minimum of 4 minutes before approaching the minefield. This delay allows faulty mines to self-destruct.

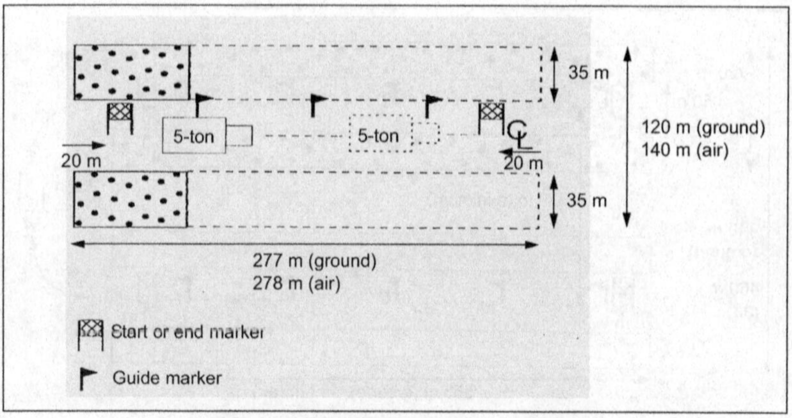

Figure 3-8. Volcano disrupt and fix minefields

Turn and block minefields (Figure 3-9, page 3-22) are emplaced using the same basic procedures as those used for disrupt and fix minefields. However, turn and block minefields use two strip centerlines along a front of 555 meters (ground) or 557 meters (air). During site layout, centerlines are separated by at least 320 meters for both ground and air delivery. This gives a total minefield depth of 440 meters (ground) or 460 meters (air). The operator dispenses 80 canisters along each centerline (40 on each side); therefore, turn and block minefields require a total Volcano load of 160 canisters. One full load of ground or air Volcano emplaces one turn or block minefield. Wherever possible, two ground Volcanoes are employed simultaneously on turn and block minefields. When only one ground delivery system is used, the crew must wait 4 minutes after dispensing the first strip before dispensing the second strip. This allows mines that fail the arming sequence to self-destruct. For air delivery, two sorties are also optimal; but demands for sorties elsewhere in the division may preclude the simultaneous employment of two Blackhawks.

Modular Pack Mine System

The MOPMS (Figure 3-10, page 3-22) is a man-portable, 162-pound, box-shaped mine dispenser that can be emplaced anytime before dispensing mines. The dispenser contains 21 mines (17 AT and 4 AP). The mines have leaf springs along their outer circumference that are designed to push the mines into proper orientation if they land on their side.

Each dispenser contains seven tubes; three mines are located in each tube. When dispensed, an explosive propelling charge at the bottom of each tube expels mines through the container roof. Mines are propelled 35 meters from the container in a 180-degree semicircle (Figure 3-11, page 3-23). The resulting density is 0.01 mine per square meter. The safety zone around one container is 55 meters to the front and sides and 20 meters to the rear.

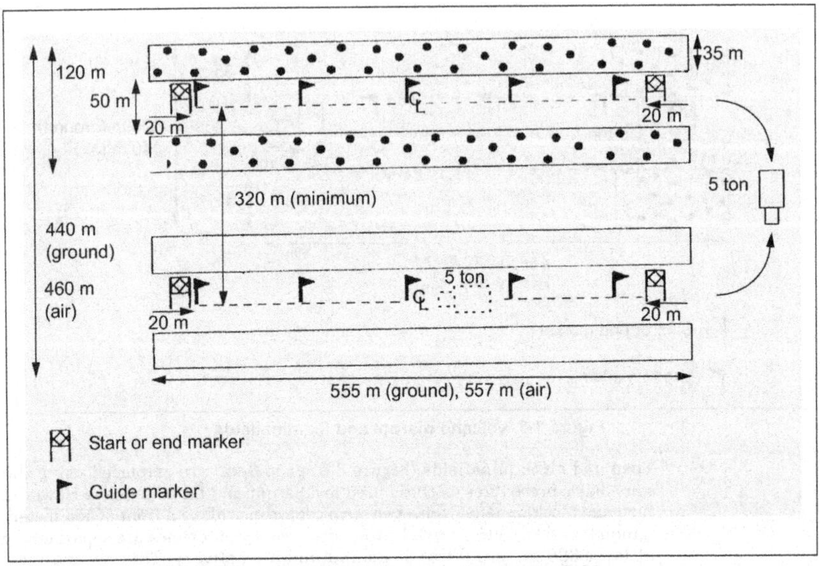

Figure 3-9. Volcano turn and block minefields

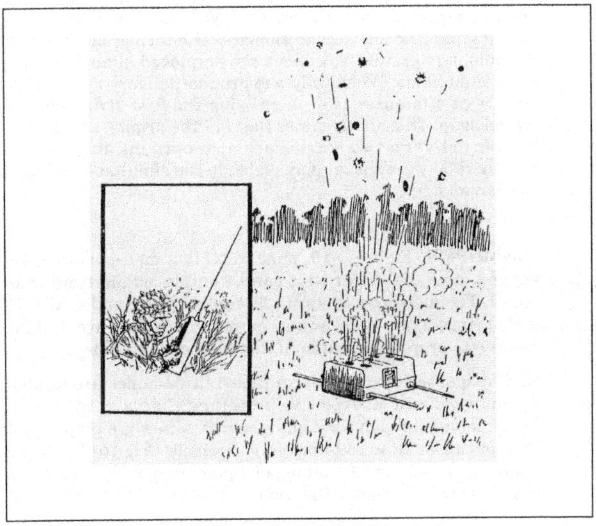

Figure 3-10. MOPMS

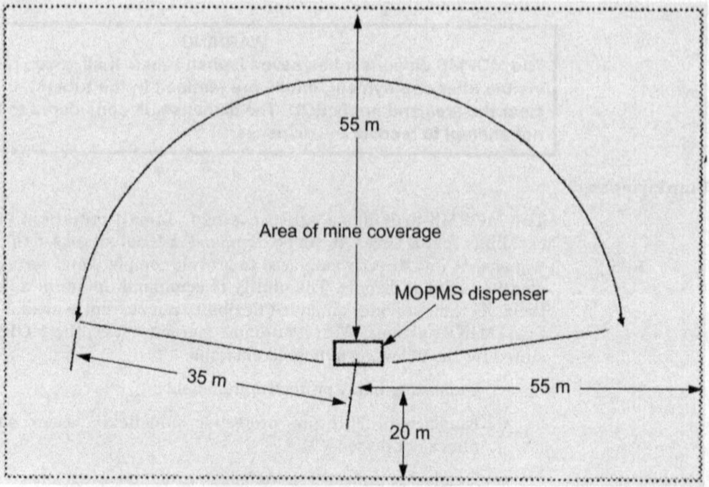

Figure 3-11. MOPMS emplacement and safety zone

Mines are dispensed on command using an M71 remote-control unit (RCU) or an electronic initiating device. Once mines are dispensed, they cannot be recovered or reused. If mines are not dispensed, the container may be disarmed and recovered for later use.

The RCU can recycle the 4-hour SD time of the mines three times, for a total duration of approximately 13 hours. Mines with a 4-hour SD time will begin to self-destruct at 3 hours and 12 minutes. All active mines must be recycled within 3 hours of the initial launch or last recycle. This feature makes it possible to keep the minefield emplaced for longer periods if necessary. The RCU can also self-destruct mines on command, allowing a unit to counterattack or withdraw through the minefield, as necessary, rather than waiting until the SD time has expired. The RCU can control up to 15 MOPMS containers or groups of MOPMS containers from a distance of 300 to 1,000 meters via separate pulse-coded frequencies. Coded frequencies defeat threat electronic countermeasures directed against the system.

If the M71 RCU is unavailable, a direct wire link is used in conjunction with an M32, M34, or M57 blasting machine. By using the M32 10-cap blasting machine, one MOPMS dispenser can be detonated at a maximum range of 1,000 meters. The M34 50-cap blasting machine can detonate one MOPMS at a maximum range of 3,000 meters. (Due to internal resistance, the maximum range is decreased by 400 meters for each additional MOPMS connected in series.) The M57 claymore-type FD can fire only one MOPMS at a maximum range of 100 meters. When controlled by direct wire, MOPMS dispensers cannot be command-detonated, and the SD time cannot be recycled.

WARNING

The MOPMS dispenser has seven launch tubes. If all seven tubes are not visible after deployment, mines are jammed in the tube(s). In this event, clear the area and notify EOD. The dispenser is considered to be UXO; do not attempt to recover the dispenser.

Employment

The MOPMS provides a self-contained, on-call minefield emplacement capability for all forces. It can be command-detonated, reused (if mines are not dispensed), and directly emplaced to provide complete and certain coverage of small or critical targets. The ability to command-detonate mines or extend their SD time provides an added flexibility not currently available with other SCATMINE systems. With its unique characteristics, the MOPMS is ideally suited for the following minefield missions:

- Emplacing hasty protective minefields.
- Emplacing deliberate protective minefields (cases emplaced, but mines not dispensed).
- Emplacing nuisance minefields (trails, crossing sites, landing zones [LZs], drop zones [DZs], and road junctions).
- Emplacing tactical disrupt and fix minefields.
- Closing gaps and lanes in existing minefields.
- Temporarily closing counterattack routes.
- Supporting ambushes.
- Supporting military operations in built-up areas (MOBA) operations.

When the MOPMS is used to close lanes, the container is positioned and dispensed by personnel in an overwatch position from a safe standoff. The MOPMS is ideally suited for creating a small disrupt obstacle in support of engineers executing a reserved demolition target. Engineers prepare the reserved target for demolition and emplace several MOPMS units on the enemy side, just out of target range. When the last forward element passes through the target, the firing party detonates the charges. If something goes wrong or the firing party needs more time, MOPMS mines can be dispensed to disrupt the enemy before it reaches the target.

The MOPMS provides light and special forces with a versatile, compact system for emplacing nuisance minefields. It can be used in low-, mid-, and high-intensity conflicts and in a variety of environments. The MOPMS cannot be transported long distances by hand because of its weight, so its use is limited.

Emplacement

MOPMS dispensers are issued as standard Class V munitions and are drawn from an ASP on a mission-by-mission basis. RCUs are organizational issues of equipment and are assigned to engineer and combat arms units. Due to the weight of the system, it will normally be transported by vehicle, as close as possible to the emplacement site, where it can easily be hand-emplaced by four soldiers using the four foldout carrying handles.

To ensure that the minefield will be dispensed in the proper location, the container should be carefully sited by the noncommissioned officer in charge (NCOIC). Several containers can be used together to provide a greater area of coverage or a higher mine density. If mines are not dispensed immediately, containers should be camouflaged and, if possible, buried. When placed in sand or snow, brace the containers to prevent them from moving during mine dispensing. Designate a firing point that gives the operator clear observation of the area to be mined. Firing systems must be inspected according to MOPMS operating instructions. If mines are dispensed immediately, remove empty containers to avoid revealing the minefield location.

The MOPMS can be employed to emplace disrupt and fix tactical minefields. Emplacement procedures are the same as for protective minefields above. However, MOPMS containers are arranged in a specific pattern to achieve the necessary depth, front, and density. Once the minefield is marked (to include the safety zone), MOPMS containers are arranged as shown in Figure 3-12 for a disrupt minefield. The safety zone is 55 meters from the front and sides and 20 meters from the rear of the container. The disrupt minefield uses four MOPMS containers that are spaced 70 meters apart to give a minefield front of 280 meters. Other MOPMS containers are offset from the baseline by 35 meters to give the minefield a depth of 70 meters. All containers are fired using the same RCU or FD.

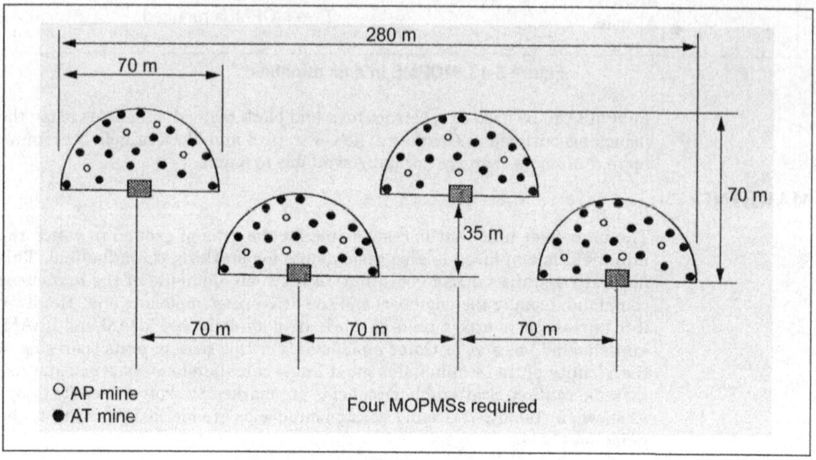

Figure 3-12. MOPMS in a disrupt minefield

Figure 3-13 illustrates the arrangement of MOPMS containers for a fix minefield. The basic layout is the same as the disrupt minefield; however, the fix minefield has one additional MOPMS that is placed 70 meters forward of the baseline to act as an IOE. This gives the same 280-meter minefield front but increases the minefield depth to 115 meters.

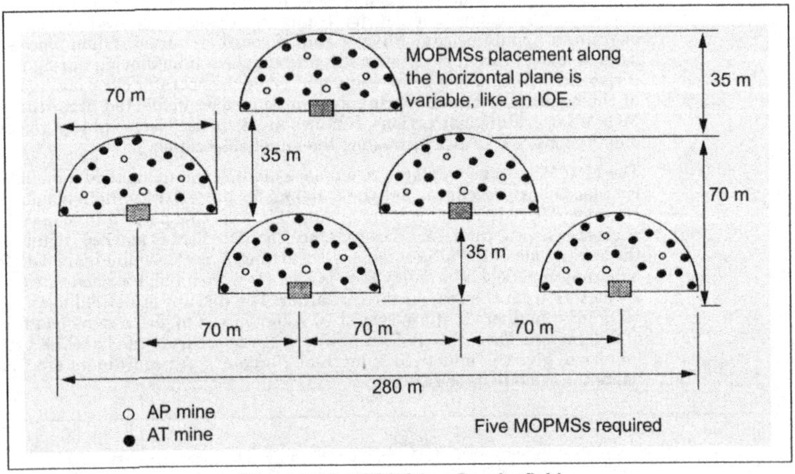

Figure 3-13. MOPMS in a fix minefield

MOPMS can be used to construct turn and block tactical minefields using the principles outlined in Chapter 2; however, turn and block minefields require more containers than are normally available to a unit.

MARKING

The maneuver unit that is responsible for the area of ground in which the minefield is emplaced is also responsible for marking the minefield. This normally requires direct coordination between elements of the maneuver command (usually the engineer) and the delivering/emplacing unit. However, it is unrealistic to expect units to mark artillery-delivered ADAM and RAAM, air-delivered Volcano, or Gator minefields. For this reason, units operating in the vicinity of these minefields must know calculated safety zones and use extreme caution. Scatterable minefields are marked to protect friendly troops as shown in Table 3-7. Ground Volcano minefields are marked according to the guidelines below.

Table 3-7. Marking scatterable minefields

Minefield Location	Marking
Enemy forward area	Unmarked
Friendly forward area	Sides and rear marked
Friendly rear area	All sides marked

SAFETY ZONES

A safety zone is an area where a stray or outlying mine has a chance of landing and laying to rest. The commander must prevent friendly forces from maneuvering into the safety zone during the minefield's life cycle. Depending on its specific location on the battlefield, the safety zone may be marked with a fence.

The safety zone around a Volcano minefield is shown in Figure 3-14.

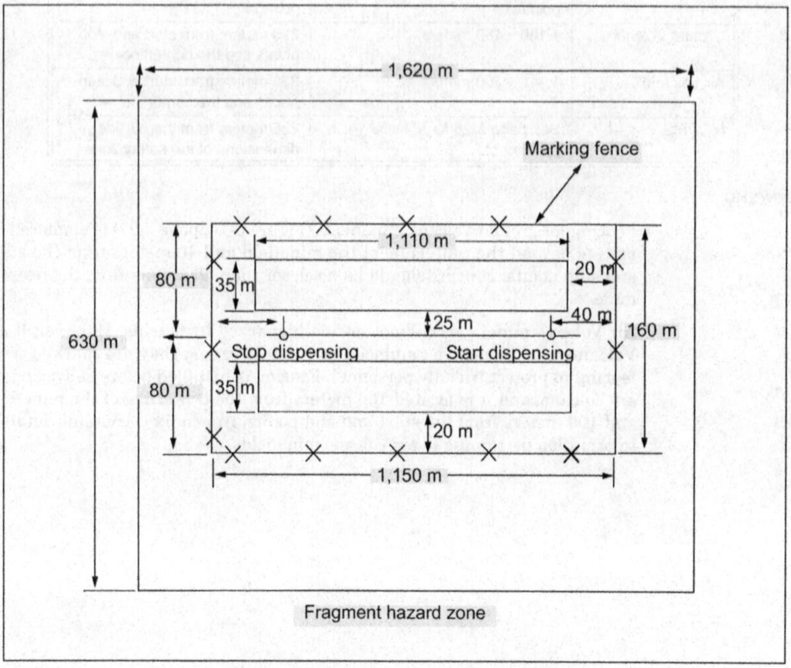

Figure 3-14. Ground Volcano minefield

FRAGMENT HAZARD ZONES

If an AT mine that is oriented on its side self-destructs, the EFP can theoretically travel 640 meters. This is the maximum fragment hazard zone; however, the chances of being struck are negligible at this distance. Tests indicate that the acceptable risk distance is 235 meters from the outer edges of the minefield's safety zone. This fragment hazard zone is also associated with the Gator and MOPMS AT mines. When the MOPMS is used for protective minefield missions, commanders must be made aware of the fragment hazard zone.

Use Table 3-8 to determine safety zones and fragment hazard zones.

Table 3-8. Safety and fragment hazard zones

System	Safety Zone	Fragment Hazard Zone
ADAM/RAAM	500 to 1,500 meters from aim point(s) (depends on delivery factors)	235 meters from the outside dimensions of the safety zone
Gator	925 x 475 meters from aim point(s)	1,395 x 945 meters from aim point(s)
Ground Volcano	1,150 x 160 meters	235 meters from start and stop points and the centerline
Air Volcano	1,315 x 200 meters	235 meters from start and stop points and the centerline
MOPMS	See page 3-28 for specific placement.	235 meters from the outside dimensions of the safety zone

FENCING

Fencing for ground Volcano minefields (Figure 3-14, page 3-27) is emplaced 80 meters beyond the centerline of the minefield and 40 meters from the start and stop points. Fencing should be no closer than 20 meters from the nearest mine.

Air Volcano minefields are not normally marked by fencing. However, if air Volcano minefields are emplaced in friendly areas, they are marked with fencing to protect friendly personnel. Fencing is installed before delivering an air Volcano, and it is located 100 meters from the centerline of the minefield and 100 meters from the start and end points. Appendix D contains detailed information pertaining to air Volcano minefields.

Chapter 4

Special-Purpose Munitions

Special-purpose munitions are hand-emplaced and used to create an expedient obstacle, enhance existing ones, and attack specific types of targets. The commander can employ these munitions to support his scheme of maneuver, to mass firepower, and to disrupt or destroy enemy forces in depth. Special considerations must be made in the planning process to effectively employ special-purpose munitions.

M18A1 CLAYMORE

The M18A1 claymore munition (Figure 4-2) is a fragmentation munition that contains 700 steel balls and 682 grams of composition C4 explosive. It weighs 1.6 kilograms and can be detonated by command **(Korea Only: or trip wire)**. It is activated by electric or nonelectric blasting caps that are inserted into the detonator well. The claymore projects a fan-shaped pattern of steel balls in a 60-degree horizontal arc, at a maximum height of 2 meters, and covers a casualty radius of 100 meters. The forward danger radius for friendly forces is 250 meters. The backblast area is unsafe in unprotected areas 16 meters to the rear and sides of the munition. Friendly personnel within 100 meters to the rear and sides of the munition should be in a covered position to be safe from secondary missiles. If the M18A1 is employed in a minefield for 72 hours or more, the minefield must be fenced on all sides.

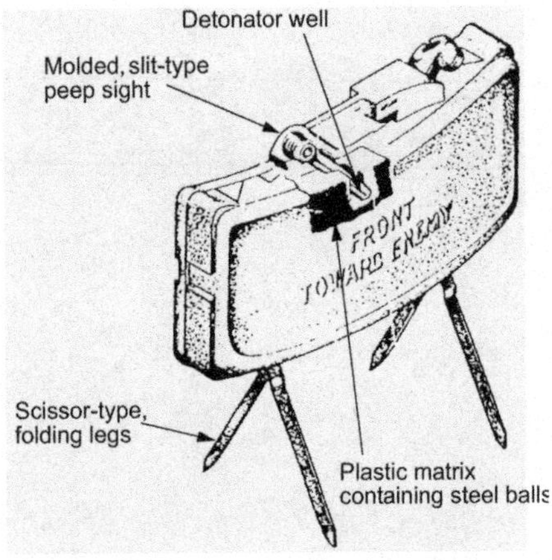

Figure 4-2. M18A1 claymore

When employing the M18A1 claymore with other munitions or mines, separate the munitions by the following minimum distances:

- 50 meters in front of or behind other M18A1s.
- 3 meters between M18A1s that are placed side by side.
- 10 meters from AT or fragmentation AP munitions.
- 2 meters from blast AP munitions.

SELECTABLE LIGHTWEIGHT ATTACK MUNITION

The selectable lightweight attack munition (SLAM) (Figure 4-3) is a multipurpose munition with an antitamper feature. The SLAM is compact and weighs only 1 kilogram, so it is easily portable. The SLAM is intended for use against APCs, parked aircraft, wheeled or tracked vehicles, stationary targets (such as electrical transformers), small fuel-storage tanks (less than 10,000-gallon), and ammunition storage facilities. The EFP warhead can penetrate 40 millimeters of homogeneous steel.

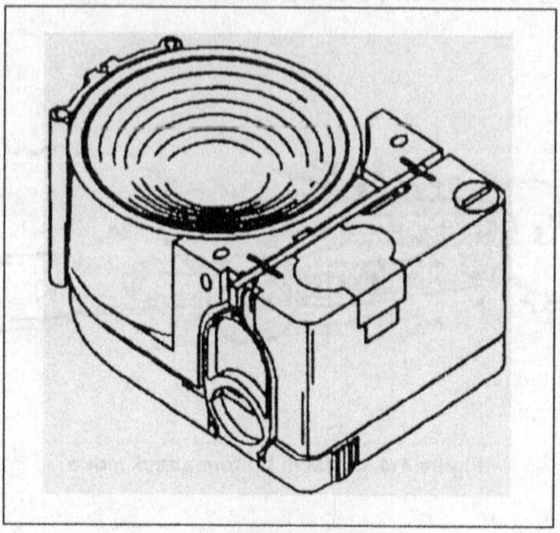

Figure 4-3. SLAM

The SLAM has two models—one is self-neutralizing (M2) and the other is self-destructing (M4):

- The M2 is solid green and has no labels, brands, or other distinguishing marks. This device is used by SOF and is not available to other units.
- The M4 is green with a black warhead (EFP) face. This device is normally used by units designated as light, airborne, air assault, crisis response, and rapid deployment.

See Appendix B for a description of major SLAM components.

OPERATING MODES

The SLAM has four possible employment methods—bottom attack, side attack, timed demolition, and command detonation.

Bottom Attack

The SLAM has a built-in magnetic sensor, so it can be used as a magnetic-influenced munition against trucks and light armored vehicles (Figure 4-4, page 4-4). It can be concealed along trails and roads where target vehicles operate and can be camouflaged with dry leaves, grass, and so forth without affecting EFP performance. Mud, gravel, water, and other debris that fill the EFP cup have minimal impact on EFP formation and effectiveness as long as the debris does not extend beyond the depth of the EFP cup. The magnetic sensor is designed to trigger detonation when it senses a vehicle's overpass. For the EFP to form properly, it needs a minimum of 13 centimeters from the point of emplacement to the target. The bottom-attack mode is active when the selector switch is set to *4, 10, or 24 HOURS* and the passive infrared sensor (PIRS) cover is in place. The SLAM will self-destruct (M4) or self-neutralize (M2) if the selected time expires before the SLAM is detonated by a vehicle.

Figure 4-4. SLAM in bottom-attack mode

Side Attack

The SLAM is equipped with a PIRS that was specifically developed for the side-attack mode (Figure 4-5). The PIRS detects trucks and light armored vehicles by sensing the change in background temperature when vehicles cross in front of the PIRS port. The PIRS is directional and aligned with the EFP when the device is aimed. The side-attack mode is active when the SLAM selector switch is set to *4, 10, or 24 HOURS* and the PIRS cover is removed to expose the PIRS. The SLAM will self-destruct (M4) or self-neutralize (M2) if the selected time expires before it is detonated by a vehicle.

Timed Demolition

The SLAM's built-in timer will trigger detonation at the end of a selected time (Figure 4-6). The timed-demolition mode is active when the SLAM selector switch is set to *15, 30, 45, or 60 MINUTES*. In this mode, the magnetic sensor and the PIRS are inoperable, and the SLAM will detonate after the selected time has expired.

Command Detonation

This mode provides manual warhead initiation using standard military blasting caps and a priming adapter (Figure 4-7). The command-detonation capability bypasses the SLAM's fuse and safing and arming (S&A) assembly.

Figure 4-5. SLAM in side-attack mode

Figure 4-6. SLAM in timed-demolition mode

Figure 4-7. SLAM in command-detonation mode

ANTITAMPER FEATURE

The SLAM has an antitamper feature that is only active in the bottom- and side-attack modes. The SLAM will detonate when an attempt is made to change the selector switch's position after arming.

M93 HORNET

The M93 Hornet (Figure 4-8) is an AT/antivehicular off-route munition made of lightweight material (35 pounds) that one person can carry and employ. The Hornet is a nonrecoverable munition that is capable of destroying vehicles by using sound and motion detection methods. It will automatically search, detect, recognize, and engage moving targets by using top attack at a standoff distance up to 100 meters from the munition. It is employed by combat engineers, rangers, and SOF.

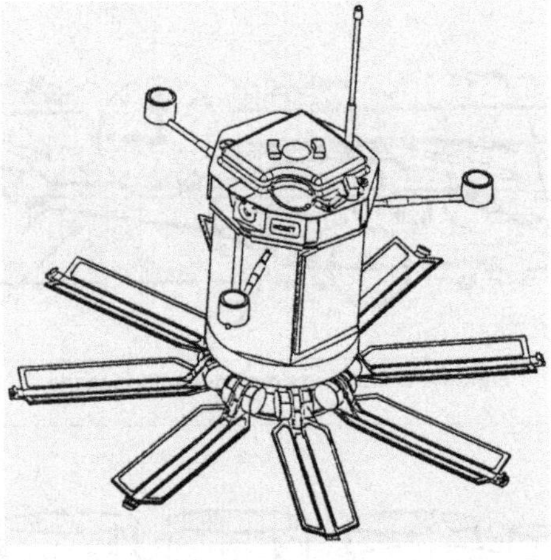

Figure 4-8. M93 Hornet

The RCU is a hand-held encoding unit that interfaces with the Hornet when the remote mode is selected at the time of employment. After encoding, the RCU can be used to arm the Hornet, reset its SD times, or destroy it. The maximum operating distance for the RCU is 2 kilometers.

High winds, heavy rain, snow, ice, extreme cold, and extreme heat reduce the Hornet's ability to detect targets at maximum range. Radio-frequency (RF) jamming devices (such as the hand-emplaced, expandable jammer [HEXJAM]), limit the Hornet's communication capabilities if they are placed in the munition field, but they will not affect the Hornet's ability to engage targets and will not damage the system. RF jamming devices affect the remote arming of current Hornet systems using the MOPMS RCU, and they will affect future Hornet's two-way communications capability with the Centurion remote control device.

See Appendix B for a description of Hornet components.

EMPLOYMENT CONSIDERATIONS

The Hornet's active battery pack is inserted during prearming and has an estimated life of 4 hours. The active battery pack powers the munition from the time it is inserted until the end of the safe-separation time, when the built-in reserve battery is activated. To prevent munitions from becoming duds, do not prearm them too early. Allow adequate time for travelling to the obstacle site, emplacing mines, throwing arming switches, and expiration of safe-separation times.

Once the Hornet is armed and the self-test is performed, the munition will remain active until its SD time expires or until it is encountered. The SD time (4 hours, 48 hours, 5 days, 15 days, or 30 days) is determined by the mission and the commander's intent. The munition will self-detonate after the SD time has expired.

Hornet munitions have an employed life of 60 days in the prearmed mode (remote arming) and 30 days in the armed mode. If the temperature exceeds 100°F, the employed life drops to 15 days in the prearmed mode and 30 days in the armed mode.

EMPLOYMENT ROLES

Combat engineers or maneuver forces under engineer supervision emplace Hornets in close operations; SOF or rangers emplace Hornets in deep operations. Hornets will be employed throughout the entire depth of the battle space to support Army operations.

Close Operations

In close operations, the Hornet can be—

- Used to fix the enemy and weaken it along its AA.
- Emplaced as an offensive-support weapon system because of its quick emplacement time and wide attack area.
- Employed rapidly along exposed flanks during a maneuver as a situational obstacle to disrupt the enemy's counterattacks.

- Used as a stand-alone tactical obstacle or as a reinforcement to conventional obstacles.
- Used to disrupt and delay the enemy, allowing long-range weapons to engage more effectively.

Deep Operations

In deep operations, the Hornet can be—

- Emplaced along key routes in gauntlet obstacles to disrupt and delay threat second-echelon forces, resupply operations, and key lines of communication (LOC).
- Used at C^2 and logistics sites to disrupt enemy operations.

Rear Operations

In rear operations, the Hornet can be emplaced (unarmed) along key routes in preparation for possible retrograde operations.

Early-Entry Operations

In early-entry operations, the Hornet can be—

- Used as an additional antiarmor weapon to supplement light forces.
- Used along high speed AAs in gauntlet obstacles to buy time and space.

TACTICAL EMPLACEMENT

There are four basic emplacement scenarios for the Hornet.

Conventional Minefield Reinforcement

The Hornet can be used to reinforce a conventional turn, block, or fix minefield (Figure 4-9).

Platoon engineers emplace the conventional minefield first, and then they traverse the safe lane that is perpendicular to the minefield. The Hornets are employed in two staggered rows, spaced 100 meters apart, 50 to 100 meters from the front edge (on the enemy side) of the conventional minefield. It is also recommended that a row of Hornets be placed 50 meters behind the minefield to reduce the enemy's breaching capability. (This row will be emplaced after the safe lane is closed.) The emplacing vehicles work toward the safe lane.

Two squads employ Hornets in two rows of ten each. One or more soldiers provide security. Under the supervision of a noncommissioned officer (NCO), four soldiers in each squad vehicle start prearming the Hornets, if necessary. They—

- Rotate the handle.
- Remove the cover.
- Insert the active battery pack and verify functionality via a solid status light.
- Reinstall the active battery-pack cover.

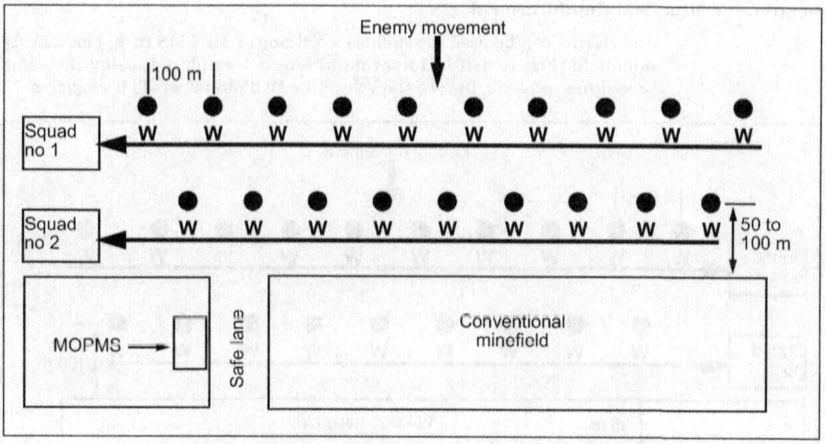

Figure 4-9. Hornet reinforcing a conventional minefield

- Select the SD time.
- Encode the Hornet with the M71 RCU and verify functionality via a flashing status light.
- Reinstall the cover.

Each emplacement vehicle moves to the first Hornet emplacement site in each row. The emplacing soldier and the arming soldier dismount. The emplacing soldier is handed a Hornet from the vehicle. He emplaces the Hornet at the designated site and returns to the vehicle.

The arming soldier rotates the handle on the Hornet, removes the cover and the safety and handling (S&H) band, rotates the SD switch to *U*, and pushes the arm switch to *ARM*. He then returns to the vehicle, taking the cover and the S&H band with him. The vehicle travels to the next Hornet emplacement site.

After all the Hornets in the two leading rows have been emplaced and armed, the emplacing vehicles exit through the safe lane and usually secure it with a MOPMS. The emplacement vehicles must be at least 475 meters (safe standoff distance) from the nearest Hornet within 30 minutes. Before sending a remote arming signal, vehicles must wait at least 36 minutes after the arming switch is thrown on the last Hornet emplaced. If a rear row is required, it is emplaced at this time. The Hornets are then remotely armed with the M71 RCU, when required. They are now capable of covering the minefield by fire and engaging threat tracked vehicles.

Scatterable Minefield Reinforcement

The Hornet can be used to reinforce a Volcano or MOPMS turn, block, or fix minefield (Figure 4-10). Hornet munitions are emplaced, using the same procedures as above, **before** the Volcano or MOPMS minefield is emplaced.

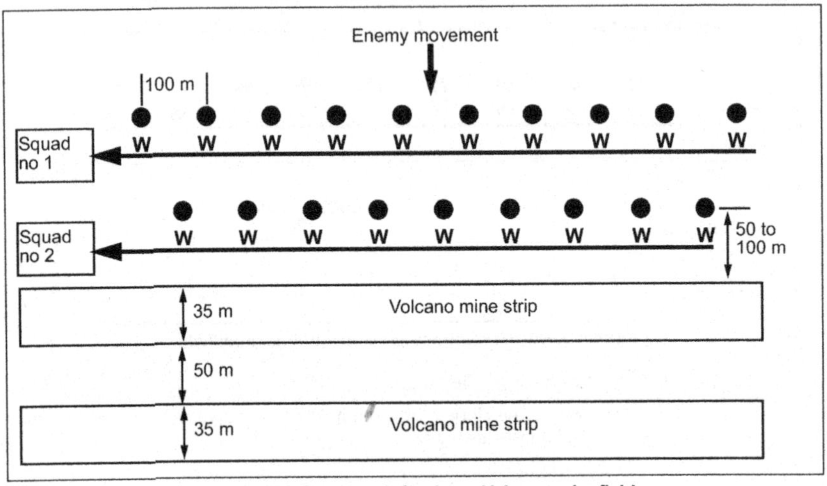

Figure 4-10. Hornet reinforcing a Volcano minefield

To ensure that the Volcano dispensing vehicle has sufficient time to reach the safe standoff distance (475 meters), Volcano dispensing should start no later than 30 minutes (minus the Volcano dispensing time) after the first Hornet is armed. This allows Hornet emplacing squads to be finished or nearly finished before the Volcano dispenser begins emplacing the minefield.

Area-Disruption Obstacle

When the X-pattern is employed, the Hornet is very effective as a disrupting obstacle (Figure 4-11). An area-disruption obstacle is employed to disrupt the enemy's approach prior to the start of the direct-fire battle. It causes disruption and attrition of the advancing threat force and encourages follow-on forces to seek an alternate route. Therefore, multiple area-disruption obstacles will typically be employed to adequately cover the cross-country AA. This requires coordinated action among multiple squads.

An engineer platoon emplaces a Hornet area-disruption obstacle. The obstacle typically consists of 20 Hornets (five clusters of four Hornets each) employed in an X-pattern over a 1- by 1-kilometer area. Individual Hornets are emplaced 100 meters apart. Emplacing this obstacle must be done as a dispense-and-roll operation to ensure that the emplacing vehicles can reach the safe standoff distance (475 meters) from any armed Hornets.

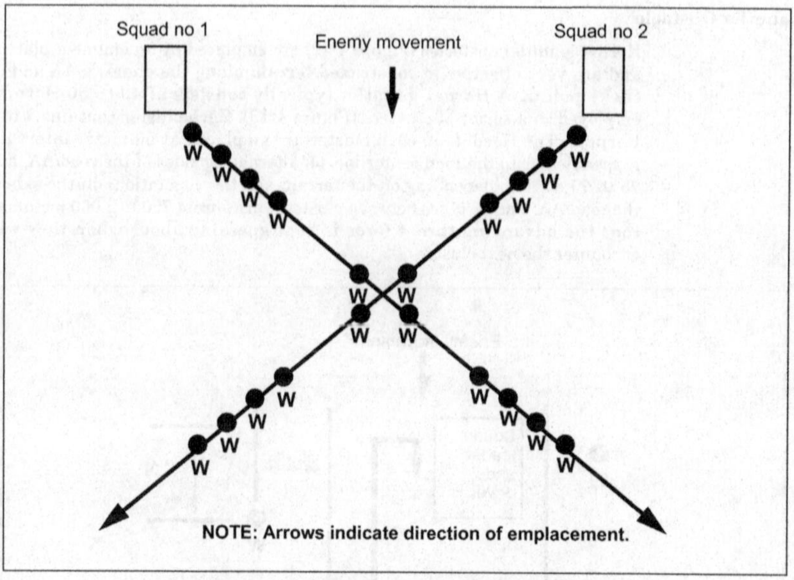

Figure 4-11. Hornet area-disruption obstacle

Area-disruption obstacles are normally armed by remote, but they can be manually armed under the following conditions:

- METT-TC requires rapid emplacement and arming.
- Terrain reconnaissance determines that there are no major impediments (rough terrain, vegetation) to maneuver.
- Emplacement is done during daylight hours (mission-oriented protective posture [MOPP] level 0 only).

Hornets are prearmed the same as above. Two squads lay the Hornets in unison, starting with the two emplacement sites closest to the enemy. Each squad drives in a straight line, crossing paths at the middle of the X, and emplaces ten Hornets.

A soldier in the back of each emplacing vehicle throws the arming switch and sets the Hornet down or drops it off (base down) the back of the vehicle. After all the Hornet clusters are emplaced, squad vehicles quickly travel to the 475-meter safe standoff distance (no further than 2 kilometers) to prepare for remote arming. Hornets can be remotely armed 36 minutes after the arming switch is thrown on the last Hornet emplaced. If manual arming is used, Hornets automatically arm at the end of their safe-separation time (5 to 6 minutes after the arming switch is thrown).

Gauntlet Obstacle

Hornet gauntlet obstacles (Figure 4-12) are emplaced by an engineer platoon and are very effective in constricted terrain along the enemy's AA and at choke points. A Hornet gauntlet typically consists of 40 to 50 Hornets employed in a series of clusters (Figure 4-13). Each cluster contains 3 to 6 Hornets. The Hornets in each cluster are emplaced at 50-meter intervals, perpendicular to the road centerline, on alternating sides of the road/AA, and 25 to 50 meters (depending on the terrain and the vegetation) off the side of the road/AA. The distance between clusters varies from 750 to 2,000 meters so that the advancing threat force is kept guessing about when they will encounter the next cluster.

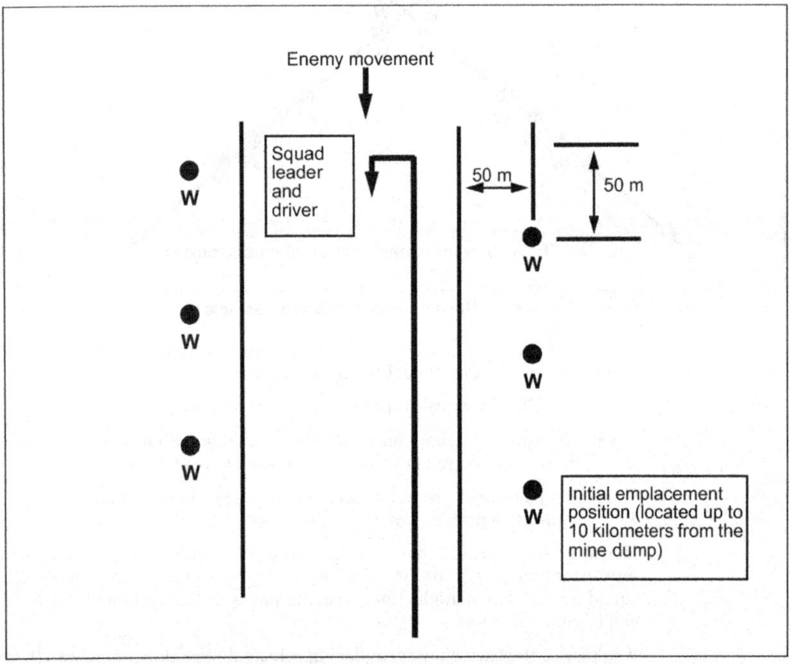

Figure 4-12. Hornet gauntlet obstacle (one cluster)

Before laying any Hornets, the munitions are prearmed as above. Soldiers also set the target switch to *HVY* for clusters closest to the enemy, so that the Hornets will only engage heavy tracked vehicles. The intent is to make threat forces commit to a route they perceive to be clear.

Hornets are emplaced beginning on the friendly side of the cluster. The first engineer squad emplaces Hornet munitions beginning with the cluster closest to the enemy. The emplacement vehicle drives even with the first Hornet

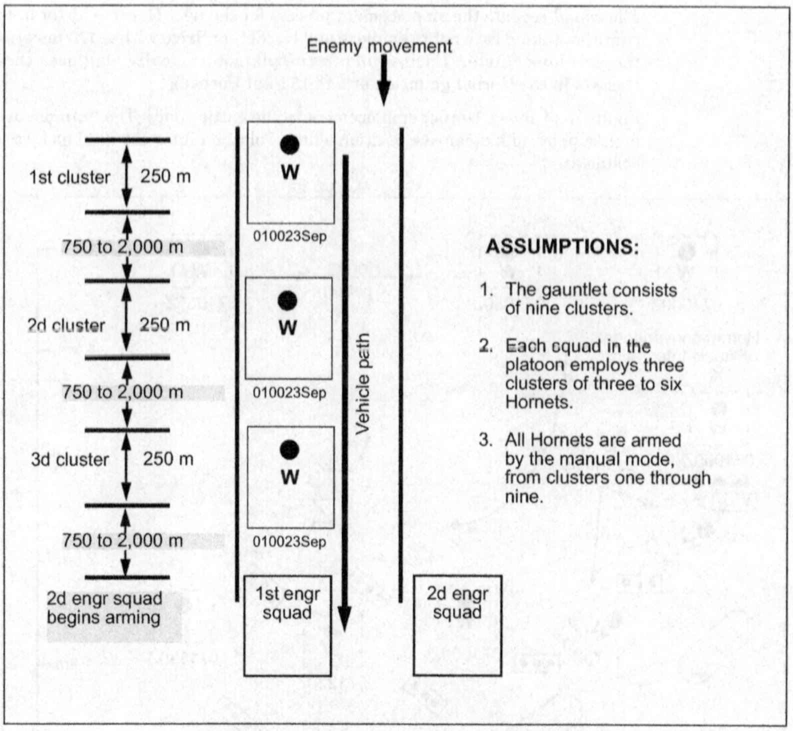

Figure 4-13. Hornet gauntlet obstacle (platoon)

emplacement site. The emplacing soldier dismounts, and a soldier in the vehicle hands him a Hornet. The emplacing soldier then proceeds to the Hornet emplacement site. The vehicle travels to a point even with each subsequent emplacement site. A soldier deploys at each emplacing site to lay one of the remaining Hornet munitions in the cluster. The vehicle then turns around and stops even with the last Hornet (on the enemy side) in the cluster.

Upon reaching the Hornet employment location, each emplacing soldier removes the cover and the S&H band, rotates the SD switch to *U*, and on the command (audible or visual signal) of the NCOIC, pushes the arm switch to *ARM*. Once the Hornets' arming switches are thrown, soldiers return to the road, taking the covers and the S&H bands with them, and wait to be picked up by the emplacement vehicle. After all the soldiers are in the emplacing vehicle, the driver quickly travels to the safe standoff distance (475 meters). The Hornet munitions in the first cluster will arm at the end of the safe-separation time (5 to 6 minutes).

The squad repeats the emplacement process for the next Hornet cluster in the gauntlet, taking care not to emplace any Hornets or drive within 475 meters of the previous cluster. Each squad in the platoon typically emplaces three clusters in the Hornet gauntlet, or 9 to 18 total Hornets.

Figure 4-14 shows Hornet emplacement in the battle space. The example used is in support of a defensive position where Volcano mines are used as tactical obstacles.

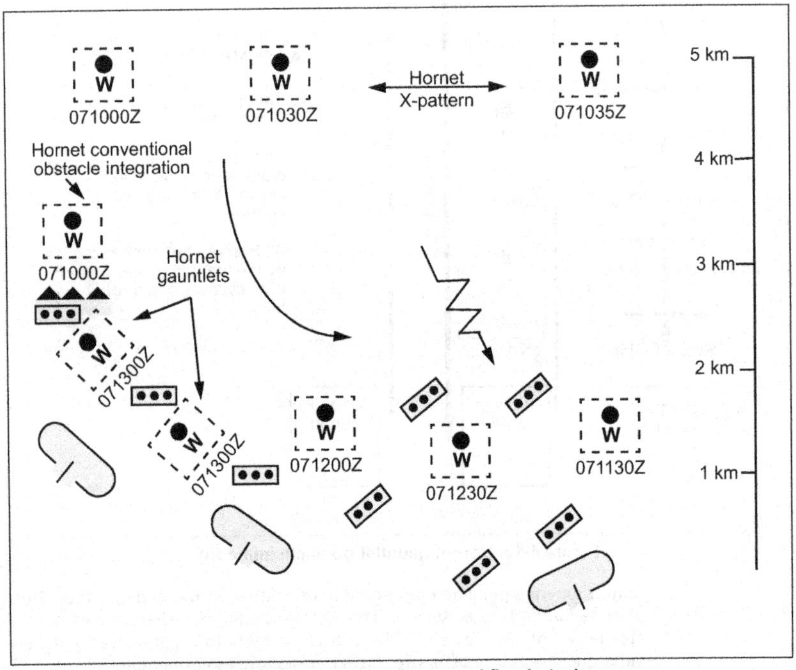

Figure 4-14. Hornet-enhanced turn-and fix-obstacle groups

Deep-Battle Interdiction Weapon

SOF or ranger units emplace Hornet munitions in the deep battle area as interdiction weapons. A typical mission requires a unit ranging in size from a six-man team to an entire company. The number of Hornets carried by the unit depends on the mission and the mode of insertion (vehicle, aircraft, or dismounted troops); a man can normally carry only one Hornet. Hornets are typically used to support a raid against an enemy position or complex and at bridges or choke points along high-speed AAs used by advancing second-echelon forces or for resupply. In these roles, the Hornets are employed similar to the clusters in a gauntlet obstacle.

Camouflage and Concealment

The best camouflage and concealment for the Hornet is tall grass and brush. The Hornet can be partially buried if the terrain or the vegetation does not provide effective natural camouflage and concealment. Placing the Hornet in a hole degrades its performance, so it should only be done when Hornets cannot be covered by fires or protected from tampering by dismounted enemy. The following conditions must be met:

- The depth of the hole must not exceed 4 inches, because the acoustic sensors must be above ground level.
- The hole must not restrict the Hornet's ability to rotate and tilt its body and to fire the sublet. To meet this requirement, the hole must be at least 36 inches wide and flat enough to support the munition. Although the Hornet should be placed on a flat surface if possible, it can operate on slopes up to 15 degrees.

Munitions placed at ground level should be no closer to obstructions than the distances shown in Table 4-1.

Table 4-1. Hornet minimum emplacement distances

Maximum Obstruction Height	Minimum Employment Distance from Obstruction
1 m	3 m
2.4 m	5 m
6.5 m	15 m
25 m	25 m

When the Hornet is emplaced and concealed, remove all indicators of excess soil and camouflage material before performing the arming sequence.

Recording and Marking

When the Hornet munition field is completed, the OIC will identify an NCO to be the recorder. The NCO will collect data from the NCOICs of the emplacing squads and complete DA Form 1355 as outlined in Chapter 8. The OIC will ensure that the DA Form 1355 is completed timely and accurately.

Marking the Hornet munition field will be completed as prescribed in Chapter 2. The fence will be no closer than 150 meters from the nearest Hornet munition. Marking must be completed before emplacing the munitions.

This chapter implements STANAG 2990.

Chapter 5

Conventional Mines

Conventional mines are hand-emplaced mines that require manual arming. This type of mine laying is labor-, resource-, and transport-intensive. Soldiers emplace conventional mines within a defined, marked boundary and lay them individually or in clusters. They record each mine location so that the mines can be recovered. Soldiers can surface lay or bury conventional mines and may place AHDs on AT mines.

ANTITANK MINES

The M15, M19, and M21 AT mines are used by US forces. They are shown in Figure 5-1, and their characteristics are listed in Table 5-1, page 5-2.

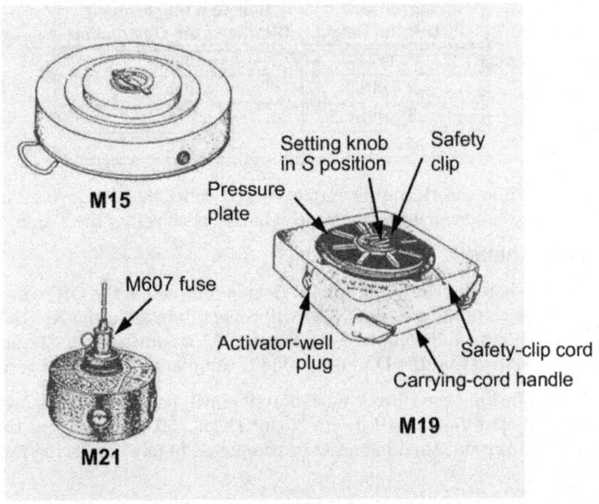

Figure 5-1. AT mines

M15

The M15 AT mine is 337 millimeters in diameter and 125 millimeters high. It weighs 13.5 kilograms and contains 9.9 kilograms of Composition B explosive. The primary fuse well is on the top center of the mine; secondary fuse wells are on the side and bottom. The M15 can contain the following fuses:

Table 5-1. Characteristics of AT mines

Mine	DODIC	Fuse	Warhead	AHD	Explosive Weight	Mine Weight	Mines per Container
M15 w/ M603 fuse	K180	Pressure	Blast	Yes	9.9 kg	13.5 kg	1
M15 w/ M624 fuse	K180 (mine) K068 (fuse)	Tilt rod	Blast	Yes	9.9 kg	13.5 kg	1
M19	K250	Pressure	Blast	Yes	9.45 kg	12.6 kg	4
M21	K181	Tilt rod or pressure	SFF	Yes*	4.95 kg	7.6 kg	4

*Conventional AHDs will not couple with this mine; however, the M142 multipurpose FD can be emplaced under this mine.

- M603 fuse. When the M603 fuse is employed on the primary fuse well, the M15 is a track-width mine that is activated by 158 to 338 kilograms on the pressure plate. This produces an M-Kill.
- M624 fuse. When the M624 fuse (with tilt rod) is employed on the primary fuse well, the M15 is a full-width mine that is activated by a deflection of 20 degrees or 1.7 kilograms of pressure to the tilt rod. Depending on the armor, this produces an M-Kill or a K-Kill.

M19

The M19 AT mine is a low-metallic, square-shaped mine that measures 332 by 332 millimeters and is 94 millimeters high. It weighs 12.6 kilograms and contains 9.45 kilograms of Composition B explosive, a tetryl booster pellet, and an M606 integral fuse. When the setting knob on the pressure plate is in the *S* (safe) position, the mine cannot function by action of the main fuse. After the safety clip has been removed and the setting knob turned to the *A* (armed) position, a force of 157.5 to 225 kilograms on the pressure plate depresses the Belleville spring and begins the firing chain. A standard FD may be used with the M2 activator in any of the secondary fuse wells on the sides or the bottom of the mine. When the M19 is employed, it is difficult to detect because of its plastic construction. It produces an M-Kill with a blast effect.

M21

The M21 AT mine is 230 millimeters in diameter and 206 millimeters high. It weighs 7.6 kilograms and has 4.95 kilograms of Composition H6 explosive. The mine is activated by 1.7 kilograms of pressure against a 61-centimeter-long rod on the end of the M607 fuse. It uses an M-S plate to produce a K-Kill. The M21 with tilt rod must be buried or staked (use three stakes at the 12, 4, and 8 o'clock positions) so that enemy vehicles will not tip the mine over. Without the tilt rod, the mine is activated by 130.5 kilograms of pressure on the M607 fuse and produces an M-Kill by blast effect.

ANTIPERSONNEL MINES

The M14 and M16 AP mines are used by US forces on the Korean peninsula. They are also used by many other countries. The M16 AP mine is likely to be seen in a modified form. These mines are shown in Figure 5-2, and their characteristics are listed in Table 5-2.

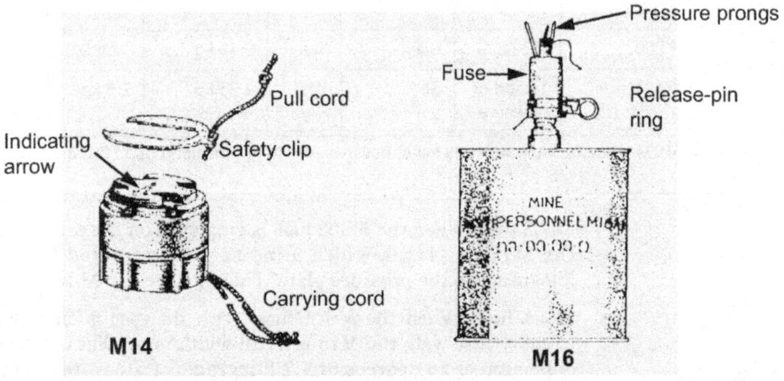

Figure 5-2. AP mines

Table 5-2. Characteristics of AP mines

Mine	DODIC	Fuse	Warhead	AHD	Explosive Weight	Mine Weight	Mines per Container
M14	K121	Pressure	Blast	No	28.4 g	99.4 g	90
M16-series	K092	Pressure or trip wire	Bounding frag	No	450 g	3.5 kg	4

M14

The M14 AP mine is a low-metallic blast mine consisting of a main charge (28.4 grams of tetryl) and a plastic fuse with a steel firing pin. It is cylindrical in shape (56 millimeters in diameter and 40 millimeters high) and weighs 99.4 grams. The pressure plate has an indented, yellow arrow that points to the *A* or *S* position on top of the fuse body. A force of 11.5 to 13.5 kilograms depresses the pressure plate and causes the Belleville spring to drive the firing pin into the detonator. The M14 is not designed to kill, but to incapacitate. The M14 AP mine has been modified by gluing a metal washer to the bottom of the mine. The modification was directed to improve the detectability of the mine. Unmodified mines are not authorized for use by US forces.

M16

The M16 AP mine is a bounding fragmentation mine that consists of a mine fuse (M605), trinitrotoluene (TNT) explosive, a propelling charge, and a projectile that are contained in a sheet-steel case. The mine is 103 millimeters in diameter, 199 millimeters high (including the fuse), and weighs 3.5 kilograms. The principal difference between the M16, M16A1, and M16A2 versions are in the construction of the detonators and boosters. The casualty radius is 27 meters for the M16 and M16A1 and 30 meters for the M16A2. A pressure of 3.6 to 9 kilograms applied on one or more of the three prongs of the M605 fuse or a pull of 1.4 to 4.5 kilograms on the trip wire will activate the mine.

EMPLACING MINES

The method used to lay and conceal each type of mine depends on the method of mine operations, the type of ground in which the mine is to be laid, and the type of ground cover available for camouflage.

Standard-pattern mine laying is laborious and time-consuming, but it is more effective and flexible than row mine laying and allows better mine concealment. Standard-pattern mine laying is well suited for protective minefields, and it can be used in terrain where the nature of the ground makes row mine laying impractical.

To achieve the maximum effect, mines must be laid where they cannot be seen and where a vehicle or a person exerts enough pressure to detonate them. The following rules should be applied to achieve the maximum effects of mines:

Mines With Prongs

Korea Only: If the mine is activated by its prongs, it should be buried flush with the ground so that only the tips of the mechanism are exposed (Figure 5-3). A mine buried in this manner is held firmly upright. The target exerts a direct, downward pressure rather than a sideways thrust. The mine is protected from damage and is difficult to see. If it is buried more deeply, it becomes unreliable because the layer of spoil may prevent the mine mechanism from operating.

If the mine is activated by a trip wire, it should be buried so that the trip wire is at least 2 to 3 centimeters above the ground (Figure 5-4).

Mines With Pressure Plates

Mines with pressure plates will function when completely buried as long as the cushion of earth above them is not too thick. AT mines are normally buried with the top of the mine approximately 5 centimeters below ground level.

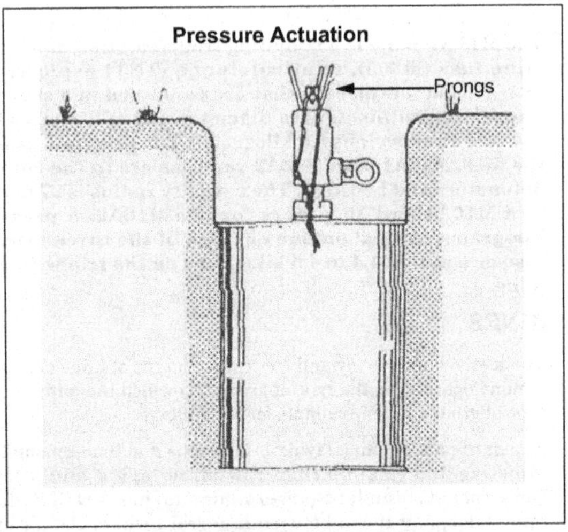

Figure 5-3. Prong-activated AP mine

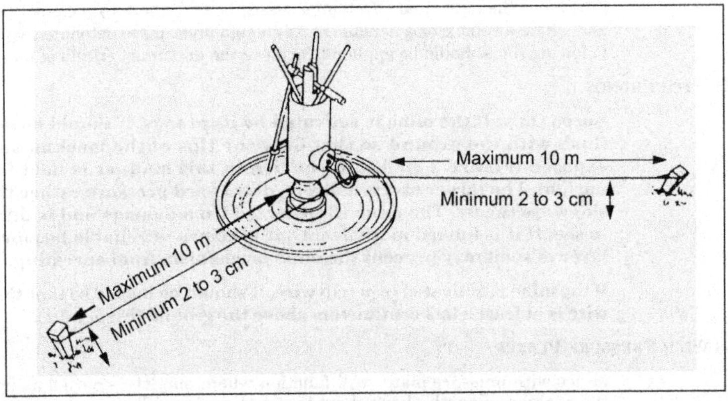

Figure 5-4. Trip-wire-activated AP mine

Korea Only: AP mines are usually placed in a hole and covered with camouflage material. If the hole is only slightly larger than the mine, the weight of the target may be supported by the shoulder of the hole, and the mine will fail to activate. Such bridging action can be avoided if the hole is dug much wider than the mine (Figure 5-5).

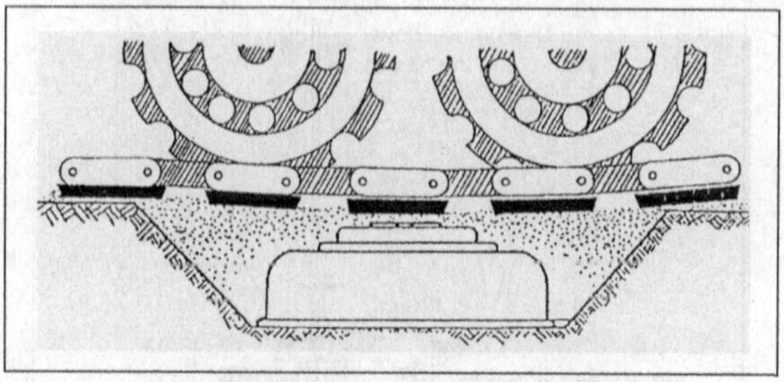

Figure 5-5. Buried mine with pressure plate

Mines With Tilt Rods

Tilt-rod fuses normally require the body of the mine to be buried and the tilt-rod assembly to be clear of the ground (Figure 5-6). A tilt-rod fuse is preferred in areas where vegetation is sufficient to conceal the extension rod. Camouflage materials are carefully used to prevent premature detonation or interference with the normal functioning of the fuse. Extension rods are camouflaged before the mine is armed. If tilt rod mines are surface-laid, they must be staked.

Bearing Boards

High pressure is required to activate AT mines. When burying a mine in soil that has a low bearing pressure (such as soft sand or clayey soil), it may be necessary to place a board or another bearing plate under the mine. Otherwise, the mine may not detonate when it is forced down.

Concealment

After digging the hole for a mine, place the spoil in a sandbag to reduce the evidence of mining. If a sandbag is not available, heap the spoil. Camouflage all traces of digging after the mine is laid. If the ground cover is turf or other matted, root material, remove spoil that cannot be hidden. Cut the sod in an X, I, or U shape in the area where the mine is to be placed; lay the mine; and then roll the sod back in place to camouflage the mine. Loose earth over a mine will eventually consolidate, so the mine location should have a small mound immediately after laying (Figure 5-7). Ensure that the mound is inconspicuous and that it blends with the surrounding area. It is very important that you make a final check after concealing each mine so that you can correct faults progressively, because they cannot be corrected later.

AT mines in standard-pattern minefields should be buried. However if conditions dictate, mines with a single-impulse fuse may be laid on the surface. Mines with double-impulse fuses should always be buried, because if

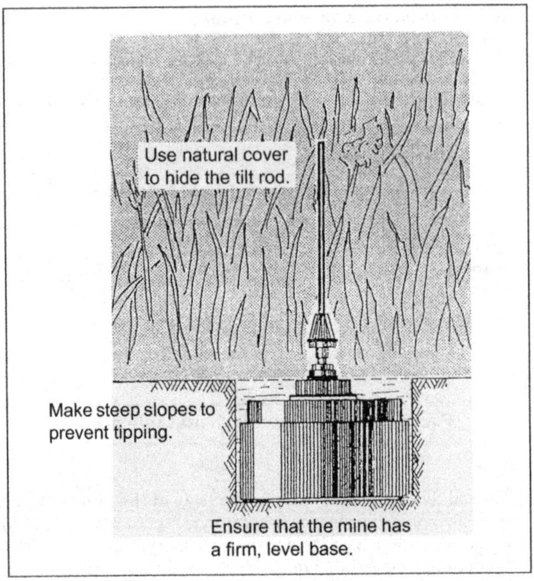

Figure 5-6. Buried mine with tilt rod

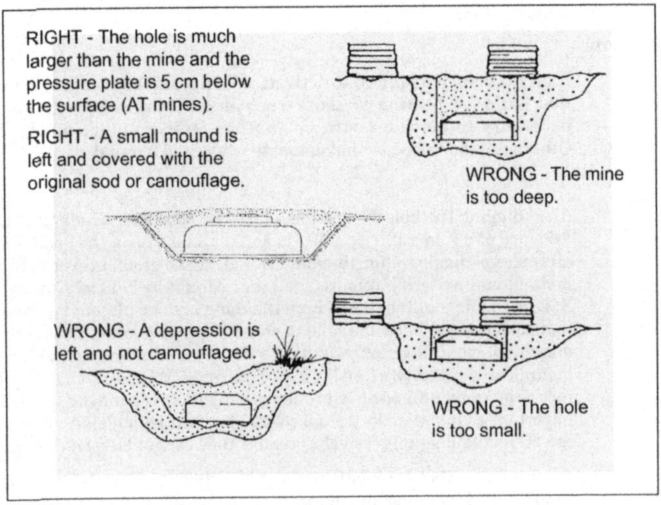

Figure 5-7. Buried and concealed mines

they are surface-laid, they may be physically damaged when pressure is exerted by a tracked vehicle. Buried mines also have some resistance to countermeasures, but surface-laid mines have none. Consideration must also be given to sympathetic detonation of AT mines (Table 5-3). US conventional mines do not have integral AHDs, so allow extra time to lay mines with AHDs.

Table 5-3. Sympathetic detonation chart

Type	M16	M15	M19
Surface-laid	NA	4.0 m	4.0 m
Buried flush	1.5 m	2.4 m	5.5 m
Buried 5 cm	NA	1.5 m	4.8 m

The difficulty of burying mines in very rocky ground and the necessity for surface laying will have a bearing on which mines are suitable. For example, small, blast-type AP mines are hard to detect and easy to camouflage. They are much easier to camouflage than larger fragmentation mines. The type of AT mine used will make little difference, because the mine's size will always make camouflage very difficult.

Maneuver Assistance

During large mine-laying operations, engineers seldom have sufficient manpower to carry out all minefield tasks. Other combat arms units must often provide work parties. Engineers must be capable of organizing, controlling, and supervising combined arms work parties. They must also instruct them in new equipment and techniques. Work parties may be integrated with engineers or given certain tasks that are within their capabilities.

When laying a standard-pattern minefield, consider supplementing work parties with other combat arms soldiers to perform the following:

- Executing Class IV/V supply point or mine dump missions. Soldiers uncrate and prepare mines and remove empty boxes and residue.
- Laying. Soldiers position mines within strips and dig holes.
- Marking. Soldiers construct the perimeter fence and emplace mine signs.

Unpacking, preparing, and loading mines are the most time-consuming tasks when laying a row minefield; and they are ideal tasks for other combat arms soldiers.

This chapter implements STANAG 2036.

Chapter 6

Row Mining

Row mining is a means of emplacement for tactical minefields. For example, a typical tactical minefield could contain several rows of regularly spaced mines.

USE

Row mining is not a new idea. It has been used since the beginning of modern mine warfare and is very effective. It is especially effective in support of maneuver-oriented doctrine. Row mining is faster than standard-pattern mining and improves the maneuver commander's flexibility by providing him an obstacle that requires less manpower effort.

Mines may be surface-laid or buried, and they are often laid directly from a slow-moving vehicle. This reduces the time and the personnel required to emplace a minefield. Row mining can be used as a tactical or situational obstacle. Minefields are usually emplaced at or near the FLOT, along flank AAs, to support security operations. Speed and efficiency make row mining a desirable option, and row mining supports current doctrine.

RULES

Rules governing authority, reporting, recording, and marking are generally the same for row minefields as they are for other minefields. Row mining is simply a method of laying mines.

The most important factor in row mining is the requirement for strict C^2. Row mining is potentially the most hazardous form of mine laying. It entails vehicles and personnel moving in and around mines without the safety of a centerline strip. Leaders must place extreme emphasis on safety because the laying procedure is very rapid.

Most of the rules governing row mining are defined in STANAG 2036. A summary of those rules and some additional rules that apply are shown below.

- Rows.
  - There are two types of mine rows—regular and short. Short rows are described under IOE rules below.
  - Regular rows are marked and recorded. They are designated by letters (A, B, and so forth), with Row A being closest to the enemy.
  - The minimum distance between rows of AT mines is 8 meters.
  - **Korea Only: The minimum distance between any row and a row containing AP mines is 15 meters.**

- The distance between the start row marker and the first mine in a row is the mine spacing for that row.
- Start and end row markers are permanent markers and must be made of detectable material.

- Clusters.
  - Clusters are placed on the row centerline and directed toward the enemy side.
  - A cluster in row mining usually consists of one AT mine **(Korea Only: but it may also contain AP mines)**.
  - Cluster composition must remain the same throughout the row.
  - **Korea Only: Different types of AP mines may be used in a cluster.**
  - **Korea Only: The total number of mines in one cluster will not exceed five; no more than one will be an AT mine.**
  - The type of AT mine may vary from one cluster to another.
  - **Korea Only: A cluster of AP mines can be laid in a 2-meter semicircle on the enemy side of the baseline.**
  - When a cluster contains a mine that is equipped with an AHD, the mine is armed before the AHD is armed. The cluster is not armed until all personnel are at least 60 meters away.
  - Omitted clusters do not contain mines. They are recorded on DA Form 1355 (see Chapter 8).
  - Clusters are omitted within lanes, within gaps, in areas less than 2 meters from boundaries and lanes, and in areas where the terrain (trees, rocks) prohibits emplacement.
- IOE.
  - The IOE is located on the enemy side of the minefield.
  - The IOE baseline must be at least 15 meters from Row A.
  - IOE mines are buried.
  - IOE short rows are labeled at start (I1) and end (I1E) points.
  - IOE short rows must be at least 15 meters apart.
- **Korea Only: Trip wires.**
  - **Trip wires can be used in regular rows, but only one mine per cluster can be actuated by a trip wire.**
  - **No more than two trip wires can be used on a mine.**
  - **Trip wires are not considered AHDs.**
  - **Trip wires must be at least 2 meters from a minefield lane, a cluster, another trip wire, an IOE short row, or a minefield perimeter fence.**

— **Trip wires can only be used with AP fragmentation mines.**

- Lanes.
  - — Lanes are sited before laying begins. Lane locations should not be obvious.
  - — Clusters must be at least 2 meters from lane edges.
  - — The number of lanes must be sufficient to ensure that no one lane is overused and turned into an obvious track.
  - — Sufficient mines must be stockpiled so that the responsible unit can seal lanes suspected of being located by the enemy.
- General.
  - — The spacing between mines or clusters can vary from 4 to 10 meters but must remain constant within the row.
  - — Mines and clusters must be at least 15 meters from the perimeter fence.
  - — If the distance between a mine or a cluster and any turning point is less than the spacing for that row, omit that mine or cluster. The mine immediately following a turning point is always located at the mine spacing for that row.
  - — The minefield has two landmarks located to the rear, never to the extreme side or front.
  - — Global-positioning systems (GPSs) can only be used to determine the coordinates for minefield landmarks and reference points (RPs).

WARNING
Do not use GPSs to chart or record minefield perimeter coordinates or to determine safe routes through or around existing minefields.

- — If landmarks are more than 200 meters away from the last row or are out of the direct line of sight, intermediate row markers or landmarks are placed at least 75 meters from the last end row marker.
- — Landmarks can be used for more than one minefield. This must be recorded in the remarks block of DA Form 1355.
- — Back azimuths are not used to record the minefield.
- — Measurements are in meters.

LOGISTICS

CALCULATIONS

To simplify the calculation process, a minefield requirements computation work sheet (Figure 6-1, pages 6-5 through 6-8) has been developed. This work sheet is provided to the platoon leader or sergeant as a step-by-step guide to the mathematics involved in the logistical computation process. Properly completed, the work sheet provides the number of mines to order (by type), the number of regular strips to be emplaced, cluster composition, the estimated man-hours required to install the minefield, the amount of fencing and marking material required, the number of truckloads required to carry the mines, and the number of rolls of engineer tape required.

Step-by-step procedures for completing the work sheet are shown in Figure 6-2, pages 6-9 through 6-14. Each step is explained in the example to facilitate the understanding of the logic behind the calculations.

Use the following steps to determine the number of AT mines required for a row minefield when not using the standard row minefields discussed later in this chapter. Round the resulting numbers up to the nearest whole number.

Step 1. Determine the number of mines required.

$$density \times front = number\ of\ mines$$

Step 2. Determine the number of mines per row.

$$front \div mine\ Spacing = number\ of\ mines\ per\ row$$

Step 3. Determine the number of rows.

$$number\ of\ mines \div number\ of\ mines\ per\ row = number\ of\ rows$$

Step 4. Determine the actual number of mines.

$$number\ of\ mines\ per\ row \times number\ of\ rows = number\ of\ mines$$

Step 5. Determine the number of mines to request (includes a 10 percent resource factor).

$$number\ of\ mines \times 1.1 = number\ of\ mines\ to\ request$$

Step 6. Determine the number of vehicle loads by using Table 2-8, page 2-45.

Step 7. Determine the fencing and marking material required.

Sample Problem: Your platoon has been tasked to emplace a 400-meter row minefield with a density of 0.5-0-0 (AT-AP fragmentation-AP blast). You have decided to space the mines 6 meters apart. Determine the number of M15 mines to order and the number of 5-ton dump trucks (with sideboards) required to deliver the mines.

- Step 1. 0.5 × 400 meters = 200 mines
- Step 2. 400 ÷ 6 = 66.6 = 67 mines per row
- Step 3. 200 ÷ 67 = 2.98 = 3 rows
- Step 4. 67 × 3 = 201 mines
- Step 5. 201 × 1.1 = 221.1 = 222 mines
- Step 6. 222 ÷ 204 = 1.08 = 2 5-ton trucks
- Step 7.
  - — Concertina: ([400 × 2] + [200 × 2] + 160) × 1.4 = 1,904 meters of concertina
  - — Pickets: 1,904 ÷ 15 = 126.9 = 127 pickets
  - — Signs: 127 pickets = 127 signs

MINEFIELD REQUIREMENTS COMPUTATION WORK SHEET			
GIVEN			
Desired density	AT _____	APF _____	APB _____
IOE representative cluster	AT _____	APF _____	APB _____
Front	_____ meters		
Depth	_____ meters		
Percentage of AHDs	_____%		
Type of mines	AT _____	APF _____	APB _____
Type of truck/trailer	_________		
Lanes/gaps/traffic tapes	_____		
Trip-wire safety tapes	_____		
PART 1. NUMBER OF MINES			
A. IOE live clusters = front ÷ 9	_____ ÷ 9 = _____ (round up)		
	AT	APF	APB
B. IOE representative cluster × Number of IOE clusters = Number of mines in IOE	_____ × _____ _____	_____ × _____ _____	_____ × _____ _____
C. Desired density × Minefield front = Mines in regular strips	_____ × _____ _____	_____ × _____ _____	_____ × _____ _____
D. Subtotal of mines (line B + line C)	_____	_____	_____
E. 10% excess factor =	× 1.10	× 1.10	× 1.10
Total number of mines to order (round up for each)	_____	_____	_____
PART 2. NUMBER OF REGULAR STRIPS			
A. Add desired density	AT _____	+ APF _____	+ APB _____ = _____
B. 0.6 × line A above	0.6 × _____ = _____ (round up)		
C. 3 × AT desired density	3 × _____ = _____		

Figure 6-1. Minefield requirements computation work sheet

D. Number of regular strips required = highest number of line B or C _____

PART 3. NUMBER OF AHDs

% AHDs × total number of AT mines _____

PART 4. STRIP CLUSTER COMPOSITION

A. Desired density

AT: 3 × _____ = _____ APF: 3 × _____ = _____ APB: 3 × _____ = _____

B. Cluster composition table

STRIP	AT	APF	APB	STRIP TOTAL (cannot exceed 5)
A	_____	_____	_____	_____
B	_____	_____	_____	_____
C	_____	_____	_____	_____
D	_____	_____	_____	_____
E	_____	_____	_____	_____
F	_____	_____	_____	_____
G	_____	_____	_____	_____
H	_____	_____	_____	_____
I	_____	_____	_____	_____
TOTAL (Cannot exceed desired density × 3 as computed in A above)	_____	_____	_____	_____

PART 5. NUMBER OF MAN-HOURS FOR INSTALLATION

Number of mines ÷ emplacement rate = mines per man-hour

Number of AT mines:______ ÷ 4 = _____ (round up)

Number of APF mines:______ ÷ 8 = _____ (round up)

Number of APB mines:______ ÷ 16 = _____ (round up)

_____ + _____ + _____ × 1.2 = _____ man-hours (round up)

Figure 6-1. Minefield requirements computation work sheet (continued)

PART 6. AMOUNT OF FENCING AND MARKING MATERIAL

A. Concertina wire or single-strand barbwire

([front × 2] + [depth × 2] + 160) × 1.4 = meters of concertina or single-strand barbwire required

([_____ × 2] + [_____ × 2] + 160) × 1.4 = _____ (round up)

Number of pickets = amount of concertina or single-strand barbwire ÷ 15

_____ ÷ 15 = _____ (round up)

- OR -

B. Double-strand barbwire

([front × 2] + [depth × 2] + 160) × 2.8 = meters of double-strand barbwire required

([_____ × 2] = [_____ × 2] + 160) × 2.8 = _____ (round up)

Number of pickets = amount of double-strand barbwire ÷ 30

_____ ÷ 30 = _____ (round up)

C. Number of signs = number of pickets = _____

PART 7. NUMBER OF TRUCKLOADS

AT mines

_____ cases per truck × _____ mines per case = _____ mines per truck

_____ mines required ÷ _____ mines per truck = _____ truckloads of AT mines

APF mines

_____ cases per truck × _____ mines per case = _____ mines per truck

_____ mines required ÷ _____ mines per truck = _____ truckloads of APF mines

APB mines

_____ cases per truck × _____ mines per case = _____ mines per truck

_____ mines required ÷ _____ mines per truck = _____ truckloads of APB mines

Total truckloads

_____ AT truckloads + _____ APF truckloads + _____ APB truckloads =

_____ total truckloads required (round up)

Figure 6-1. Minefield requirements computation work sheet (continued)

PART 8. AMOUNT OF ENGINEER TAPE

A. Minefield boundaries depth × 2 = _____ × 2 = _____

B. Regular strips front × number of regular strips = _____ × _____ = _____

C. IOE front + (number of IOE clusters × 3) = _____ + (_____ × 3) = _____

D. Lanes and gaps depth × 2 × number of lanes and gaps = _____ × 2 × _____ = _____

E. Traffic tapes depth × number of traffic tapes _____ × _____ = _____

F. Trip-wire safety tape front × number of regular strips with trip wire _____ × _____ = _____

G. Subtotal (lines A + B + C + D + E + F)

_____ + _____ + _____ + _____ + _____ + _____ = _____ meters (round up)

H. Number of rolls to order (line G × 1.2)

_____ × 1.2 = _____ meters

_____ meters ÷ 170 meters per roll = _____ rolls of engineer tape (round up)

PART 9. SANDBAGS

A. Number of clusters in IOE (from 1A) = _____

B. Number of clusters in minefield = number of clusters in IOE × 3 × number of regular strips (from 2D) = _____

C. Total number of clusters (line A + line B) = _____

D. Number of sandbags = number of clusters × 3 sandbags per cluster (line C × 3) = _____

Figure 6-1. Minefield requirements computation work sheet (continued)

Basic information pertaining to the minefield is normally determined by the engineer company commander or the staff engineer. It is provided to the OIC or NCOIC of the emplacing unit during the mission briefing. In this example, the following guidance is given to the emplacing unit:

Desired density	AT 1	APF 4	APB 8
IOE representative cluster	AT 1	APF 2	APB 2
Front	200 meters		
Depth	300 meters		
Percentage of AHDs	10%		
Type of mines	AT M15	APF M16A2	APB M14
Type of truck/trailer	5-ton dump (with sideboards)		
Lanes/gaps/traffic tapes	1 lane, 1 traffic tape (foot troops)		
Trip-wire safety tapes	3		

The rest of this work sheet is completed by using the above information.

The regular strip has a cluster density of one cluster every 3 meters. The IOE has a cluster density of one-third that of a regular strip, or one cluster every 9 meters. Therefore, to obtain the number of clusters in the IOE, the length of the strip is divided by 9. Decimals are rounded up to the next higher whole number.

PART 1. NUMBER OF MINES

Step 1.

IOE live clusters 200 ÷ 9 = 23 (rounded up)

The representative cluster composition for the IOE clusters is established and provided by the commander based on METT-TC factors. The number of clusters in the IOE is multiplied by the cluster composition to determine the number of mines, by type, in the entire IOE.

Step 2.

	AT	APF	APB
IOE representative cluster ×	1	2	2
Number of IOE clusters =	23	23	23
Number of mines in IOE	23	46	46

The minefield front multiplied by the desired density determines the number of mines in the minefield.

NOTE: The desired density pertains only to the regular strips and does not take into account the number of mines in the IOE which were calculated in Step 2.

Figure 6-2. Step-by-step procedures for completing the minefield requirements computation work sheet

Step 3.

Desired density ×	1	4	8
Minefield front =	200	200	200
Mines in regular strips	200	800	1,600

The number of mines required for the IOE (Step 2) is added to the number of mines in the regular strips (Step 3).

Step 4. Subtotal of mines

(Step 2 + 3)	223	846	1,646

Ten percent is added to the total number of mines required to allow for damaged items and irregularities in terrain and strip length. This is accomplished by multiplying the total number of mines (Step 4) by 1.1. Decimals are rounded up to the next higher whole number.

Step 5.

10% excess factor =	1.1	1.1	1.1
Total number of mines to order	246	931	1,811

These figures represent the total number of mines, by type, required for the entire minefield. When ordering by the case rather than by individual mines, the total should be divided by the number of mines per case and rounded up to the next whole case. (See Table 2-8, page 2-45.)

PART 2. NUMBER OF REGULAR STRIPS

Step 1.

Add desired density AT 1 + APF 4 + APB 8 = 13

Each regular mine strip has a cluster every 3 meters; therefore, its density is one-third cluster per meter of front. A total density of 13 mines per meter of front in the previous example would equal 3 × 13 or 39 mines per 3 meters of front. Clusters may contain a maximum of five mines, so the resulting figure must be divided by 5. In short, to determine the minimum number of regular strips required, the total density must be multiplied by three-fifths (3 meters between clusters and five mines per cluster). For ease of calculation, three-fifths is converted to the decimal 0.6. Decimals are rounded up to the next highest whole number.

Step 2.

0.6 × Step 1 0.6 × 13 = 8 (rounded up)

The calculations to determine the minimum number of regular strips previously described are not suitable when the ratio of AT to AP mines is greater than 1:4. For example, if the desired density is 1-1-1, the total density is 3. The minimum number of strips would be 3 × 3/5 = 1.8, rounded up to 2 strips. However, because of the restriction on the number of AT mines per cluster, it is impossible to obtain a density of 1 AT mine per meter of front with only 2 strips. A minimum of 3 regular strips is required. The alternative means of determining the number of regular strips is founded by multiplying the AT desired density by 3.

Figure 6-2. Step-by-step procedures for completing the minefield requirements computation work sheet (continued)

Step 3.

3 × AT desired density 3 × 1 = 3

The number of regular strips calculated by the first method and the alternative method are compared, and the higher figure is used as the minimum number of regular strips. The 8 determined by the 3/5 rule is larger than the 3 determined by the alternative method. Therefore, the minimum number of regular strips in the example is 8.

Step 4.

Number of regular strips required = highest number of Step 2 or 3 = 8

PART 3. NUMBER OF AHDs

0.1 × 223 = 23 (rounded up)

PART 4. STRIP CLUSTER COMPOSITION

The cluster composition table is prepared by the OIC of the laying unit to control the allocation of mines to a regular strip. The cluster composition remains constant within a particular strip, but it may vary among different strips. As the mines are allocated by strip, no more than 1 AT mine can be placed in each representative cluster, and each cluster can have a maximum of 5 mines.

A tabular format is prepared (see Figure 6-1, page 6-6) to facilitate the distribution of mines by emplacement personnel. Note that each component of the desired density is multiplied by 3. The number 3 is always used regardless of the minimum number of regular strips because it is the number of mine strips required to give a minefield density of 1 mine per meter of front when a cluster contains only 1 mine of each type. Each mine strip has a cluster every 3 meters; therefore, it has a density of one-third mine per meter when a cluster contains 1 of each type of mine.

Step 1. Desired density

AT: 3 × 1 = 3 APF: 3 × 4 = 12 APB: 3 × 8 = 24

The total of each column in the table cannot exceed the number of mines above. For example, with an APF desired density of 4, 3 × 4 = 12, so the total APF mines in the representative cluster composition for each of the regular strips cannot exceed 12.

PART 5. NUMBER OF MAN-HOURS FOR INSTALLATION

Remember, the total number of mines includes the mines in regular strips and the mines in IOE short strips. The laying rates are—

AT mines: 4 per man-hour.

APF mines: 8 per man-hour.

APB mines: 16 per man-hour.

Figure 6-2. Step-by-step procedures for completing the minefield requirements computation work sheet (continued)

The number of man-hours required for each mine type is computed and rounded up. These amounts are totalled and a 20 percent excess factor is included by multiplying the total by 1.2. The resulting figure is the total number of man-hours required for emplacement and represents straight work time only. It does not take into account the time for transportation to and from the emplacement site, meals, and breaks; limited visibility; or NBC conditions. The commander should use his judgment and past experience to determine the time required for transportation, meals, and breaks. When working under limited visibility or NBC conditions, the total man-hours (after the excess factor has been included) should be multiplied by 1.5.

In this example, a total of 357 man-hours is required as determined below. Note that each decimal is rounded to the next higher whole number.

Number of mines ÷ emplacement rate = mines per man-hour

Number of AT mines	246 ÷ 4 = 62 (rounded up)
Number of APF mines	961 ÷ 8 = 121 (rounded up)
Number of APB mines	1,811 ÷ 16 = 114 (rounded up)
Total	62 + 121 + 114 = 297 × 1.2 = 357 (rounded up)

PART 6. AMOUNT OF FENCING AND MARKING MATERIAL

Standard-pattern minefields must be marked and fenced. The amount of fencing required depends on whether barbwire (single- or double-strand) or concertina is used.

The amount of wire for a single-strand barbwire or a single-strand concertina fence is calculated with the following formula:

([front × 2] + [depth × 2] + 160) × 1.4

The amount of wire for a double-strand barbwire fence is calculated with the following formula:

([front × 2] + [depth × 2] + 160) × 2.8

Step 1.

Concertina Wire or Single-Strand Barbwire

([200 × 2] + [300 × 2] + 160) × 1.4 = 1,624

The number of pickets required is determined by dividing the total amount of fence by 15.

Number of pickets = amount of fence ÷ 15   1,624 ÷ 15 = 109 (rounded up)

Step 2.

Double-Strand Barbwire

([200 × 2] + [300 × 2] + 160) × 2.8 = 3,248 (rounded up)

The number of pickets required is determined by dividing the total amount of fence by 30.

Number of pickets = amount of fence ÷ 30   3,248 ÷ 30 = 109 (rounded up)

Figure 6-2. Step-by-step procedures for completing the minefield requirements computation work sheet (continued)

The number of minefield marking signs is equal to the number of pickets.

NOTE: These calculations determine the marking and fencing materials required for the minefield perimeter only. Additional materials may be required for lanes and gaps.

PART 7. NUMBER OF TRUCKLOADS

The number of vehicles required depends on the type and amount of mines as well as the type of vehicles available. The total mines, by type, required is divided by the haul capacity of available vehicles to determine the number of truckloads required to transport the mines.

In this example, crated M15 AT mines, M16A2 APF mines, and M14 APB mines are hauled in M930 5-ton dump trucks (with sideboards). (See Table 2-8, page 2-45).

AT mines:

246 mines required ÷ 204 mines per truck = 1.2 truckloads of AT mines

APF mines:

931 mines required ÷ 888 mines per truck = 1.05 truckloads of APF mines

APB mines:

1,811 mines required ÷ 13,770 mines per truck = 0.13 truckloads of APB mines

Total truckloads:

1.2 AT truckloads + 1.05 APF truckloads + 0.13 APB truckloads = 2.38 truckloads = 3 truckloads (rounded up) required

PART 8. AMOUNT OF ENGINEER TAPE

An extensive amount of engineer tape is used to mark the initial layout of a standard-pattern minefield. Engineer tape comes in 170-meter rolls and is used to mark several portions of the minefield.

NOTE: In this example, only one lane and one roll of traffic tape is required.

Step 1.

Minefield boundaries depth × 2 300 × 2 = 600

Step 2.		
Regular strips	front × number of regular strips	200 × 8 = 1,600
Step 3.		
IOE	front + (number of IOE clusters × 3)	200 + (23 × 3) = 269
Step 4.		
Lanes and gaps	depth × 2 × number of lanes and gaps	300 × 2 × 1 = 600
Step 5.		
Traffic tape	depth × number of traffic tapes	300 × 1 = 300

Figure 6-2. Step-by-step procedures for completing the minefield requirements computation work sheet (continued)

Step 6.		
Trip-wire safety tape	front × number of regular strips with trip wire	200 × 3 = 600
Step 7.		
Total of Steps 1 through 6 = 600 + 1,600 + 269 + 600 + 300 + 600 = 3,969 meters		
Step 8.		
Add 20% excess	total amount of engineer tape required for the minefield × 1.2 = 3,969 × 1.2 = 4,762.8 = 4,763 (rounded up)	
Step 9.		
Total number of rolls	total amount of engineer tape, in meters, from Step 8 ÷ 170 meters per roll = 4,763 meters ÷ 170 = 28.02 = 29 rolls	
PART 9. SANDBAGS		
To determine the number of sandbags for the removal of spoil.		
Step 1.		
Number of clusters in IOE (Part 1, Step 1) = 23		
Step 2.		
Number of clusters in minefield = number of clusters in IOE × 3 × number of regular strips (Part 1, Step 4)		
23 × 3 × 8 = 552		
Step 3.		
Total number of clusters (add Steps 1 and 2) = 575		
Step 4.		
Number of sandbags = number of clusters × 3 sandbags per cluster		
575 x 3 = 1,725		

Figure 6-2. Step-by-step procedures for completing the minefield requirements computation work sheet (continued)

TASK ORGANIZATION

To maximize the efficiency of the row-mining process, the platoon leader must task-organize his platoon. The organization of the task, as a whole, is intricate and places great demands on the leader. Leave nothing to chance when planning and executing a row minefield, because each situation is different. Make allowances for transporting, handling, and controlling the mines. The officer in charge (OIC) and the squad leaders must be able to exercise control throughout the task under all conditions. Always observe safety.

Organize the platoon into four parties—siting and recording, marking, mine dump, and laying.

Siting-and-Recording Party

The platoon leader directs the party and is responsible for siting, recording, and reporting the minefield. This party consists of one or two soldiers and a vehicle to carry material. (If a vehicle is not available, increase the party to three soldiers.) Because siting is usually done in daylight, the party must take appropriate physical-security measures. The party starts well ahead of the actual laying, sets out control markers, and avoids using sharp turns. The party marks the vehicle traffic routes to and from the minefield rows.

When siting is complete, the OIC identifies one member of the party to be the recorder and assigns the remaining soldiers to other tasks. The recorder collects data from the laying party NCOIC and completes DA Form 1355 as outlined in Chapter 8. The OIC ensures that the DA Form 1355 is completed timely and accurately.

Marking Party

This party is composed of an NCOIC and personnel who are not working as members of other teams. After the minefield is sited, the marking party emplaces fence posts, wire, and marking signs.

Mine-Dump Party

This party is controlled by the platoon sergeant (PSG) and is composed of personnel who are not working as members of other teams. The mine-dump party accounts for all strip packages that arrive from other sources, sets up vehicle mine sets at the mine dump, and hauls supplies as required. The PSG places row packages in a location that maximizes speed and provides concealment for minefield emplacement, and he also ensures that the mine dump is prepared for night operations. The party marks the mine dump's entrance and exit and the routes to them. The PSG verifies the strip feeder reports with the squad leaders upon the completion of each row and passes the reports to the recording party. The PSG is not required to stay at the mine dump continuously; he has the flexibility to move around the area to perform other activities.

The mine-dump party creates vehicle sets by setting aside the number of mines and fuses that are required by each laying vehicle. The party loosens and then hand-tightens arming and shipping plugs, helps load the mines onto laying vehicles, and disposes of residue. Soldiers may also assist the marking party and provide local security. For initial vehicle loads, the mine-dump party may be assisted by the laying party.

Laying Party

This party consists of an NCOIC, four soldiers, and a vehicle to carry the mines. The NCOIC controls the movement of each laying vehicle. He directs each vehicle to start and stop laying and controls immediate-action drills. The NCOIC initiates a strip feeder report with the PSG or the mine-dump NCOIC, receives azimuths from the siting party, and directs his element to the correct row. He is responsible for replacing the temporary row markers with permanent markers and for ensuring that mines are laid according to the azimuths, mines are spaced correctly, and the strip feeder report is accurate.

The NCOIC ensures that the end row marker is emplaced at the completion of each row, and he closes the strip feeder report with the PSG.

NOTE: Using tilt-rod fuses requires additional soldiers to stake mines, insert fuses, and arm mines.

When laying three rows at once, each laying party consists of an APC or an organic squad vehicle and a carrier, sapper, and digging team.

Carrier Team

This team is comprised of the APC driver and the track commander (TC). They ensure that the APC maintains the proper speed and stays on the proper course.

Sapper Team

This team is composed of the squad leader and the remaining squad members. It provides personnel to lay and arm mines. Each soldier carries wrenches and fuses. The squad leader supervises laying and tasks personnel who are not needed for laying to other parties.

Digging Team

The digging team buries mines. It consists of an NCOIC and several soldiers (may be soldiers from supported maneuver units) who are equipped with suitable digging tools. Increase the arming party by two to speed up the laying process or task personnel who are not needed to other parties. **NOTE: If mines are surface-laid, there is no digging team.**

SITE LAYOUT

Once the platoon leader has coordinated the location of the minefield(s) with the maneuver commander, siting in the minefield can begin. Siting is the first step in the actual laying process and is done for safety and control. Although the minefield may be emplaced at night or during limited visibility, the siting party should site the minefield under favorable conditions, preferably during daylight. Siting consists of identifying landmarks; establishing routes; and emplacing start, end, and intermediate row markers. Actual control measures (stakes or pickets) should not stand out to such an extent that they give away the minefield orientation, but they must be easily discernible to the laying party.

Certain features, like thick woods and deep, wide streams, are natural obstacles. Mine rows should be laid to reinforce terrain and increase the effectiveness of the minefield.

Mine Rows

Mine rows are labeled with a letter and should be laid in order. Row A is nearest the enemy, followed by rows B, C, D, E, and so forth. When laying tactical minefields, each row has permanent start and end row markers. Intermediate markers may be required, depending on the row length and the terrain. Platoon leaders determine the number of laying vehicles to be employed. The preferred technique is to use three vehicles so that three rows can be laid simultaneously. Using more than three vehicles is beyond the platoon's C^2 capabilities and is not considered. The distance between rows is determined by the following factors:

- Depth and density of the minefield.
- Terrain.
- Suitability of the ground.
- Desired obstacle intent.

NOTE: Rows are spaced 50 meters apart in standardized row minefields (discussed later in this chapter).

Mine Spacing

The minefield OIC determines the mine spacing. The desired density, the availability of laying vehicles, the number of rows, and the possibility of sympathetic detonation (Table 5-3, page 5-8) affect the distance between mines. **NOTE: Mines are spaced 6 meters apart in standardized row minefields (discussed later in this chapter).**

Control Measures

Control measures are temporary markers that are used to guide vehicles and troops during row-mining operations. Markers are constructed of different materials for different uses. For example, use VS17 panels on poles for start and end row markers, and use M133 hand-emplaced minefield marking set (HEMMS) poles with flags for intermediate markers. Use the following temporary markers:

- Start row (does not replace the mandatory permanent marker).
- Start laying (first intermediate marker after the start row marker).
- Intermediate (used between the last row marker and the next visible point, not more than 100 meters away).
- Change of direction or turning point (actually consists of three markers—warning, turning point, and new direction).
- Stop laying.
- End row (does not replace the mandatory permanent marker).

The following materials may be used to construct temporary markers:

- U-shaped pickets.
- HEMMS poles.
- Wooden posts.
- Steel rods.
- Engineer tape.
- VS17 panels.

Control measures for laying mines at night require lights or infrared (IR) equipment as follows:

- Chem-lights placed in U-shaped pickets or hand-held.

- Directional flashlights taped in U-shaped pickets or hand-held.
- HEMMS lights used with U-shaped pickets or poles.
- Lights from a minefield marking set number 2.
- IR reflectors.

NOTE: The use of control measures should be incorporated into unit standard operating procedures (SOPs).

Procedures

The minefield OIC arrives on the site with the siting-and-recording party. He selects Landmark 1 and then sites the left (or right) boundary fence and start row markers (all start and end row markers are permanent markers). The siting-and-recording party takes distances and azimuths to be used in preparing the recording form. If the tactical situation permits and the marking party is ready, emplacement of the fence should begin.

If the minefield is to have an IOE row, the siting-and-recording party proceeds across the IOE and establishes I1, I1E, I2, I2E, and so on until it reaches the end. Personnel proceed down the right (or left) boundary and emplace start row marker A1. Proceeding from A1 to A2, they place intermediate markers as required. Personnel use different colored markers to identify each row (for example: Row A, red light; Row B, green light; Row C, blue light). For IR markings, they use multiple horizontal IR light sources that are spaced at least 6 inches apart (for example: Row A, one light; Row B, two lights; Row C, three lights). When they reach A2, they emplace an end row marker and repeat the procedure from B1 to B2, C1 to C2, and so on until they emplace all the required control measures (Figure 6-3). The siting-and-recording party establishes Landmark 2 and the left (or right) rear fence location. Personnel also assist the PSG in siting mine dumps near the minefield.

MINE-LAYING VEHICLES

Soldiers normally lay row minefields from a tactical vehicle. Consider vulnerability, capacity, and trafficability when selecting a vehicle. Before emplacing the minefield and preparing the vehicle for mine laying, drive it in a random pattern across the minefield site. The random pattern deceives the enemy by masking the actual laying pattern. Load enough mines so that each vehicle can complete at least one entire row before reloading, but do not stack fused mines more than two-high.

LAYING A ROW MINEFIELD

The following drills demonstrate how to lay a minefield:

Drill 1

Squad vehicles arrive on the site and proceed down the left (or right) boundary of the minefield to their assigned row. (A separate party must be detailed to install the IOE.) At the start row marker, the squad vehicle moves into position and prepares to lay mines. The squad leader for Row A directs Vehicle 1 to move out.

Mines are laid on the ground at the required spacing, along the temporary markers positioned by the siting-and-recording party.

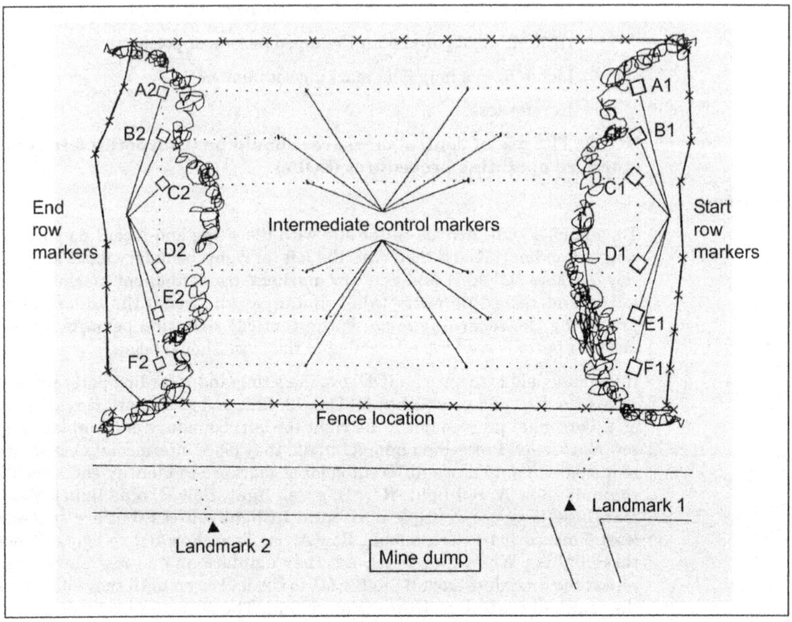

Figure 6-3. Site layout

As mines are laid, the arming party moves behind the vehicle and arms the mines. Personnel remove temporary markers installed by the siting-and-recording party and replace the end row markers with permanent markers.

When Vehicle 1 moves a safe distance (approximately 25 meters) along Row A, Vehicle 2 begins to lay mines on Row B. When Vehicle 2 moves a safe distance along Row B, Vehicle 3 begins to lay mines on Row C.

The marking party continues to emplace the left and right boundary fences (Figure 6-4a, page 6-20).

The IOE party exits the minefield outside the left (or right) boundary after it completes the IOE.

When Vehicles 1 and 2 finish their assigned rows, they move past the end row marker and execute a left (or right) turn and wait for Vehicle 3 to complete its row. All vehicles move in column down the left (or right) boundary to the mine dump, load the next row's mines, and then move to their next assigned row. The process of laying and arming mines is repeated (Figure 6-4b, page 6-21).

After the minefield is laid, all the vehicles exit down the left (or right) boundary and out the rear. The marking party completes the rear boundary fence, and the recording party completes DA Form 1355. The OIC or PSG ensures that the DA Form 1355 is complete and accurate and signs it.

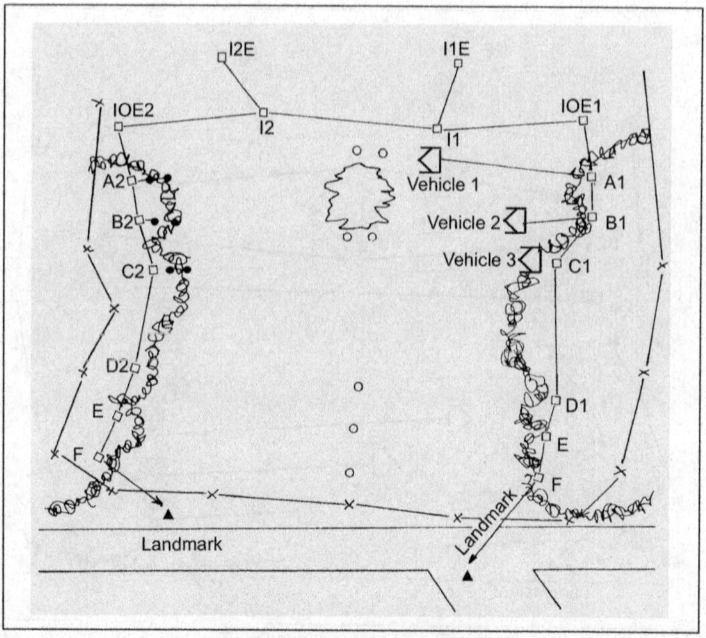

Figure 6-4a. Laying a minefield

Drill 2

This drill may be used to speed up mine laying; however, strict C^2 is vital to ensure security and safety. This method is difficult to use when the terrain is rugged or when weather or visibility is subject to change.

The drill is conducted by three squad vehicles, each laying one row. Row B has turning points and Rows A and C do not. If the minefield has six rows, Row E has turning points and Rows D and F do not. The squad leader (laying leader) in Row B (and Row E, if required) is in charge of the overall laying.

Squad vehicles arrive on the site and proceed down the left (or right) boundary of the minefield to their assigned row. Squad vehicles move into position at start row markers and prepare to lay mines.

The laying leader directs Vehicle 1 to move out on Row A. The sapper team lays mines on the ground at the required spacing. If an IOE is required, the Row A team emplaces the IOE concurrently with Row A and at the same spacing. When Vehicle 1 reaches the IOE short-row start marker, the laying party lays mines along an azimuth designated by the laying leader (Figure 6-5). After the IOE short row is laid, Vehicle 1 returns to Row A and continues laying mines.

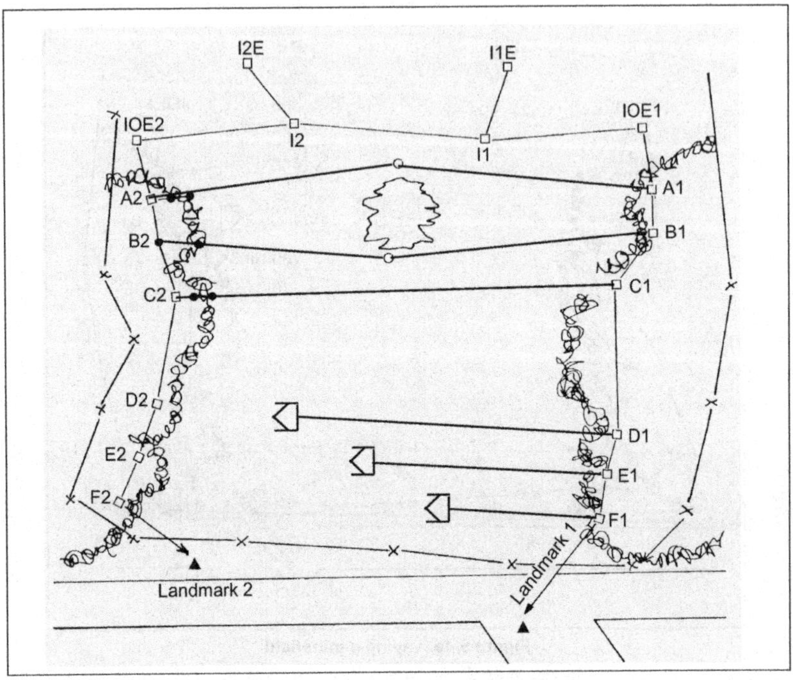

Figure 6-4b. Laying a minefield (continued)

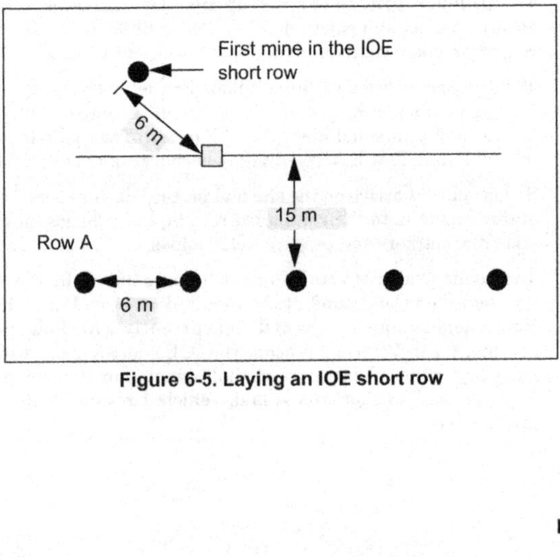

Figure 6-5. Laying an IOE short row

After all the mines are laid, the arming party moves behind the vehicle and arms the mines. Personnel remove temporary markers and replace start and end row markers with permanent markers. The arming party must be distinguishable from everyone else. The last member of the arming party should wear a colored vest or carry a specific colored chem-light. No one is allowed behind the last member of the arming party.

The NCOIC completes a strip feeder report (Figure 6-6) and gives it to the recording party. The strip feeder report includes the number of mines laid, the type of mines laid, azimuths of IOE strips and turning points, AHDs emplaced (by cluster number), and any other information (such as omitted mines) the platoon leader requires.

STRIP FEEDER REPORT FOR STRIP/ROW ______					
Type of Mine	Number of Mines	AHDs by Cluster	IOE-Strip Azimuth	Turning-Point Azimuth	Remarks

Figure 6-6. Sample strip feeder report

Vehicle 1 moves down Row A and lays mines until the laying leader directs it to stop. (The laying leader chooses vehicle stops to coincide with the locations of turning points.) The laying leader then directs Vehicle 3 to begin laying mines along Row C. Vehicle 3 lays mines on Row C until the laying leader directs it to stop (somewhere well past Vehicle 1).

Vehicle 2 moves toward Vehicle 1 and begins to lay mines on Row B. He lays mines to within 15 meters of Vehicle 1. Vehicle 2 then turns toward Vehicle 3 and lays mines on Row B to within 15 meters of Vehicle 3. Vehicle 2 then turns back toward Vehicle 1 and continues to lay mines in this pattern until Row B is laid.

Figure 6-7 shows vehicle positions when using the above method to lay a row minefield.

NOTES:

1. At night or during low visibility, Vehicle 1 has two red flashlights and Vehicle 3 has two green flashlights. The flashlights are held side by side, and pointed toward Vehicle 2. The driver of Vehicle 2 moves forward until he is within 15 meters of the lights or until the light holder turns the lights off.

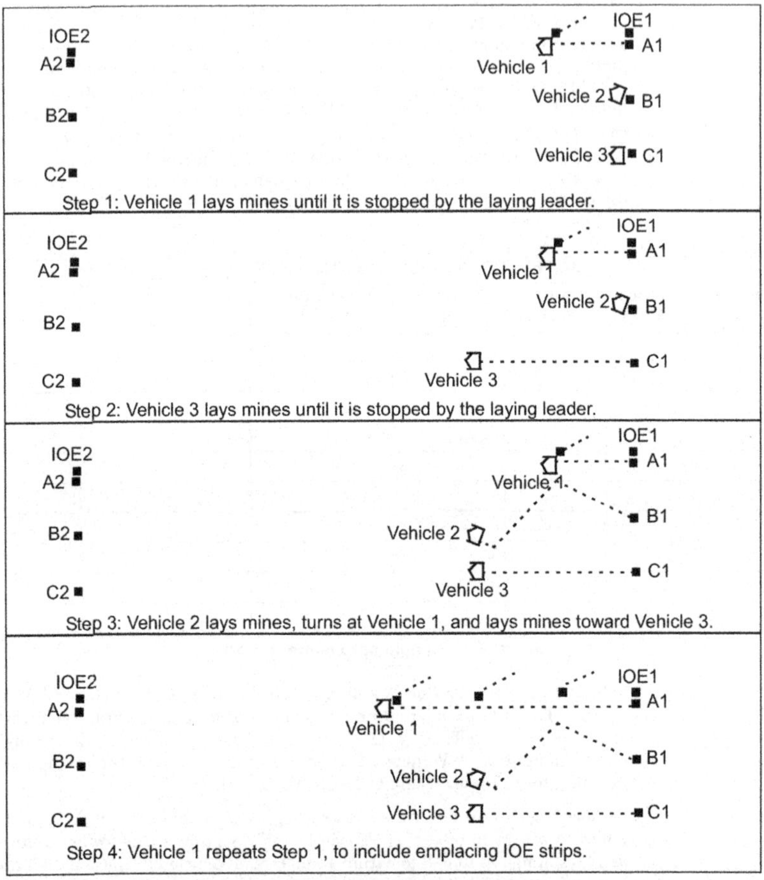

Figure 6-7. Laying a row minefield

2. If the platoon leader feels that low visibility or other reasons preclude the use of vehicle positions as turning points, he may have the siting party emplace turning-point markers (three intermediate markers) for Vehicle 2 to use as a guide. In this event, the three vehicles emplace mines simultaneously.

After Vehicles 1 and 2 finish their assigned rows, they move past the end row marker, execute a left (or right) turn, and wait for Vehicle 3 to complete its row. All the vehicles move in column down the left (or right) boundary to their next assigned row, if there is one, and continue to lay and arm mines. This process is repeated until the entire minefield is laid. All the vehicles then exit the minefield down the left (or right) boundary and out the rear. The marking party completes the rear boundary fence, and the recording party completes DA Form 1355. The OIC or PSG ensures that the DA Form 1355 is complete and accurate and signs it.

IMMEDIATE-ACTION DRILL

If the enemy attacks the platoon during minefield emplacement, the laying party should execute the following actions:

- Sapper teams enter vehicles and recover spacing sandbags.
- Vehicle 1 exits the minefield by making a wide turn around the front of the other two vehicles.
- Vehicle 2 follows by making a wide turn around the front of Vehicle 3.
- Vehicle 3 exits the minefield.
- The three squads conduct the immediate-action drill as ordered by the platoon leader.

SQUAD DRILL

During row mining, the squad in each laying vehicle performs the following actions:

- Squad leader.
  - — Directs the squad to start laying mines.
  - — Supervises mine arming and placing.
  - — Allocates a vehicle, if possible, to help remove spoil from the site.
- Carrier team.
  - — Moves the APC to the row start point.
  - — Lowers the APC ramp until it is horizontal or opens the rear door. (If using the APC ramp to distribute mines, chains the ramp open to support the weight.)
  - — Moves the APC at a low speed (3 to 5 kph) in a straight line toward the row end point.
- Sapper team.
  - — Soldier 1 ties the rope to the end of the lowered ramp or the tow pintle.
  - — Soldier 2 ties the partially filled sandbag on the other end of the rope. (The rope length from the end of the ramp door to the sandbag is the correct spacing between mines [Figure 6-8]).
  - — Soldier 3 (squad leader) positions the team members. Soldier 1 is at the rear of the compartment, Soldier 2 sits on the edge of the APC ramp or open door, and Soldier 4 walks behind the APC.
  - — Soldier 1 fuses a mine and passes it to Soldier 2. **(Korea Only: If AP mines are also laid, they are given out simultaneously.)**
  - — Soldier 2 records all the mines issued.
  - — Soldier 2 places the fused mine on the ground when the sandbag tied to the rope is even with the previously laid mine.
  - — Soldier 3 (squad leader) walks behind the vehicle and supervises mine laying.

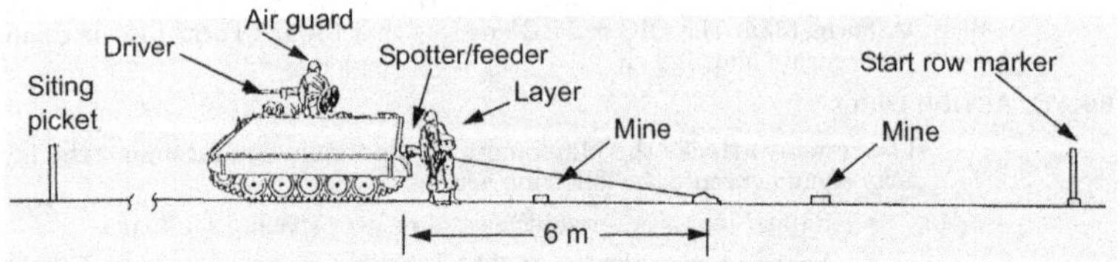

Figure 6-8. Measuring distances between mines with sandbags

- Soldier 4 walks behind the vehicle and arms mines.
- After the mine row is armed and camouflaged, Soldier 4 buries pins, clips, and shipping plugs 30 centimeters to the rear of the start row marker.
- The sapper team repeats the above steps until the end of the row is reached.

• Digging team, if needed. (The NCOIC selects the mine to be buried by each soldier and supervises the operation.)

- Follows the laying party along the friendly side of the row.
- Digs in mines but leaves them exposed until arming is complete.
- **Korea Only: Arms AP mines in a cluster before arming AT mines.**

Marking, Recording, and Reporting Row Minefields

Marking procedures for row minefields are the same as those for other minefields (see Chapter 2).

Row minefields are recorded on DA Form 1355 (Figures 6-9a and 6-9b, pages 6-26 and 6-27). Reporting procedures for intent, initiation, status, and completion reports are detailed in Chapter 8.

STANDARDIZED TACTICAL ROW MINEFIELDS

The specific composition of a tactical row minefield depends on METT-TC factors and available resources. To aid in standardization of platoon techniques, four compositions have been developed to match desired obstacle effects. Using standardized minefields facilitates planning the obstacle type, size, and logistical requirements. It is imperative that the design and the effect of these minefields are well understood. They are an integral part of combined arms obstacle doctrine and form the cornerstone of engineer obstacle operations.

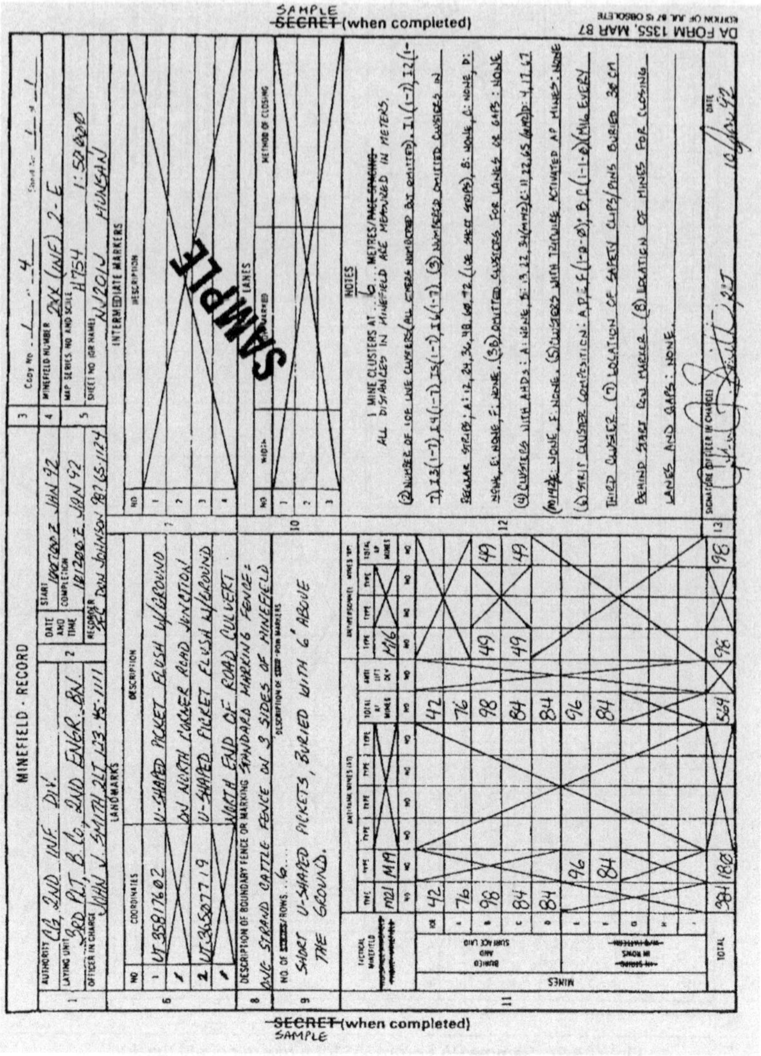

Figure 6-9a. Sample DA Form 1355 for a row minefield (front)

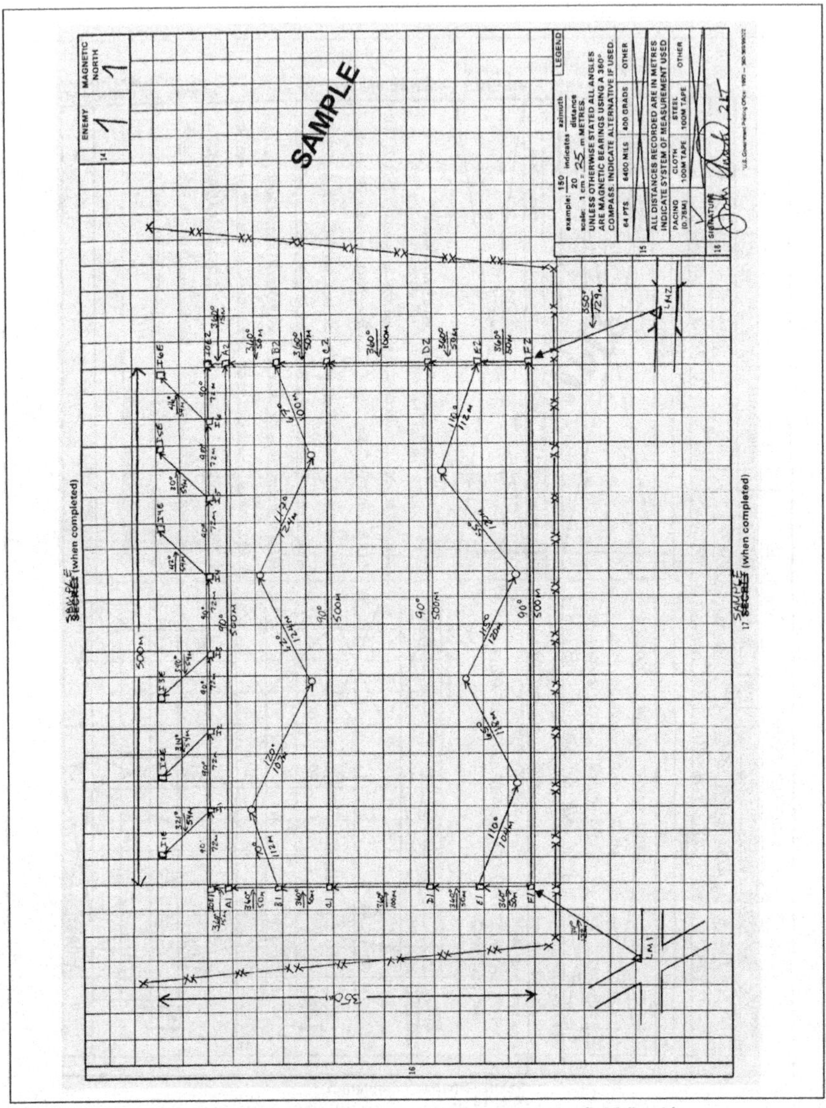

Figure 6-9b. Sample DA Form 1355 for a row minefield (back)

DISRUPT AND FIX

Disrupt and fix row minefields (Figure 6-10) are similarly constructed, but a fix minefield has an IOE and does not have AHDs.

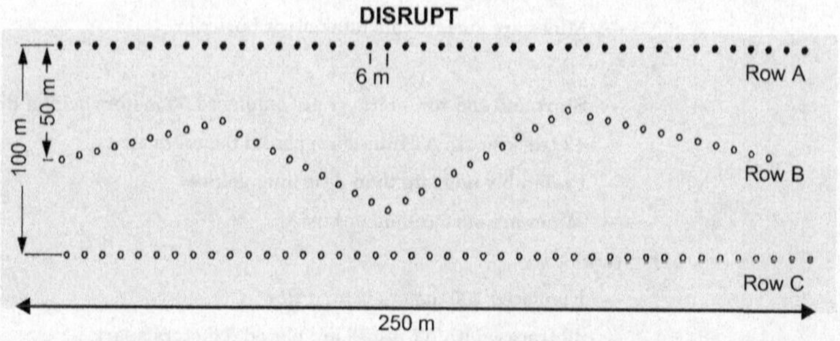

	Disrupt	Fix
Front	250 m	250 m
Depth	100 m	120 m
Rows of AT full-width mines	1	1
Rows of AT track-width mines	2	2
IOE	No	Yes
AHD	No	No
Platoon hours required	1.5	1.5
Full-width mines	42	63
Track-width mines	84	84
Density	0.5	0.6

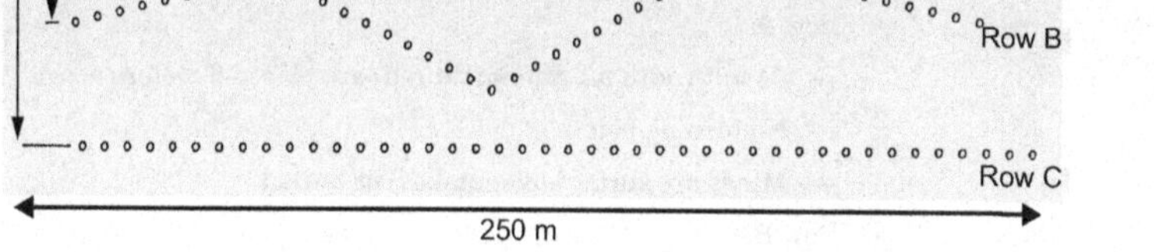

Figure 6-10. Standardized disrupt and fix row minefields

Disrupt and fix row minefields are emplaced as follows:

- Row A.
  - — 42 full-width AT mines (tilt-rod) are placed 6 meters apart.
  - — No turning points.
  - — Mines are surface-laid (staked) or buried.
- Row B.
  - — Start and end row markers are emplaced 50 meters behind Row A.
  - — 42 track-width AT mines are placed 6 meters apart.
  - — Preferably no more than 3 turning points.
  - — Mines are surface-laid or buried.
- Row C.
  - — Emplaced 100 meters behind Row A.
  - — 42 track-width AT mines are placed 6 meters apart.
  - — No turning points.
  - — Mines are surface-laid or buried.
- IOE (fix minefield only).
  - — 3 IOE short rows.
  - — A separate IOE baseline.
  - — The IOE baseline is on the enemy side, 15 meters from Row A.
  - — 7 full-width AT mines are placed 6 meters apart on each IOE short row.
  - — Mines are buried.
  - — The first IOE short row is placed 48 meters from the IOE end marker, the next short row is 84 meters from the first short row, and the third short row is 84 meters from the second short row.

TURN

A turn minefield (Figure 6-11, page 6-30) consists of four rows of full-width mines and two rows of track-width mines.

A turn row minefield is emplaced as follows:

- Row A.
  - — 84 full-width AT mines (tilt-rod) are placed 6 meters apart.
  - — No turning points.
  - — Mines are surface-laid (staked) or buried.
- Row B.
  - — Start and end row markers are emplaced 50 meters behind Row A.
  - — 84 full-width AT mines (tilt-rod) are placed 6 meters apart.

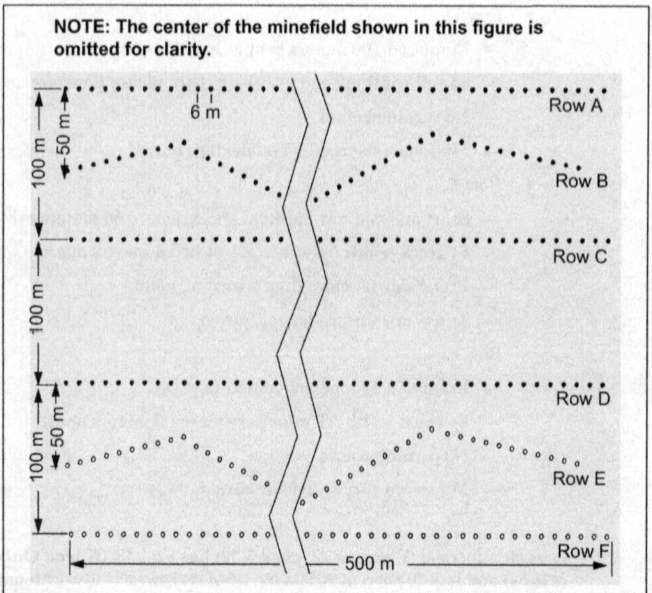

Front	500 m
Depth	300 m
Rows of AT full-width mines	4
Rows of AT track-width mines	2
IOE	No
AHD	No
Platoon hours required	3.5
Full-width mines	336
Track-width mines	168
Density	1.0

Figure 6-11. Standardized turn row minefield

— Preferably no more than 5 turning points.

— Mines are surface-laid (staked) or buried.

- Row C.

  — Emplaced 100 meters behind Row A.

  — 84 full-width AT mines (tilt-rod) are placed 6 meters apart.

  — No turning points.

  — Mines are surface-laid (staked) or buried.

- Row D.
  - — Emplaced 100 meters behind Row C.
  - — 84 full-width AT mines (tilt-rod) are placed 6 meters apart.
  - — No turning points.
  - — Mines are surface-laid (staked) or buried.
- Row E.
  - — Start and end row markers are emplaced 50 meters behind Row D.
  - — 84 track-width AT mines are placed 6 meters apart.
  - — Preferably no more than 5 turning points.
  - — Mines are surface-laid or buried.
- Row F.
  - — Emplaced 100 meters behind Row D.
  - — 84 track-width AT mines are placed 6 meters apart.
  - — No turning points.
  - — Mines are surface-laid or buried.

BLOCK

A block minefield (Figure 6-12, page 6-32) has an IOE **(Korea Only: and AP mines)** and has 20 percent AHDs in two of its rows of full-width mines. AHDs are placed in Rows B and C for the best effect. **(Korea Only: A block minefield also requires 84 M16 or M14 AP mines and has a density of 0.17 M16/M14 AP mine per linear meter.)**

A block row minefield is emplaced as follows:

- Row A.
  - — 84 full-width AT mines (tilt-rod) are placed 6 meters apart.
  - — No turning points.
  - — Mines are surface-laid (staked) or buried.
- Row B.
  - — Start and end row markers are emplaced 50 meters behind Row A.
  - — 84 full-width AT mines (tilt-rod) are placed 6 meters apart.
  - — Preferably no more than 5 turning points.
  - — Mines are surface-laid (staked) or buried.
- Row C.
  - — Emplaced 100 meters behind Row A.
  - — 84 full-width AT mines (tilt-rod) are placed 6 meters apart.
  - — No turning points.

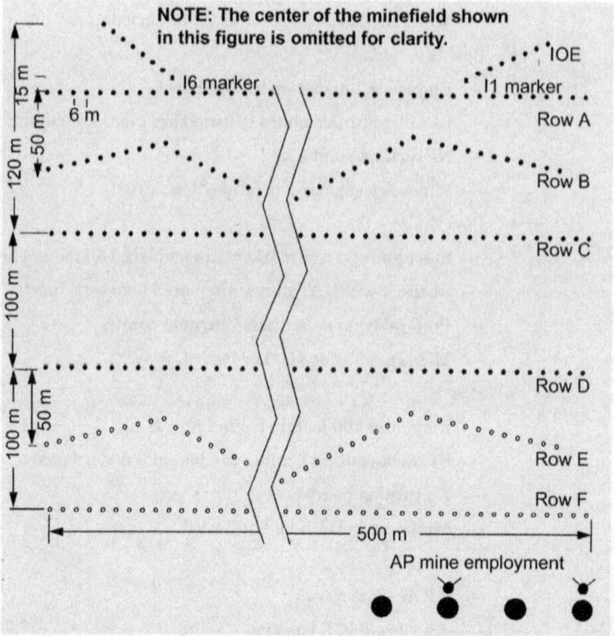

Front	500 m
Depth	320 m
Rows of AT full-width mines	4
Rows of AT track-width mines	2
IOE	Yes
AHD	Yes
Platoon hours required	5
Full-width mines	378
Track-width mines	168
AT mine density	1.1
Korea Only: AP frag mines	**84**
Korea Only: AP mine density	**0.17**

Figure 6-12. Standardized block row minefield

- Mines are surface-laid (staked) or buried.

- Row D.
  - Emplaced 100 meters behind Row C.
  - 84 full-width AT mines (tilt-rod) are placed 6 meters apart.
  - No turning points.
  - Mines are surface-laid (staked) or buried.
- Row E.
  - Start and end row markers are emplaced 50 meters behind Row D.
  - 84 track-width AT mines are placed 6 meters apart.
  - Preferably no more than 5 turning points.
  - Mines are surface-laid or buried.
- Row F.
  - Emplaced 100 meters behind Row D.
  - 84 track-width AT mines are placed 6 meters apart.
  - No turning points.
  - Mines are surface-laid or buried.
- IOE.
  - 6 IOE short rows.
  - A separate IOE baseline.
  - 7 full-width AT mines are placed 6 meters apart on each IOE short row.
  - Mines are buried.
  - The first IOE short row is placed 72 meters from the IOE end marker, and the five subsequent IOE short rows are placed at 72-meter intervals on the IOE baseline.
- **Korea Only: AP mines.**
  - **AP mines are placed on two rows of full-width AT mines.**
  - **AP mines are placed in a cluster around AT mines.**
  - **One AP mine is placed in front of every other AT mine.**

HASTY PROTECTIVE ROW MINEFIELDS

Hasty protective row minefields are temporary in nature and are used as part of a unit's defensive perimeter. Units usually use mines from their basic load. If time permits, mines should be buried to increase their effectiveness, but they can be surface-laid. The minefield can be easily recovered because AHDs, nonmetallic mines, or low-metallic mines are not used. Mines are employed outside the hand-grenade range but within the range of small-caliber weapons. All mines are picked up by the emplacing unit upon leaving the area, unless enemy pressure prevents mine retrieval or the minefield is being transferred to a relieving commander.

RULES

The brigade commander has the initial authority to employ hasty protective row minefields. This authority may be delegated to the battalion or company commander on a mission-by-mission basis.

Most of the rules governing hasty protective row mining are defined in STANAG 2036. A summary of those rules and some additional rules that apply are shown below.

- Rows.
  - — Rows are marked and recorded. They are designated by letters (A, B, and so forth) with Row A being closest to the enemy.
  - — The minimum distance between rows of AT mines is 8 meters.
  - — **Korea Only: The minimum distance between any row and a row containing AP mines is 15 meters.**
  - — The distance between the start row marker and the first mine in a row is the mine spacing for that row.
  - — An IOE row is not used.
  - — Start and end row markers are permanent markers and must be made of detectable material.
- General.
  - — The spacing between mines is at least 4 meters to prevent sympathetic detonation. There is no maximum distance between mines within the row.
  - — Mines are at least 15 meters from the perimeter fence.
  - — The minefield must be fenced on all sides if M18A1 AP mines are employed and the minefield will be in place for more than 72 hours.
  - — The minefield has at least one landmark that is located to the rear, never to the extreme side or front.
  - — The minefield has an easily identifiable RP (tree, stump, stake).
  - — Back azimuths are not used to record the minefield.
  - — Measurements are in meters.
  - — AHDs, nonmetallic mines, or low-metallic mines are not used.
  - — Minefields are recorded on DA Form 1355-1-R (see Chapter 8).

SITE LAYOUT

Requesting and receiving permission to lay mines is the first step when emplacing a hasty protective row minefield. The next step is conducting a thorough leader's reconnaissance of the proposed minefield area. Identify mine locations that cover likely AAs, enhance key weapon systems, and cover dead space. Establish an easily identifiable RP between the minefield and the unit position. From the RP, visualize mines as running in rows parallel to the unit position (Figure 6-13). After the RP is established and the minefield is visualized, complete recording and emplace the mines, but do not arm them. This procedure simplifies recording and makes retrieval quicker and safer.

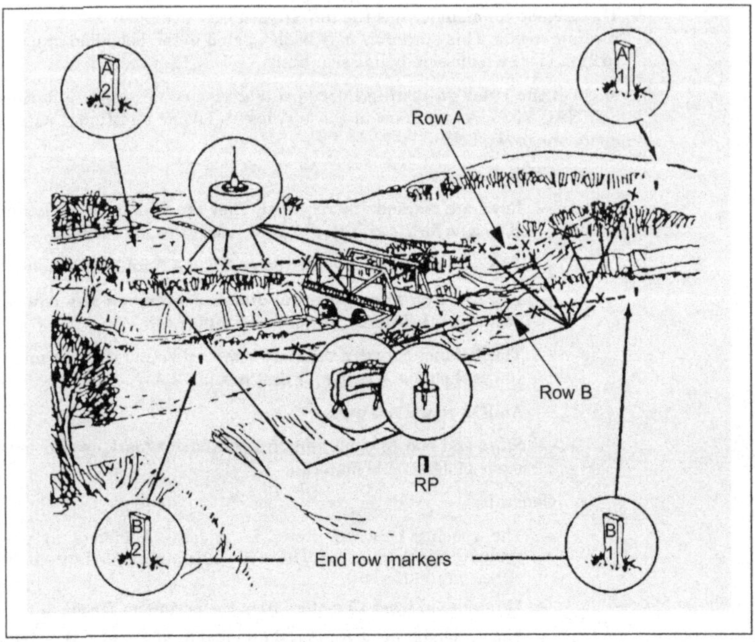

Figure 6-13. Site layout

Mine Rows

The row closest to the enemy is designated as Row A; succeeding rows are designated B, C, D, and so on. The ends of rows are shown by two end row markers. They are labeled with the letter of the row and the number 1 for the right end of the row or the number 2 for the left end of the row. The rows are numbered from right to left, facing the enemy. The marker should be an easily identifiable object, such as a wooden stake with a nail or a steel picket so that it can be found with an AN/PSS-12 mine detector.

Laying Procedures

From the RP, the leader measures the magnetic azimuth, in degrees, to a selected point on the right side (facing the enemy) of the tentative minefield. He paces off the distance and records it in meters. This point (B1) marks the beginning of the second row. The leader places a marker at B1 and records the azimuth and the distance on DA Form 1355-1-R.

From B1, the leader measures the azimuth and distance to a second point on the right side of the minefield (facing the enemy). He places a marker at this point (A1), and records the information.

The leader measures the distance and the azimuth from A1 to the location of the first mine in that row. The distance (or spacing) from the end row marker to the first mine is the mine spacing for that row. After the leader records the location, the mine is emplaced, but it is not armed.

The distance and azimuth are measured from the first mine to the second mine, and so on, until all the mines are emplaced and the locations are recorded. This procedure is repeated for the second row. As each mine is recorded, it is assigned a number to identify it in the minefield record.

When the last mine location is recorded for a row, the distance and the azimuth are measured from that point to another arbitrary point, A2 or B2. A marker is placed here in the same manner as A1 and B1. Next, the distance and azimuth from the RP to B2 and from B2 to A2 are measured and recorded.

When all the mines have been placed and recorded, the leader measures the distance and the azimuth between the RP and a permanent landmark that can be found on the map. He records the information on DA Form 1355-1-R. The landmark is used to assist others in locating the minefield if it is transferred or unexpectedly abandoned.

Mines can be armed after recording is complete. Mines nearest the enemy are armed first, allowing soldiers to safely work their way back to the unit position. Pins and clips are buried 30 centimeters behind row markers, the RP, or any easily identifiable, accessible location. Record the location of the pins and clips in the remarks section of DA Form 1355-1-R. The leader then reports the completion of the minefield to higher headquarters.

If the minefield is transferred to another unit, the transferring unit leader briefs the gaining unit leader. The gaining unit leader signs and dates the mines-transferred block on the DA Form 1355-1-R. The form is destroyed when the minefield is removed. If the minefield is abandoned unexpectedly, the DA Form 1355-1-R is forwarded to higher headquarters.

Mine Removal

When removing mines from a hasty protective row minefield, the leader first determines the best method to use:

- If the minefield has been under constant observation from the time it was laid and has not been tampered with, the squad leader directs the personnel who laid the mines to pick up the same mines. He uses DA Form 1355-1-R preceded by a mine detector to determine the types of mines to be removed and where they are located.
- If the minefield has not been under constant observation, may have been tampered with, or the personnel who laid the mines are not available or do not remember the location of the mines, the squad leader uses DA Form 1355-1-R and a clearance team as outlined in Chapter 11 to locate and remove mines.

The leader retrieves safety devices, shipping plugs, and other items that accompanied the emplaced mines. Using the azimuths and distances provided on the DA Form 1355-1-R, the removal team starts at the RP and moves to B1. They then move from B1 to the mine and remove the mine. If B1 is destroyed, the team moves from the RP to B2. The team then shoots a back azimuth (subtract 180 degrees) from the recorded azimuth from B2 to the first mine and removes the mine. Personnel continue this process until all the mines have been removed. The stakes at Al, B1, A2, and B2 are necessary because it is safer to find a stake than to find an armed mine.

The removal team observes basic safety precautions by maintaining 30 meters between personnel, not running, and moving only in cleared areas. The team starts with the row closest to the defender and works toward the enemy. Personnel—

- Check the sides and bottoms of the mines for AHDs, and disarm or mark the mines as they are found.
- Replace all pins, clips, and other safety devices before the mines are removed from the ground.
- Turn arming dials to *SAFE* or *UNARMED;* or if mines have screw-type fuses, remove the fuses and take them away from the mines.
- Lift the mines from the holes after they have been rendered safe.
  - If a mine was put in place and kept in sight by the individual who removes it, he lifts it directly from the hole after rendering it safe.
  - If a mine has not been kept in sight, the individual attaches a 60-meter rope or wire to the mine, takes cover, and pulls the mine from the hole.
- As each mine is removed, place a tick mark beside it on the DA Form 1355-1-R.
- Assemble all the mines in one location for accountability.

NOTE: AHDs are not used in hasty protective row minefields. However, as a safety precaution, consider all mines to be equipped with AHDs until proven otherwise.

The leader confirms the removal of the mines and accounts for the number of mines, by type, as recorded on the DA Form 1355-1-R. The leader may find it necessary to confirm an exploded mine to account for all the mines.

To confirm a mine explosion that was not witnessed, identify the crater or traces of burnt soil made by the detonated mine and place a tick mark beside the mine number on the DA Form 1355-1-R. Ensure that the crater found in the vicinity of the mine was caused by a land mine and not by artillery. A mine crater is normally circular, and it shows traces of burnt soil. The impact and the soil dispersion of artillery is normally elongated.

The squad leader confirms that each mine is disarmed and safe. The removal team cleans and repacks serviceable mines for future use and destroys the others; they repack serviceable mines in their original containers and store them according to the unit SOP and local regulations. The removal team removes and stores the row markers. The leader submits a report to his higher headquarters stating that the minefield has been removed and that the area is clear.

This chapter implements STANAG 2036.

Chapter 7

Standard-Pattern Minefields

Emplacing standard-pattern minefields is laborious and time-consuming, but it allows better mine concealment than row mining. Standard-pattern laying is well suited for protective and nuisance minefields. It can be used in terrain where the nature of the ground makes row mining impracticable.

To achieve their maximum effect, mines must be laid so that they cannot be seen and so that a vehicle's wheel or track or a person's foot exerts enough pressure to detonate them.

The method used to lay mines depends on the mine operation, the type of ground in which the mine is to be laid, and the type of ground cover that is available for camouflage.

COMPONENTS

MINE STRIPS

The mine strip is the foundation of a standard-pattern minefield. If a mine strip was laid in one straight line, the enemy could easily locate the mines; therefore, mine strips are laid in several segments as shown in Figure 7-1, page 7-2.

When siting, laying, and recording mine strips, all measurements are expressed in meters. Directions are recorded as magnetic azimuths, in degrees.

MINE CLUSTERS

The cluster is the basic unit of a minefield. It consists of one to five mines that are laid within a 2-meter-radius semicircle (Figure 7-2, page 7-3). When clusters are placed in a mine strip, they are numbered progressively and may consist of—

- One AT mine.
- **Korea Only: One AT mine and one to four AP mines.**
- **Korea Only: One to five AP mines.**

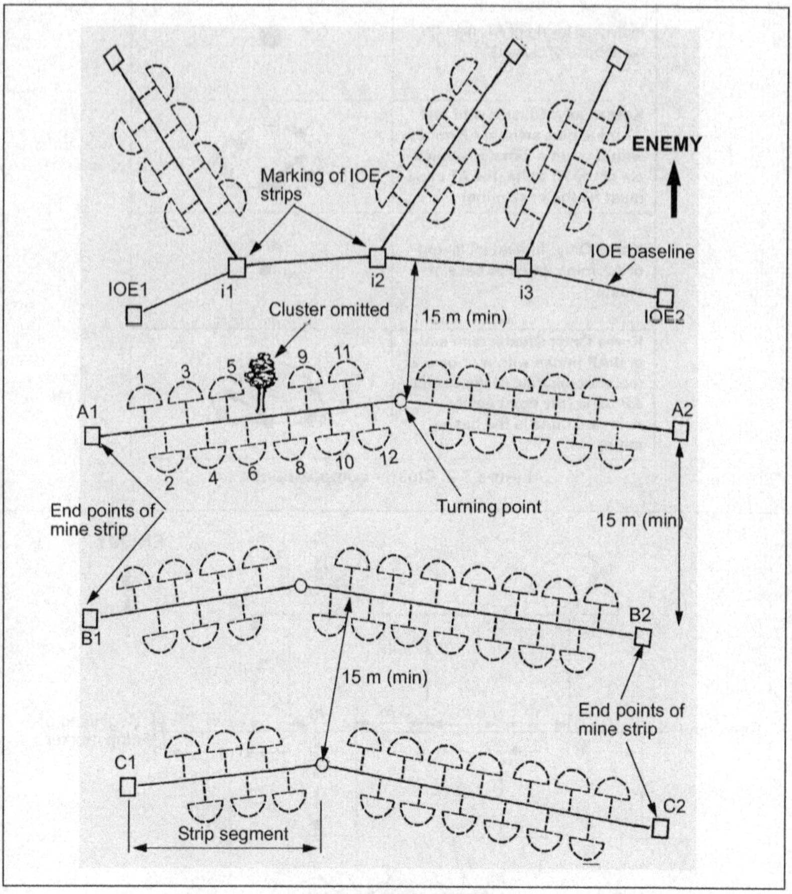

Figure 7-1. Minefield layout

Clusters are placed at 6-meter intervals, center to center, to form rows. Two parallel rows, 6 meters apart, form a mine strip. The arrangement of clusters in a mine strip is shown in Figure 7-3.

Rules for Positioning Clusters Within a Strip

The first cluster is placed on the enemy side of the strip centerline, 6 meters from the beginning-of-strip marker. The following clusters are numbered consecutively. Odd-numbered clusters are always on the enemy side of the

Individual laying of AT mine (to be at base of cluster)	
Korea Only: Cluster with one AT mine plus several AP mines within or on a 2-meter semicircle of the AT mine (the AT mine must be the base mine)	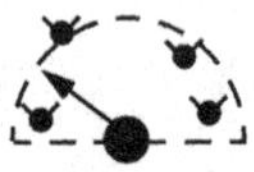
Korea Only: Individual laying of AP mine (to be at base of cluster)	
Korea Only: Cluster with several AP mines within or on a 2-meter semicircle of the center AP mine (the most easily detected mine is the base mine)	

Figure 7-2. Cluster compositions

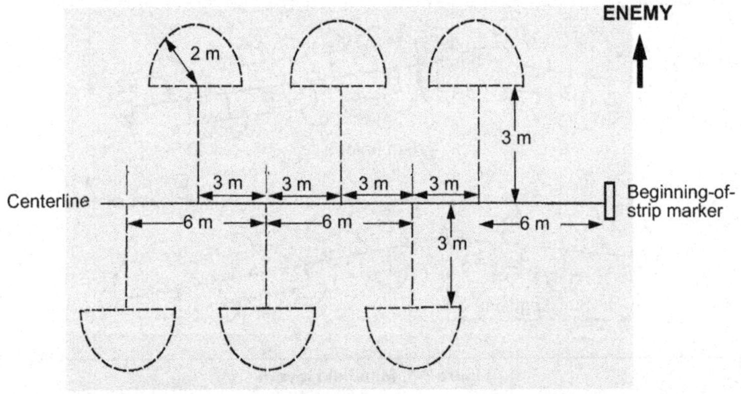

Figure 7-3. Arrangement of clusters in a mine strip

strip centerline. The direction of laying follows the numbering (A1 to A2, B1 to B2, and so on).

The IOE is normally the first part of the minefield encountered by the enemy. It consists of a baseline from which short strips are extended (Figure 7-4, page 7-4). Short strips along the IOE deceive the enemy on the minefield's pattern, spacing, and size. IOE placement and composition are largely dictated by the time allowed for laying the minefield, terrain conditions at the laying site, and the tactical situation.

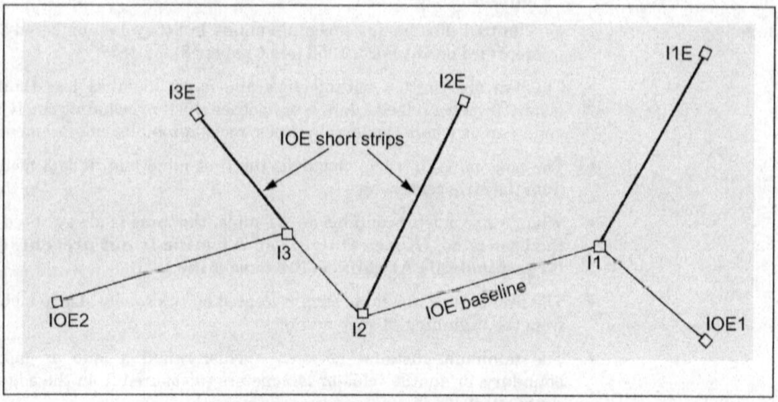

Figure 7-4. IOE baseline with short strips

The IOE baseline extends from one end point (IOE1) to another end point (IOE2). The laying direction is indicated by end-of-strip markers. Laying always begins at IOE1. Intermediate or turning points are marked in consecutive order beginning with I1. On the enemy side of the IOE baseline, short strips are extended from turning points at irregular angles. They are identified by turning-point markers.

Turning points should be no more than 45 degrees from the last azimuth. The length of short strips is not standard. A marker is emplaced at the end of each short strip. Markers are numbered in consecutive order beginning with I1E for recording purposes. No trip wires are used in the IOE, but AHDs may be employed.

Korea Only: AP mines actuated by trip wires are placed on the enemy side of each regular strip. No more than one mine per cluster uses trip wires, and no more than two trip wires extend from the mine. Trip wires are angled toward the enemy and should be at least 2 meters from the cluster, the lane border, and the minefield boundary. Trip wires are only used with AP fragmentation mines; they are not considered to be AHDs.

STANDARD-PATTERN MINEFIELD RULES

The following rules apply to tactical and protective standard-pattern minefields. They do not apply to nuisance minefields.

Clusters

- A cluster is a 2-meter-radius semicircle located 3 meters off the strip centerline.
- There are two types of clusters—live and omitted.

— A live cluster contains one AT mine. **(Korea Only: A live cluster contains as many as five mines, only one can be an AT mine).**

— Omitted clusters do not contain mines, but they are numbered and recorded on DA Form 1355 (see Chapter 8).

- Clusters are omitted within lanes and gaps; in areas less than 2 meters from boundaries, lanes, or another cluster (including the IOE); and in areas where the terrain (trees, rocks) prohibits emplacement.
- The base mine in a live cluster is the first mine laid. It is 3 meters from the strip centerline.
- When a live cluster contains an AT mine, the mine is always used as the base mine. **(Korea Only: If an AT mine is not present, the largest metallic AP mine is the base mine.)**
- The first cluster in a mine strip is located on the enemy side, 6 meters from the beginning-of-strip marker.
- The minimum distance between a cluster and a lane, a gap, a boundary, or another cluster is 2 meters (measured from the edge of the cluster).
- Cluster composition is the number of mines, by type, in any cluster in a specific group.
- Cluster composition remains the same through the entire mine strip and is recorded on DA Form 1355.
- **Korea Only: The types of AP mines may vary within a cluster.**
- The cluster boundary must at least 15 meters from the minefield perimeter fence.
- Clusters are numbered, beginning with the first cluster on the enemy side. Odd-numbered clusters are always on the enemy side of the strip, and even-numbered clusters are always on the friendly side.

Regular Strips

- A regular strip (sometimes referred to as a *lettered strip*) consists of a strip centerline and two rows of clusters (Row 1, enemy side; Row 2, friendly side).
- Regular strips are marked and recorded. They are designated by letters (A, B, and so forth), with Strip A being closest to the enemy.
- A standard-pattern minefield contains at least three regular strips.
- The minimum distance between strip centerlines is 15 meters; there is no maximum distance.
- Safety tapes are used to ensure that personnel installing trip wires do not move forward into armed clusters. A safety tape is used behind each regular strip. Safety tapes are 8 meters from the strip centerline (3 meters from the outer edge of the cluster).

- The marking of end points indicates the direction of laying (for example, A1 to A2).

IOE

- The IOE consists of a baseline from which short strips are extended.
- The IOE is located on the enemy side of the minefield.
- The number of clusters in an IOE is approximately one-third the number used in a regular strip.
- The first cluster along a short strip is placed on the enemy side and must be at least 6 meters from the IOE baseline; the cluster boundary must be at least 2 meters from the IOE baseline (Figure 7-5). If the short strip is exactly parallel to the enemy direction of travel, the NCOIC designates the enemy side of the strip.
- The IOE baseline is labeled at the beginning (IOE1) and end (IOE2) according to the direction mines are emplaced.
- Short strips are labeled at turning points (I1) and at the end (I1E).
- The IOE contains a safety tape that is 2 meters behind the IOE baseline and runs parallel with it.
- The IOE baseline is at least 15 meters from any point of the strip centerline of a regular strip; there is no maximum distance.
- An IOE short strip is at least 15 meters from another IOE short strip; there is no maximum distance.
- AHDs may be employed.

IOE Short Strips

- Short strips originate from turning points along the IOE baseline.
- The number and length of short strips depend on the tactical situation and the resources available.

Trip Wires (Korea Only)

- **Trip wires are not used in an IOE.**
- **Trip wires may be used in regular strips, but only one mine per cluster may be actuated by a trip wire.**
- **Trip wires are employed no closer than every third cluster.**
- **No more than two trip wires can be used on one mine.**
- **Trip wires are used only on the enemy side of the strip.**
- **Trip wires are not considered AHDs.**
- **Trip wires are located at least 2 meters from a lane, a safety tape, a cluster, another trip wire, the IOE baseline, and the minefield perimeter fence.**
- **Trip wires can only be used with AP fragmentation mines.**

Turning Points

- Clusters must be at least 3 meters from turning points.

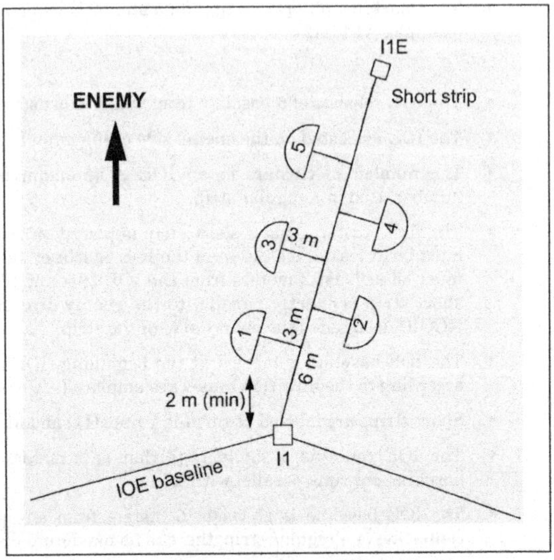

Figure 7-5. Clusters on an IOE short strip

- The first cluster after a turning point is laid on the opposite side of the strip centerline from the last cluster before the turning point.
- The angle of any given turning point cannot exceed 45 degrees from the last azimuth. (This ensures a minimum distance of 2 meters between cluster boundaries in the same row.)

Lanes

- Minefield lanes (Figure 7-6, page 7-8) are used by dismounted patrols and vehicles.
- Lanes are sited before laying begins.
- Lane locations should not be obvious.
- Clusters are not laid within 2 meters of lane edges.
- Lanes are zigzagged, not straight.
- Lanes cross the strip centerline at approximately right angles.
- Direction changes will not exceed 45 degrees. (This ensures that long vehicles will be able to negotiate turns.)
- The number of lanes must be sufficient to ensure that no one lane is overused and turned into an obvious track.
- Sufficient mines are stockpiled so that the responsible unit can close lanes suspected of being located by the enemy.
- Recommended minefield lane widths are—
  — Footpath: 1 meter.

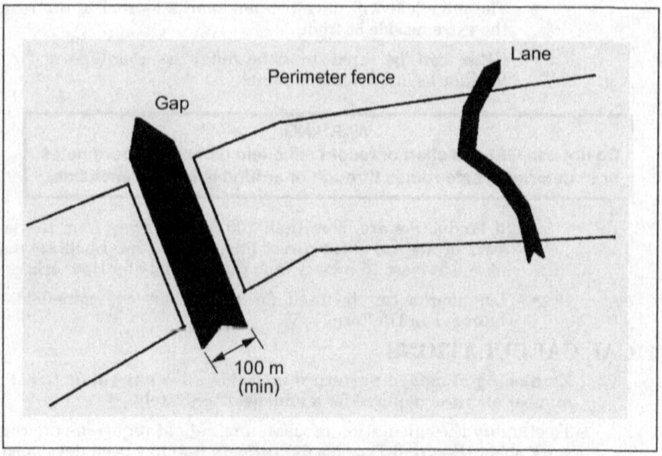

Figure 7-6. Minefield lanes and gaps

— One-way vehicle lane: 8 meters.

— Two-way vehicle lane: 16 meters.

Gaps

- Minefield gaps (Figure 7-6) are left so that friendly forces can pass through the minefield in tactical formation.
- Gaps must be at least 100 meters wide.
- Gaps are sited before laying begins.
- Gaps are located along recognizable features (fences, tracks, creeks).
- Gaps should run straight through a minefield and not contain bends.
- Sufficient mines must be stockpiled so that the responsible unit can close gaps when necessary.
- Gaps should closely resemble the rest of the minefield so that they will not be discovered by the enemy.
- The ground within a gap should be disturbed with tracks to represent the passage of a mine-carrying vehicle.
- Signs of mine laying (digging, scattered spoil, crates) should be visible in a gap.

General

- The farthest extremities of a regular strip determine the minefield front.
- Minefield depth is measured from the first to the last strip and includes the IOE (if applicable).
- Strips can be laid left to right or right to left.
- Back azimuths are not used to record the minefield.

- The minefield will have two landmarks located to the rear, never to the extreme side or front.
- GPSs can be used to determine the coordinates for minefield landmarks and reference points.

WARNING
Do not use GPSs to chart or record minefield perimeter coordinates or to determine safe routes through or around existing minefields.

- If landmarks are more than 200 meters away from the last regular strip or are out of the direct line of sight, intermediate markers are placed at least 75 meters from the last end-of-strip marker.
- Landmarks can be used for more than one minefield. They are recorded on DA Form 1355.

LOGISTICAL CALCULATIONS

Emplacing standard-pattern minefields allows the unit to calculate the number of mines required for a minefield accurately.

To simplify the calculation process, a minefield requirements computation work sheet (Figure 6-1, pages 6-5 through 6-8) has been developed. Step-by-step procedures for completing the work sheet are shown in Figure 6-2, pages 6-9 through 6-14.

CLUSTER CALCULATION

Accuracy is essential when emplacing a standard-pattern minefield. The following check system has been designed so that the minefield OIC can accurately record the number of mines laid. Compare the strip feeder report with the—

- Number of pins and clips returned by the laying party.
- Mine tally sheet.
- Cluster computation.

EXAMPLE: You receive a strip feeder report for Strip A. There are 26 clusters in the strip. The strip has three segments (30, 42, and 21 meters) and two turning points.

Step 1. Add the total length of the strip as determined by the recording party.

30 + 42 + 21 = 93 meters

Step 2. The first and last cluster in the strip are located 6 meters in from the end-of-strip markers. Subtract 12 from Step 1.

93 - 12 = 81 meters

Step 3. Clusters are not located on turning points. Multiply the number of turning points by 3 and subtract it from Step 2.

81 - (3 × 2) = 75 meters

Step 4. Divide Step 3 by the cluster spacing.

75 ÷ 3 = 25 clusters

Step 5. Add one cluster, because when a line is divided, there is one more interval than spacing.

25 + 1 = 26 clusters

From these computations, the minefield OIC is able to cross-check the information on the strip feeder report. In this example, Strip A should have a total of 26 clusters, less any omitted clusters.

PLATOON ORGANIZATION

The platoon is the basic unit used to install a standard-pattern minefield. Orders to the laying unit OIC specify the proposed location, the length, and the mine type and density. The platoon organization and equipment are shown in Table 7-1.

Table 7-1. Platoon organization and equipment

Personnel	Officer	NCO	EM	Equipment
Supervisory	1	1	0	Officer: Map, lensatic compass, notebook, and minefield record forms NCO: Map, lensatic compass, and notebook
Siting party	0	1	3	Stakes or pickets, sledgehammers, engineer tape on reels, and nails to peg tape
Marking party	0	1	2	Rolls of barbwire or concertina, marking signs, lane signs, wire cutters, gloves, sledgehammers, pickets, and picket pounder
Recording party	0	1	2	Sketching equipment, lensatic compass, minefield record forms, map, and metric tape
First laying party	0	1	6-8	Notebook for squad leader, picks, shovels, and sandbags
Second laying party	0	1	6-8	Same as first laying party
Third laying party	0	1	6-8	Same as first laying party
Totals	1	7	25-31	

The OIC performs a map study, and if the situation permits, conducts a ground reconnaissance of the site. He determines locations for mine strips, landmarks, fences, mine dumps, and approaches. Using the minefield requirements computation work sheet (Figure 6-1, pages 6-5 through 6-8), the OIC determines the required number of mines and other materials. He arranges for mines to be delivered and organizes the platoon into siting, laying, recording, marking, and mine-dump parties.

Siting Party

The siting party places boundary stakes or pickets as strip markers at the beginning and end of each mine strip and at the points where strips change direction. It lays tape on the centerlines of each strip, lane, and traffic path. After siting is complete, the siting party augments other parties.

Laying Parties

One laying party is responsible for installing, arming, and camouflaging all the mines on a strip or a portion of a strip. Each laying party is then assigned additional strips.

Recording Party

The recording party obtains the necessary reference data, prepares DA Form 1355 (see Chapter 8), and when needed, installs intermediate markers.

Marking Party

The marking party erects fences and signs to mark minefield boundaries and lanes. After marking is complete, the marking party augments other parties.

Mine-Dump Party

The mine-dump party, controlled by the PSG, is composed of personnel who are not working as members of other teams. It accounts for all Class IV and V supplies that arrive from other sources, organizes mine sets at the mine dump, and hauls additional supplies as required. The PSG is not required to stay at the mine dump continuously; he has the flexibility to move around the area to perform other activities.

Mine-Emplacement Procedures

Mine-emplacement procedures are shown in Figure 7-7. In the illustration, the minefield is laid from right to left.

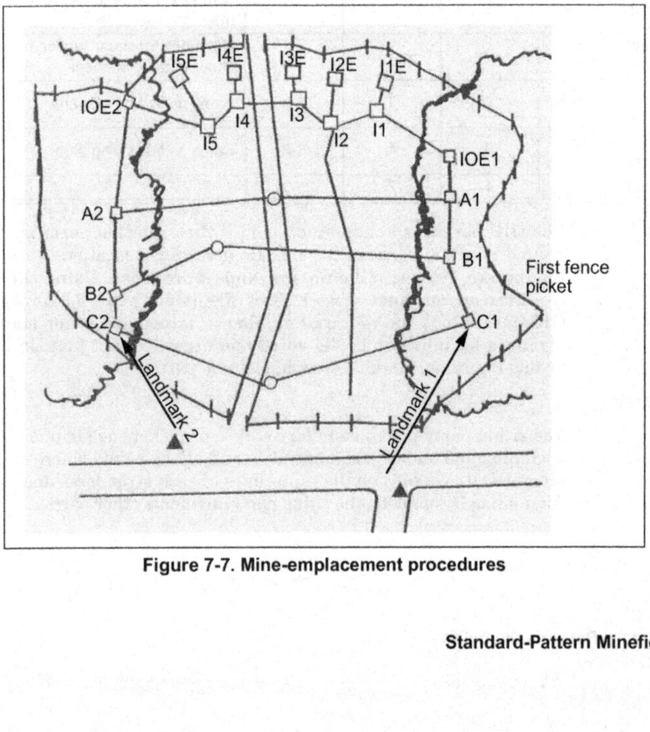

Figure 7-7. Mine-emplacement procedures

The OIC arrives at the site with the siting and marking parties. He goes to the right or left (depending on the laying direction) rear boundary of the minefield. This part of the minefield is the farthest from the enemy. The OIC indicates the starting point of the rear strip (this is Strip C in a three-strip minefield), and the siting party emplaces a beginning-of-strip marker.

The OIC designates a starting point for the marking party that is at least 15 meters to the right of the boundary stake. He indicates where the minefield marking fence should be placed. The marking party immediately begins to install fence pickets, working in a counterclockwise direction.

When all the pickets are installed, the marking party encircles the field with a single strand of barbwire (at waist height) and emplaces mine signs. Personnel emplace a second strand of barbwire if required.

From the boundary stake of Strip C, the OIC moves in the direction of the enemy and establishes the starting point of Strip B. Strip centerlines should not be parallel nor less than 15 meters apart. Two members of the siting party emplace a beginning-of-strip marker at the starting point of Strip B. The remaining two members lay tape between the two strips and fasten the tape to the ground at frequent intervals to prevent movement. This procedure is followed until the beginning-of-strip markers of the three regular strips (C, B, and A) and the IOE on the right-hand side of the minefield have been installed.

At the IOE beginning-of-strip marker (IOE1), the OIC gives the siting party a sketch of the minefield and instructions on siting the IOE baseline and strip centerlines. The NCOIC and one other member of the siting party immediately begin setting stakes to indicate the IOE baseline. The centerline laying team lays tape on the IOE baseline, leaving tape reels where tapes run out. Short strips extending from the IOE baseline are established at the same time. Each short strip starts with a marker that is designated as I1, I2, and so forth and ends with a marker that is designated as I1E, I2E, and so forth. Turning points are not used on short strips.

Upon reaching the IOE end-of-strip marker (IOE2), the NCO moves away from the enemy side, establishes the left boundary stake of Strip A, stakes out Strip A, and repeats the procedure until all strip centerlines are taped. All stakes are driven flush with the ground.

While the IOE is being taped, the recording party obtains reference data for the DA Form 1355. Working behind the siting party, the recording party starts from Landmark 1 (designated by the OIC) and proceeds to C1. After C2 has been sited, the recording party proceeds from Landmark 2 to C2 to establish the distance and the azimuth. Finally, the recording party ties C1 and C2 to both landmarks in case one of the landmarks is removed or destroyed. The amount of detail obtained by the recording party depends on the tactical classification of the minefield and any special orders. Aerial photographs taken of the minefield before the tracing tape is removed become valuable supplements to the DA Form 1355.

As soon as laying parties arrive at the site with mines, they establish mine dumps a minimum of 150 meters apart and 50 meters behind the field. They uncrate and stack AT mines and place fuses and detonators in separate boxes.

Fuse types are not mixed. **(Korea Only: AP mines are left in their crates, and the crate lids are removed.)**

When the siting party completes the centerline staking, it installs lane tapes and traffic tapes, respectively. Lane tapes are used by tactical vehicles and patrols. Traffic tapes are used by laying personnel to assist in camouflage and to reduce the amount of traffic on strip centerlines. Traffic tapes are laid perpendicular to the minefield trace at 100-meter intervals.

Mine Emplacement

The laying party must know the cluster composition of the strip, the location of any omitted cluster, and future lane locations. When the centerline tape for a regular strip has been installed, the NCOIC designates all but two members of the laying party to emplace mines in the ground. The remaining two soldiers, usually the most experienced, are designated as fusers and are responsible for arming mines. Layers carry the maximum load of mines to be used as base mines in the clusters. Fusers carry the fuses and the detonators.

The NCOIC moves to the right or left (depending on the laying direction) beginning-of-strip marker of the strip and organizes the layers into one column to his rear, directly on the centerline. He measures 6 meters along the centerline for the first cluster and, pointing perpendicular from the centerline and in the direction of the enemy, indicates the placement of the base mine. The first layer on the enemy side places a mine on the ground, 3 meters from the centerline.

The NCO measures 3 more meters and indicates the placement of the second base mine on the opposite (friendly) side of the strip. The first layer on that side places a base mine on the ground. As the initial load of mines is laid, each layer returns to the nearest mine dump for another load. Fusers follow behind layers and insert mine fuses, but they do not arm the mines. This procedure is followed until the end-of-strip marker on the far side of the minefield is reached.

Korea Only: The NCO tells layers the number and type of mines to be placed next to the base mine in each cluster. As AP mines are being placed, the NCO proceeds along the strip and ensures that the proper number of AP mines is placed in each cluster. The NCO places a spool of trip wire next to the mines that are to be activated by trip wire.

When all the mines are positioned in clusters, one layer is assigned to dig the holes for all the mines in a cluster. He places the spoil from the holes in sandbags and leaves the sandbags beside the base mine in each cluster. The layer checks the positioning of the mines in the holes, removes the mines from the holes, and places the mines beside the holes. **(Korea Only: The layers anchor trip wires with nails or stakes and wrap the loose ends of trip wires around the fuses.)**

When digging has progressed at least 25 meters from the first mine laid, the arming procedure begins. Fusers arm all the mines in a cluster, beginning with the mine farthest from the centerline and work backward. They place all the mines in the holes **(Korea Only: attach trip wires)** and arm and camouflage the mines. They place filled sandbags on the centerline of the strip, opposite the base mine. Fusers keep their back toward the centerline. Other personnel must remain at least 25 meters from the fusers (Figure 7-8).

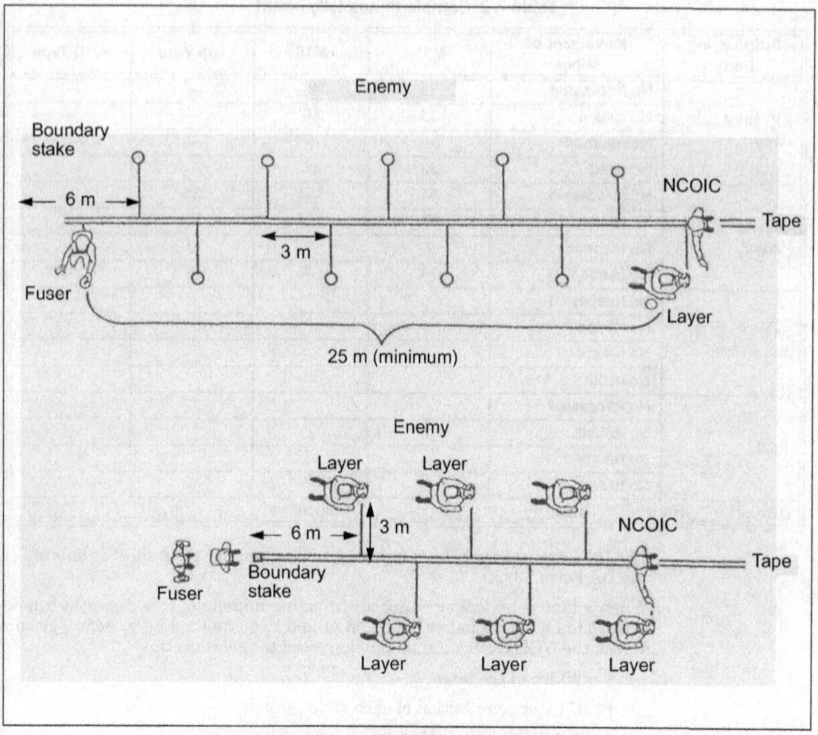

Figure 7-8. Laying and fusing mines

Mines located in lanes are not initially buried. They are placed aside to prevent confusion when counting clusters. The mines can be buried after the lane is closed. Upon completing the arming operation, fusers give the safety clips to the NCO, who verifies that all the mines have been armed and camouflaged. The NCO checks the strip and ensures that sandbags, tape, and debris have been picked up. The NCO gives the safety clips to the PSG, who buries them 30 centimeters to the rear of the beginning-of-strip marker.

All mines and other explosive items are recorded upon issue. They are summarized on a mines tally sheet (see Table 7-2). If more than one mine dump is established, a mines tally sheet is kept at each dump and the information is later transferred to a master tally sheet.

Table 7-2. Sample mines tally sheet

Strip/Laying Party	Movement of Mines	M15	M16	Trip Wire	AHD Type
IOE strip/ Party 1	No forecasted	23	46	—	—
	No issued	23	46	—	—
	No returned	—	—	—	—
	No used	23	46	—	—
Strip A/ Party 2	No forecasted	81	154	35	—
	No issued	81	154	35	—
	No returned	7	6	9	—
	No used	74	148	26	—
Strips B, C, D, and so forth	No forecasted				
	No issued				
	No returned				
	No used				
Total	No forecasted				
	No issued				
	No returned				
	No used				
Date ______ Rank _____ Name __________________ Signature ____________________					

The PSG ensures that the number of mines used per tally sheet is entered on the DA Form 1355.

When a lane is no longer required through a minefield, it is closed by a lane-closure team that consists of one NCO and two other soldiers. Before closure starts, the NCO checks the minefield record to ascertain the—

- Width of the lane.
- Cluster composition of each strip.
- Total number of mines required.
- Number of strips that intersect the lane.
- Azimuth of each strip.
- Distance between strips, along the lane centerline.
- Location of the mine dump.

The procedures for lane closure is as follows (see Figure 7-9, page 7-16):

- The lane-closure team moves along the safe-lane centerline until it arrives at the strip safe-lane marking picket.
- The NCO lays out a strip centerline tape and a tape along both sides of the safe lane to mark its boundaries.
- The team lays clusters.
- The team recovers the tapes.

- The above steps are repeated at successive strips.
- The NCO amends the DA Form 1355.

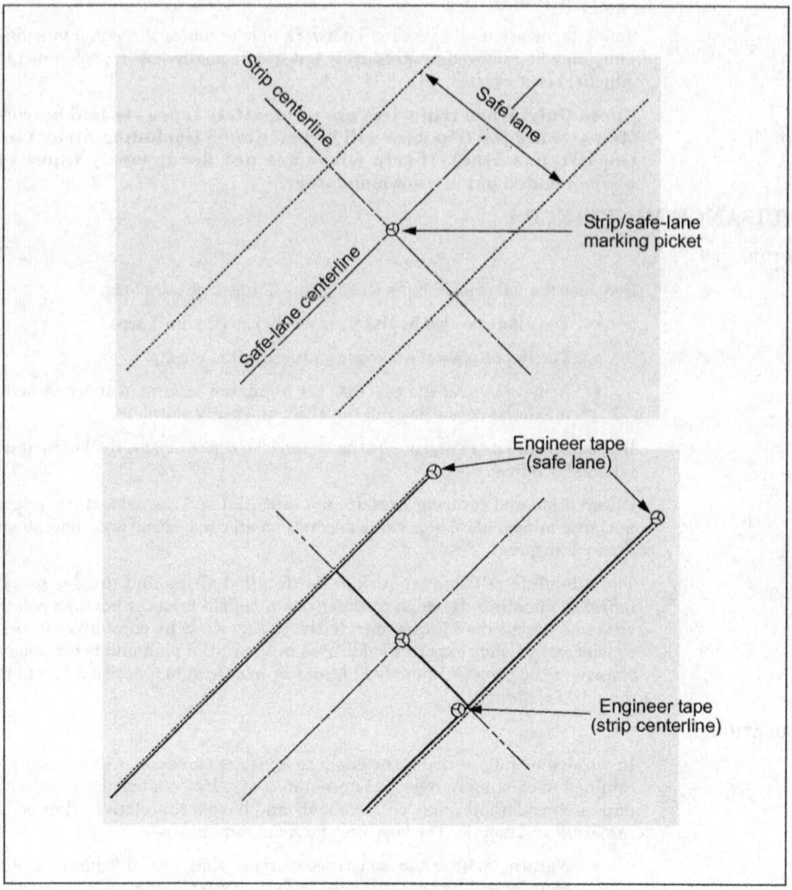

Figure 7-9. Lane closure

NOTE: When gaps must be closed, fences are temporarily erected along the side boundaries. They are removed later to avoid the indication of a passage through the minefield.

NCOICs do not act as working members of their parties; they ensure that—

- No one moves back into a mined area.
- Any irregularity, such as an omitted cluster, is recorded.

- All safety devices are recovered and checked against the minefield record.

Safety tapes are used to create a network of safe routes through a minefield. They may be removed progressively but are normally left in place until the minefield is complete.

Korea Only: When trip wires are used, safety tapes are laid between strips where the trip wire will be positioned (including Strip A and the IOE baseline). If trip wires are not used, safety tapes are recommended but are not mandatory.

NUISANCE MINEFIELDS

SITING

Consider the following factors when siting a nuisance minefield:

- The effort needed by the enemy to bypass a mined area.
- The importance of an area or a route to the enemy.
- Achievement of the goal (use the minimum amount of effort needed to maximize casualties and the effect on enemy morale).

Remember, the more ingenuous the methods of concealment, the longer it will take to lay mines.

Observation and covering fires are not essential and are seldom feasible for nuisance minefields. Their value depends on effective siting and concealment to cause surprise.

The minefield OIC is responsible for detailed siting and the design of a nuisance minefield. He must consider the minefield from the enemy's point of view and assess the courses open to the enemy when he encounters it. Such considerations may expose weaknesses in the initial plan and bring about a change to the proposed minefield layout or may lead to a decision to site the minefield elsewhere.

LOCATION

In wooded or hilly terrain, the enemy's logistics transport will normally be confined to existing routes. Nuisance mines at selected sites along roads can impose considerable delay on the enemy and have a cumulative effect on his resources and morale. The best sites for axial mining are—

- Natural defiles and constructed areas that are difficult to bypass (cuts, embankments, causeways, fords, forest tracks, built-up areas).
- In the vicinity of road craters, AT ditches, and any obstacles that have to be cleared.
- Around culverts.
- Demolished bridges, particularly the home bank, including likely adjacent crossing places and alternative building sites.
- Likely assembly areas.
- Covered approaches or dead space.

- In the vicinity of enemy-needed fuel, supplies, and engineering materials that cannot be destroyed or removed.
- Railroads (in or near culverts, bridges, sharp turns, tunnels, steep grades). Mines should be laid where enemy trains cannot bypass the mined area on branches or spurs.

LAYING

There is no requirement for recording the precise location of individual nuisance mines. Recording mine positions that are laid to a pattern is easy and quick. Pattern laying should be used when it can be done without prejudicing concealment.

When the number of mines to be laid on the site makes it impracticable or undesirable to lay mines in a pattern, they may be scatter-laid unless otherwise directed. Scatter laying by hand is useful in road blocks, bridge abutments, and craters when it would be difficult and wasteful to lay mines in a pattern. Again, scatter-laying along routes to be denied to the enemy will add considerably to the delay imposed. All available types of AT mines **(Korea Only: and AP mines)** are used to make nuisance minefields complex and difficult to remove. Combinations of mine types should be varied constantly so that each minefield presents a clearance problem. Deeply buried mines can be included; however, they take much longer to lay. These mines may be worthwhile around craters, where the enemy is likely to need earthmoving equipment.

If a nuisance minefield is laid in a standard pattern, standard procedures are followed. If mines are selectively positioned, procedures must be tailored to suit the situation. In all occasions, however, the following rules should be observed:

- The intended position for each mine is clearly marked on the ground before laying begins.
- Laying parties work in pairs, and each pair is detailed to lay specific mines.

INSPECTION AND MAINTENANCE

Mines that are left in the ground for an extended period of time may deteriorate and malfunction for one or more of the following reasons:

- Moisture may have entered the igniter or body of the mine and either neutralized the explosive or corroded the metal parts. Such actions may be aggravated by local factors (soil acidity, wide temperature swings).
- Frost or heat may have subjected the mine to mechanical strain and caused distortion.
- Insects or vegetation may have caused obstructions.
- Animals may have turned mines over or detonated them.

Technical inspections should only be made by experienced engineers or EOD personnel. When a minefield deteriorates below the operating level, additional mine strips and/or rows must be added to restore its effectiveness. They are sited to the front or rear of the existing minefield to increase its depth. New mine strips and/or rows are treated as a separate, additional minefield.

Technical inspections of minefields are normally done at three-month intervals, and personnel work in pairs. They are done more frequently during extreme weather conditions.

HANDOVER

Minefield handover is an extremely important task. The following items need to be addressed between emplacing and overwatching units:

- Discussing minefield compositions.
- Discussing the minefield configuration and walking and/or riding the minefield trace.
- Discussing lane closure, if applicable.
- Training the unit on how to the close lane, if applicable.
- Discussing obstacle protection against enemy dismounted patrols.
- Signing over the written report (Figure 2-26, page 2-54).
- Discussing indirect fires.
- Reporting completion of handover to higher headquarters.
- Forwarding a copy of the written report and the DA Form 1355 to higher headquarters.

This chapter implements STANAG 2036.

Chapter 8

Reporting and Recording

Reporting and recording mine and minefield/munition field information is critical to the success of a unit's mission. It not only provides tactical data for the commander, but it also provides force-protection information to subordinate and adjacent units.

MINEFIELD/MUNITION FIELD REPORTS

A minefield/munition field report is an oral, electronic, or written communication that concerns friendly or enemy mining activities. The report format is specified by the local command. It is submitted by the emplacing unit commander through operational channels to the G3/S3 of the authorized headquarters. The headquarters integrates the report with terrain intelligence and disseminates it along with tactical intelligence. The report is sent by the fastest, most secure means available. Figure 8-1, page 8-2, summarizes the minefield/munition field report flow at the division level and below.

REPORT OF INTENTION

When planning to emplace a minefield/munition field, the unit must submit a report of intention to notify their higher headquarters. The report doubles as a request when it is initiated at levels below emplacement authority. The report includes—

- Tactical purpose of the minefield/munition field.
- Estimated number and type of mines to be emplaced.
- Location.
- Proposed start and completion times.
- Type of minefield/munition field.
- Whether mines are surface-laid or buried.
- Whether AHDs are used.
- Location and width of lanes and gaps.

Conventional minefields/munition fields that are part of an OPLAN or a general defense plan (GDP) approved by the authorizing commander do not require a report of intention because inclusion in an OPLAN or a GDP implies an intention to lay.

REPORT OF INITIATION

A report of initiation is mandatory. It informs higher headquarters that emplacement has begun and that the area is no longer safe for friendly movement and maneuver. The report specifies the time emplacement began

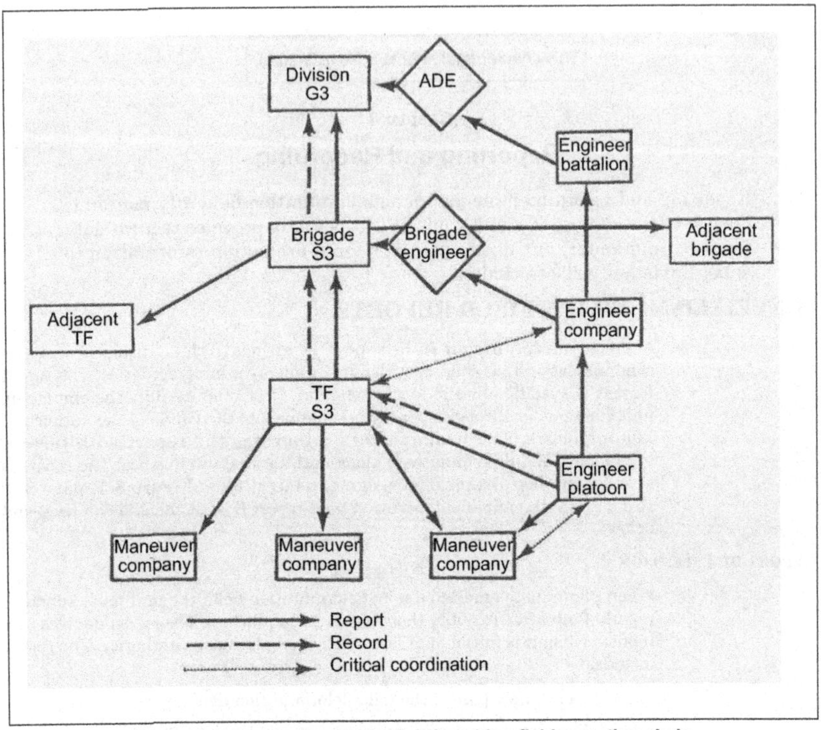

Figure 8-1. Conventional minefield/munition field reporting chain

and identifies the location and the target number of minefields/munition fields.

Report of Completion

A report of completion is usually an oral report to the authorizing commander. It indicates that the minefield/munition field is complete and functional. A report of completion is immediately followed by a completed DA Form 1355 or 1355-1-R.

Report of Transfer

Minefield/munition field responsibility is transferred from one commander to another in a report of transfer (Figure 2-26, page 2-54). This report is signed by the transferring and receiving commanders. It includes a certificate stating that the receiving commander was shown or otherwise informed of all the mines within the transferring commander's zone of responsibility. The report states that the receiving commander assumes responsibility for those mines.

The report of transfer is sent to the next higher commander who has authority over the transferring and receiving commanders.

REPORT OF CHANGE

A report of change is made immediately upon any change or alteration to a previously reported minefield/munition field. It is sent to the next higher commander and through channels to the headquarters that keeps the written minefield/munition field record. A report of change is made by the commander responsible for surveillance and maintenance of the minefield/munition field.

PROGRESS REPORTS

During the emplacing process, the commander may require periodic reports on the amount of work completed.

MINEFIELD/MUNITION FIELD RECORDS

Most conventional minefields/munition fields are recorded on DA Form 1355, hasty protective row minefields/munition fields are recorded on DA Form 1355-1-R.

The laying unit prepares the minefield/munition field record. The OIC signs the form and forwards it to the next higher command as soon as possible. Once the information is entered on the form, the form is classified SECRET, NATO SECRET, or SECRET-ROKUS. (When used for training, the record is marked SAMPLE.) The number of copies prepared depends on the type of minefield/munition field emplaced and the local procedures. Unit SOPs should provide advanced guidance on how minefield/munition field information is to be passed to higher, lower, and adjacent commands. Minefield/munition field records are circulated on a need-to-know basis. Large minefields/munition fields are recorded on two or more DA Forms 1355.

Once the minefield/munition field is removed, the date of removal, the time removal was complete, and the name of the unit that removed the minefield/munition field will be placed in the remarks block of the form. The form will then be forwarded to the unit's central control cell for mine clearance information. The central control cell will retain the form for two years and then destroy it.

When changes are made to an existing minefield/munition field, a new record must be prepared. This record is marked REVISED. It shows the minefield/munition field as it is after changes. The original minefield/munition field number remains unchanged. Some changes which require a new record are—

- Relocation of mines in safe lanes.
- Relocation of safe lanes.
- Changed lane or minefield/munition field marking.
- Inclusion of the minefield/munition field into a larger minefield/munition field system.
- Removal or detonation of mines.
- Addition of mines to the minefield/munition field.

Conventional minefield/munition field records are forwarded through operational channels to theater army headquarters (TAHQ) where they will be maintained on file by the theater engineer. If a TAHQ has not been established, minefield/munition field records are maintained on file with the assistant corps engineer of the AO where the minefield/munition field is located.

MINEFIELD RECORD

DA Form 1355 consists of a single sheet that is printed on both sides and folded in half. The front side contains blocks of tabular data, the inside is a graph of 1-centimeter squares for a scaled sketch of the field, and the back side includes a computation formula and instructions. The scale for plotting minefields/munition fields depends on the size of the field. To avoid using two sheets for the sketch, adjust the scale so that one form will support the sketch. For very large minefields/munition fields, two sheets may be required. The system of measurement and the scale size must be indicated in the legend block. A second form may be used to support any additional information in the notes block. Any blocks or lines not used on the form must be crossed out to avoid unauthorized entries on the form.

An example of a completed DA Form 1355 is shown in Figures 8-2a through 8-2c, pages 8-5 through 8-7.

The following step-by-step instructions are provided for completing DA Form 1355:

Block 1. Enter complete data on the laying authority and the laying unit. The OIC block should include rank, name, and social security number (SSN).

1	AUTHORITY: CG 2nd INF DIV
	LAYING UNIT: B Co 2nd ENGR BN 1st PLT
	OFFICER IN CHARGE: 1LT R. YOUNG 762-01-1352

Block 2. Enter the date-time group (DTG) for start and completion times. The recorder block includes rank, name, and SSN.

2	DATE AND TIME	START 090630Z JAN 90
		COMPLETION 091800Z JAN 90
	RECORDER SFC F. LING 550-72-1332	

Block 3. Enter the copy and the sheet number. The number of copies prepared depends on the unit SOP and the minefield/munition field classification.

3	Copy No. 1 of 4	Sheet No. 1 of 1

NOTE: The minefield/munition field record is forwarded by the laying unit. One copy is retained by the overwatching unit, one copy by the next higher command, one copy by the unit central control cell (for mine clearance information), and one copy by the proper national territorial authority.

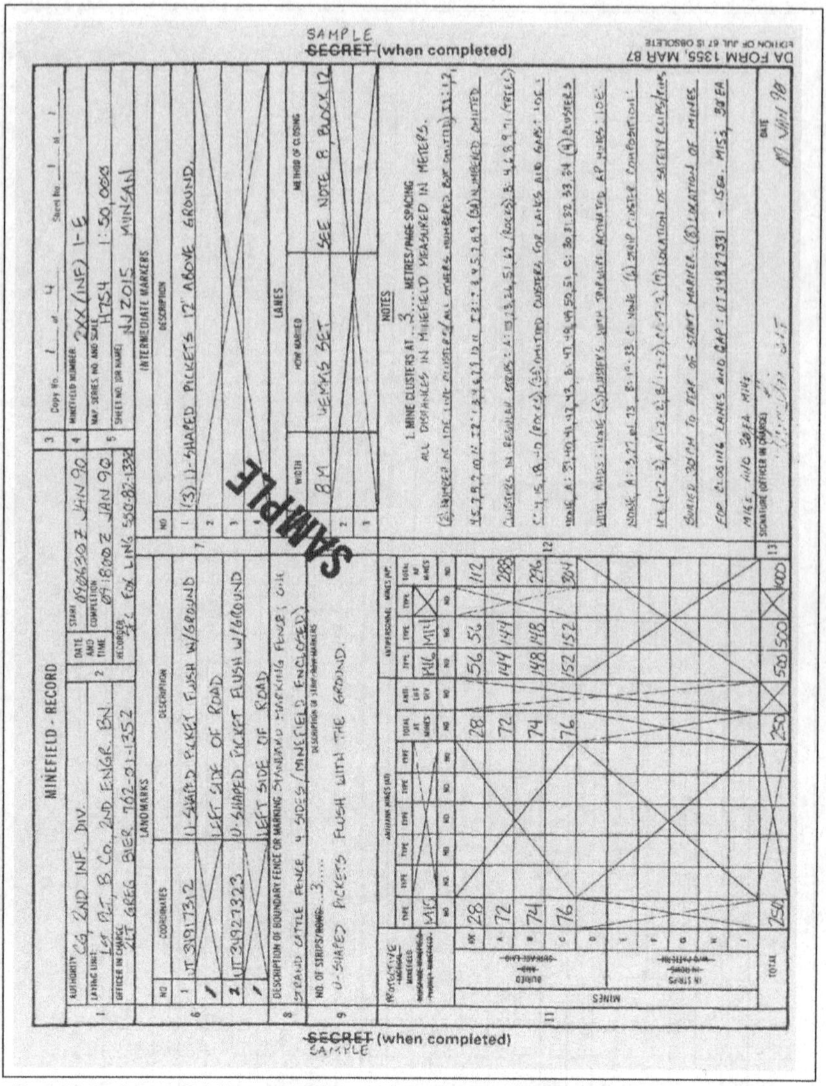

SAMPLE
~~SECRET~~ (when completed)

MINEFIELD - RECORD

1 AUTHORITY: CG, 2ND INF. DIV.
LAYING UNIT: 1st PLT, B Co., 2ND ENGR. BN.
OFFICER IN CHARGE: 2LT GREG BIER 762-01-1352

2 DATE AND TIME: START 090630Z JAN 90 / COMPLETION 091800Z JAN 90
RECORDER: SFC FOX LING 580-82-1332

3 Copy No 1 of 4 — Sheet No 1 of 1

4 MINEFIELD NUMBER: 2XX (INF) 1-E
MAP SERIES NO AND SCALE: H754 1:50,000

5 SHEET NO (OR NAME): NJ 2015 MUNSAN

6 LANDMARKS

NO	COORDINATES	DESCRIPTION
1	UT 31917312	U-SHAPED PICKET FLUSH W/GROUND
/		LEFT SIDE OF ROAD
2	UT 31927323	U-SHAPED PICKET FLUSH W/GROUND
/		LEFT SIDE OF ROAD

7 INTERMEDIATE MARKERS

NO	DESCRIPTION
1	(3) U-SHAPED PICKETS 12" ABOVE GROUND.
2	
3	
4	

8 DESCRIPTION OF BOUNDARY FENCE OR MARKING: STANDARD MARKING FENCE; ONE STRAND CATTLE FENCE 4 SIDES/(MINEFIELD ENCLOSED)

9 NO. OF STRIPS/~~ROWS~~: 3
DESCRIPTION OF STRIP/~~ROW~~ MARKERS: U-SHAPED PICKETS FLUSH WITH THE GROUND.

10 LANES

NO	WIDTH	HOW MARKED	METHOD OF CLOSING
1	8M	HEMMS SET	SEE NOTE 8, BLOCK 12
2			
3			

11 PROTECTIVE ~~TACTICAL~~ MINEFIELD ~~NUISANCE MINEFIELD~~ ~~PHONEY MINEFIELD~~

MINES

		AT TYPE M15	TYPE	TYPE	TYPE	TYPE	TOTAL AT MINES	ANTI-LIFT DEV	AP TYPE M16	AP TYPE M14	TYPE	TOTAL AP MINES
		NO	NO	NO	NO	NO	NO	NO	NO	NO	NO	NO
~~SURFACE LAID~~ ~~AND~~ BURIED	RX	28					28		56	56		112
	A	72					72		144	144		288
	B	74					74		148	148		296
	C	76					76		152	152		304
	D											
~~W/O PATTERN~~ ~~IN ROWS~~ IN STRIPS	E											
	F											
	G											
	H											
	I											
TOTAL		250					250		500	500		1000

12 NOTES

1. MINE CLUSTERS AT 3 METRES/PACE SPACING
ALL DISTANCES IN MINEFIELD MEASURED IN METERS.
(2) NUMBER OF IDL LINE CLUSTERS/ALL OTHERS NUMBERED BUT OMITTED) I1:1,2, 4,5,7,8,9,10,11 I2: 1,3,4,6,7,9,10,11 I3: 1,3,4,5,7,8,9 (3A) NUMBERED OMITTED
CLUSTERS IN REGULAR STRIPS: A: [illegible] B: [illegible] (ROCKS) B: 4,6,8,9,11 (TREES)
C: 4,15,18,40 (ROCKS) (3B) OMITTED CLUSTERS FOR LANES AND GAPS: IDL:
NONE A: 39,40,41,42,43 B: 47,48,49,50,51 C: 30,31,32,33,34 (4) CLUSTERS
WITH AHDS: NONE (5) CLUSTERS WITH TRIPWIRES ACTIVATED AP MINES: IDL:
NONE A: 3,27,61,73 B: 10,33 C: NONE (6) STRIP CLUSTER COMPOSITION:
IDL (1-2-2), A (1-2-2), B (1-2-2), C (1-2-2) (7) LOCATION OF SAFETY CLIPS/PINS
BURIED 30CM TO REAR OF START MARKER. (8) LOCATION OF MINES
FOR CLOSING LANES AND GAP: UT34827331 - 15EA. M15; 30EA
M16, AND 30EA M14

13 SIGNATURE (OFFICER IN CHARGE): [illegible signature] DATE: 10 JAN 90

DA FORM 1355, MAR 87 — EDITION OF JUL 67 IS OBSOLETE

~~SECRET~~ (when completed)
SAMPLE

Figure 8-2a. Sample DA Form 1355 (front side) for a standard-pattern minefield/munition field

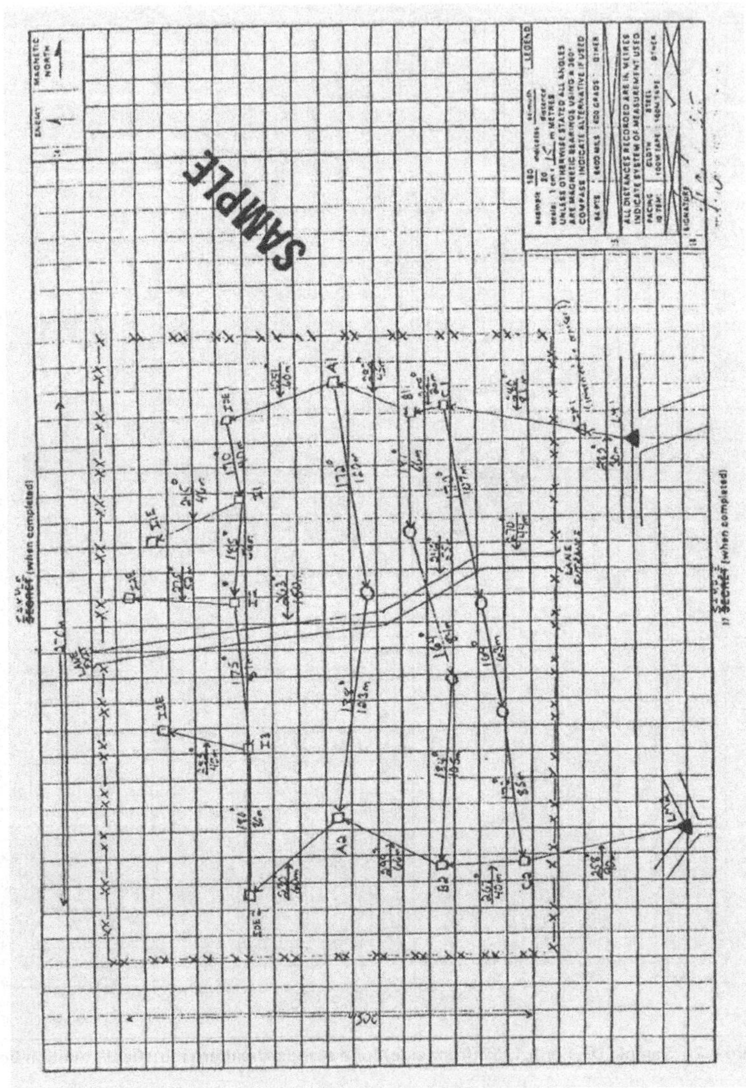

Figure 8-2b. Sample DA Form 1355 (inside) for a standard-pattern minefield/munition field

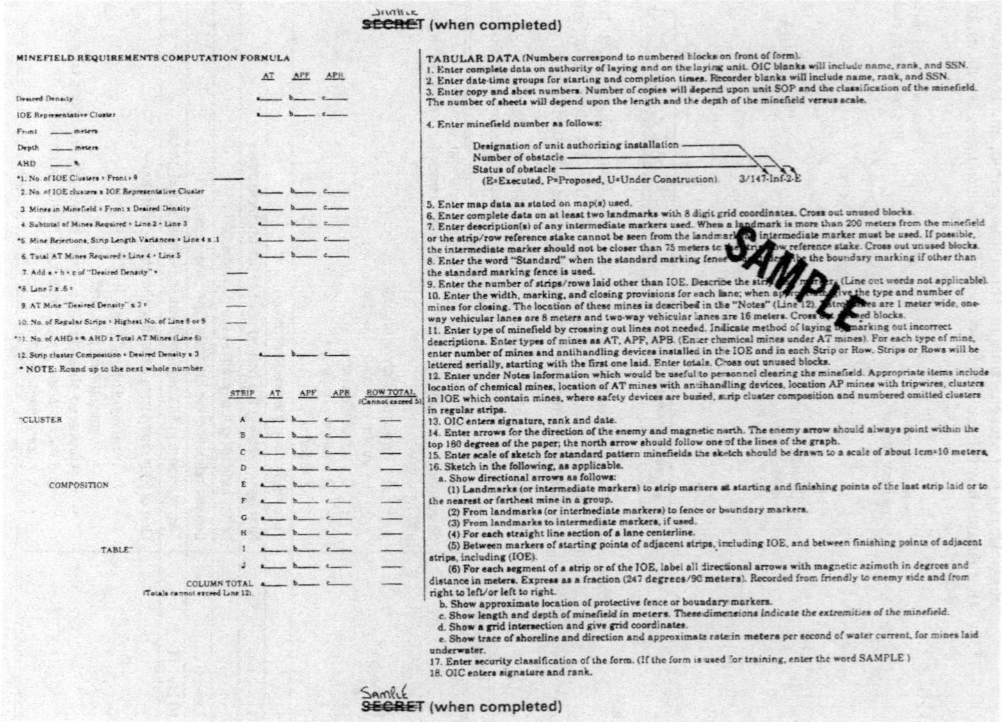

~~SECRET~~ Sample (when completed)

MINEFIELD REQUIREMENTS COMPUTATION FORMULA

		AT	APF	APB
Desired Density		a___	b___	c___
IOE Representative Cluster		a___	b___	c___
Front ___ meters				
Depth ___ meters				
AHD ___ %				
*1. No. of IOE Clusters = Front ÷ 9	___			
2. No. of IOE clusters x IOE Representative Cluster		a___	b___	c___
3. Mines in Minefield = Front x Desired Density		a___	b___	c___
4. Subtotal of Mines Required = Line 2 + Line 3		a___	b___	c___
*5. Mine Rejections, Strip Length Variances = Line 4 x .1		a___	b___	c___
6. Total AT Mines Required = Line 4 + Line 5		a___	b___	c___
7. Add a + b + c of "Desired Density" =	___			
*8. Line 7 x .6 =	___			
9. AT Mine "Desired Density" x 3 =	___			
10. No. of Regular Strips = Highest No. of Line 8 or 9	___			
*11. No. of AHD = % AHD x Total AT Mines (Line 6)	___			
12. Strip cluster Composition = Desired Density x 3		a___	b___	c___

* NOTE: Round up to the next whole number.

	STRIP	AT	APF	APB	ROW TOTAL (Cannot exceed 5)
"CLUSTER	A	a___	b___	c___	___
	B	a___	b___	c___	___
	C	a___	b___	c___	___
	D	a___	b___	c___	___
COMPOSITION	E	a___	b___	c___	___
	F	a___	b___	c___	___
	G	a___	b___	c___	___
	H	a___	b___	c___	___
TABLE"	I	a___	b___	c___	___
	J	a___	b___	c___	___
COLUMN TOTAL (Totals cannot exceed Line 12)		a___	b___	c___	___

TABULAR DATA (Numbers correspond to numbered blocks on front of form).
1. Enter complete data on authority of laying and on the laying unit. OIC blanks will include name, rank, and SSN.
2. Enter date-time groups for starting and completion times. Recorder blanks will include name, rank, and SSN.
3. Enter copy and sheet numbers. Number of copies will depend upon unit SOP and the classification of the minefield. The number of sheets will depend upon the length and the depth of the minefield versus scale.

4. Enter minefield number as follows:

Designation of unit authorizing installation
Number of obstacle
Status of obstacle
(E=Executed, P=Proposed, U=Under Construction). 3/147-Inf-2-E

5. Enter map data as stated on map(s) used.
6. Enter complete data on at least two landmarks with 8 digit grid coordinates. Cross out unused blocks.
7. Enter description(s) of any intermediate markers used. When a landmark is more than 200 meters from the minefield or the strip/row reference stake cannot be seen from the landmark, an intermediate marker must be used. If possible, the intermediate marker should not be closer than 75 meters to the strip/row reference stake. Cross out unused blocks.
8. Enter the word "Standard" when the standard marking fence is used. Describe the boundary marking if other than the standard marking fence is used.
9. Enter the number of strips/rows laid other than IOE. Describe the strip/row markers (Line out words not applicable).
10. Enter the width, marking, and closing provisions for each lane; when appropriate, give the type and number of mines for closing. The location of these mines is described in the "Notes" (Line 12). Footpath lanes are 1 meter wide, one-way vehicular lanes are 8 meters and two-way vehicular lanes are 16 meters. Cross out unused blocks.
11. Enter type of minefield by crossing out lines not needed. Indicate method of laying by marking out incorrect descriptions. Enter types of mines as AT, APF, APB. (Enter chemical mines under AT mines). For each type of mine, enter number of mines and antihandling devices installed in the IOE and in each Strip or Row. Strips or Rows will be lettered serially, starting with the first one laid. Enter totals. Cross out unused blocks.
12. Enter under Notes information which would be useful to personnel clearing the minefield. Appropriate items include location of chemical mines, location of AT mines with antihandling devices, location AP mines with tripwires, clusters in IOE which contain mines, where safety devices are buried, strip cluster composition and numbered omitted clusters in regular strips.
13. OIC enters signature, rank and date.
14. Enter arrows for the direction of the enemy and magnetic north. The enemy arrow should always point within the top 180 degrees of the paper; the north arrow should follow one of the lines of the graph.
15. Enter scale of sketch for standard pattern minefields the sketch should be drawn to a scale of about 1cm=10 meters.
16. Sketch in the following, as applicable.
a. Show directional arrows as follows:
(1) Landmarks (or intermediate markers) to strip markers at starting and finishing points of the last strip laid or to the nearest or farthest mine in a group.
(2) From landmarks (or intermediate markers) to fence or boundary markers.
(3) From landmarks to intermediate markers, if used.
(4) For each straight line section of a lane centerline.
(5) Between markers of starting points of adjacent strips, including IOE, and between finishing points of adjacent strips, including (IOE).
(6) For each segment of a strip or of the IOE, label all directional arrows with magnetic azimuth in degrees and distance in meters. Express as a fraction (247 degrees/90 meters). Recorded from friendly to enemy side and from right to left/or left to right.
b. Show approximate location of protective fence or boundary markers.
c. Show length and depth of minefield in meters. These dimensions indicate *the extremities of the minefield.*
d. Show a grid intersection and give grid coordinates.
e. Show trace of shoreline and direction and approximate rate in meters per second of water current, for mines laid underwater.
17. Enter security classification of the form. (If the form is used for training, enter the word SAMPLE.)
18. OIC enters signature and rank.

SAMPLE

~~SECRET~~ Sample (when completed)

Figure 8-2c. Sample DA Form 1355 (back side) for a standard-pattern minefield/munition field

Block 4. The minefield/munition field obstacle numbering system (Table 8-1) consists of 11 characters and 1 obstacle status symbol. It shows the type of obstacle, the belt and the zone in which it is located, and the headquarters that established the zone. Abbreviations for obstacle types are shown in Table 8-2.

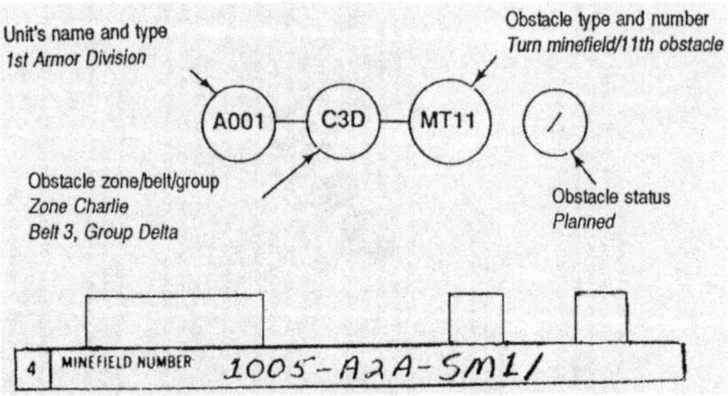

Example: Obstacle number I005-A2A-SM21/ indicates that the 5th Infantry Division planned the obstacle in Zone A. It is the 21st obstacle in Group A, Belt 2, and has not been executed. The obstacle is a MOPMS.

Table 8-1. Minefield/munition field obstacle numbering system

Character	Description
1 through 4	Alphanumeric description of the headquarters type and numerical designation that established the obstacle zone; character 1 designates the unit type with a letter: • A = armor division or brigade • I = infantry division or brigade • C = cavalry division • R = cavalry regiment • Z = corps
5	Letter indicating the obstacle zone
6	Number indicating the belt number in the obstacle zone
7	Letter indicating the group in the obstacle belt
8 and 9	Letters indicating the obstacle type (see Table 8-2)
10 and 11	Two numbers indicating the obstacle number in the group
12	One of four characters indicating the obstacle status— • / = planned obstacle • — = obstacle being prepared • + = prepared obstacle (This sign is for reserve demolition targets and may indicate a readiness state of safe or armed.) • X = completed obstacle

Table 8-2. Abbreviations for obstacle types

B—Bridge Demolition		**W—Wire Obstacle**	
BA	Abutment	WA	Double apron
BS	Span	WB	Booby-trapped
BC	Abutment and span	WF	Tanglefoot
M—Minefield/Munition Field		WG	General-purpose, barbed tape
MD	Disrupt	WN	Nonstandard
MT	Turn	WR	Road block
MF	Fix	WT	Triple standard
MB	Block	**R—Road Crater**	
MN	Nonstandard	RH	Hasty
MP	Protective	RD	Deliberate
MQ	Nuisance	RM	Mined
MS	Standard pattern	**M—Miscellaneous**	
S—Scatterable Minefield/Munition Field		AD	AT ditch
SA	ADAM	AR	Rubble by CEV gun
SB	Gator	AB	Rubble by blade
SR	RAAM	AT	Abatis
SF	ADAM and RAAM	AE	Rubble by explosives
SM	MOPMS	AM	Movable MOBA obstacle (car, bus)
SV	Volcano	AN	Expedient nonstandard
H—Hand-Emplaced Munitions		AL	Log crib, log obstacle
HH	Hornet	AP	Post obstacle (hedgehog, tetrahedron)
HS	SLAM	AH	Log hurdle

Block 5. Enter the map data as stated on the map(s) used.

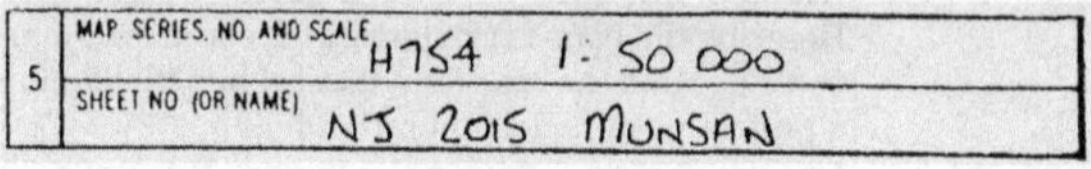

Block 6. Enter the grid coordinates and a description of at least two landmarks. If the landmarks are roads, trails, or routes, enter their name or number. This makes identification easier when removing the minefield/munition field. When recording minefields/munition fields, GPSs can only be used to determine the coordinates for minefield/munition field landmarks and RPs.

WARNING
Do not use GPSs to chart or record minefield/munition field perimeter coordinates or to determine safe routes through or around existing minefields/munition fields.

6	LANDMARKS		
	NO	COORDINATES	DESCRIPTION
	1	UT 34917312	U-SHAPED PICKET FLUSH WITH GROUND
	2		NEXT TO ROAD
	2	UT 34927323	U-SHAPED PICKET FLUSH WITH GROUND
	4		NEXT TO ROAD

Block 7. Enter the description of intermediate markers, if applicable. When a landmark is more than 200 meters from the minefield/munition field, or a strip or row reference stake cannot be seen from the landmark, an intermediate marker must be used. If possible, the intermediate marker is at least 75 meters from the strip or row reference stake.

7	INTERMEDIATE MARKERS	
	NO	DESCRIPTION
	1	3 U-SHAPED PICKETS 12" ABOVE GROUND
	2	
	3	
	4	

Block 8. Enter the word STANDARD when a standard marking fence is used. Describe the boundary marking if a standard marking fence is not used. (Use two sides and the rear for a tactical minefield/munition field; use four sides for a protective minefield/munition field.)

8	DESCRIPTION OF BOUNDARY FENCE OR MARKING STANDARD
	4 SIDES MINEFIELD FULLY ENCLOSED

Block 9. Enter the number of strips or rows laid. (Do not include the IOE.) Describe the strip or row markers. Cross out words that do not apply.

9	NO. OF STRIPS/~~ROWS~~: ...3.....	DESCRIPTION OF STRIP/~~ROW~~ MARKERS
	U-SHAPED PICKETS FLUSH WITH GROUND	

Block 10. Enter the width, the marking, and the provisions for each lane. When appropriate, give the types of mines and the number of each type of mine for closing. (The location of these mines is described in Block 12.)

10	LANES			
	NO	WIDTH	HOW MARKED	METHOD OF CLOSING
	1	8m	HEMMS SET	15xM15, 30xM16, 30xM14
	2			
	3			

Block 11. Enter the type of minefield/munition field by crossing out the lines that are not needed. Indicate the method of laying by crossing out incorrect descriptions. Enter the types of mines and the number of each type of mine. Also enter the number of AHDs installed in the IOE and in each row. Letter the strips or rows sequentially, starting with the first one laid. Enter totals.

11 MINES	Protective ~~TACTICAL~~ MINEFIELD ~~NUISANCE MINEFIELD~~ ~~PHONEY MINEFIELD~~	ANTITANK MINES (AT)								ANTIPERSONNEL MINES (AP)			
		TYPE M15	TYPE	TYPE	TYPE	TYPE	TYPE	TOTAL AT MINES	ANTI-LIFT DEV	TYPE M16	TYPE M14	TYPE	TOTAL AP MINES
		NO	NO	NO	NO	NO	NO	NO	NO	NO	NO	NO	NO
BURIED ~~AND~~ ~~SURFACE LAID~~	IOE	28						28		56	56		112
	A	72						72		144	144		288
	B	74						74		148	148		296
	C	76						76		152	152		304
	D												
IN STRIPS ~~IN ROWS~~ ~~W/O PATTERN~~	E												
	F												
	G												
	H												
	J												
TOTAL		250						250		500	500		1000

If the type of munition field is Hornet, enter "Hornet" above the word "tactical" and line out "nuisance minefield" and "phoney minefield." (See Figures 8-3a and 8-3b, pages 8-12 and 8-13.)

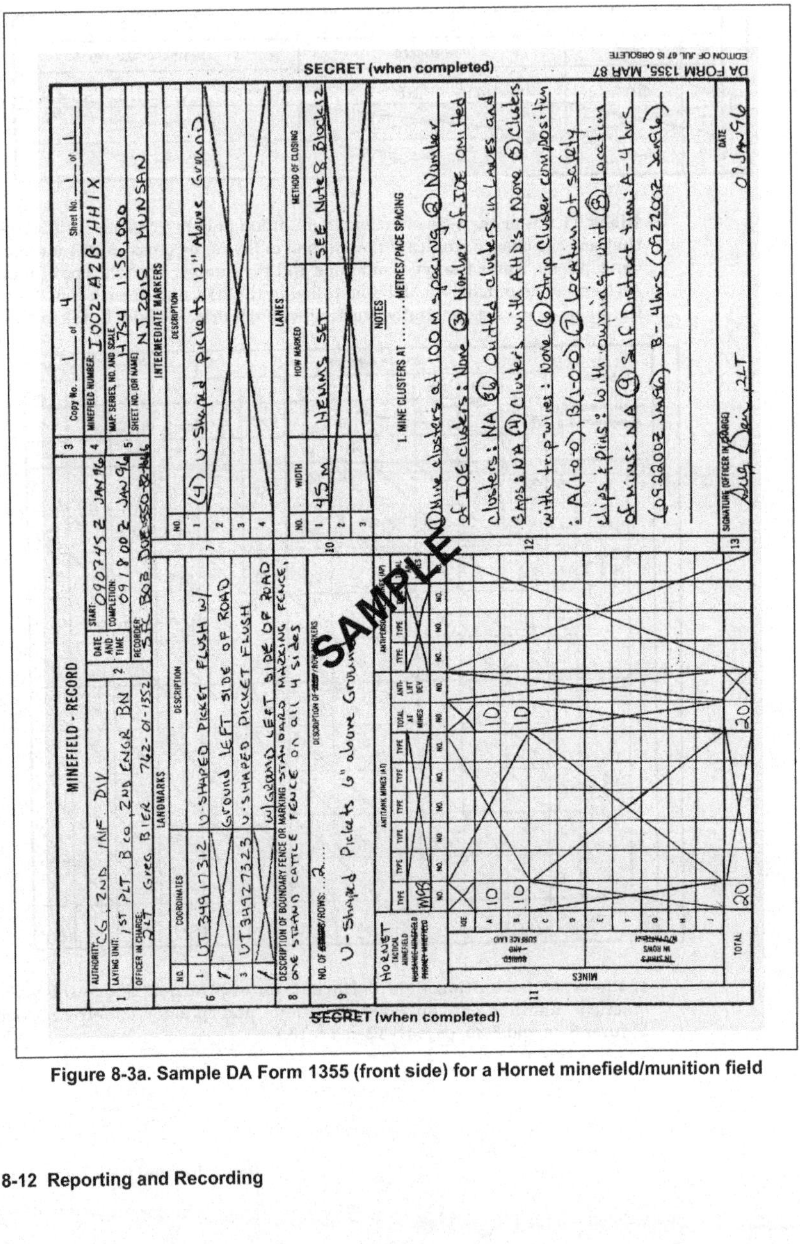

Figure 8-3a. Sample DA Form 1355 (front side) for a Hornet minefield/munition field

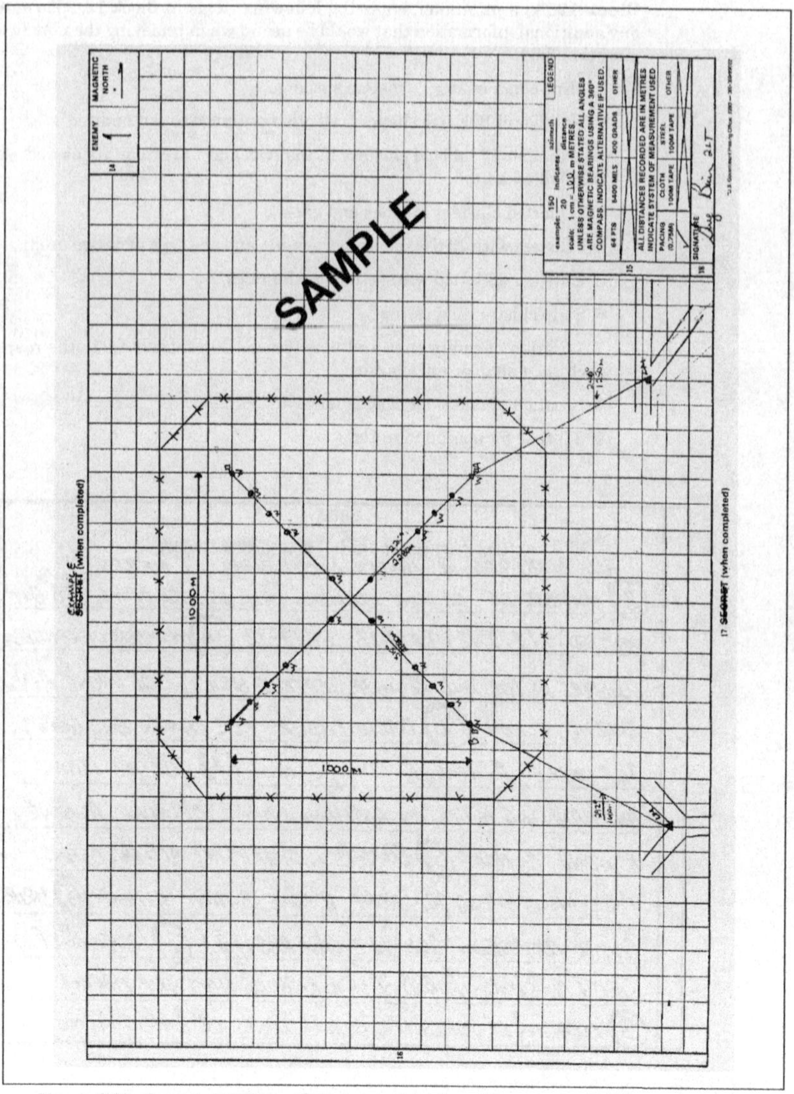

Figure 8-3b. Sample DA Form 1355 (back side) for a Hornet minefield/munition field

Block 12. As a minimum, enter the following items in Block 12. Also enter any additional information that would be useful when removing the minefield/ munition field.

1. Mine clusters at ____ meters spacing.

2. Number of IOE live clusters (all others numbered but omitted).

3a. Number of omitted clusters in the IOE and in regular strips and why they were omitted.

3b. Omitted clusters in lanes and gaps.

4. Clusters with AHDs (identify the mine and the type of device used).

5. Clusters with trip-wire-actuated AP mines.

6. Strip cluster composition.

7. Location of safety clips and pins (buried 30 centimeters to the rear of each start strip or row marker).

8. Location of mines for closing lanes and gaps.

9. SD time for special munitions.

NOTES

12

1. MINE CLUSTERS AT ..3.... METRES/[redacted] SPACING

ALL DISTANCES IN MINEFIELD MEASURED IN METERS

(2) NUMBER OF IOE LIVE CLUSTERS (ALL OTHERS NUMBERED BUT OMITTED); I1: 1-12; I2: 1-13, I3: 1-12 (3A) NUMBER OF OMITTED CLUSTERS IN IOE AND REGULAR LETTERED STRIPS; IOE: NONE, A: NONE B: NONE, C: NONE (3B) OMITTED CLUSTERS FOR LANES AND GAPS; IOE: NONE, A: NONE, B: NONE, C: NONE (4) CLUSTERS WITH ANTI HANDLING DEVICE ON ANTITANK MINE; IOE: NONE, A: NONE, B: NONE, C: NONE (5) CLUSTERS WITH TRIPWIRE ACTIVATED ANTI-PERSONNEL MINES; IOE: NONE, A: NONE, B: NONE, C: NONE (6) STRIP CLUSTER COMPOSITION; IOE: 1-2-2, STRIP ABC: 1-2-1 (7) LOCATION OF SAFETY CLIPS/PINS: BURIED 30cm TO REAR OF START STRIP MARKER

(8) LOCATION OF MINES FOR CLOSING LANES OR GAPS: NONE

Block 13. The emplacing unit OIC signs and dates the form.

13	SIGNATURE (OFFICER IN CHARGE)	DATE
	Roger P. Young 1LT	10 Jan 90

The front side of DA Form 1355 is now complete. Continue on the inside as follows:

Block 14. When completing the sketch, enter arrows for the direction of the enemy and the magnetic north. The enemy arrow will always point within the top 180 degrees of the form; the north arrow will follow one of the graph lines.

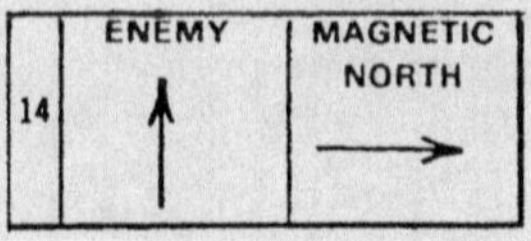

Block 15. If a compass was not available, enter what was used in the information block. Indicate the system of measurement and the scale used.

15	example: 150/20 indicates azimuth/distance scale: 1 cm = 15 m METRES. UNLESS OTHERWISE STATED ALL ANGLES ARE MAGNETIC BEARINGS USING A 360° COMPASS. INDICATE ALTERNATIVE IF USED.			LEGEND
	64 PTS	6400 MILS	400 GRADS	OTHER
	X	X	X	X
	ALL DISTANCES RECORDED ARE IN METRES INDICATE SYSTEM OF MEASUREMENT USED			
	PACING (0.75M)	CLOTH 100M TAPE	STEEL 100M TAPE	OTHER
	X	X	✓	X

Block 16. Enter the information shown below on the sketch:

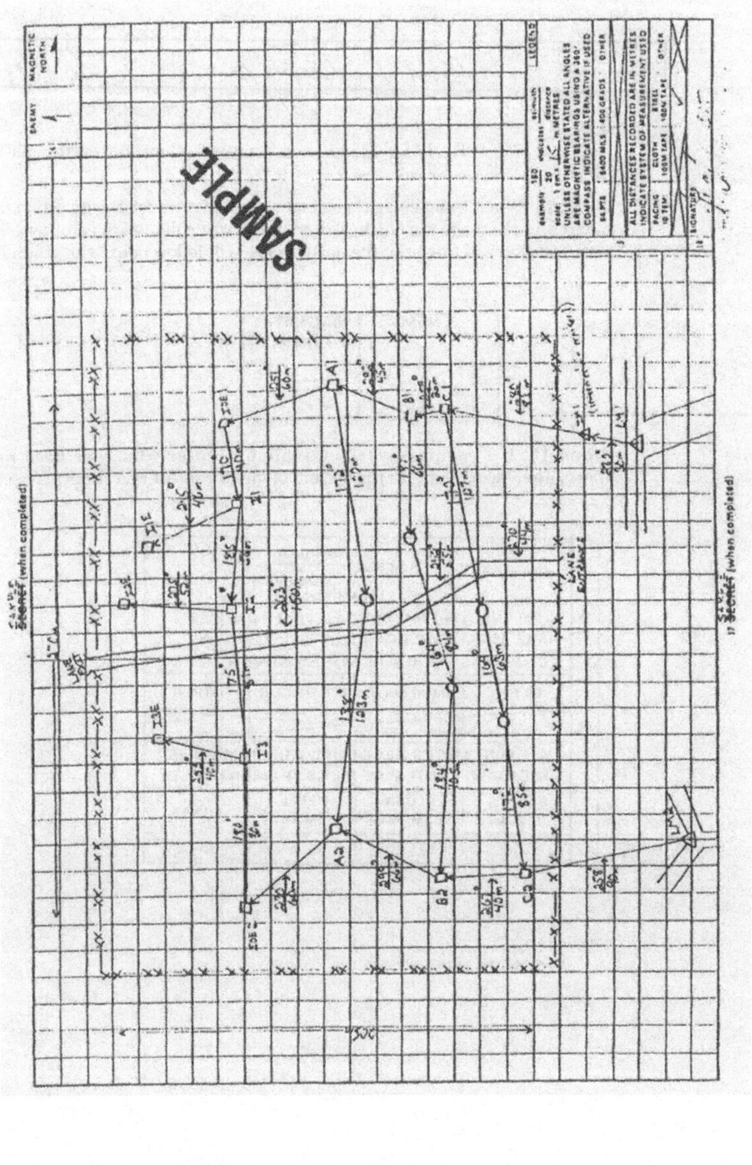

Block 17. Enter the security classification of the form. If the form was used for training, enter the word SAMPLE.

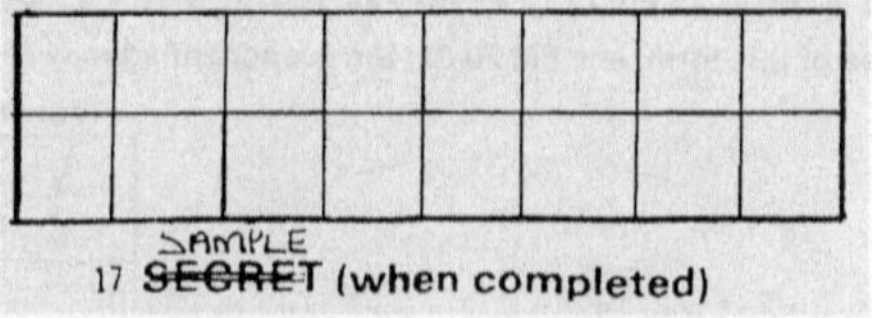
17 ~~SECRET~~ SAMPLE (when completed)

Block 18. The emplacing unit OIC signs the signature block.

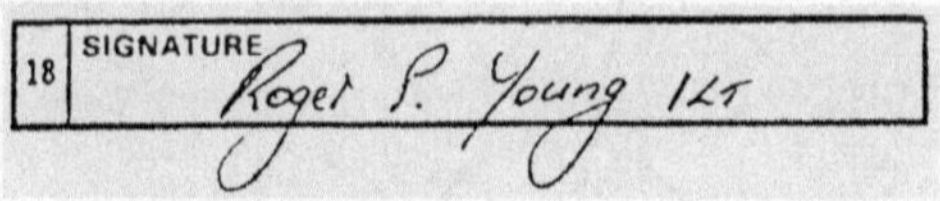
18 SIGNATURE Roger P. Young 1LT

Hasty Protective Row Minefield Record

Hasty protective row minefields/munition fields are recorded on DA Form 1355-1-R (Figure 8-4, page 8-18). DA Form 1355-1-R will be printed on 8½- by 11-inch paper. A copy for local reproduction is available in the back of this FM. DA Form 1355-1-R is also available on the Army Electronic Library (AEL) CD-ROM (EM 0001) and the USAPA web site (www.usapa.army.mil).

Use the following formula to determine the scale used on DA Form 1355-1-R:

Distance from RP to the farthest point in the minefield + 10 meters ÷ 4 = scale

Example: 90 meters + 10 meters = 100 meters ÷ 4 = 25 meters

The number 4 is a constant and represents the four concentric rings on DA Form 1355-1-R. Ten is added as a safety margin to ensure that the minefield/munition field sketch is entirely contained within the largest ring. The distance between rings is 2 centimeters; therefore, the scale used in this example is 2 centimeters = 25 meters.

HASTY PROTECTIVE ROW MINEFIELD RECORD

For use of this form, see FM 20-32; the proponent agency is TRADOC.

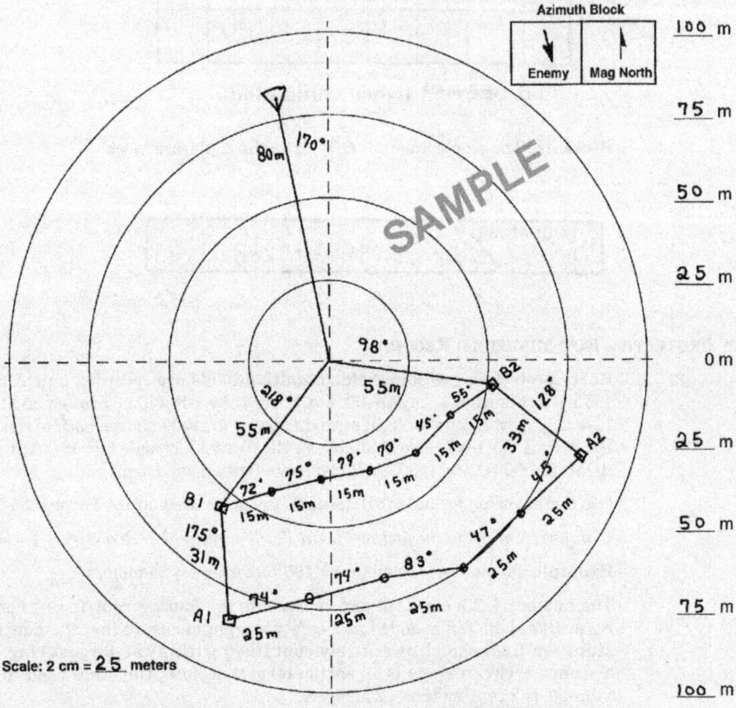

Scale: 2 cm = 25 meters

Tabular Block

Row	Type	Actuation	Mine Number
A	M-15	Pressure	1, 4
A	M-21	Tilt-Rod	2, 3
B	M-15	Pressure	1, 3, 5
B	M-15	Tilt-Rod	2, 4

Remarks Points: A1, A2, B1, B2 are marked with orange tent pegs flush to the ground.

Identification Block

Unit 2nd PLT. C. Co. 317TH Eng. BN.	
Reference Point Tree stump right side of the road w/white eng. tape at NA 14503505	
Remarks Landmark: Center of road junction at NA 13403409	
Map and Sheet No. Talbot #5568	
Name of OIC 2nd LT. Roy Bean 123-45-7611	
Signature Roy Bean	Time and Date 1400Z/1 JUL 01
Mines Removed	
Mines Transferred	

Figure 8-4. Sample DA Form 1355-1-R

The following step-by-step instructions are provided for completing DA Form 1355-1-R (see Figure 8-5):

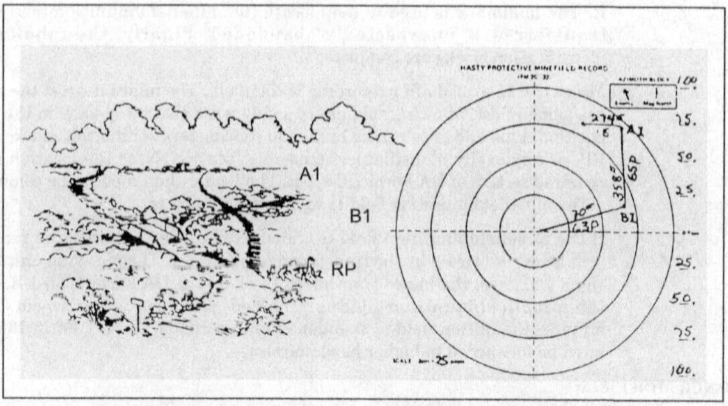

Figure 8-5. Hasty protective row minefield/munition field record

From the RP, the leader measures the magnetic azimuth, in degrees, to a specific point on the right side (facing the enemy) of the tentative minefield/ munition field. He determines the distance and records it in meters. He identifies the method of measurement in the remarks block of DA Form 1355-1-R (for example, distances were paced out and multiplied by 0.75 to convert the measurement to meters). This point, called B1 (if there are two rows), marks the beginning of the second row. A marker is placed at B1, and the leader records the azimuth and the distance.

The leader measures the azimuth and the distance from B1 to a second point on the right side of the minefield/munition field (facing the enemy). A marker is placed at this point and recorded as A1.

The leader measures the distance and the azimuth from A1 to the location of the first mine in that row and records the information.

The distance (or spacing) from the end row marker to the first mine is the spacing between all the mines in that row. After the location is recorded, the mine is emplaced, but it is not armed.

The leader measures the distance and the azimuth from the first mine to the second, and so on, until all the mines are emplaced and their locations are recorded. This procedure is repeated for the second row. As each mine is recorded, it is assigned a number to identify it in the minefield/munition field record. When the last mine location is recorded for a row, the distance and the azimuth are measured from that point to another arbitrary point, A2 or B2. A marker is placed here in the same manner as A1 and B1. Next, the distance and the azimuth from the RP to B2 and from B2 to A2 are measured and recorded.

When all the mines have been emplaced and recorded, the leader measures the distance and the azimuth between the RP and a permanent landmark that can be found on the map. He records the information on DA Form 1355-1-R. The landmark is used to help locate the minefield/munition field if it is transferred or unexpectedly abandoned. Finally, the tabular and identification blocks are completed.

Mines can be armed after recording is complete. The mines nearest the enemy are armed first, allowing soldiers to safely work their way back to the unit's position. Pins and clips can be buried 30 centimeters behind row markers, the RP, or any easily identifiable, accessible location. Note the location in the remarks section of DA Form 1355-1-R. The leader then reports the completion of the minefield/munition field to higher headquarters.

If the minefield/munition field is transferred to another unit, the receiving unit leader is briefed by the transferring unit leader. The receiving unit leader signs and dates the mines transferred block on the DA Form 1355-1-R. When the minefield/munition field is removed, the form is destroyed. If the minefield/munition field is abandoned unexpectedly, the DA Form 1355-1-R must be forwarded to higher headquarters.

Nuisance Minefield

It is unnecessary to record the precise location of individual nuisance mines that are laid in a pattern. Recording the position of a minefield/munition field that is laid in a pattern (either a standard pattern or one adopted for the situation) is easy and quick. However, the number of mines to be laid on site may make it impracticable or undesirable to lay mines in a pattern. In this case, they may be laid individually and their positions accurately recorded.

Figures 8-6a and 8-6b, pages 8-21 and 8-22 provide an example of a completed nuisance minefield/munition field record.

SCATTERABLE MINEFIELD/MUNITION FIELD REPORTING AND RECORDING

The speed and responsiveness of SCATMINE employment require accurate, uniform, and timely reports. All known information on scatterable minefields/ munition fields is simply and rapidly reported to all affected units. Although SCATMINES have SD capability, they are still recorded; and the information is disseminated to prevent casualties to friendly forces.

Since the locations of individual SCATMINEs are unknown, they cannot be recorded in detail. For most systems, a safety zone is calculated from one or more aim points. For example, a RAAM minefield/munition field is recorded based on the target location (the grid coordinates given to the firing battery). The size of the minefield/munition field depends on the number of rounds fired, the number of aim points, and the angle of fire. An artillery-delivered minefield/munition field is recorded by plotting it on a map, based on the aim point and the safety zone specified in the scatterable minefield/munition field report and record that was prepared by the emplacing unit. A ground-Volcano minefield/munition field can be recorded more accurately by plotting each of the minefield/munition field corner points rather than an aim point.

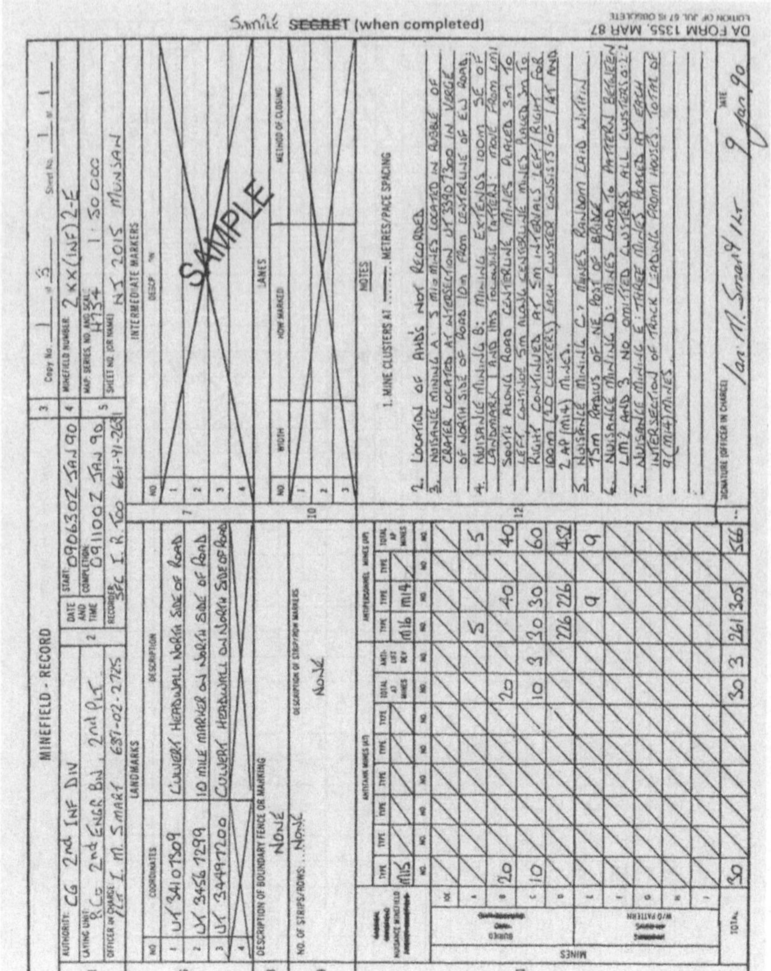

SAMPLE SECRET (when completed)

MINEFIELD - RECORD

AUTHORITY	CG 2nd INF DIV
LAYING UNIT	B Co 2nd ENGR BN, 2nd PLT
OFFICER IN CHARGE	1LT I. M. SMART 681-02-2725
DATE AND TIME — START	090630Z JAN 90
COMPLETION	091100Z JAN 90
RECORDER	SFC I. R. TOO 661-91-2631
Copy No.	1 of 3
Sheet No.	1 of 1
MINEFIELD NUMBER	2 KX(INF) 2-E
MAP: SERIES, NO AND SCALE	H734 1:50 000
SHEET NO. (OR NAME)	NJ 2015 MUNSAN

LANDMARKS

NO	COORDINATES	DESCRIPTION
1	UT 34107309	CULVERT HEADWALL NORTH SIDE OF ROAD
2	UT 34561299	10 MILE MARKER ON NORTH SIDE OF ROAD
3	UT 34497200	CULVERT HEADWALL ON NORTH SIDE OF ROAD
4		

INTERMEDIATE MARKERS (NO / DESCP) — crossed out, SAMPLE

DESCRIPTION OF BOUNDARY FENCE OR MARKING: NONE

NO. OF STRIPS/ROWS: NONE

DESCRIPTION OF STRIP/ROW MARKERS: NONE

LANES (NO / WIDTH / HOW MARKED / METHOD OF CLOSING) — crossed out

MINES — NUISANCE MINEFIELD

	AT M15	Total AT	Anti-lift dev	AP M16	AP M14	Total AP
				5		5
	20	20			40	40
	10	10	3	30	30	60
				226	226	452
					9	9
TOTAL	30	30	3	261	305	566

NOTES

1. MINE CLUSTERS AT .... METRES/PACE SPACING
2. LOCATION OF AHDS NOT RECORDED
3. NUISANCE MINING A: 5 M16 MINES LOCATED IN RUBBLE OF CRATER LOCATED AT INTERSECTION UT 33907300 IN VERGE OF NORTH SIDE OF ROAD 10m FROM CENTERLINE OF E-W ROAD.
4. NUISANCE MINING B: MINING EXTENDS 100m SE OF LANDMARK 1 AND HAS FOLLOWING PATTERN: MOVE FROM LM1 SOUTH ALONG ROAD CENTERLINE. MINES PLACED 3m TO LEFT, CONTINUE 5m ALONG CENTERLINE. MINES PLACED 3m TO RIGHT. CONTINUES AT 5m INTERVALS LEFT/RIGHT FOR 100m (20 CLUSTERS). EACH CLUSTER CONSISTS OF 1 AT AND 2 AP (M14) MINES.
5. NUISANCE MINING C: MINES RANDOM LAID WITHIN 15m RADIUS OF NE POST OF BRIDGE.
6. NUISANCE MINING D: MINES LAID IN PATTERN BETWEEN LM2 AND 3. NO OMITTED CLUSTERS. ALL CLUSTERS 0:2:2
7. NUISANCE MINING E: THREE MINES PLACED AT EACH INTERSECTION OF TANK LEADWAY FROM HOUSES. TOTAL OF 9 (M14) MINES.

SIGNATURE (OFFICER IN CHARGE): I. M. Smart 1LT — DATE: 9 Jan 90

DA FORM 1355, MAR 87 — EDITION OF JUL 67 IS OBSOLETE

SAMPLE SECRET (when completed)

Figure 8-6a. Sample DA Form 1355 (front side) for a nuisance minefield/munition field

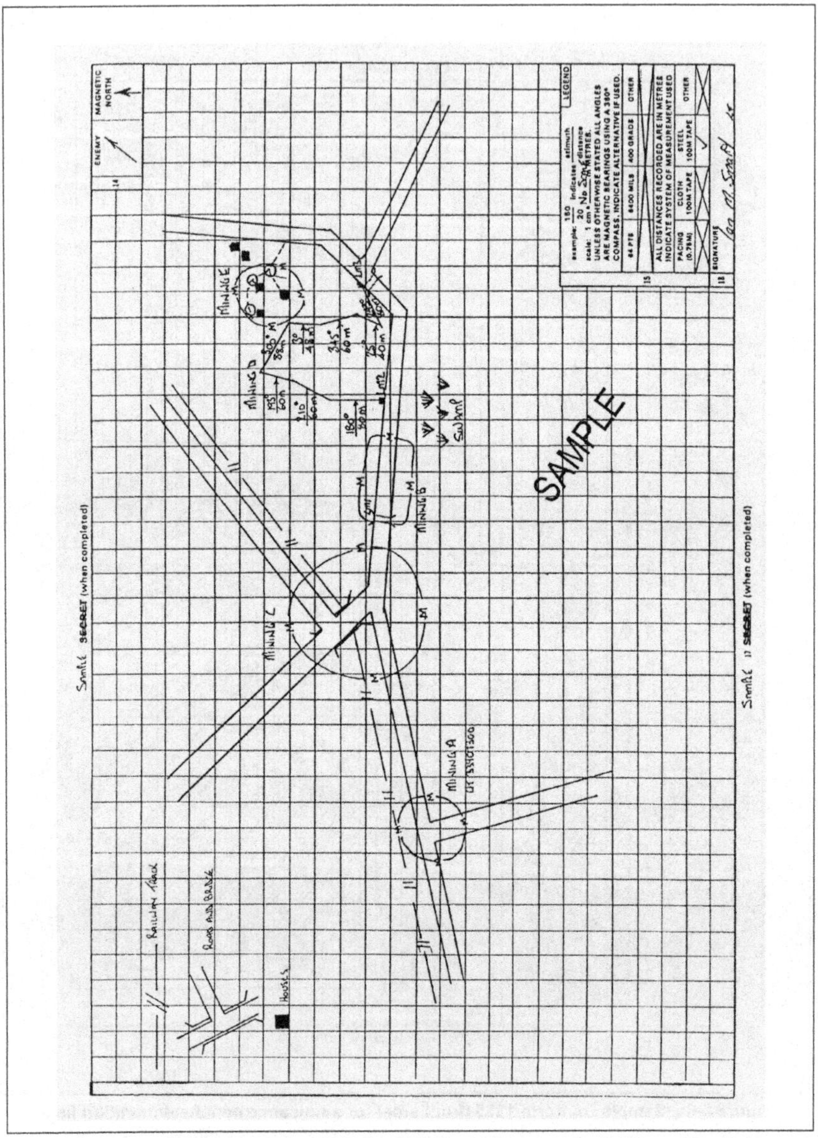

Figure 8-6b. Sample DA Form 1355 (inside) for a nuisance minefield/munition field

To facilitate reporting and recording of scatterable minefields/munition fields, a simple, uniform procedure is used. This procedure combines the report and the record into one document (Figure 8-7) that is applicable for all delivery systems.

Line No	Information Required	Data
1	Approving authority	Enter the approving authority, such as CDR 3AD.
2	Target/obstacle number	If the minefield/munition field is part of an obstacle plan, enter the obstacle number, such as 2XXX0157. This number represents II Corps, target number 157. If the minefield/munition field is not a part of an obstacle plan or does not have a number, then leave this line blank or enter NA.
3	Type of emplacing system	Enter the type system that emplaced the minefield/munition field, such as artillery or Volcano.
4	Type of mines	Enter AP for antipersonnel mines and AT for antitank mines. If both types of mines are used, enter AP/AT.
5	Life cycle	Enter the DTG the minefield/munition field was emplaced and the DTG the last mine SDs.
6-14	Aim point/corner points of minefield/munition field	If the system used to emplace the minefield/munition field uses a single aim point to deliver the mines, enter that aim point, such as MB 10102935. If the system has distinct corner points (Volcano), enter those corner points, such as MB 17954790, MB 18604860, MB 18504890, and MB 18054895.
15	Size of safety zone from aim point	If an aim point is given in Line 6, enter the size of the safety zone from that aim point. **Example:** Artillery emplaces a minefield/ munition field from aim point MB 10102935, and the safety zone is 1,000 x 1,000 m. Enter 500 m so that personnel plotting or receiving the information can plot the coordinates and go 500 m in each direction from the aim point to plot the safety zone.
16	Unit emplacing mines/ report number	Enter the unit emplacing mines and the report number, such as BCO 23 ENGR BN 4. Reports should be numbered consecutively. This would be the fourth minefield/munition field that B Company has emplaced.
17	Person completing report	Enter the person's name completing the report, such as SFC Jones.
18	DTG of report	Enter the DTG of the report, such as 160735ZOCT90.
19	Remarks	Include any other items the reporting unit may feel are important.

Figure 8-7. Scatterable minefield/munition field report and record work sheet

In addition to the scatterable minefield/munition field report and record, the SCATMINWARN (a sample is shown in Figure 8-8, page 8-24) notifies effected units that SCATMINEs will be emplaced. These two reports are the only reports used with scatterable mines.

A completed scatterable minefield/munition field report and record for an ADAM/RAAM artillery mission is shown in Figure 8-9, page 8-24. Note that on line 6, only one grid coordinate is given. It is the aim point used when the mission was fired. Also note that the 500-meter distance from the aim point (line 15) designates a safety zone that is 1,000 by 1,000 meters.

Line	Message
Alpha	Emplacing system
Bravo	AT (Yes or No)
Charlie	AP (Yes or No)
Delta	4 aim or corner points
Echo	Grid coordinates of aim points/corner points and size of the safety zone
Foxtrot	DTG of the life cycle

Figure 8-8. Sample SCATMINWARN

Line No	Information Required	Data
1	Approving authority	CDR 3AD
2	Target/obstacle number	2XXX0157
3	Type of emplacing system	Artillery
4	Type of mines	AT/AP
5	Life cycle	0816102-082020OCT90
6	Aim point/corner points of minefield/munition field	MB 10102935
7		
8		
9		
10		
11		
12		
13		
14		
15	Size safety zone from aim point	500 m
16	Unit emplacing mines/report number	2/48FA/2
17	Person completing report	SFC Hollins
18	DTG of report	061645ZOCT90
19	Remarks	NA

Figure 8-9. Scatterable minefield/munition field report and record for an ADAM/RAAM artillery mission

The SCATMINWARN provides affected units with the necessary warning to plan and execute their operations. The information is kept to a minimum to ensure rapid dissemination. The report may be sent orally, digitally, or hard copy. It is sent before or immediately after the mines have been emplaced. A completed SCATMINWARN for an artillery mission is shown in Figure 8-10.

Line	Message
Alpha	Artillery
Bravo	Yes
Charlie	Yes
Delta	One
Echo	MB 10102935 500 m
Foxtrot	081610Z-081920ZOCT90

Figure 8-10. Sample SCATMINWARN for an artillery mission

MINEFIELD/MUNITION FIELD OVERLAY SYMBOLS

The symbols contained in Figure 8-11, pages 8-26 through 8-30, are extracted from FM 101-5-1 and are provided for posting mine data on maps and overlays.

Description	Symbol
Minefield/Munition Fields	
Korea Only: AP mine	
AT mine	
AT mine with AHD	
Directional mine (arrow points in direction of main effect)	
Mine cluster	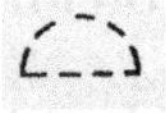
Mine, type unspecified	
Trip wire	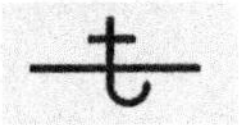
Control Measures	
Zone	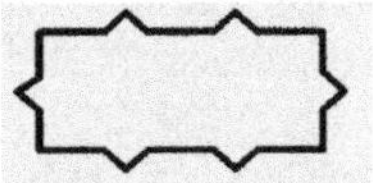
Belt	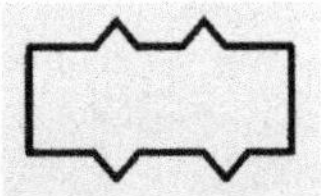
Restrictions	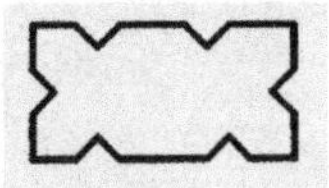

Figure 8-11. Minefield/munition field overlay symbols

Description	Symbol
Block effect	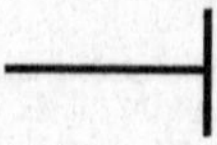
Turn effect	
Disrupt effect	
Fix effect	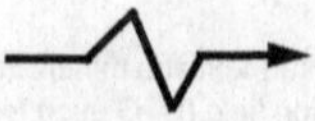
Conventional	
A planned minefield/munition field consisting of unspecified mines	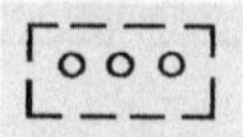
A completed minefield/munition field consisting of unspecified mines	
Scatterable minefield/munition field (DTGs used for SD times)	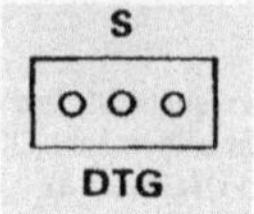
Conventional AP minefield/ munition field reinforced with SCATMINEs	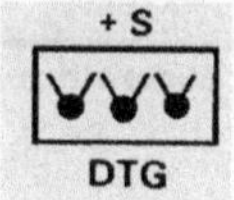
Tactical AP row minefield/ munition field (outline drawn to scale)	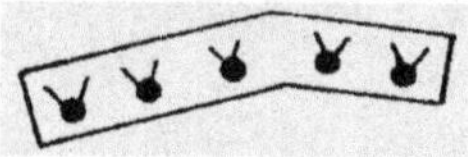

Figure 8-11. Minefield/munition field overlay symbols (continued)

Description	Symbol
Tactical minefield/munition field of scatterable AT mines, effective until 101200Z	
Completed AT minefield/munition field (drawn away from the location and connected by a vector)	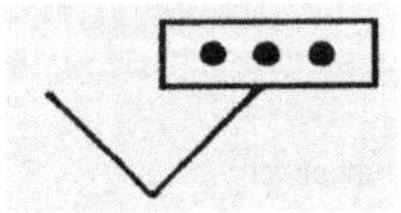
Executed Volcano minefield/ munition field (DTG used for SD time)	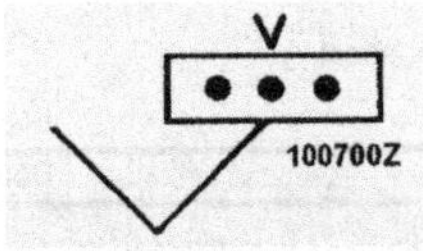
Lane in conventionally laid AT minefield/munition field	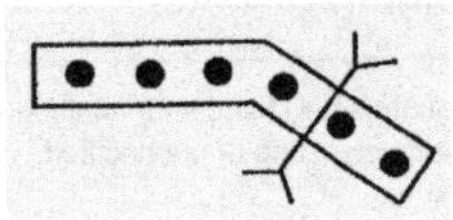
Gap in conventionally laid AT minefield/munition field (DTG opened to DTG closed)	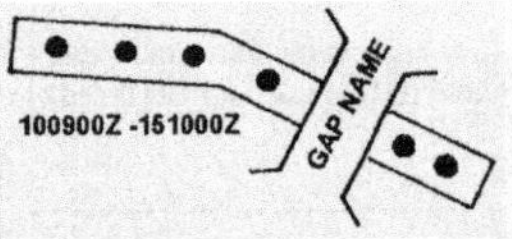

Figure 8-11. Minefield/munition field overlay symbols (continued)

Description	Symbol
AT ditch reinforced with AT mines	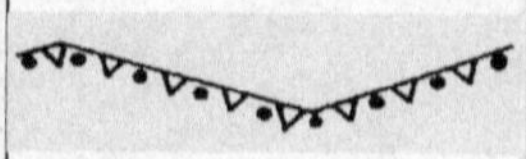
UXO	
UXO area	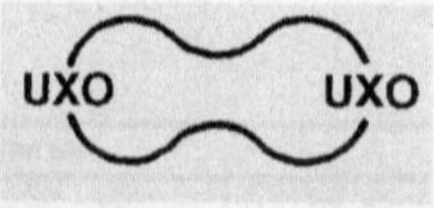
Nuisance	
Nuisance minefield/munition field	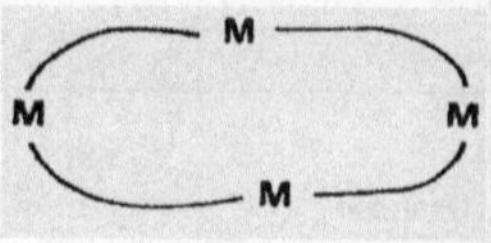
Demolished crossroads with nuisance mines	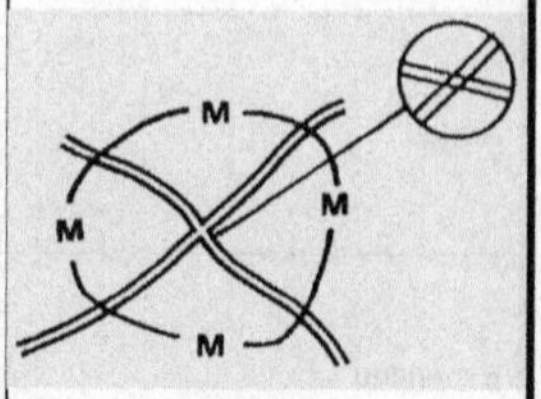
Phony	
Phony minefield/munition field	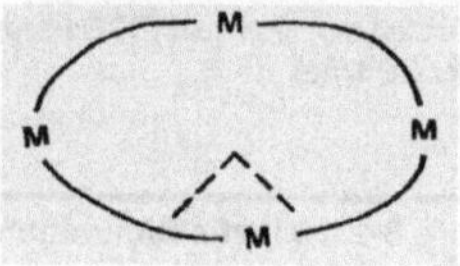

Figure 8-11. Minefield/munition field overlay symbols (continued)

Description	Symbol
Phony minefield/munition field, fenced	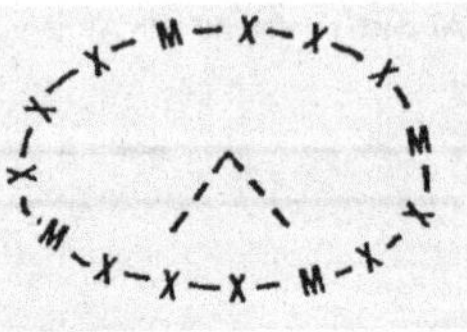
Hornet Symbology	
Planned	
Unarmed	
Armed	
Expended	
Armed munition field (DTG used for SD time)	

Figure 8-11. Minefield/munition field overlay symbols (continued)

PART TWO

Counteroperations

This part of the manual provides overall guidance for conducting counteroperations by US forces. The types of breaching and clearing operations conducted, the tasks performed, and the equipment required are described in detail. Responsibilities and planning considerations are outlined for each operation.

Chapter 9

Countermine Operations

Countermine operations are undertaken to breach or clear a minefield. All the tasks fall under breaching or clearing operations and include detecting, reporting, reducing, proofing, and marking.

DEFINITIONS

OBSTACLE

The term *obstacle* is used often in this chapter because the same breaching and clearing operations are used for minefields and other obstacles. For the purpose of this manual, breaching and clearing tactics, techniques, and procedures (TTP) focus solely on minefields.

REDUCTION

Reduction is the act or actions taken against an obstacle that diminishes its original effect. For example, creating a lane in a minefield would yield a reduction of the minefield obstacle.

BREACHING

Breaching is the employment of TTP to project combat power to the far side of an obstacle. It is a synchronized combined arms operation that is under the control of the maneuver commander.

AREA CLEARANCE

Area clearance is the total elimination or neutralization of an obstacle or portions of an obstacle. Clearing operations are not conducted under fire. They are usually performed by follow-on engineer forces after a breaching operation or anytime in a friendly AO where an obstacle is a hazard or hinders movement.

ROUTE CLEARANCE

Route clearance is the removal of mines along preexisting roads and trails.

MINE NEUTRALIZATION

Mine neutralization occurs when a mine is rendered incapable of firing on passage of a target. The mine may still be dangerous to handle.

PROOFING

Proofing is done by passing a mine roller or other mine-resistant vehicle through a lane as the lead vehicle. It verifies that a lane is free of mines.

DEMINING

Demining is the complete removal of all mines and UXO within a geopolitical boundary after hostilities cease.

BREACHING OPERATIONS

Breaching is a synchronized combined arms operation that is under the control of the maneuver commander. FM 3-34.2 provides combined arms commanders and staffs with doctrine TTP that are needed to successfully overcome obstacles. Breaching operations make maneuver possible in the face of enemy obstacle efforts. Since obstacles may be encountered anywhere, maneuver forces integrate breaching operations into all movement plans.

When possible, enemy minefields are bypassed to maintain the momentum and conserve critical countermobility assets. However, when making the decision to bypass rather than breach, consider the likelihood of friendly units being channelized into kill zones. Bypassing is done by maneuvering around a minefield or, if aviation assets are available, moving over the minefield. When maneuvering around an obstacle, attempt to locate a portion of the force in overwatch positions to cover the bypass of the main element. Even when the decision is made to conduct a breach, scouts should continue to reconnoiter for bypass routes.

The first step in understanding breaching operations is to know the obstacle breaching theory. Knowing the theory behind breaching operations equips the engineer and the maneuver commander with fundamentals that are needed to integrate breach into the tactical planning, preparation, and execution of an operation.

Successful breaching operations are characterized by the application of the following tenets of breaching:

INTELLIGENCE

In any operation where enemy obstacles interfere with friendly maneuver, obstacle intelligence (OBSTINTEL) becomes a priority intelligence requirement (PIR). Finding enemy obstacles or seeing enemy obstacle activity validates and refines the S2's picture of the battlefield. OBSTINTEL helps determine enemy intentions, plans, and strength. The force engineer is the unit's expert on enemy countermobility, and he assists the S2 in templating enemy obstacles and analyzing OBSTINTEL.

When collecting OBSTINTEL, reconnaissance is a combined arms activity that includes engineers. An engineer squad moves with scouts or the patrol and conducts dismounted reconnaissance of templated or discovered obstacles. Additional information on reconnaissance can be found in FM 5-170. Reconnaissance teams gather the following OBSTINTEL information from the reconnaissance:

- Minefield location. Plot the perimeter location on a large-scale map and refer to recognizable landmarks.

- Perimeter description. Describe how the perimeter is fenced. If it is unfenced, describe how it is marked. If it is unmarked, show how it was recognized.
- Nuisance mines. If you discover a nuisance mine forward of the minefield's outer edge, remember, there may be others. Assembly areas might also be mined.
- Types of mines. Indicate whether mines are AT or AP or have unknown fuses (self-neutralized or SD). If possible, recover specimens of unknown or new mines and note the details.
- Details of any other devices. Describe booby traps, trip wires, and flares.
- Laying method. Indicate whether mines are buried or surface-laid.
- Density and pattern. Include the mine spacing and the number of mine rows; estimate the mine density based on this information.
- Minefield depth. Provide the distance between strips or rows and describe markers.
- Safe lanes and gaps. Plot the location of suspected safe lanes and gaps and describe their markings.
- Ground conditions. Include information on general ground conditions.
- Other obstacles. Plot the location and the construction of other obstacles.
- Enemy defenses. Describe the enemy's location and size. Include the location of enemy direct-fire weapons.

NOTE: Engineers engaged in reconnaissance for OBSTINTEL should rarely, if ever, be used to reduce obstacles during the reconnaissance.

Each reconnaissance team commander submits a detailed OBSTINTEL report to the next higher headquarters when the reconnaissance is complete. A sample OBSTINTEL report is shown in Figure 9-1.

Letter Designation	Explanation
A	Map sheet(s).
B	Date and time the information was collected.
C	Type of minefield (AT, AP, or mixed).
D	Grid references of minefield extremities, if known.
E	Depth of minefield.
F	Estimated time required to clear the minefield.
G	Estimated material and equipment required to clear the minefield.
H	Routes for bypassing the minefield, if any.
I-Y	Grid references of lanes (entry and exit) and width of lanes, in meters.
Z	Additional information such as types of mines used, unknown mines, or types of booby traps.

Figure 9-1. Sample OBSTINTEL report

FUNDAMENTALS

Suppress, obscure, secure, and reduce (SOSR) are the breaching fundamentals that must be applied to ensure success when breaching against a defending enemy. This TTP will always apply but may vary based on specific situations.

Suppress

Suppression is the focus of all available fires on enemy personnel, weapons, and equipment to prevent effective fires on friendly forces. The purpose of suppression is to protect forces who are reducing the obstacle and maneuvering through it and to soften the enemy's initial foothold.

Obscure

Obscuration hampers enemy observation and target acquisition, and it conceals friendly activities and movement. Obscuration smoke deployed on or near the enemy position minimizes the enemy's vision.

Secure

The force secures the breaching site to prevent the enemy from interfering with reduction and to prevent the enemy's passage through lanes created during reduction. In general, enemy tactical obstacles are secured by fires, and protective obstacles are secured by force.

Reduce

Reduction means creating lanes through or over the obstacle to allow passage of the attacking force. The lanes must be sufficient to allow the force to cross and accomplish the mission. (Table 9-1 provides information on lane widths.) The unit reducing the minefield marks the minefield and lane locations and reports their conditions to higher headquarters.

Table 9-1. Lane widths

Lane	Width	Remarks
Assault footpath	1 meter	Allows passage of dismounted troops so that they may continue an attack or secure the far side of a minefield while vehicle lanes are being breached
Initial lane	4.5 meters	Allows passage of breaching and assault forces (initial lanes are widened and marked as soon as the tactical situation allows)
Intermediate lane	4.5 meters	Upgrades sign posting and further defines exit markers
Full lane	10 meters	Allows passage of two-way traffic

ORGANIZATION

The commander organizes the force to accomplish SOSR breaching fundamentals quickly and effectively. This requires him to organize support, breach, and assault forces with the necessary assets to accomplish their missions.

Support Force

The support force is primarily responsible for eliminating the enemy's ability to interfere with the breaching operation. It must isolate the battlefield with fires and suppress enemy fires that are covering the breach location. This involves massive direct and indirect fire to destroy enemy vehicles and personnel who are able to bring fire on the breach force.

Breach Force

The breach force creates lanes that enable the assault force to pass through the obstacle and continue the attack. It is also responsible for marking lanes (length and entry points) to speed passage of assault and follow-on forces. The breach force is a combined arms force. It includes engineers, breaching assets, and enough maneuver force to provide local security. The breach force must be able to deploy and begin reducing the obstacle as soon as enemy fires have been suppressed. It must be capable of creating a minimum of one lane for each assaulting company or two lanes for an assaulting TF. At a minimum, the lanes must be marked and their locations and conditions reported to higher headquarters and follow-on units as prescribed in the unit's SOP. The commander should expect a 50 percent loss of mobility assets in close combat.

Assault Force

The assault force destroys or dislodges the enemy on the far side of the obstacle. It secures the far side by physical occupation in most breaching operations. The assault force may be tasked to assist the support force with suppression while the breach force reduces the obstacle. The assault force must be sufficient in size to seize objectives that eliminate fires on the breaching site.

MASS

Breaching is conducted by rapidly applying concentrated force at a designated point to crack the obstacle and rupture the defense. Massed combat power is directed against an enemy weakness. Smoke and terrain help isolate the enemy force that is under attack. The commander also masses engineers and breaching equipment to reduce the obstacle. The breach force is organized and equipped to use several different reduction techniques in case the primary technique fails (a key vehicle is destroyed or casualties render dismounted engineers ineffective). Additional reduction assets are present to handle the unexpected (50 percent over the requirement are normally positioned with the breach force).

SYNCHRONIZATION

Breaching operations require precise synchronization of the SOSR breaching fundamentals by support, breach, and assault forces. Failure to synchronize effective suppression and obscuration with reduction and assault can result in rapid, devastating losses of friendly troops in the obstacle or the enemy's fire sack. The commander cannot adequately synchronize his force's application of combat power in the short time available to him when he encounters an obstacle. The number of decisions that he must make while under fire in an unclear situation will rapidly overwhelm him. Even with a force trained to execute a combined arms breach, synchronizing all the necessary tasks remains a complex, difficult process. The commander uses the following principles to ensure synchronization through proper planning and force preparation:

- Detailed reverse planning.
- Clear subunit instructions.
- Effective C^2.
- A well-rehearsed force.

CLEARING OPERATIONS

Clearing is the total elimination or neutralization of mines from an area. It is not usually conducted under enemy fire, but it can be conducted by engineers during war or after hostilities as part of nation assistance.

A limited clearing operation can be conducted by follow-on engineers after the breaching force has reduced the minefield and secured the area. Engineers initially improve existing breach lanes by widening and marking them, and they also clear and mark new lanes through the minefield. The clearing operation supports the continued passage of forces.

A clearing operation is also conducted to eliminate all the mines in a minefield (previously identified, reported, and marked in a friendly AO) that hinders mobility or is a hazard to friendly forces.

UPGRADE OF BREACH LANES

Lane clearance is more deliberate than lane breaching and normally takes longer. Follow-on engineers upgrade breach lanes to improve existing lanes through minefields and to create new lanes. This clearing operation is intended to further reduce the minefield so that follow-on units can pass through it quickly.

The breach force that initially reduced the obstacle and marked the lanes turns over the lanes to follow-on engineers. Follow-on engineers can expect lane widths of 4.5 meters. The total number of lanes depends on the size of the lead assault force. Two to four assault lanes are normal if the lead unit was brigade-size.

If forces continue to pass through existing lanes while further reduction and clearance is conducted, follow-on engineers first begin reducing new lanes. At a minimum, two lanes are required for an assaulting TF and one lane is required for an assaulting company/team.

A limited amount of mechanical breaching assets is available for clearing operations. Follow-on engineers will probably not have tank-mounted mine-clearing blades (MCBs) or mine-clearing rollers (MCRs). The main mechanical clearing asset is an armored dozer with a mine rake. Mine-clearing line charges (MICLICs) are used if available. Engineers conducting clearing operations—

- Ensure that lanes are a minimum of 100 meters apart.
- Reduce additional lanes by using the equipment and techniques outlined in Chapter 10.
- Widen lanes to 10 meters to allow two-way traffic.
- Mark breach lanes by using the original marking system or the division SOP. (Marking procedures are outlined in Chapter 10.)
- Emplace entrance, exit, and left and right lane markers to provide day and night capability.

Traffic control is critical during lane reduction and when shifting lanes to improve existing lanes. Engineers conducting reduction and clearance may also provide guides at the lanes. Control procedures are outlined in FM 90-13-1.

To eliminate the danger of forces entering the minefield adjacent to lanes, the minefield is marked with fencing (barbwire or concertina) and mine markers.

Marking is emplaced across the front, on both sides, between lanes, and to the left and right of the crossing site as far out as practical.

Engineers may also help remove damaged vehicles from minefield lanes. Recovery vehicles should be available near lanes for this purpose.

AREA CLEARANCE

Clearing operations are done when engineers receive a mission to clear an area of mines or to clear a specific minefield in a friendly AO. The minefield was reported and may already be marked on all sides. The worst case would be if the minefield was reported but not marked and its limits were unknown. The engineer unit receiving the mission bases plans on available information and prepares equipment based on the estimate. Detailed techniques and procedures for area and route clearance operations are outlined in Chapter 11.

Actions at the minefield begin with a thorough reconnaissance to identify the minefield limits and the types of mines. This is a time-consuming process that is hazardous to shortcut. Identified limits are marked with an expedient system of single-strand barbwire or concertina. In this situation, since all mines must be destroyed, the unit takes a systematic approach to clearing mines. The procedure depends on the types of mines and whether the mines are buried or surface-laid.

If mines are magnetic- or seismic-fused, mechanical assets are used. Pressure mines can be destroyed by using hand-emplaced explosives. When a manual procedure is used, eliminate trip wires on AP mines with grapnel hooks before moving forward to detect mines.

Using the manual procedure, engineers visually detect mines or detect them with mine detectors and probes. They also mark mines for destruction by explosives. Chapter 11 contains information on minesweeping procedures.

After the mines are destroyed, engineers proof used lanes and routes to ensure that all the mines were eliminated. This is done by using a mine roller or another blast-resistant device. Proofing is discussed further in Chapter 10.

DEMINING

Demining is the complete removal of all mines and UXO to safeguard the civilian population within a geopolitical boundary after hostilities cease. It is an extremely manpower- and time-intensive operation and is sometimes contracted. Although not a formal Army mission or function, SOFs may provide special expertise in training demining organizations, acting as advisors, and taking the lead in providing clearance equipment or techniques that can be useful in demining operations. Demining TTP are outlined in TC 31-34.

Chapter 10

Minefield Reduction

Reduction is the physical creation of a lane through a minefield. It is a fundamental of breaching operations as discussed in Chapter 9 and in FM 3-34.2. A number of tasks (detecting, reporting, reducing, proofing, and marking) directly support or are included in minefield reduction.

DETECTING

Detection is the actual confirmation and location of mines. It may be accomplished through reconnaissance, or it may be unintentional (such as a vehicle running into a mine). Mine detection is used in conjunction with intelligence-gathering operations, minefield bypass reconnaissance, and breaching and clearing operations. There are four types of detection methods—visual, physical (probing), electronic, and mechanical.

VISUAL

Visual detection is part of all combat operations. Personnel visually inspect the terrain for the following minefield indicators:

- Trip wires.
- Signs of road repair (such as new fill or paving, road patches, ditching, culvert work).
- Signs placed on trees, posts, or stakes. Threat forces mark their minefields to protect their own forces.
- Dead animals.
- Damaged vehicles.
- Disturbances in previous tire tracks or tracks that stop unexplainably.
- Wires leading away from the side of the road. They may be firing wires that are partially buried.
- Odd features in the ground or patterns that are not present in nature. Plant growth may wilt or change color, rain may wash away some of the cover, the cover may sink or crack around the edges, or the material covering the mines may look like mounds of dirt.
- Civilians. They may know where mines or booby traps are located in the residential area. Civilians staying away from certain places or out of certain buildings are good indications of the presence of mines or booby traps. Question civilians to determine the exact locations.
- Pieces of wood or other debris on a road. They may be indicative of pressure or pressure-release FDs. These devices may be on the surface or partially buried.

- Patterns of objects that could be used as a sighting line. The enemy can use mines that are fired by command, so road shoulders and areas close to the objects should be searched.

PHYSICAL

Physical detection (probing) is very time-consuming and is used primarily for clearing operations, self-extraction, and covert breaching operations. Detection of mines by visual or electronic methods should be confirmed by probing. Use the following procedures and techniques when probing for mines:

- Roll up your sleeves and remove your jewelry to increase sensitivity. Wear a Kevlar helmet, with the chin strap buckled, and a protective fragmentation vest.
- Stay close to the ground and move in a prone position to reduce the effects of an accidental blast. When moving into a prone position—
  - Squat down without touching your knees to the ground.
  - Scan forward up to 2 meters and to the sides up to 3 meters for mine indicators.
  - Probe the area around your feet and as far forward as possible.
  - Kneel on the ground after the area is found to be clear, and continue probing forward until you are in a prone position.
- Use sight and touch to detect trip wires, fuses, and pressure prongs.
- Use a slender, nonmetallic object as a probe.
- Probe every 5 centimeters across a 1-meter front.
- Gently push the probe into the ground at an angle that is less than 45 degrees.

DANGER

Use extreme caution when probing. If the probe is pushed straight down, its tip may detonate a pressure fuse.

- Apply just enough pressure on the probe to sink it slowly into the ground.
- If the probe encounters resistance and does not go into the ground freely, carefully pick the soil away with the tip of the probe and remove the loose dirt by hand. Care must be taken to prevent functioning the mine.
- When you touch a solid object, stop probing and use two fingers from each hand to carefully remove the surrounding soil and identify the object.
- If the object is a mine, remove enough soil to show the mine type and mark its location. Do not attempt to remove or disarm the mine. Use explosives to destroy detected mines in place, or use a grappling hook and rope to cause mines to self-detonate. Do not use metal grappling hooks on magnetic-fused mines.

Probing is extremely stressful and tedious. The senior leader must set a limit to the time a prober can actually probe in the minefield. To determine a reasonable time, the leader must consider METT-TC factors, weather conditions, the threat level, the unit's stress level, and the prober's fatigue level and state of mind. As a rule, 20 to 30 minutes is the maximum amount of time that an individual can probe effectively.

ELECTRONIC

Electronic detection is effective for locating mines, but this method is time-consuming and exposes personnel to enemy fire. In addition, the suspected mines must be confirmed by probing.

AN/PSS-12 Mine Detector

The AN/PSS-12 mine detector (Figure 10-1) can only detect metal, but most mines have metal components in their design. The detector can locate and identify plastic or wooden mines by a slight metallic signature. Employment and operation procedures for the AN/PSS-12 are discussed in Appendix F, and technical data is available in TM 5-6665-298-10. The detector is hand-held and identifies suspected mines by an audio signal in the headphones.

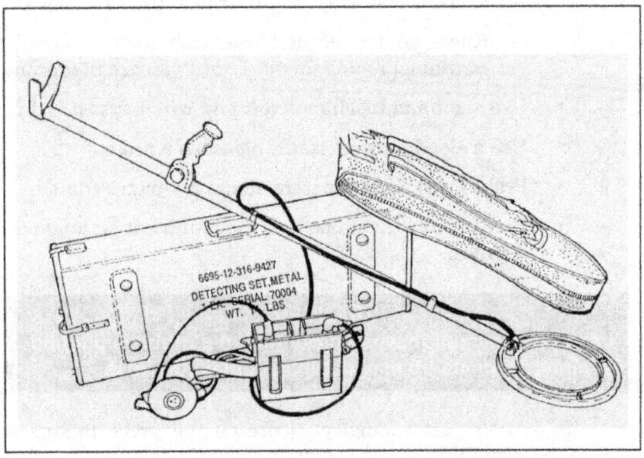

Figure 10-1. AN/PSS-12 mine detector

As in probing, consideration must be taken for the maximum amount time an individual can operate the detector. The leader considers METT-TC factors, weather conditions, the threat level, the unit's stress level, and the individual's fatigue level and state of mind. As a rule, 20 to 30 minutes is the maximum amount of time an individual can use the detector effectively.

Airborne Standoff Minefield Detection System

The Airborne Standoff Minefield Detection System (ASTAMIDS) (Figure 10-2, page 10-4) provides US forces with the capability to detect minefields rapidly. Environmental conditions must be favorable for aircraft and ASTAMIDS operations. ASTAMIDS can be mounted on a UH-60 Blackhawk helicopter, an unmanned aerial vehicle (UAV), or a fixed-wing aircraft. The system detects and classifies thermal and other anomalies as suspected minefields along routes or in areas of interest. ASTAMIDS can be used to protect advancing forces and can operate in concert with air and ground units in reconnaissance missions.

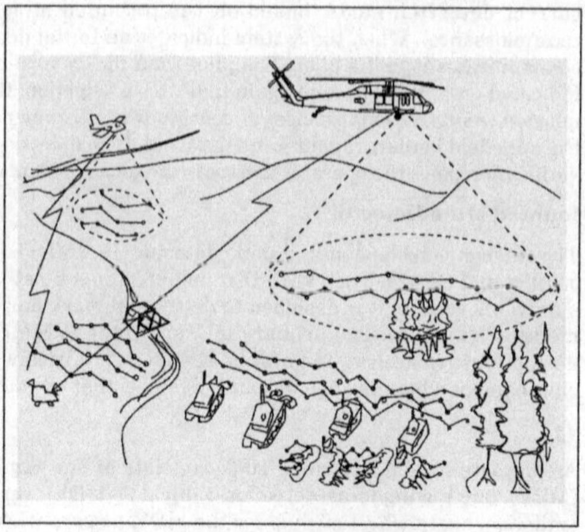

Figure 10-2. ASTAMIDS

System Components

ASTAMIDS hardware and software components consist of a sensor with associated electronics and the minefield-detection algorithm and processor (MIDAP). Surrogate equipment includes an air-data package (GPS, radar altimeter, inertial measurement unit [IMU]), a power supply, a work station(s), a digital data recorder, mounting racks, and a modified floor for the specific aircraft.

Operators view the data displayed on the monitors, communicate with the aircrew, and perform other functions (such as changing data tapes and producing reports). The aircrew must maintain an altitude of 300 feet and an airspeed of approximately 70 knots for the system to detect mines accurately within the sensor's ground swath (approximately 215 feet wide). The system has a 2-hour operational capability, based on standard flight time for the mission profile.

Employment Concept

ASTAMIDS is a fast method for detecting tactical minefields. When it is employed by aviation elements in support of maneuver units, close coordination between aviation and ground units assures that minefield detection is reported accurately and quickly. ASTAMIDS is not as precise as ground detection systems, but it is accurate enough to help mitigate the dangers inherent with minefields. It can be used in both friendly and enemy territories. The use of a Blackhawk ASTAMIDS in areas of threat observation and fire must be planned and coordinated very carefully, because a helicopter is extremely vulnerable while flying the mission profile required for detection (steady altitude, speed, and path).

Once airborne and at its start point, the ASTAMIDS system is placed in the correct detection mode, based on the intended mission (route or area reconnaissance). When the system indicates an initial detection, the operator communicates it to the pilot. The pilot then flies a verification pass over the indicated area. If the system again indicates a detection, the pilot resumes the mission (route reconnaissance) or continues the survey pattern to determine the minefield borders (area reconnaissance). If no detection is indicated on the verification pass, the operator instructs the pilot to resume the flight plan.

Interim Vehicle-Mounted Mine Detector

The interim vehicle-mounted mine detector (IVMMD) is used in all levels of conflict and OOTW. The IVMMD is mounted on a blast- and fragmentation-survivable vehicle; it is designed to detect and mark buried and surface-laid, metallic AT mines. The primary mission of the IVMMD is to detect mines during route clearance. The system should not be used when operating in an environment where the enemy employs mines that are not pressure-fused.

System Components

A complete IVMMD (Figure 10-3) consists of one mine-detection vehicle (MDV), one towing/mine-detection vehicle (T/MDV), three mine detonation trailers, a spare-wheel module for the MDV, a spare-wheel module for the T/MDV, and a container of spare parts.

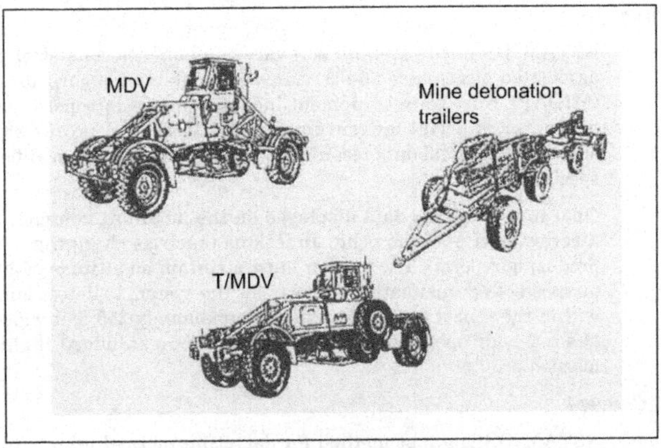

Figure 10-3. IVMMD components

The MDV's only mission is to detect mines. It can negotiate vertical slopes up to a 20 percent grade. The MDV employs a 4-cylinder engine and a manual transmission to propel the 4.8-ton vehicle with a 3-meter-wide detection array. The detection array consists of two separate induction coils (one for the left side and one for the right side) that detect magnetic fields below the vehicle.

The detection array is suspended between the two axles of the vehicle. When the detector encounters a metallic object, the operator is notified by an audible signal in the earphone. A visual signal appears on the dashboard that denotes which side of the array detected the object. The operator then stops the vehicle, backs it up, and reencounters the metallic object. (The MDV has two detection modes—the locate mode is used to identify the metal object, and the pinpoint mode is used to find the center of the object.) When the operator encounters the strongest signal, he activates the marking system (a nozzle mounted on the rear frame and centered on each detection array) that deploys a water-based ink onto the roadway.

The T/MDV has a 6-cylinder engine and the same detection and marking system as the MDV. The T/MDV tows three detonation trailers. The recommended maximum operating speed while towing the trailers is 20 kph. The T/MDV (with trailers) can negotiate vertical slopes up to a 20 percent grade; however, going down such slopes is difficult. The T/MDV must be in first gear, and the trailer brakes must be deployed to decrease the speed of the vehicle when going down a slope.

The mine detonation trailers are very heavy, and they are specifically designed to apply heavy ground pressure that initiates pressure-activated mines. Each trailer has two axles of different lengths so that the three trailers provide a full 3-meter-wide proofing capability behind the T/MDV. If a mine detonates underneath the trailers, the wheel bolts are designed to sheer so that repair is limited to replacing a single wheel.

The detection array is suspended between the two axles of the MDV. Although the vehicle is designed to produce very little ground pressure, it will detonate most pressure-fused mines, depending on the sensitivity of their fuses.

- The MDV produces 27.9 pounds per square inch (psi) of ground pressure when the tires are inflated to 14.5 psi and 21.8 psi of ground pressure when the tires are inflated to 8.7 psi.
- The T/MDV produces 49.8 psi of ground pressure when the tires are inflated to 29 psi and 23.4 psi of ground pressure when the tires are inflated to 8.7 psi.

Employment Concept

The IVMMD is used to support route-clearance operations. Clearance operations ensure that LOC are safe for the passage of personnel and equipment. The IVMMD should not be used during hours of limited visiblity, because it hampers the operator's ability to see surface-laid mines and visual signatures that indicate mining activities.

MECHANICAL

The track-width mine roller is a mechanical minefield-detection system. It is most effectively deployed to lead columns on route movement, but it can be used to precede tactical formations. In column movement, unit vehicles travel a narrow path, and one or two mine rollers can effectively detect mines in the path. Mine rollers can also be used to detect minefields in front of deployed tactical formations; however, more than one roller is required for a good probability of detection.

REPORTING

Intelligence concerning enemy minefields is reported by the fastest means available. Spot reports (SPOTREPs) are the tactical commander's most common source of minefield intelligence. They originate from patrols that have been sent on specific minefield reconnaissance missions or from units that have discovered minefield information in the course of their normal operations. The information is transmitted to higher headquarters.

REDUCING

Minefield reduction and clearing equipment is broken down into explosive, mechanical, electronic, and manual. Combat engineers and the operators of breach assets practice and become proficient in these reduction means. They integrate them into the breach drills of the units they support. The team applies different TTP to breach drills and prepares and rehearses them as part of the TF plan.

EXPLOSIVE

M58A4 Mine-Clearing Line Charge

The MICLIC (Figure 10-4, page 10-8) is a rocket-propelled, explosive line charge. It is used to reduce minefields that contain single-impulse, pressure-activated AT mines and mechanically activated AP mines. It clears a 14- by 100-meter path. The MICLIC has a 62-meter standoff distance from the launcher to the detonation point. The MICLIC's effectiveness is limited against prong AP mines, magnetically activated mines (including some SCATMINEs), top-attack mines, side-attack mines, and mines containing multiple-impulse or delay-time fuses. It also has little effect on other obstacles, such as log and concrete barriers, antivehicular ditches, and walls. The shock effect and the psychological impact of the detonation make the MICLIC a useful weapon in a close fight or in MOBA.

The MICLIC is mounted on a rubber-tired trailer, or two MICLICs can be mounted on an armored vehicle-launched bridge (AVLB), with the bridge downloaded, using a fabricated I-beam frame (procedures for mounting the MICLIC on the AVLB are outlined in TM 9-1375-215-14&P). This is called an armored vehicle-launched MICLIC (AVLM) (Figure 10-5, page 10-8), and it is the preferred system because no trailer is involved to hinder the mobility of the vehicle.

Towing vehicles for the trailer-mounted MICLIC are a combat engineer vehicle (CEV), an M113 APC, M2 and M3 Bradleys, an M9 armored combat earthmover (ACE), a 5-ton wheeled vehicle, and a 2½-ton wheeled vehicle. The trailer limits the MICLIC's mobility in rough terrain and degrades the maneuverability of the towing vehicle, thereby increasing vulnerability. Since the MICLIC is critical to the breach, it is kept under the protection of the force and is moved to the breach site along easily trafficable, covered, and concealed routes. This effectively prevents the towing vehicle from performing any other task (firing or maneuvering) or serving as an engineer squad vehicle unless MICLIC employment is the squad's only mission. This is an important consideration when selecting the towing vehicle because this vehicle must be solely dedicated to the mission.

Figure 10-4. MICLIC

Figure 10-5. AVLM

The MICLIC can be fired from within an armored vehicle without exposing soldiers to fires; however, the prefiring preparations must be done in advance at a covered and concealed location near the breach site. The initiating cable is brought into the vehicle through the hatch, which must be left ajar, or through the portal of the periscope, which has been removed. Therefore, the crew is not afforded nuclear, biological, chemical (NBC) protection. When the MICLIC is fired from a wheeled vehicle, however, the crew must move to a covered position outside the backblast area. The special-purpose cable on the firing control switch is long enough to allow adequate standoff.

The vehicle operator must be proficient in all aspects of preparing and deploying the MICLIC, including the critical aspect of selecting the optimum breach site. Although the operator will be directed to the breach site by the engineer platoon leader or the breach force commander, ensuring that he can independently accomplish the task will simplify the operation and greatly enhance its likelihood of success. The towing vehicle and the operator must be selected well in advance and be dedicated solely to the task. The operator must be included in all rehearsals and planning sessions and, if possible, during leaders' reconnaissances.

Each MICLIC trailer transports and fires one charge, and then it must be reloaded. The AVLM can fire both MICLICs before reloading. The loaded charge container weighs 1,283 kilograms, so a lifting device such as a 5-ton wrecker or a HEMTT is needed. Reloading, which can be done by an experienced crew in about 20 minutes, entails loading a rocket on the rail and lifting a new charge container onto the launcher. The reloading operation must be done in a covered and concealed location.

The exact limits and depth of an enemy minefield are seldom known before the breach. This is particularly true when the situation is unclear, and the minefield is encountered simultaneously with enemy contact. The first and only indication that a unit is in a minefield may be when a vehicle encounters a mine. The leading edge of the minefield still may be an uncertainty, because the vehicle could have hit a mine in an interior row. The number of MICLICs needed to clear a single lane through a minefield depends on the minefield depth:

- Clearing a lane through a minefield less than 100 meters deep requires one MICLIC (Figure 10-6, page 10-10). The leading edge of the minefield is identified and, if possible, confirmed by reconnaissance. The MICLIC is deployed from a minimum standoff distance of 62 meters from the leading edge of the minefield.
- Clearing a lane through a minefield more than 100 meters deep or of uncertain depth requires two or more MICLICs (Figure 10-7, page 10-10). If the leading edge cannot be identified, the MICLIC is deployed 100 meters from the possible edge or stricken vehicle. When the first MICLIC is detonated, a second MICLIC moves 25 meters into the first MICLIC's path and fires its charge. This extends the lane an additional 87 meters. Additional MICLICs are used for minefields of extreme depth, and each one moves down the lane 25 meters into the path created by the previous charge.

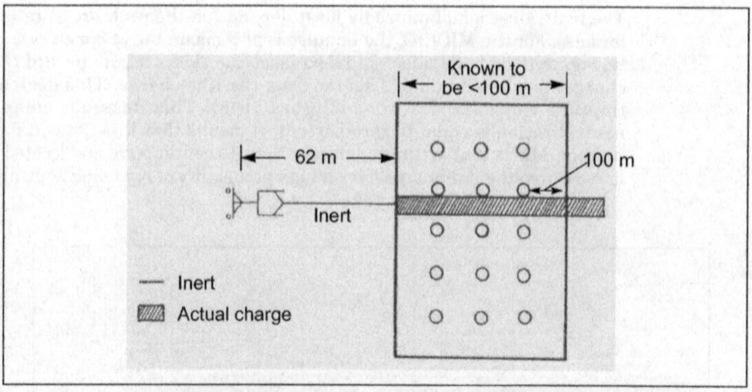

Figure 10-6. MICLIC employment in a minefield less than 100 meters deep

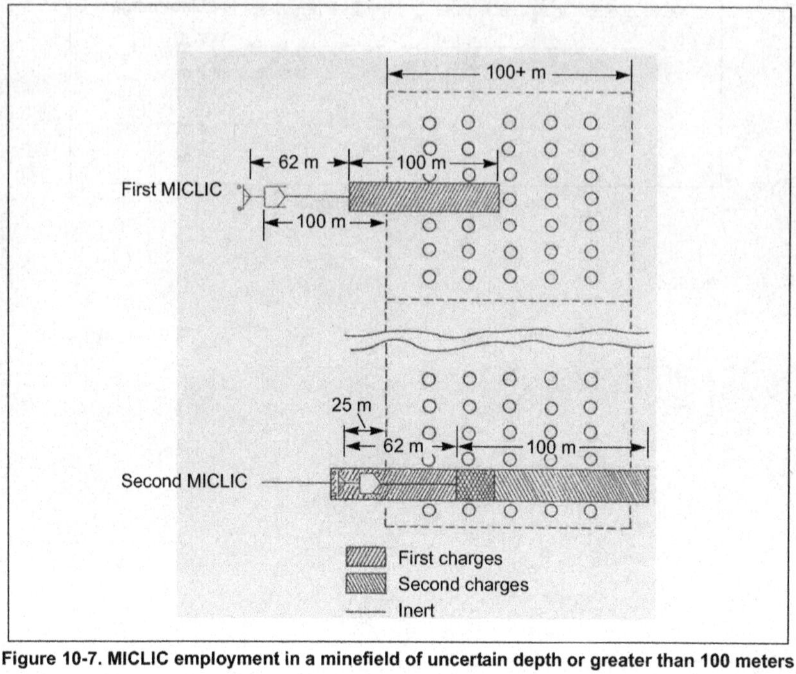

Figure 10-7. MICLIC employment in a minefield of uncertain depth or greater than 100 meters

The neutralization of mines by blast depends on the peak pressure and the impulse. For the MICLIC, the impulse is at a maximum of 3 meters from the line charge (on both sides) and decreases the closer it gets toward the line charge, to a minimum of 1 meter from the line charge. This decrease on impulse causes a *skip zone* (Figure 10-8). This does not mean that neutralization is equal to zero percent; it means that it is not equal to 100 percent. Mines that are buried deeper than 10 centimeters and located 1 to 2 meters from the line charge have a high probability of not being neutralized.

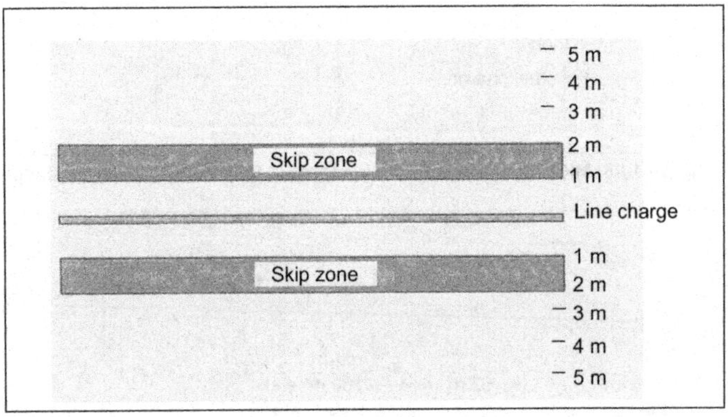

Figure 10-8. Skip zone

Antipersonnel Obstacle Breaching System

The Antipersonnel Obstacle Breaching System (APOBS) (Figure 10-10) is a man-portable device that is capable of quickly creating a footpath through AP mines and wire entanglements. The APOBS is normally employed by combat engineers, infantry soldiers, or dismounted armored cavalry personnel. The APOBS provides a lightweight, self-contained, two-man, portable line charge that is rocket-propelled over AP obstacles from a standoff position away from the edge of the obstacle.

For dismounted operations, the APOBS is carried in 25-kilogram backpacks by no more than two soldiers for a maximum of 2 kilometers. One backpack assembly consists of the rocket-motor launch mechanism, containing a 25-

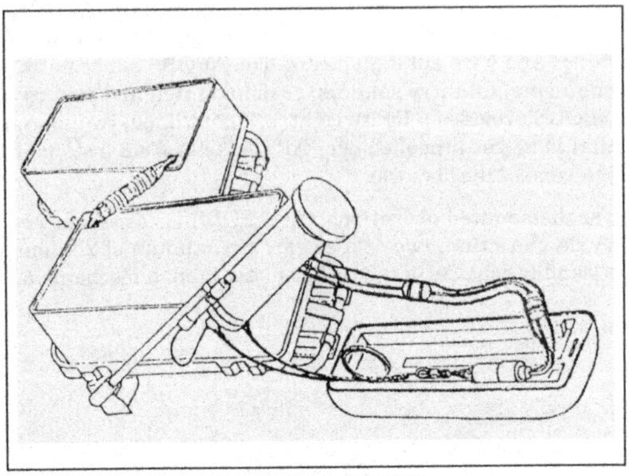

Figure 10-9. APOBS

meter line-charge segment and 60 attached grenades. The other backpack assembly contains a 20-meter line-charge segment and 48 attached grenades. The total weight of the APOBS is approximately 54 kilograms. It is capable of breaching a footpath that is approximately 0.6 by 45 meters and is fired from a 25-meter standoff.

M1A1/M1A2 Bangalore Torpedo

The bangalore torpedo (Figure 10-11) is a manually emplaced, explosive-filled pipe that was designed as a wire breaching device, but it is also effective against simple pressure-activated AP mines. It is issued as a demolition kit and consists of ten 1.5-meter tubes. Each tube contains 4 kilograms of high explosives and weighs 6 kilograms. The kit clears a 1- by 15-meter lane.

Figure 10-10. Bangalore torpedo

The bangalore torpedo is used by dismounted infantry and engineer troops. An individual soldier or a pair of soldiers connects the number of sections needed and pushes the torpedo through the AP minefield before priming it. A detailed reconnaissance is conducted before employing the bangalore torpedo to ensure that trip wires have not been used.

The bangalore torpedo generates one short impulse and is not effective against pronged, double-impulse, or pressure-resistant AP and AT mines.

> **WARNING**
> **Do not modify the bangalore torpedo. Cutting the bangalore in half or performing any other modification could cause the device to explode.**

MECHANICAL

MCBs and MCRs are fielded as armor battalion sets that contain 12 MCBs and 4 MCRs. Blades clear lanes through minefields, while rollers are used to detect minefields and proof lanes created by other means. Rollers are not a good primary system for lane reduction because multiple mine detonations destroy the roller system and the vehicle pushing it. (The roller is designed to resist two conventional-mine or three scatterable-mine strikes, depending on the mine type.)

The CEV, the ACE, and dozer blades were not designed for breaching minefields and should be employed only as a last resort, because using them is extremely hazardous to the crew and equipment. However, CEVs or ACEs can effectively clear a lane through AP scatterable minefields because they sustain little or no damage and offer protection to the crew. When using a dozer to clear a path through minefields, the operator is exposed to mine effects. Before clearing begins, the operator's cabin should be sandbagged or *up-armored* and the lane should be cleared of trip wires. When using an engineer blade to clear a path through a scatterable minefield, use the skim technique (Figure 10-12). Start skimming 100 meters from the suspected minefield leading edge.

Mine-Clearing Blade

The MCB (Figure 10-13) is used to remove land mines from the minefield. It consists of a blade arrangement with scarifying teeth to extract mines, a moldboard to cast mines aside, and leveling skids to control the depth of the blade.

The MCB lifts and pushes mines, which are surface-laid or buried up to 31 centimeters deep, to the side of the track-width lanes. The blade has three depth settings—21, 25, and 31 centimeters. The blade creates a 1.5-meter cleared path in front of each track. Figure 10-14, page 10-16, shows inside clearance distances between tracks of common track vehicles in relationship to the uncleared area left by the MCB.

Mines armed with AHDs, magnetic fuses, or seismic fuses may be activated when they are lifted by the blade; and they may disable the blade. Mines lifted by the blade are left in the spoil on each side of the furrowed path and remain a hazard until they are removed. Double-impulse mines that are lifted into the spoil on the side have a probability of functioning into the hull of the plowing vehicle. The skid shoe for each blade exerts adequate pressure to activate

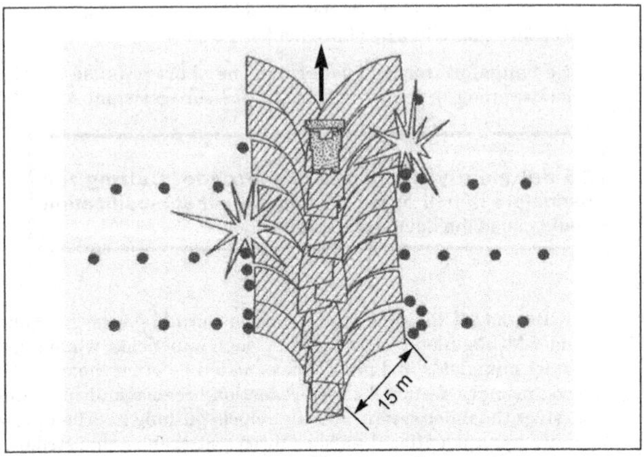

Figure 10-11. Skim technique

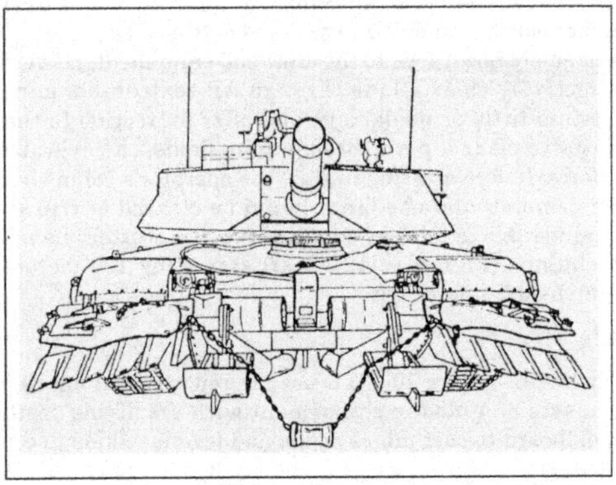

Figure 10-12. MCB

most single-impulse mines, which effectively clears a section of the centerline by explosive detonation. This action may disable the blade. Multiple-impulse pressure fuses encountered by the skid shoe are not defeated. A dog-bone assembly between the blades defeats tilt-rod mines. The improved dog-bone assembly (IDA) projects a magnetic signature and defeats tilt-rod and magnetic mines.

The MCB weighs approximately 3,150 kilograms and can be mounted on an M1 tank without special preparation or modification. Mounting requires lift capability and takes up to an hour, so it must be mounted well in advance of the mission. It is not easy to mount or transfer the MCB to another tank under battlefield conditions.

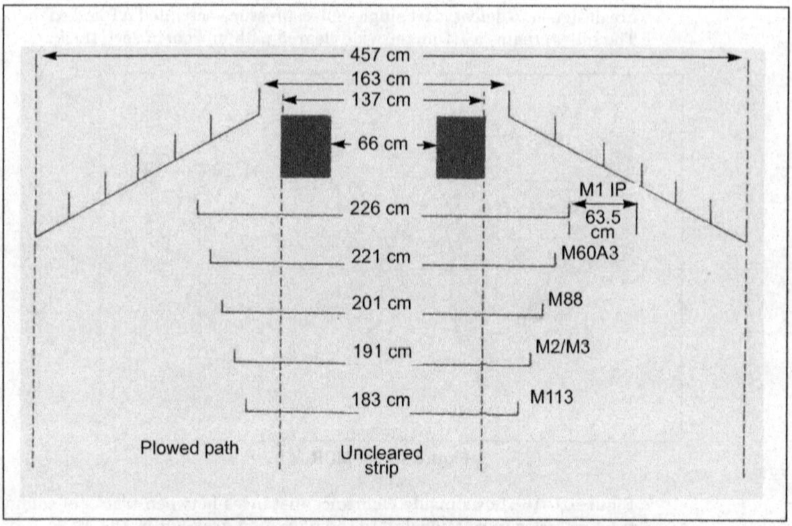

Figure 10-13. Mine-blade width compared to track-vehicle widths

Once mounted, an electric motor raises and lowers the blade. When it is in the raised position, it minimally effects the M1's maneuverability and speed. This will not greatly impact the employment of the weapon system except when the blade is in operation. The MCB is also equipped with an emergency, quick-disconnect feature.

The M1 should perform plowing operations from 8 to 10 kph, depending on soil conditions. It cannot maneuver but must continue in a straight path through the minefield to avoid damaging the blade. The main gun must be traversed to the side during plowing because mine detonation under the blade may cause the gun to be thrown violently into the air, damaging the tube. The area selected for the lane must be relatively flat and free of rocks or other obstacles.

The operator begins plowing approximately 100 meters from the estimated minefield leading edge. He creates a lane extending another 100 meters beyond the estimated minefield far edge to ensure that the lane extends through the entire minefield. Multiple vehicles crossing the breach will deepen the cut made by the MCB, and pressure-fused mines left in the uncleared strip will be dangerous. The uncleared strip should be cleared as soon as possible.

Mine-Clearing Roller

The MCR (Figure 10-15) consists of a roller assembly, a mounting kit, and a hand winch kit. The roller assembly weighs approximately 9,072 kilograms and consists of two push beams mounted to the front of the tank. The rollers are designed to defeat most single-pulse, pressure-activated AT and AP mines. The roller creates a 1.1-meter-wide cleared path in front of each track.

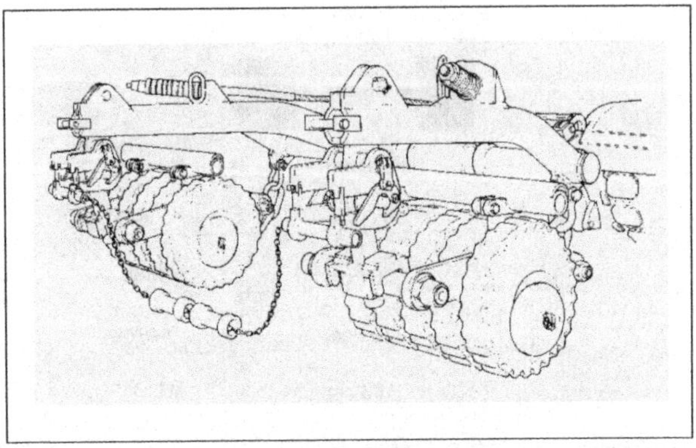

Figure 10-14. MCR

Figure 10-16 shows inside clearance distances between tracks of common track vehicles in relationship to the uncleared area left by the MCR. A dog-bone-and-chain assembly between the rollers defeats tilt-rod mines. The IDA can be fitted to the roller. The roller is designed to withstand multiple mine explosions before damage; however, this depends on the size of the mines. Large blasts may destroy the roller or the vehicle or injure the crew.

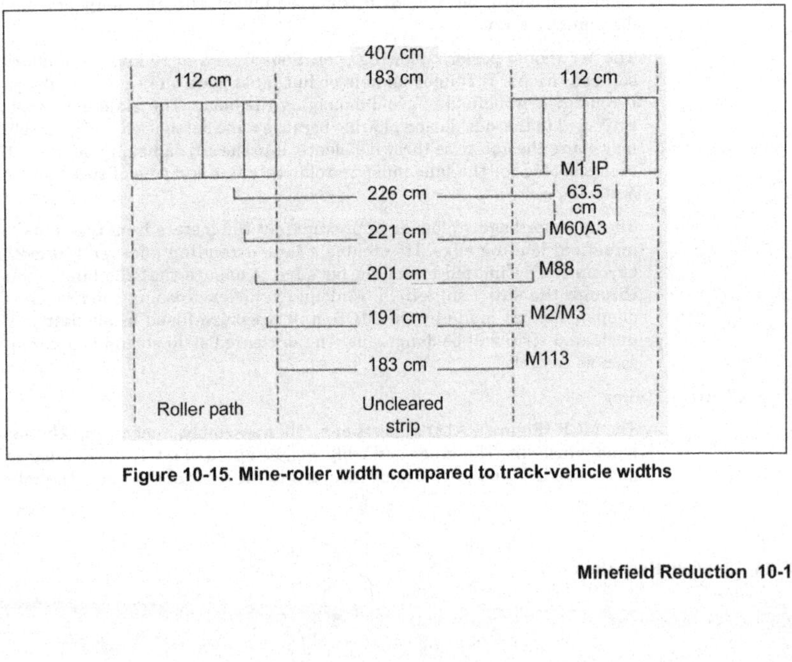

Figure 10-15. Mine-roller width compared to track-vehicle widths

The roller can be mounted on an M1 or M60 tank that is modified with a permanently attached mine-roller mounting kit. Mounting the roller to a tank is a cumbersome, time-consuming operation because it is very difficult under battlefield conditions and requires lift capability. The roller tank is limited to a speed of 5 to 15 kph. When employed in a suspected minefield, the MCR must travel in a relatively straight path, because tight turns may cause the roller to deviate from the path of the track and leave the tank vulnerable to mines. Ground fluctuations, bumps, and berms may cause the roller to lift from the ground and miss mines.

The MCR is not designed to negotiate gaps on its own; however, it can be used on AVLB caution crossings. In this situation, the curbing from the bridge is removed. To prevent damage to the bridge's hydraulic line, the tank driver uses a strap to lift the dog bone and chain when crossing the bridge. The main gun must be traversed to the side when a mine encounter is possible or imminent, because a mine blast can throw the roller or parts of the roller violently into the air and damage the tube. The main gun should only be fired from a temporary halt.

When the situation and the mission permit, MCRs may be employed as lead vehicles to detect minefields. This is most viable when the supported element is traveling in a column. The roller may also be used to lead a supported element traveling in a tactical formation other than a column, but it is less effective than other methods because—

- Vehicles not directly behind the roller may encounter mines passed by the roller.
- The roller may travel well into or completely through a widely spaced minefield without encountering a mine, thus giving the formation a false sense of security.
- A mine encountered by the roller may not be on the leading edge of the minefield.
- The roller vehicle is extremely vulnerable because it can only use its weapon system from a temporary halt.

Rollers are best used to proof lanes in obstacles that are breached by other means, such as a MICLIC or an MCB. A roller pulling a trailer-mounted MICLIC can proof a lane created by a MICLIC that was launched by another vehicle. The roller then fires the second MICLIC and proofs its own lane.

If rollers participate in a deliberate breach operation or if the force incorporates rollers into a hasty breach plan, rollers must be mounted before rehearsals. Unmounted rollers that not being used for the mission are carried in the TF formation on M916 tractor trailers. Rollers require lift capability (such as an M88), a secure location, and 30 to 60 minutes to mount on a tank that is fitted with a mounting kit.

Panther

The M60 Panther (Figure 10-17) is one of several developmental countermine systems used by US forces during operations *Joint Endeavor* and *Joint Task Force Eagle*. The Panther is a remotely controlled vehicle with mine rollers, and it is used to proof lanes and assembly areas. The system consists of a turretless M60 tank, Israeli mine rollers, an antimagnetic actuating device, and an RCU that is mounted in a separate vehicle. Additionally, a remote video camera allows the operator to see the road ahead.

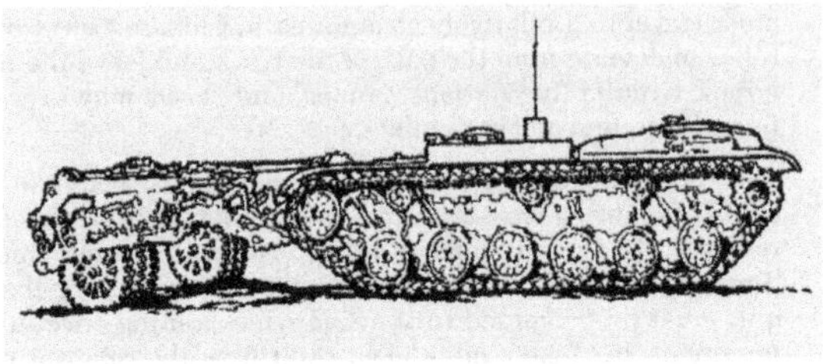

Figure 10-16. Panther

During route clearance or proofing operations, the Panther is the lead vehicle on the route. It is followed closely by an armored control vehicle, usually an M113. The control vehicle contains the Panther operator, the RCU, and the monitor. The monitor displays the route being proofed or cleared through a camera mounted on the Panther. The Panther is controlled from the commander copula or troop hatch of the control vehicle. The control vehicle should be approximately 200 to 300 meters behind the Panther, and its hatches should be secured open. Crew members in the control vehicle should be wearing Improved Body Armor System, Individual Countermine (IBASIC) protective garments.

Mine rollers can be raised for limited travel while mounted on the Panther. If the distance is excessive, the rollers must be transported on a cargo carrier. Rollers must be adjusted before every mission to ensure that they have contact with the ground and that their weight is uniformly distributed. To ensure proper coverage and overlap of rollers, at least three passes should be conducted. Passes should have a minimum of 30 centimeters overlap. Inside roller distances are the same as the MCR.

MiniFlail

The MiniFlail (Figure 10-18. page 10-20) is a remotely operated, line-of-sight, AP-mine and UXO neutralization system that was developed for use by US light forces. It can clear at a rate of 1,200 square meters per hour. The MiniFlail detonates or disables AP mines from a safe operating distance. The MiniFlail neutralizes by striking objects with a rotating chain assembly, called a *flail*, and clears a foot path approximately 1.1 meters wide. The system neutralizes AP mines and UXO by detonation, mechanical destruction, or displacement from the cleared lane. The MiniFlail is approximately 1.3 meters wide, 1.3 meters high, 3 meters long, and weighs 1,100 kilograms. The system is operated by a hand-held controller that has a maximum range of 300 meters. It is fully armored with a material similar to Kevlar, and the tires are filled with foam. The flail is a self-articulating, hydraulically powered shaft with 84 chains; each chain is 0.5 meter long.

Figure 10-17. MiniFlail

M1 Grizzly

The Grizzly (Figure 10-19) provides a hasty capability for breaching complex obstacles of mines, wire, posts, rubble, and tank ditches to create a lane for other vehicles to follow. The Grizzly's primary features are a full-width, 4.2-meter MCB and a power arm. The power arm has a reach of 9 meters and a bucket capacity of 1.2 cubic meters. Its primary missions are to reduce berms and fill AT ditches.

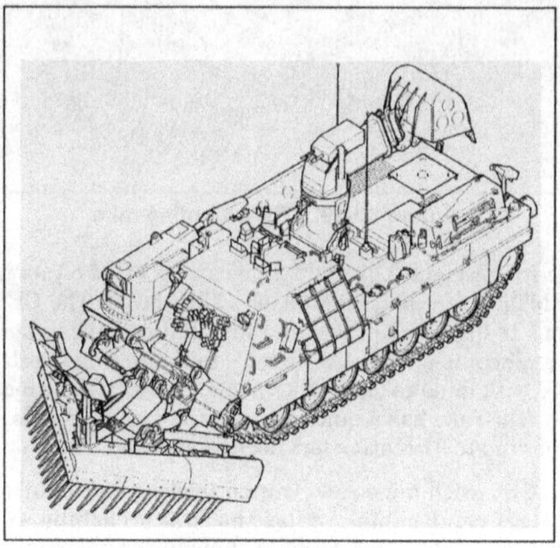

Figure 10-18. Grizzly

The system is armed with an M240 7.62-millimeter machine gun and an MK19 40-millimeter grenade launcher. The Grizzly has an M1-series tank chassis.

The Grizzly lifts and pushes mines, which are surface-laid or buried up to 31 centimeters deep, to the side of full-width lanes. The blade has multiple depth settings, depending on the mission, and it creates a 4.2-meter-wide cleared path. When plowing, the Grizzly is restricted to less than 10 kph, depending on soil conditions. The operator begins plowing approximately 100 meters from the estimated minefield leading edge. He creates a lane extending another 100 meters beyond the estimated minefield far edge to ensure that the lane extends through the entire minefield.

The Grizzly has integrated digital features to enhance battlefield awareness. Some of the digital features are thermal and video cameras, ground-speed sensors, terrain-mapping sensors, and an integrated commander's control station.

Combat Engineer Vehicle with Full-Width Mine Rake

NOTE: The CEV with full-width mine rake will be used in Korea only.

This system consists of a wedge-shaped rake that is mounted to a CEV blade (Figure 10-20).

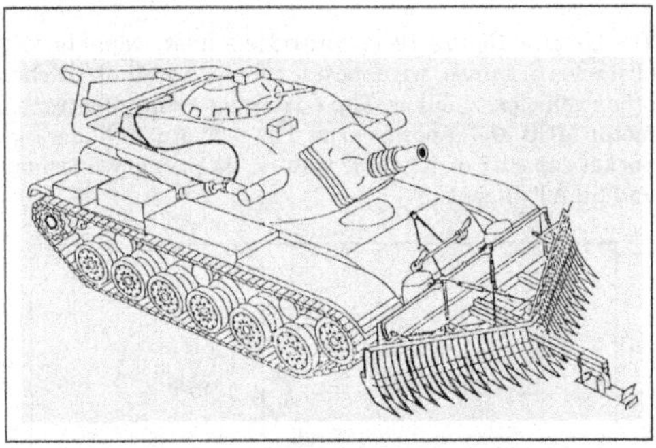

Figure 10-19. CEV with mine rake

The rake weighs 2,025 kilograms and is lifted off its transport vehicle with a HEMTT, a wrecker, an M88, or a CEV boom. The CEV crew uses basic-issue items to install the rake, and installation takes approximately 30 minutes. The rake has a skid shoe to maintain a raking depth of 31 centimeters. The CEV with mine rake provides a vehicle-width clearance (4.5 meters) at 5 to 10 kph. The rake has a quick-disconnect feature. It lifts surface-laid and buried mines (up to 31 centimeters deep) and pushes them off to both sides.

The CEV with mine rake is used to clear lanes during minefield breaching. While it can be employed as the first breaching asset into a minefield, a MICLIC should be used first to eliminate as many mines as possible. The rake is then used to proof the lane. The system can pull a MICLIC and fire it before proofing. Raking begins 100 meters before the minefield and continues 100 meters beyond the suspected limit. The CEV maintains a straight course through the minefield. If the skid shoe is damaged, the operator reduces speed and manually controls the blade depth. This is very difficult and risky.

Mine-Clearing/Armor-Protection Kit

The mine-clearing/armor-protection kit (MCAP) consists of two parts—the mine-clearing rake and the armor protection.

The rake uses a tine that is mounted on a diagonal beam. The rake assembly is designed to sift through the soil, lift out mines, and windrow buried and surface-laid mines to the right of the vehicle. The system clears a 30-centimeter-deep path through a minefield. The rake has a skid shoe that acts as a depth control guide for the operator.

The armor protects the crew against mine blast, small-arms fire, and artillery fire. Protection is also provided for the engine, the fuel tank, and exposed hydraulic cylinders and lines. Ballistic glass blocks are provided at each vision port to permit unrestricted view and operation of the vehicle and the equipment.

The MCAP is mounted on a D7 dozer to perform minefield breaching and lane widening. Proofing the lane must be conducted after the dozer has cleared the lane. Some AP mines may still be left in the lane.

Electronic

The Field-Expedient Countermine System (FECS) is a series of copper coils that fit over the front of tracked and wheeled vehicles. Power is supplied by the vehicle's battery. The coil emits a large magnetic signature that detonates magnetically fused mines located 2 to 5 meters in front of the vehicle. The FECS is designed to defeat magnetically influenced mines only and must be used with other countermine systems.

Manual

When stealth is required or advanced mechanical equipment is unavailable, manual breaching procedures can be used. Engineers use hand-emplaced explosives, grapnel hooks attached to ropes, probes, mine detectors, and hand-emplaced marking equipment to manually breach obstacles. This is the only method that works in all situations and under all conditions because certain types of terrain, weather, and sophisticated fuses can severely degrade the effectiveness of rollers, plows, and line charges.

Surface-Laid Minefields

The enemy possesses a significant mechanical, mine-burying capability. It has the capacity and the propensity for the labor-intensive effort required to bury mines by hand; however, the enemy often lays mines on the surface. Buried mines are usually found in a highly prepared defense that requires a deliberate breach operation. Training and execution of surface and buried minefield breaches should always assume the presence of AHDs and trip wires until proven otherwise.

From covered positions, the engineers first use grapnel hooks to check for trip wires in the lane. The limited range of the tossed hook requires the procedure to be repeated through the estimated width of the obstacle. A demolition team then moves through the lane. The team places a *line main* down the center of the lane, ties the line from the explosives into the line main, and places blocks of explosives next to surface-laid mines. After the mines are detonated, the team makes a visual check to ensure that all of the mines were cleared before directing a proofing roller and other traffic through the lane.

Manual procedures must be well-practiced. Members of the demolition team are assigned special tasks, such as grappler, detonating-cord man, and demolitions man. All of the members should be cross-trained on all the procedures. Demolitions are prepared for use before arriving at the breach site. An engineer platoon uses squads in series through the minefield to clear a lane for a company team. The platoon must rehearse reduction procedures until execution is flawless, quick, and technically safe. During the breach, the engineer platoon will be exposed in the lane for 5 minutes or more depending on the mission, the minefield depth, and the platoon's level of training.

Buried Minefields

Manually reducing a buried minefield is extremely difficult to perform as part of a breaching operation. It is usually part of a clearance operation. If the mine burrows are not easily seen, mine detectors and probes must be used to locate the mines. The mines are then destroyed by hand-emplaced charges. As an alternative, the mines can be removed by using a grappling hook and, if necessary, a tripod (Figure 10-21). Using a tripod provides a vertical lift on the mine, making it easier to pull the mine out of the hole.

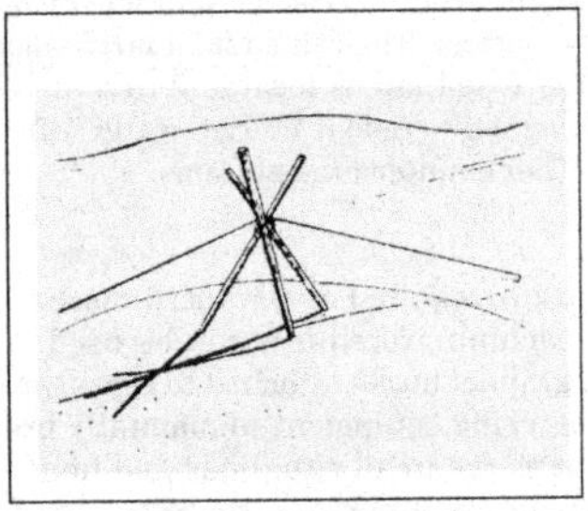

Figure 10-20. Tripod

The platoon leader organizes soldiers into teams with distinct, rehearsed missions including grappling, detecting, marking, probing, and emplacing demolitions and detonating cord. The platoon is exposed in the obstacle for long periods of time.

Grappling Hook

The grappling hook (grapnel) is a multipurpose tool that is used for manual obstacle reduction. Soldiers use it to detonate mines from a standoff position by activating trip wires and AHDs. After the grapnel is used to clear the trip wires in a lane, dismounted engineers can move through the minefield, visually locate surface-laid mines, and prepare the mines for demolition. In buried minefields, soldiers grapple, then enter the minefield with mine detectors and probes.

There two types of grapnel hooks—hand-thrown and weapon-launched.

Hand-Thrown

A 60+-meter light rope is attached to the grapnel for hand-throwing. The throwing range is usually no more than 25 meters. The excess rope is used for the standoff distance when the thrower begins grappling. The thrower tosses the grapnel and seeks cover before the grapnel and rope touch the ground in case their impact detonates a mine. He then moves backward, reaches the end of the excess rope, takes cover, and begins grappling. Once the grapnel is recovered, the thrower moves forward to the original position, tosses the grapnel, and repeats the procedure at least twice. He then moves to the end of the grappled area and repeats this sequence through the depth of the minefield.

Weapon-Launched

A 150-meter light rope is attached to a lightweight grapnel that is designed to be fired from an M16A1 or M16A2 rifle using an M855 cartridge. The grapnel is pushed onto the rifle muzzle, with the opening of the retrieval-rope bag oriented toward the minefield. The firer is located 25 meters from the leading edge of the minefield, and he aims the rifle muzzle at a 30- to 40-degree angle for maximum range. Once fired, the grapnel will travel 75 to 100 meters from the firer's position. After the weapon-launched grapnel hook (WLGH) has been fired, the firer secures the rope, moves 60 meters from the minefield, moves into a prone position, and begins to grapnel. The WLGH can be used only once, but it can be reused up to 20 times for training (blanks are used to fire the grapnel for training).

Multiple grapplers can clear a lane of trip wires quickly and thoroughly, but they must time their efforts and follow procedures simultaneously, if possible. A hit on a trip wire or a pressure fuse can destroy the hook and the cord, so engineers should carry extras.

PROOFING

Proofing is done by passing a mine roller or another mine-resistant vehicle through the minefield as the lead vehicle to verify that a lane is free of mines. An MCB, a Panther, a MiniFlail, or an MCR can be used to proof lanes. If the risk of live mines remaining in the lane does not exceed the risk of loss to enemy fires while waiting, proofing may not be practical. Some mines are resistant to some breaching techniques (for example, magnetically fused mines may be resistant to the MICLIC blast), so proofing should be done when the time available, the threat, and the mission allow.

During a limited clearing operation, proof upgraded breach lanes following a breach. After the minefield is completely cleared, proof the routes used through the area.

MARKING

This section implements STANAGs 2036 and 2889.

This section provides commanders with a standard system for marking breach lanes and bypasses. It centers around a systematic, phased upgrade of lane marking. Each upgrade conforms to the tactical requirements for that phase of the attack, from initial reduction of the obstacle to the passage of larger follow-on forces, as well as the return traffic necessary to sustain the force. Additional guidelines are discussed in FM 3-34.2.

Marking breach lanes and bypasses is critical to obstacle reduction. Effective lane marking allows the commander to project forces through the obstacle quickly, with combat power and C^2. It gives the assaulting force confidence in the safety of the lane and helps prevent unnecessary minefield casualties.

There are two critical components of the lane-marking system:

- Lane-marking pattern (location of markers indicating the entrance, the lane, and the exit).
- Marking device (type of hardware emplaced to mark the entrance, the lane, and the exit).

The lane-marking system outlined in this section centers around standardized marking patterns rather than the marking device. Standardizing the marking pattern is critical to offensive operations. A common lane pattern—

- Enables cross attachments and adjacent units to recognize breach lanes easily with minimal knowledge of a particular unit's tactical SOP.
- Gives all forces a standardized set of visual cues that are needed to pass through a lane safely while maintaining their momentum.
- Facilitates quick conversion to the lane-marking requirements of STANAGs 2889 and 2036 (discussed later in this chapter).

The standard lane-marking hardware is decided by unit commanders. This gives units greater flexibility and allows them to adopt marking devices that are tailor-made for their type of unit and operational focus (such as an armored or light force, a mounted or dismounted attack, limited visibility, thermal capability). However, regardless of the type of device used, it must support the standard lane-marking pattern outlined in the following paragraphs. Therefore, commanders should consider these guidelines and examples before developing or adopting their own marking system.

LANE-MARKING TERMS

The definitions in the following paragraphs provide a common basis for discussing lane marking.

Entrance Markers

Entrance markers indicate the start of a reduced lane through an obstacle. They signify the friendly-side limit of the obstacle and the point at which movement is restricted by the lane width and path. Entrance markers are placed to the left and the right of the entrance point and spaced the width of the reduced lane. They must be visually different from handrail markers to help the force distinguish this critical point in the lane.

Handrail Markers

Handrail markers define the lane path through the obstacle and indicate the limits of the lane width. As a minimum, mounted and dismounted lanes will have a left handrail marker. Mounted and dismounted forces moving through the lane should keep the left handrail marker immediately to their left. As the operation progresses, lane marking may be upgraded to include left and right handrail markers.

Exit Markers

Exit markers indicate the far side of the reduced lane through an obstacle. Like entrance markers, exit markers must be distinguishably different from handrail markers; however, the exit may be marked the same as the entrance. Exit markers are placed to the left and the right of the exit point and spaced the width of the reduced lane. This visual reference is critical when only the left handrail is marked. The combination of entrance markers, left handrail markers, and exit markers provide the driver and the tank commander with visual cues so that they can safely pass through a reduced lane.

Entrance Funnel Markers

Entrance funnel markers augment entrance marking. The V formed by a funnel marker forces the platoon into a column and helps drivers and tank commanders make last-minute adjustments before entering a lane.

Final-Approach Markers

Final-approach markers are highly visible, robust markers that augment the visual signature of entrance funnel markers. They are critical when initial assault forces must maneuver to the breaching site. Normally, the initial assault force can observe the breaching area but cannot clearly distinguish entrance funnel markers. Final-approach markers provide the assault force commander with a highly visible RP toward which to maneuver his formation. They also signal company team commanders to begin changing from combat column to column formation, with platoons in combat column.

Far Recognition Markers

Far recognition markers are highly visible markers that are located between the final-approach marker and the friendly unit. They are primarily used when passing forces are denied direct observation of the final-approach marker due to distance, visibility, or terrain. When possible, far recognition markers should be different from the final-approach marker. Far recognition markers indicate the point at which forces begin changing their formation to posture for the passage. A single far recognition marker may serve up to two initial breach lanes. Once lanes are upgraded to two-way traffic, far recognition markers are required for each two-way lane. When a far recognition marker serves more than one lane, a guide or a traffic-control post (TCP) is collocated with the far recognition marker that is nearest to the breach.

Guides and Traffic-Control Posts

A TCP or a guide consists of a two-man team with communications means. The team assists the commander in controlling the movement of forces. When possible, military police (MP) should man TCPs. However, the commander may initially use other personnel as guides to man critical far recognition markers until the MP establish full TCPs. TCPs and guides provide the commander with a man on the ground who controls traffic flow to the appropriate lanes. When there are multiple lanes branching off a single far recognition marker, the TCP can assist in breaking parts of the formation off into various lanes. The TCP can also help modify the traffic flow when lanes have been closed for maintenance, for lane expansion, or by enemy SCATMINEs. The guide or TCP must give the assault force commander the azimuth and distance to the final-approach marker, identify the device used for the final-approach marker, and provide the level of the lane-marking pattern. For light forces, guides may physically escort passing units from the far recognition marker to the lane entrance.

LEVELS OF LANE MARKING AND PATTERNS

The three standard levels of marking for breach lanes and bypasses are initial, intermediate, and full.

Each lane-marking level provides an increase in lane signature and capability. Lane requirements change as a breaching operation matures from an initial breach to the forward passage of large combat forces.

Initial lane-marking requirements are driven by the nature of the fight through the obstacle. Marking must be rapid, providing only the bare minimum signature needed to pass small units who make up the initial assault force. This contrasts with the lane requirements of later phases of an offense where larger units are passed to subsequent objectives. Here, the lane signature must be more extensive and more visible, because it must guide larger forces over a greater distance to the lane's entrance without interruption. Two-way traffic becomes a priority for the simultaneous forward passage of combat units as well as the return traffic (such as ambulances and empty supply vehicles) that is necessary to sustain the force. Lane-marking limits must be absolutely clear to the most inexperienced driver or crewman. A fully developed lane must support two-way traffic and be completely marked.

Bypasses are not marked the same as lanes. They are marked with directional panels indicating the direction of the bypass. The limits of the mine threat must be marked to prevent friendly forces from entering the minefield. Marking the direction of the bypass and the minefield limits will enable the maneuvering element to bypass the minefield without having to unnecessarily defile through a marked lane. Further information on bypass marking can be found in FM 3-34.2.

Commanders must be aware of how the needs of the force change with the operation so that they can anticipate lane-marking and lane-capability requirements. Integrating the levels of lane marking into the overall breaching plan ensures that the unit's needs are satisfied. Forces necessary to mark, maintain, and upgrade lanes must be allocated and tasked with the mission. The phases of the scheme of maneuver and the service-support plan are the basis for analyzing lane requirements. The following paragraphs describe lane-marking patterns in detail and provide guidelines on when the commander should upgrade lane marking and lane capability.

Initial Lane Marking

Initial lane marking (Figure 10-22, page 10-28) is emplaced by the breach force immediately after the lane is reduced and proofed. It provides a signal to the assault force commander that the lane is ready for traffic. Initial lane marking is kept to a minimum, centering on markings needed to pass immediate assault forces through the lane to seize the initial foothold on the objective. Normally, the assault force can observe the breach and does not need the more visual signature of a mature lane marking. The initial lane-marking pattern has the following markers:

- Entrance.
- Exit.
- Left handrail.
- Entrance funnel.
- Final-approach.

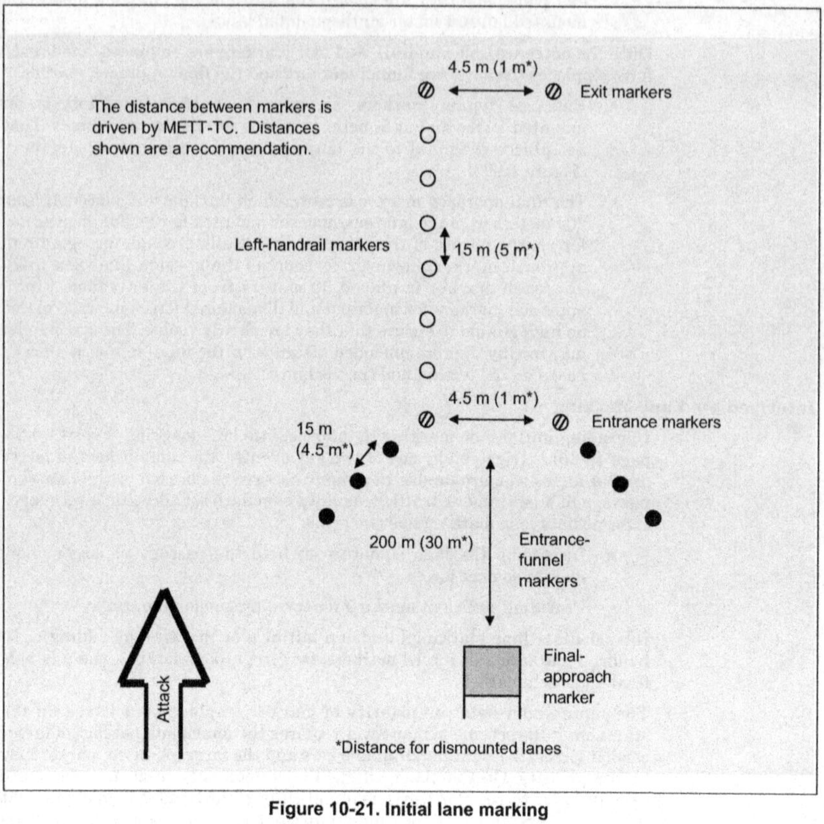

Figure 10-21. Initial lane marking

The entrance, left handrail, and exit markers are the first markers emplaced by the breach force because they define the location and the limits of the reduced lane.

- Entrance markers are placed to the left and the right of the reduced lane's entrance point, and they are spaced the width of the lane (4.5 meters for mounted lanes, 1 meter for dismounted lanes).
- Left handrail markers are placed at the left limit of the lane, along the entire path. Handrail markers are placed at 15-meter intervals for mounted forces and at 5-meter intervals for dismounted forces. Commanders may have to modify the intervals based on the terrain, the visibility, the lane length, and the lane path.
- Exit markers are placed to the left and the right of the reduced lane's exit point, and they are spaced the width of the lane (4.5 meters for mounted lanes, 1 meter for dismounted lanes).

Once the entrance, left handrail, and exit markers are emplaced, the breach force emplaces the entrance funnel markers and the final-approach marker.

- Entrance funnel markers are placed at 15-meter intervals for mounted forces and at 5-meter intervals for dismounted forces. They are placed diagonal to the lane entrance and form a 45-degree V (Figure 10-22).
- The final-approach marker is centered on the lane and placed at least 200 meters from the lane entrance for mounted forces. For dismounted forces, the nature of the attack may initially preclude using a final-approach marker; however, as soon as the mission allows, a final-approach marker is placed 30 meters from the entrance. Final-approach markers for mounted and dismounted forces must be placed on high ground to ensure that they are clearly visible. The commander may modify the recommended distance for the final-approach marker, based on the terrain and the visibility.

Intermediate Lane Marking

Upgrading initial lane marking to intermediate lane marking (Figure 10-23, page 10-30) is triggered by one of two key events—the commitment of larger combat forces who are unable to directly observe the breach or the rearward passage of sustainment traffic (casualty evacuation and vehicle recovery). Intermediate lane marking has two goals:

- Increasing the lane signature to help the passage of larger, more distant combat forces.
- Providing sufficient marking for two-way, single-lane traffic.

Intermediate lane marking builds on initial lane marking by adding right handrail markers, exit funnel markers, far recognition markers, and a farside final-approach marker.

The commander sets the priority of marker emplacement based on the situation. If the scheme of maneuver requires the immediate passage of larger combat forces, the right handrail markers and the far recognition marker may be the priority. On the other hand, if it is necessary to ground evacuate casualties or to recover vehicles, emplacing right handrail markers, exit funnel markers, and a farside final-approach marker may be required first.

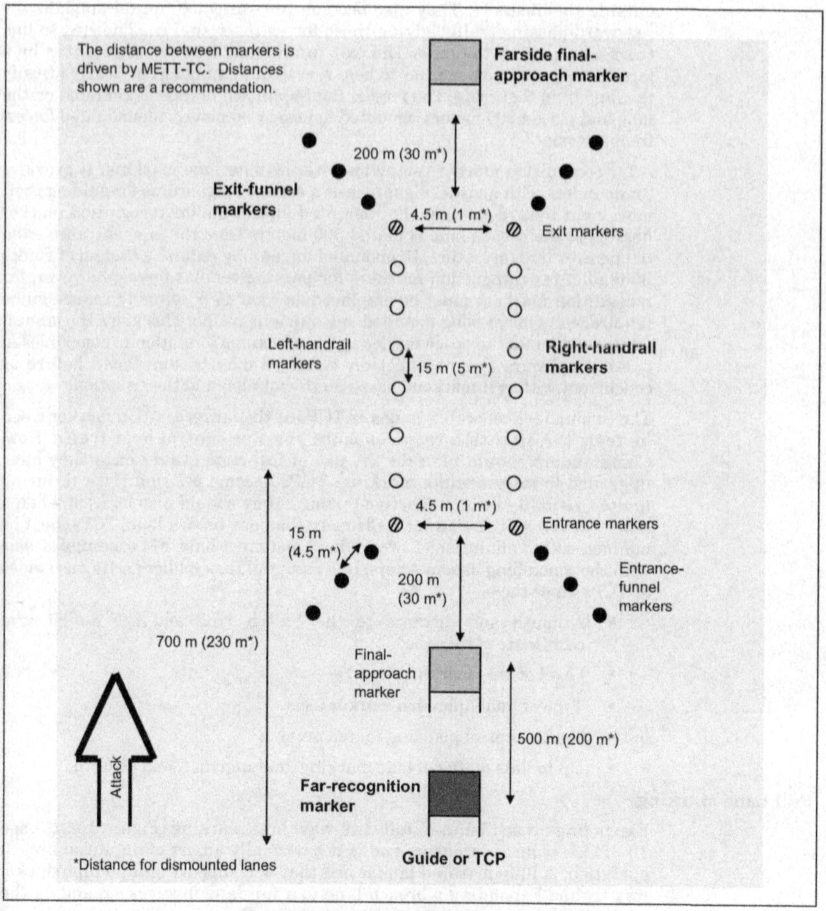

Figure 10-22. Intermediate lane marking

When upgrading to intermediate marking, the first step is to emplace the right handrail markers. Right handrail markers define the rightmost limit of the lane. They are placed the width of the lane as defined by the entrance and exit markers. The right handrail follows a path parallel to the left handrail through the obstacle. Right handrail markers are placed at the same interval as left handrail markers.

Exit funnel markers and a farside final-approach marker are emplaced to mirror the entrance markers. Exit funnel markers prevent the premature deployment of the passing force into combat formation before it is safely outside the obstacle. They also become the entrance funnel markers for rearward passing traffic, giving these forces the visual cues needed to line themselves up on the lane. The exit funnel markers are augmented by a farside final-approach marker to help rearward passing forces clearly identify the lane from their side. The farside final-approach marker is centered on the lane and placed 200 meters (mounted forces) or 30 meters (dismounted forces) from the exit.

A far recognition marker completes intermediate lane marking. It provides commanders with a visual signature or a series of signatures for guiding their movement toward the lane. For mounted forces, the far recognition marker nearest to the breach lane is placed 500 meters from the lane entrance or on the nearest terrain feature. Dismounted forces may require a system of guides instead of far recognition markers for passing combat forces; however, far recognition markers must be emplaced as soon as possible to reduce guide requirements for passing mounted sustainment traffic. This gives the assault force commander the space needed to transition his formation to companies in combat column. Far recognition markers may be emplaced before or concurrent with exit markers, based on the mission and the situation.

The commander collocates guides or TCPs at the far recognition marker when he feels the situation requires more positive control over traffic flow. Commanders should plan for the use of full-time guides once they have upgraded to intermediate marking. TCPs become mission-critical during limited visibility or in restrictive terrain. They should also be used when a single far recognition marker feeds more than one breach lane. TCPs must be manned with a minimum of two soldiers and must have FM communications with the controlling headquarters. It is essential that soldiers acting as guides or TCPs know the—

- Azimuth and distance to the breach lane and the 8-digit grid coordinate of the lane.
- Level of lane marking.
- Type of final-approach marker used.
- Traffic-control plan and march order.
- Up-to-date status of lane marking, maintenance, and so forth.

Full Lane Marking

Expanding breach lanes to full (two-way) lane marking (Figure 10-24, page 10-32) is resource-intensive and is not normally a part of an initial breach operation. A fully matured lane is one that will support uninterrupted, two-way traffic. Expanding a breach lane to a full lane involves expanding the width of the lane to accommodate two-way traffic and modifying the marking pattern to give forward and rearward passing forces the same visual signature. Upgrading to a full lane is normally assigned to follow-on engineer forces, since it is usually beyond the immediate capability of engineers with forward units.

Upgrading intermediate lane marking to full lane marking begins by temporarily closing the lane, rerouting traffic, and expanding the lane width. The initial reduced and proofed lane is always expanded to the left, in relation to the direction of the attack. Engineers reduce and proof the obstacle beginning at the left handrail to give a total lane width of 10 meters (5 meters

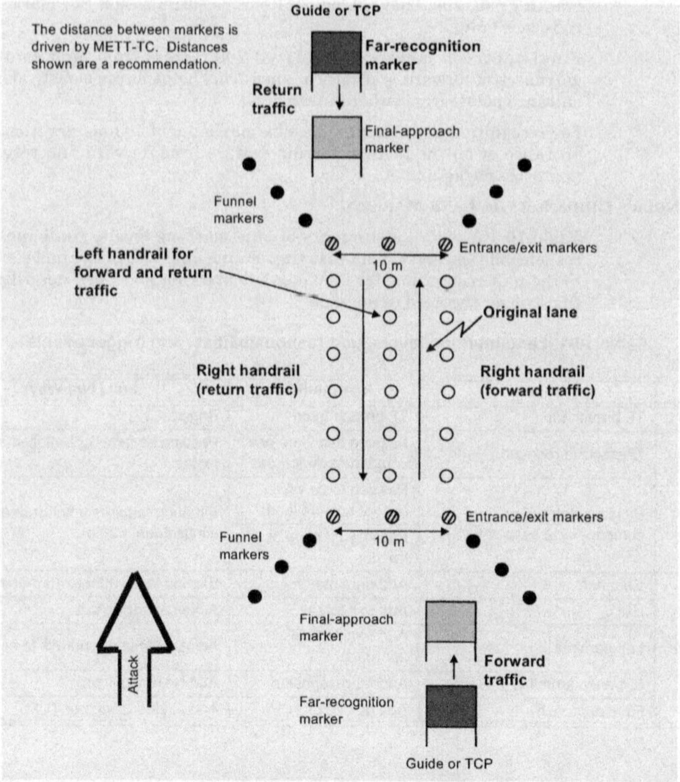

Figure 10-23. Full lane marking

each way). The expansion width requirement is the same for armored and light forces, because both forces must be able to pass mounted sustainment and combat forces during this phase.

Once the engineers expand the lane width to 10 meters, they ensure that entrance, exit, handrail, funnel, and final-approach markers are replaced on the return lane. All markings are the same as described in previous paragraphs.

The full lane-marking pattern has three entrance and three exit markers. They are placed the width of forward and return lanes and are visually different from other markers. Units must be trained to recognize that three entrance markers indicate a two-way traffic lane and that they should always use the rightmost lane.

Entrance and exit funnel markers are placed slightly different from previous marking patterns. They extend out from the entrance and exit markers on the right side only.

Final-approach markers are placed 200 meters from, and centered on, entrances of forward and return lanes. This helps forces clearly identify the entrance points from either direction.

Far recognition markers are placed a maximum of 500 meters from the lane entrance or on the nearest terrain feature from forward and return final-approach markers.

COMMANDER'S GUIDANCE FOR LANE MARKING

Table 10-1 provides a summary of lane-marking levels, guidelines on unit responsibilities, and events that trigger lane upgrade. In the table, *who* refers to the unit responsible for lane upgrade marking and *when* describes events that trigger the need to upgrade.

Table 10-1. Lane-marking levels, unit responsibilities, and trigger events

	Initial	Intermediate	Full (Two Way)
Who	TF breach force	TF breach force	Brigade
When	Obstacle is reduced	Passing battalion- or company-size forces	Passing brigade- or battalion-size forces
	Passing platoon- or company-size assault forces	Passing force which cannot see the lane	Situation requires uninterrupted sustainment traffic
		Passing TF combat trains	
Markers	Entrance	Add right handrail	Expand lane width to 10 meters
	Exit	Add exit funnel	Adjust entrance/exit
	Left handrail	Add farside final approach	Adjust left/right handrails to new width
	Entrance funnel	Add far recognition	Add far recognition
	Final approach	Add guides or TCPs	Add farside guides or TCPs

LANE-MARKING DEVICES

The majority of lane marking in the field is done by using nonstandard marking devices. When adopting a nonstandard marking device, commanders should consider the guidelines summarized in Table 10-2.

Table 10-2. Guidelines for lane-marking devices

Marker	Mounted Forces	Dismounted Forces
Handrail and funnel markers	Visible by TC and driver (buttoned up) from 50 meters	Visible by a dismounted soldier in a prone position from 15 meters
	Quick and easy to emplace, minimizing the need to expose soldiers outside the carrier	Lightweight, quick, and easy to emplace (a dismounted soldier should be able to carry enough markers for the lane and still be able to fire and maneuver)
Entrance and exit markers	Visible by TC buttoned up from 100 meters	Visible by a dismounted soldier from 50 meters
	Visually different from handrail and funnel markers	Visually different from handrail and funnel markers
	Quick and easy to emplace (may require soldiers to dismount to emplace)	Lightweight, quick, and easy to emplace
	Easily man-portable	
Final-approach and far recognition markers	Visible by TC (not buttoned up) from 500 meters	Visible by a dismounted soldier on the march from 100 meters
	Visually different from each other	Visually different from each other
	Visually alterable to facilitate traffic control through multiple lanes	Visually alterable to facilitate traffic control through multiple lanes

Figure 10-25 shows some of the devices that can be utilized for lane marking, and they are easily procured or fabricated. This is not an inclusive listing but is intended to show commanders some of the options.

Some general requirements for lane marking are—

- Markers must be able to withstand the rigors of the terrain, the weather, and the battlefield.
- Markers should be easy to modify, using minimal manpower and equipment, when visibility is limited.
- Lane-marking panels should have thermal and IR reflective marking so that they can be easily identified during limited visibility.
- Enhancements for limited visibility should be a constant source rather than a pulsating strobe. Strobes do not make the marking pattern readily apparent, particularly when approaching from an angle.

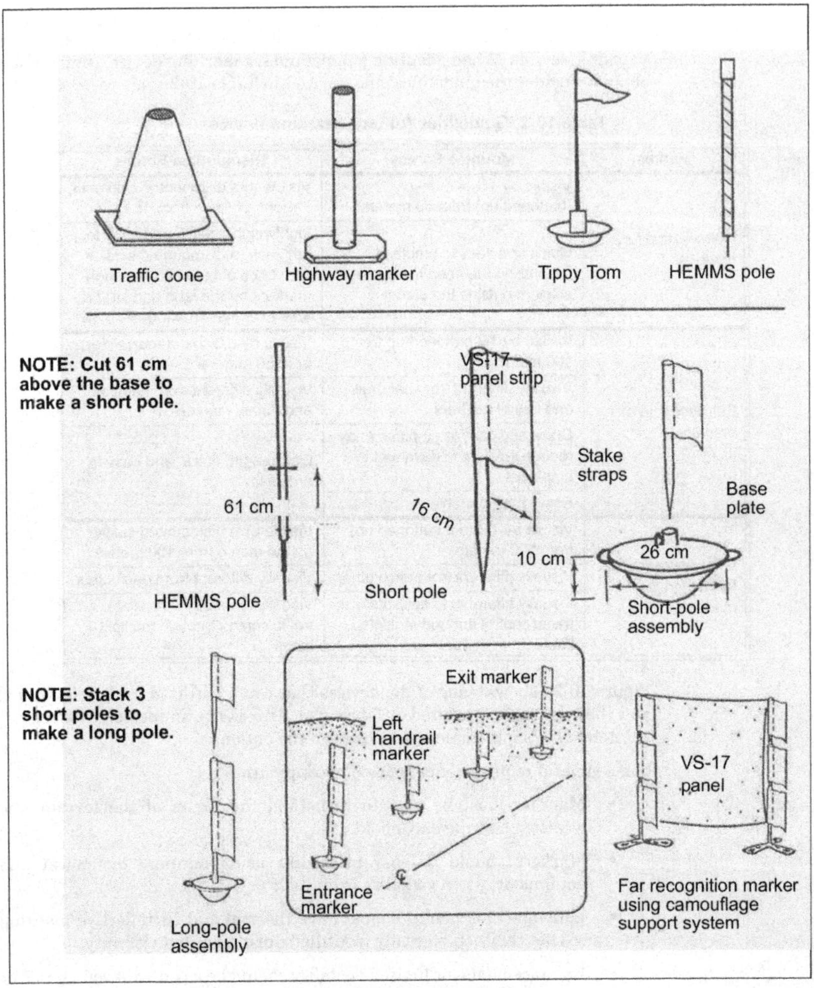

Figure 10-24. Marking devices

The following standard marking sets are available through normal supply channels:

- Minefield marking set number 2, line item number (LIN): M49096, NSN: 9905-00-375-9180.
- HEMMS, LIN: M49483, NSN: 9905-01-019-0140.

Marking Requirements of the North Atlantic Treaty Organization

The following paragraphs paraphrase the lane-marking requirements outlined in STANAGs 2889 and 2036. They also establish the procedures used by US forces to modify intermediate and full lane marking to STANAG standard. With the combined nature of warfare, commanders need to be aware of their responsibilities for marking hazardous areas, particularly breach lanes.

STANAGs 2889 and 2036 state that the type of marking device, pattern, and lighting used to mark breach lanes in forward areas is at the discretion of national authorities or the authorized commander. This gives commanders who are participating in a combined operation the flexibility to mark lanes consistent with their respective Army's standard. It also outlines minimum requirements for the lane-marking pattern before it is used by troops of other nations; however, commanders must plan for converting a lane to NATO standard as early as possible. When converting to NATO standard, the STANAG directs commanders to use lane-marking devices as stated below. Within an offensive operation, marking a lane to NATO standard will not normally occur until after the lane is matured to a full lane.

Marking Pattern and Device

The intermediate lane marking discussed earlier satisfies the minimum lane-marking pattern that must be used before forces from another country are able to pass through a lane. STANAGs 2889 and 2036 state that regardless of the marking device used, the entrance point, exit point, and left and right handrails are the minimum required lane signature. Therefore, once the lane is marked to the intermediate level, allied forces can use the lane without any additional marking.

STANAG 2889 requires that commanders convert marking devices to NATO standard as early as possible. Figure 10-26 shows a NATO standard marker. The marker is placed at right angles to the direction of travel, so that the white portion of the arrow points inward to the lane, indicating the safe side of the lane. The red portion is outward, indicating the lane limit or dangerous side of the lane. STANAG 2889 also requires that markers be large enough to be visible from 50 meters under most daylight conditions and have a field life of 60 days.

Conversion to NATO Standard Marking

To convert intermediate and full lane marking to NATO standard, affix NATO markers to long pickets and replace the existing entrance, exit, funnel, and handrail markers one for one (Figure 10-27).

Two NATO markers are used for entrance and exit markers to make them distinctly different. One NATO marker is affixed to each funnel marker and to

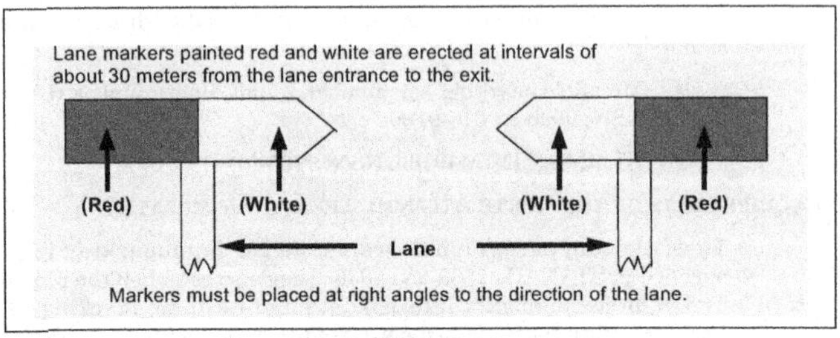

Figure 10-25. NATO standard marker

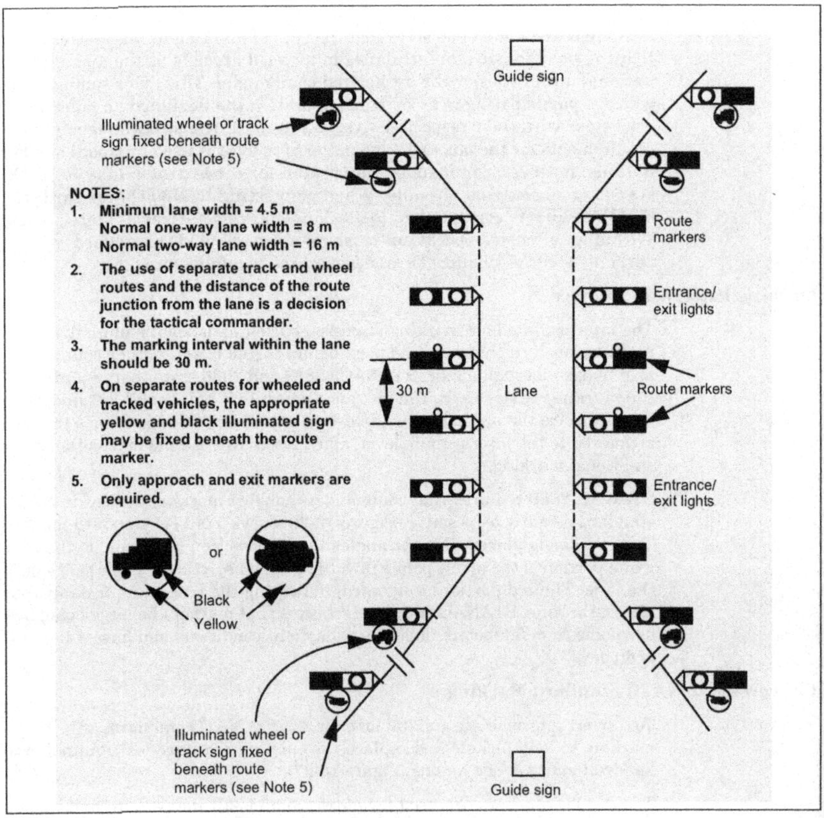

Figure 10-26. NATO lane-marking conversion

each left and right handrail marker. When converting full lane marking, the center handrail is marked with a modified NATO marker. The combination of a modified center handrail marker and directional arrows at each lane entrance provides allied forces with the signature necessary to distinguish two separate lanes. In addition, a barbwire or concertina fence (one strand minimum) is laid 1 meter above the ground to connect funnel markers, entrance markers, handrail markers, and exit pickets.

NATO uses white or green lights to illuminate markers at night (Figure 10-28). Entrance and exit markers are marked with two green or white lights placed horizontally, so that the safe and dangerous markings on them are clearly visible. One white or green light is used on funnel and handrail markers. The commander decides whether the light is placed on top of the NATO marker or placed so that it illuminates the markers. Lights must be visible from a minimum of 50 meters under most conditions and have a continuous life of 12 hours.

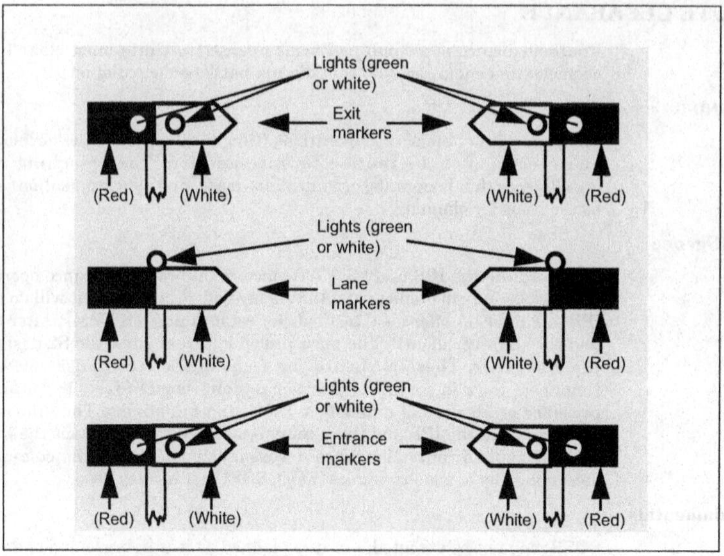

Figure 10-27. NATO standard marking for limited visibility

The mission to convert intermediate or full lane marking to NATO standard is normally assigned to corps-level engineer battalions working in the division rear area. In special cases, divisional engineer battalions may be tasked with NATO marking.

Chapter 11

Route and Area Clearance

The ability to move forces and material to any point in an AO is basic to combat power and often decides the outcome of combat operations. Maneuver relies on the availability of LOC within an AO; and during OOTW, clear LOC is essential to the movement of forces. Units must conduct route and area clearance to ensure that LOC enables safe passage of combat, combat support (CS), and CSS organizations. Clearance operations are best-suited for rear-area and stability support operations.

ROUTE CLEARANCE

Route clearance is a combined arms operation. Units must clear LOC of obstacles and enemy activity that disrupt battlefield circulation.

PLANNING

The principles of breaching operations (Chapter 9) apply to the development and execution of the route-clearance mission. The breaching tenets (intelligence, fundamentals, organization, mass, and synchronization) should be the basis for planning.

Intelligence

Incorporating the IPB and METT-TC factors into route-clearance operations will enable units to predict what the enemy will do and where it will do it. The IPB and the EBA offer ideal methods for establishing a SITEMP. After the S2 and the engineer identify the most probable threat sites, the S2 designates them as NAIs. These NAIs are the focus of the reconnaissance effort. Engineers work in concert with other reconnaissance assets to confirm the presence or absence of ambushes, UXO, and minefields. The information gathered from the IPB and the reconnaissance effort determines the method and the type of route clearance necessary. It also helps the commander determine any outside resources (EOD, SOF) that he may need.

Fundamentals

SOSR may not be executed, but it is planned as it is in breaching operations. Units must be prepared to execute SOSR fundamentals as necessary.

Organization

Task organization for a route clearance is similar to the task organization for a deliberate breach. The clearance company team is organized into breach, support, and assault forces. The breach force conducts clearing operations, the support force isolates the area being cleared, and the assault force performs security functions beyond the clearance site (traffic control points) and assists the breach force in disengagement, as required. Table 11-1 shows a sample task organization for a route clearance.

Table 11-1. Sample task organization for a route clearance

Team	Support Force	Assault Force	Breach Force
Heavy	• Mechanized infantry platoon with dismount capability • Armor platoon	• Mechanized infantry platoon • Engineer squad • Mortar section • Medical team (two ambulances) • PSYOP team • FIST • MP element	• Engineer platoon with organic vehicles • Armor platoon with plows and rollers
Light/Heavy	Two infantry platoons (light)	• Bradley platoon with dismount capability • Engineer squad • 60-mm mortar section • Medical team (two ambulances) • PSYOP team • Forward observer • MP element	• Engineer platoon with organic vehicles • Armor platoon with plows and rollers
Light	Two infantry platoons (light)	• AT/MP section with M60/MK19 mix • 60-mm mortar section • Medical team (two ambulances) • PSYOP team • Forward observer • MP element	• Engineer squad (+) • Infantry platoon (light) • AT/MP section with M60/MK19 mix

Mass

Sufficient maneuver and engineer assets must be allocated to the clearance company team. The length and the width of the route and the type of clearance to be conducted determine the size of the sweep team. Clearing a Class A military road with the deliberate sweep technique requires at least two engineer squads due to the total lane width to be cleared and the requirement for the rotation of mine-detector operators. Depending on the type of sweep operations, the commander can expect a 50 percent loss of sweep assets. Normally, as in breaching, a 50 percent redundancy of engineer assets should be allocated to the sweep team.

Synchronization

All aspects of synchronization should be implemented when planning route clearance. It is especially important that rehearsals be conducted at the combined arms level. Rehearsals should include—

- Reaction to enemy contact.
- Reaction to an ambush.
- Communications exercise.
- Fire support (obscuration smoke, immediate suppression fires, critical friendly zones for counterfire radar, and no-fire area around the clearance site).

- CSS (maneuver, casualty evacuation, marking materials, and demolitions resupply).

PLANNING CONSIDERATIONS

The purpose of breaching is to project combat power to the farside of an obstacle, and breaching usually occurs under enemy fire. Route-clearance operations focus on opening and maintaining LOC to ensure the safe passage of combat, CS, and CSS organizations. Like breaching, route-clearance operations require extensive BOS coordination. The following planning considerations should be used by brigade and battalion TFs when planning route-clearance operations:

Intelligence

- Identify choke points, bridges, tunnels, critical road junctions, and other built-up areas. These are the most suspect areas for obstacle emplacement. However, depending on the enemy's overall mission, it may not always emplace obstacles at these locations. This is especially true if the enemy's goal is to psychologically disrupt our convoys.
- Maintain a situation map with a graphics overlay that reflects the most current intelligence information.
- Maintain an incident map with a graphics overlay to facilitate a pattern analysis.
- Maintain a threat order-of-battle database, such as how the enemy will disrupt unit LOC.
- Develop a detailed R&S plan that incorporates modern battlefield techniques and systems, such as ground sensors, forward-looking airborne radar, and satellite images. As a minimum—
  - — Coordinate for UAV support, if available.
  - — Develop infiltration routes to support recon and security at likely enemy ambush sites.
  - — Develop an estimate of impact to civilians on the battlefield (COBs). COBs include local nationals, nongovernment organizations (NGOs), and private volunteer organizations (PVOs).
  - — Conduct a daily flight over the area to provide up-to-the-minute intelligence. When available, coordinate ASTAMIDS coverage.
  - — Coordinate with the USAF to periodically check the route (for example, using an AC-130 Specter gunship).
- Provide an intelligence update to company team leaders before departure. This should be in the form of a 1:50,000 enemy SITEMP overlay (confirmed and suspected/templated).
- Establish liaison with the host nation, NGOs, and SOF.

Maneuver

- Provide personnel for TCPs.

- Clear and secure flanks (at least 100 meters) and the farside of suspected and known obstacle locations.
- Close the route to US-controlled traffic during route-clearance operations to minimize the target presented to enemy forces.
- Identify and clear potential sniper positions before beginning obstacle reduction or clearance.
- Provide security for the cleared route.
- Give operational control (OPCON) of aviation assets to the route-clearance commander for clearance-support missions.
- Plan the building of static security points along the cleared route to reduce the probability of reseeding.

Fire Support

- Plan smoke for templated locations.
- Position mortars to ensure continuous coverage of the operation (move and set up with the support force).
- Prepare fires within the tactical rules of engagement.
- Ensure that the route-clearance team has a FIST coordinator. The clearance commander should locate the FIST element well forward in the order of march.
- Designate obstacle clearance sites as critical friendly zones for counterfire radar and no-fire areas.

NOTES:

1. Priority targets shift in conjunction with company team movement on the main supply route (MSR).

2. Clearance of fires is the responsibility of the maneuver commander of the sector where the target is located.

3. Adequate Q-36 radar coverage is necessary for deliberate sweep operations.

Mobility/Survivability

- Provide detailed OBSTINTEL on minefields. It must include the—
  - Description of mines or explosive devices most likely encountered.
  - Composition and pattern of obstacle(s).
  - Enemy actions or techniques used during obstacle emplacement.
- Conduct deliberate sweep operations 100 meters past the obstacle or suspected threat.
- Report, clear, and mark mines, obstacles, and explosive devices to facilitate unimpeded movement.
- Ensure that lane marking meets the standards outlined in Chapter 10 and that materials and techniques are standard throughout the route.

- Consider including road repair equipment and material as part of the sweep element (for example, a 5-ton dump truck filled with soil and an ACE to spread the soil).
- Keep all radios, electronic equipment, and aviation assets at a safe distance during reduction operations.
- Block uncleared roads and trails that branch from the route being cleared. This protects units from inadvertently traveling an uncleared route.
- Debrief the chain of command and the TF S2 on the location, the composition, and the orientation of all obstacles cleared and encountered. This assists the S2 and the engineer in IPB/EBA pattern analysis.

Air-Defense Artillery

- Consider the possibility of an air attack.
- Use the following passive air-defense measures:
  - Eliminate glare by using mud, tape, cardboard, or camouflage nets to cover headlights, mirrors, and portions of windshields.
  - Reduce dust clouds by reducing speed.
  - Plan routes that offer natural concealment.
  - Use air guards.
- Increase the distance between vehicles.
- Incorporate Stinger missile teams into the support force.

Combat Service Support

- Ensure that clearance operations are supported by a logistical/combat health support (CHS) package from the brigade support area.
- Plan for air and ground evacuation of casualties. The preferred evacuation method is by air; the routine method is by ground.
  - Conduct an air-mission brief with air ambulance assets, to include pickup zones and markers. Rehearse procedures for evacuation requests.
  - Ensure that the medical team consists of one or two ambulances. Locate the medical team with the support force.
  - Identify the ambulance exchange point along the route to be cleared.
- Ensure that all personnel wear flak vests or IBASIC (Figure 11-1, page 11-6).
- Ensure that all vehicles have tow cables in the front and the rear for extraction purposes.
- Ensure that all vehicles carrying troops have hardening (sandbags on floors and sides).

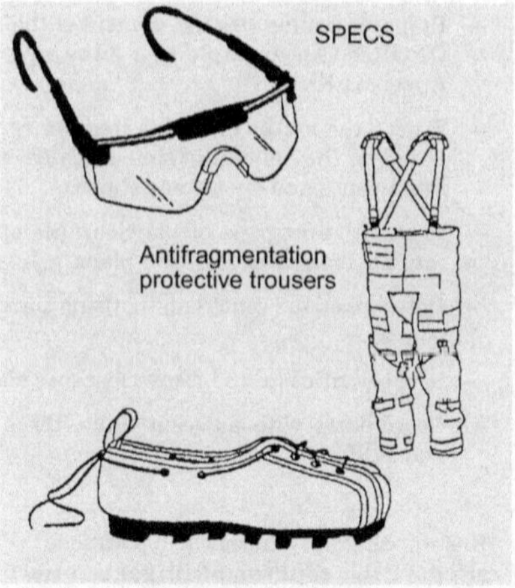

Figure 11-1. IBASIC

- Provide MP and explosive-sniffing dogs to help in clearance and provide security for convoys during and after clearing operations.

Command and Control

NOTE: The company team commander is required to operate on three separate frequencies—battalion command network, company team command network, and fire-support network.

- Designate, recognize, and include minefield indicators (Chapter 10) as part of company team rehearsals.
- Designate a reserve force (at least platoon-size) that is mechanized or air-assault capable.
- Ensure that proper rehearsals are planned and conducted according to FM 3-34.2. As a minimum, the clearance force should rehearse actions on the obstacle, actions on enemy contact, casualty evacuation, and the control of COBs.
- Ensure that the tasked unit has a clear understanding of the mission, intent, and end state. For example, the clearing unit commander should understand that his unit must clear the road width, including the shoulders, and secure the route.
- Assign clearance responsibilities to brigade and battalion assets.
- Ensure that the maneuver commander/TF S3—

- Controls the movement of all personnel and equipment along the route (travel authorization is coordinated through the S4).
- Prepares a mine risk assessment of the mission before issuing the OPORD. (An example of a mine risk assessment is shown in Appendix F.)
- Tracks the status of routes (red, amber, green) in the TF sector, based on the amount of time since the route was cleared and the intelligence and enemy situations.
- Tracks the progress of the clearance operation and integrates it into the maneuver and CSS plans.
- Determines the route length, using clearly definable start and end points.
- Sets priorities for the route-clearance element.
- Coordinates with adjacent units, the host nation, NGOs, PVOs, and SOF.

Special Operations

- Ensure that psychological operations (PSYOP)/civil affairs (CA) support the counterintelligence effort by conducting civilian interviews.
- Direct civilians along the MSR to the displaced-personnel holding areas or along the routes that the brigade has indicated for use.
- Employ PSYOP/CA teams forward to disperse civilians and provide traffic management to isolate the route during clearance operations.

TASK ORGANIZATION

A brigade or battalion TF normally conducts clearance-in-zone operations. To clear a route, the battalion TF focuses a company team as the main effort on the proposed MSR. Table 11-1, page 11-2, shows a sample task organization for a route clearance.

Support Force

This force is comprised of two maneuver platoons and the maneuver company team XO. The support force provides flank security, rear security, and protection to the breach force. It neutralizes hostile forces encountered by the company team. In rugged terrain or highly mined areas, moving the assault force on the flanks would be too risky. Aviation assets can provide flank security while ground forces provide rear security. The assault force also searches for suspected off-route mines.

Assault Force

This force is comprised of a maneuver platoon, an engineer squad, a mortar section, a medical team, a PSYOP team, an EOD team (or one that is on call), and a forward observer. The assault force's mission is the same as in a breaching operation (Chapter 9).

Breach Force

This force is comprised of a maneuver platoon (including the commander) and an engineer platoon (minus). The breach force sweeps the route and reduces mine and explosive threats. It is further task-organized into sweep teams.

A sweep team is a trained detection team that searches for mines and explosive devices. The organization of the sweep team depends on the type of sweep mission and the length, the width, and the surface composition (pavement, gravel, dirt) of the area to be swept. A platoon-size element can normally clear a 4.5-meter-wide path, and a squad-size element can normally clear a 1.5-meter-wide path. If the route is wider or time does not permit multiple passes of the route, additional engineer assets are required. Table 11-2 outlines personnel and equipment requirements for a sweep team.

Table 11-2. Personnel and equipment requirements for a sweep team

Personnel	Support Personnel	Equipment
• NCOIC • Mine-detector operators • Probers/markers • Radio operator • Demolition teams	• Medics • Vehicle operator	• One panel marker • Operational map with required maneuver graphics • Four smoke grenades (minimum) • Six mine detectors (includes three backups) and extra batteries • Two grappling hooks with 60 meters of rope each • One demolition kit or bag for each demolition man • Six probes • Mine marking material • Lane marking tape • Ten stakes

Platoon-Size Sweep Team

The normal configuration for a platoon-size sweep team is twelve soldiers in a modified column (Figure 11-2). The platoon leader supervises the entire operation. This configuration is best suited for sweeping routes in friendly territory that is not under constant surveillance.

When sweeping in areas where 100 percent coverage is required, the team should establish a clear lane from which to operate. This is conducted by the grapple hook being thrown one additional time into the suspected minefield. The cord on the grapple hook is pulled taunt, with the free end secured to a stake.

This establishes a left or right physical boundary marker for the mine-detector operator to use as a guide. The first mine-detector operator will sweep a 1.5-meter lane left or right of this cord, up to the grapple hook. At the hook, a second stake will be emplaced to hold the boundary line and the grapple hook will be thrown again and pulled for trip wires as before. A fourth throw will serve as the boundary marker through the suspected minefield. This procedure will continue until the first mine-detector operator is out of the minefield on the farside.

The boundary marker will then be secured to stakes, using the lane marking tape. A second boundary maker will be placed 1 meter to the left or right of the original boundary marker, depending on which side was swept. This will

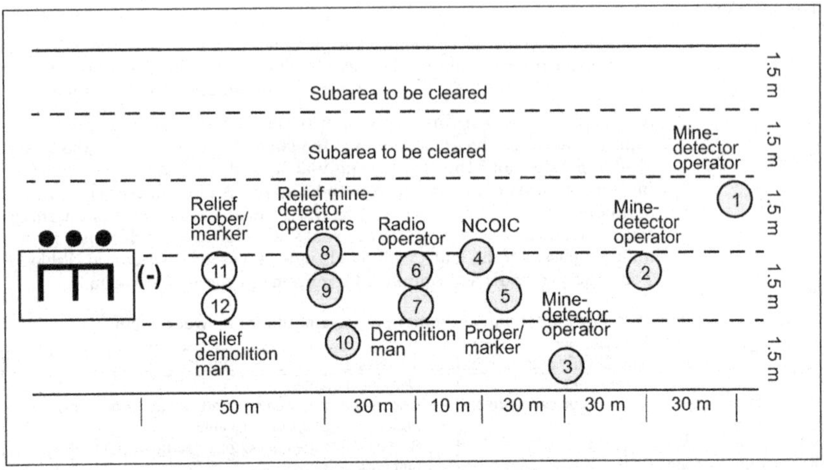

Figure 11-2. Platoon-size sweep team

establish a fixed 1-meter lane. The 0.5-meter area that was not included in the original lane is to ensure that no gaps in coverage were missed on the edge of the first lane.

The unit will then stretch a piece of marking tape out from across the clear footpath, moving it left or right 1.5 meters and staking it down on both ends. This will be repeated as many times as necessary.

Always mark lanes that have not been cleared with a piece of marking tape across the entry to that lane to prevent personnel from using the lane before it has been swept. After multiple lanes are physically marked, the unit can perform a platoon-size sweep. Lanes should be swept and cleared by working out from the original 1-meter footpath. As the cleared area becomes bigger, the marking tape is removed to allow the passage of personnel and equipment though the cleared lane.

As subsequent lanes are being swept, the marking tape will need to be secured on the outside edge of the lane. Always conduct a thorough sweep of the spot where the stake will be placed. Stake intervals will depend on terrain and atmospheric conditions.

Soldiers 1, 2, and 3 (mine-detector operators) lead the sweep team in echelon. Each sweep team covers 1.5 meters of front, and sweep teams are spaced 30 meters apart to prevent fatalities from accidental detonation by other mine-detector operators. If required, a fourth mine-detector operator can be added to the detection column.

Soldiers 4 (NCOIC) and 5 (prober/marker) follow 30 meters behind the last mine-detector operator (Soldier 3) and are centered in the cleared lane. The prober/marker is responsible for marking the cleared lane on both sides.

Soldiers 6 (radio operator) and 7 (demolition man) follow 10 meters behind Soldiers 4 and 5 and are centered in the cleared lane.

Soldiers 8, 9, 10 (relief mine-detector operators), 11 (relief prober/marker), and 12 (reserve demolition man) follow 30 meters behind Soldiers 6 and 7. If a fourth mine-detector operator is added to the column, an additional relief mine-detector operator must also be added.

The remaining platoon members help the support force or act as a reserve force, as required. They should first be integrated into the sweep team as a relief element and then moved forward as needed.

Squad-Size Sweep Team

The normal configuration for a squad-size sweep team is seven soldiers in a modified column (Figure 11-3). The squad leader supervises the entire sweep operation. This configuration is designed for sweeping routes in friendly territory that is not under constant surveillance.

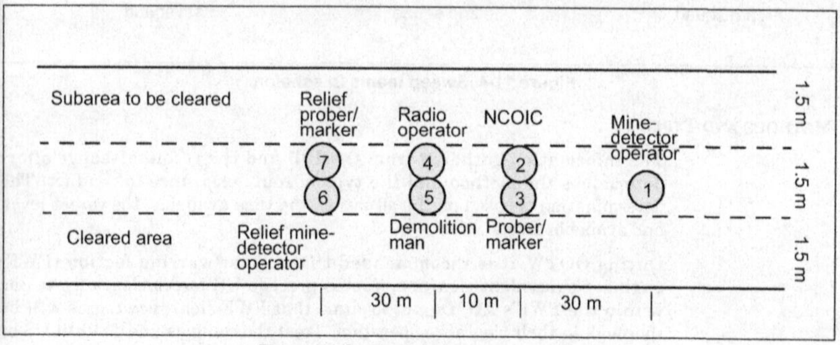

Figure 11-3. Squad-size sweep team

Soldier 1 (mine-detector operator) leads the sweep team and covers a 1.5-meter-wide path.

Soldiers 2 (NCOIC) and 3 (prober/marker) follow 30 meters behind Soldier 1 and are centered in the cleared lane. The prober/marker is responsible for marking the cleared lane on both sides.

Soldiers 4 (radio operator) and 5 (demolition man) follow 10 meters behind Soldiers 2 and 3 and are centered in the cleared lane.

Soldiers 6 (relief mine-detector operator) and 7 (relief prober/marker) follow 30 meters behind Soldiers 4 and 5. If the squad cannot use seven team members, the relief prober/marker position can be eliminated from the formation.

The engineer platoon can configure the platoon into squad-size sweep teams and place them in echelon (Figure 11-4).

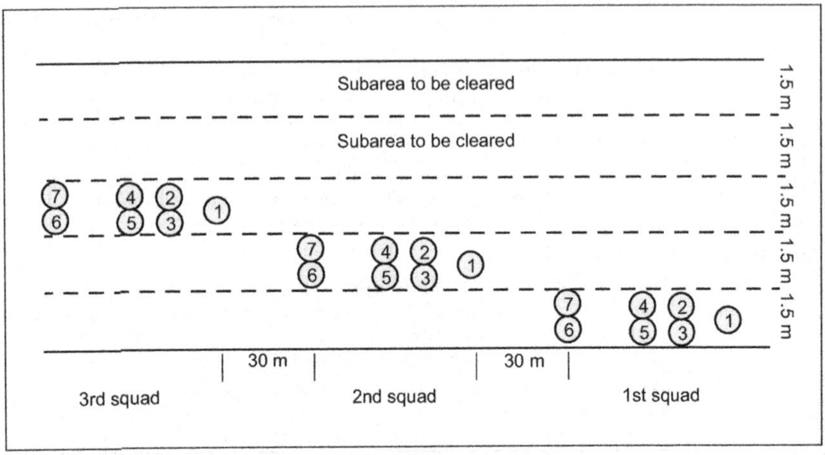

Figure 11-4. Sweep teams in echelon

METHODS AND TYPES

The information gathered from the IPB and the reconnaissance effort determines the method and the type of route clearance to conduct. The determination is based on the situation, the time available, the threat level, and available assets.

During OOTW, it is recommended that former warring faction (FWF) engineer-equivalent clearance teams precede US forces clearance teams within the FWF's AO. Do not assume that FWF clearance teams will be thorough in their clearance operation. Treat the route as unsafe until US or allied force clearance teams have proofed the route to confirm that it is cleared.

Methods

There are three methods of route clearance—linear, combat, and combined. The method employed depends on the situation, the time available, and the clearance assets available. The maneuver force should always establish static security positions at critical locations following the completion of route clearance.

Linear Clearance

In linear clearance (Figure 11-5), sweep and security teams begin route clearance at Point A and complete it at Point B. This method provides the best assurance of route coverage. Although this is an effective method, it is not the most secure method in a high-threat environment. It is also time-intensive and constrains the maneuver commander's flexibility.

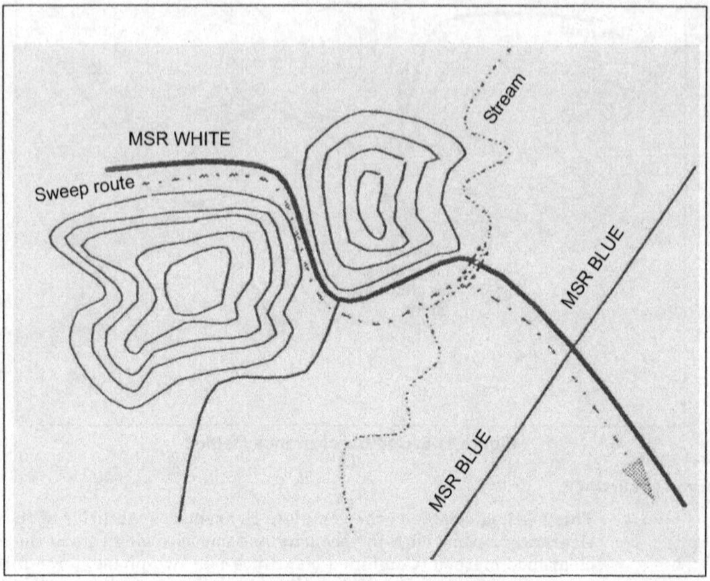

Figure 11-5. Linear clearance method

Combat Clearance

Whereas linear clearance focuses on a specific route, combat clearance (Figure 11-6) focuses on specific points along a route. As mentioned previously, IPB and EBA can identify areas for likely mine and ambush locations. These areas become NAIs or objectives for combat clearance missions. The combat clearance method divides a route into sections according to the number of suspected high-threat areas. Once the sweep element (maneuver and engineer forces) secures and sweeps these areas, the route is considered clear. Combat forces can patrol the route from these objectives to ensure that the route is secure, and if necessary, the sweep element can sweep the surrounding area if a minefield is found. Following the seizure of these objectives, the commander must assume a moderate risk that the S2 and the force engineer have identified all high-threat areas and that the route is clear of mines. Combat clearance is ideal for dismounted (light) forces since it provides the maximum use of surprise and concealment.

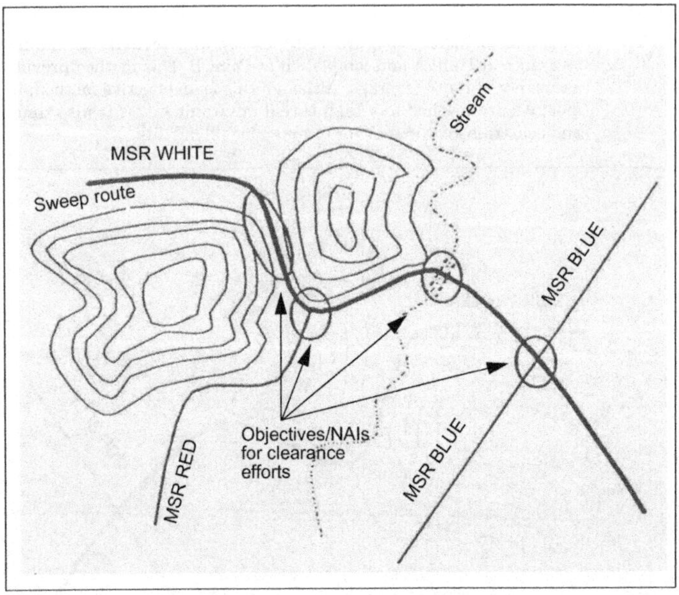

Figure 11-6. Combat clearance method

Combined Clearance

This method combines the complete clearance capabilities of the linear clearance method with the security and surprise elements of the combat clearance method. Combined clearance is a two-phase, force-intensive operation and may require a battalion-size effort, depending on the length of the route. First, identify high-threat areas through IPB/EBA and target them as NAIs and/or objectives to secure. Then, clear obstacles and enemy forces before the movement of sweep elements. The sweep team moves down the road and clears any obstacles missed or not identified during the planning process. The main advantage of this method is that the TF commander has immediately secured MSRs, allowing him to *push out* (expand forces out past the secured area and secure additional areas) and find the enemy with a degree of confidence that follow-on forces will be much safer.

Types

There are two types of sweep operations—deliberate and hasty. Deliberate and hasty clearance operations can be modified to meet the time and equipment limitations of the TF, but the commander assumes greater risk when the clearance type is modified.

Deliberate

A deliberate sweep (Figure 11-7) is very thorough and includes a complete sweep of the entire road (shoulders, culverts, ditches, and bridges). It is the most time-consuming sweep operation and relies on electronic (primary) and visual (secondary) detection systems.

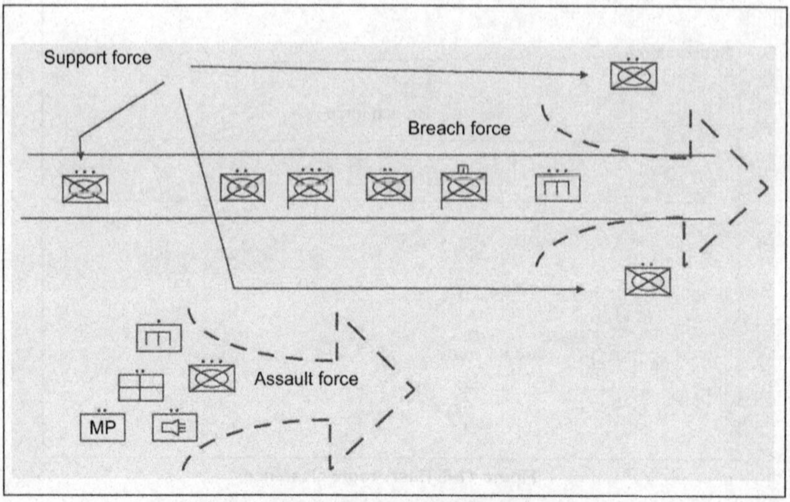

Figure 11-7. Deliberate route clearance

The platoon sweep team (Figure 11-2, page 11-9) is dismounted to focus its attention on the entire length of the route. The support force (company-size) secures at least 100 meters on the flanks and 100 meters forward to clear possible enemy direct-fire systems and overwatching elements in front of the breach force. This not only allows the breach force to focus solely on the route but also clears the area of off-route and command-detonated mines.

If enemy contact is made, the support force fixes the threat while the assault force reacts. The sweep teams withdraw to a location that provides concealment and/or security. Mechanical detection provides a third means of detection and is the method used to proof the route after the sweep team has passed through the area. The deliberate sweep includes a route reconnaissance and looks at all areas of a route, including bypasses. The deliberate sweep focuses on thoroughness rather than speed. This method is very slow and tedious and should only be used when time is not a factor; 80 to 100 meters can be covered per hour.

Hasty

A hasty sweep (Figure 11-8) consists of visual inspection, physical search or probing, and the use of mine detectors. It is the fastest, most risky method and is suited for an armored or mechanized team. It relies primarily upon visual detection (thermal sights or the naked eye) for minefield identification. The breach force looks for mines, wire, and other minefield indicators. The road surface, culverts, ditches, and bridges are inspected and searched. Visual detection is accompanied by a mechanical proofing system. Electronic mine detectors are used by sweep teams to check all suspected areas.

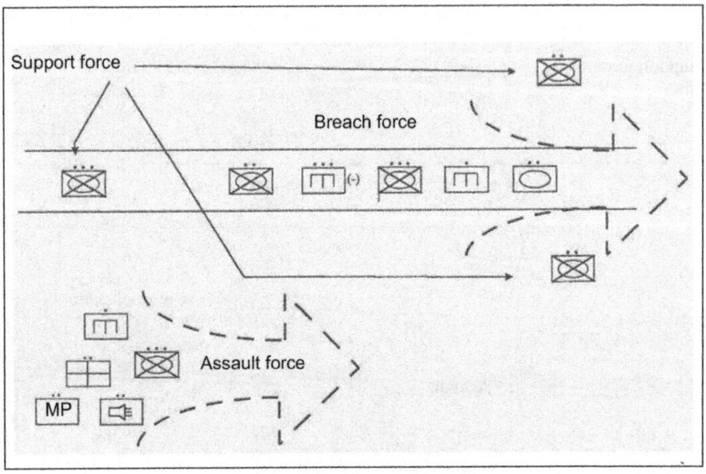

Figure 11-8. Hasty route clearance

The support force includes a maneuver platoon that provides overwatching fire and/or security. Actions upon enemy contact are the same as in a deliberate sweep. The primary objective of this technique is speed, moving approximately 3 to 5 kph. This method is extremely similar to the instride breach method when encountering minefields.

The sweep team focuses on identifying immediate risks to traffic, neutralizing those risks, and continuing on with the mission. A hasty sweep is used during the combat clearance method to validate the areas that were not deliberately cleared by the sweep team. It is also used if the METT-TC analysis does not permit a deliberate sweep or if the need for a road to be opened is urgent. Time and distance factors may be imposed. A light force may not have an MCR system but can conduct the same sweep method with an improvised roller system, or the force can use a sandbagged, 5-ton truck moving backwards as a last-resort method. Using MCRs or their equivalent is absolutely imperative due to the high risk of encountering a minefield. The mine rake or plow is not a satisfactory substitute because it destroys road surfaces.

AREA CLEARANCE

Area clearance is also a combined arms mission. Clearing operations occur when engineers receive a mission to clear an area of mines or to clear a specific minefield in a friendly AO. In most cases, the minefield has been reported and may already be marked on all sides. The engineer unit receiving the mission must base its plans on available information and prepare equipment based on the estimate.

PLANNING

Planning area-clearance operations is very similar to planning breach operations. Commanders and staffs plan and coordinate all the breaching tenets.

Intelligence

Intelligence is particularly important for discovering the types of mines and mine fuses the enemy employs. The engineer uses this information to determine which clearance and neutralization techniques offer the best chance for success and minimize the risk to the sweep teams. Intelligence also helps the commander determine the need for outside resources, such as EOD and SOF elements.

The information needed for area-clearance operations includes—

- Minefield location.
- Minefield orientation.
- Presence of wire as an obstacle.
- Location of gaps and bypasses.
- Minefield composition (buried or surface-laid mines, AT or AP mines, AHDs, trip wires, and minefield depth).
- Types of mines, employment techniques, fuses, and booby-trap configurations.
- Information on whether or not the minefield is marked and, if it is marked, the material used.
- Possibility of hostile forces in the area.

Fundamentals

You must plan to apply the four fundamentals of breaching operations (SOSR), but you may not execute all of them due to the lack of enemy presence.

Organization

Task organization is similar to that used for route-clearance operations. The breach force is the clearance element, and the support force is responsible for all security and maneuver responsibilities. There is no assault force in area-clearance operations. Table 11-3 shows a sample task organization for an area clearance. The size of the force can be tailored, based on the probability of contact.

Mass

Sufficient maneuver and engineer assets must be allocated to the clearance company team. The length and width of the route and the amount of time

Table 11-3. Sample task organization for an area clearance

Team	Support Force	Breach Force
Heavy	• Mechanized infantry platoon • Mortar section • FIST • Armor platoon	• Engineer platoon with organic vehicles • Medical team (two ambulances) • EOD team
Light/heavy	• Bradley platoon with dismount capability • 60-mm mortar section • Forward observer • One infantry platoon (light)	• Engineer platoon with organic vehicles • Medical team (two ambulances) • EOD team
Light	• AT/MP section with M60/MK19 mix • 60-mm mortar section • Forward observer • Two infantry platoons (light)	• Engineer squad (-) • Medical team (two ambulances) • EOD team

available determines the size of the sweep team. A platoon is normally used to clear a 200- by 300-meter minefield; additional assets are required to clear larger and multiple minefields.

Synchronization

All aspects of synchronization should be implemented when planning area clearance. It is especially important that rehearsals be conducted at the combined arms level. Rehearsals should include—

- Reaction to enemy contact.
- Reaction to an ambush.
- Communications exercise.
- Fire support (obscuration smoke, immediate suppression fires, critical friendly zones for counterfire radar, and no-fire area around the clearance site).
- CSS (maneuver, casualty evacuation, marking materials, and demolitions resupply).

PLANNING CONSIDERATIONS

Area-clearance BOS planning considerations are parallel to route-clearance planning considerations. Brigade and battalion TF staffs should use the planning considerations outlined on pages 11-3 through 11-7 plus the ones outlined below:

- Intelligence. Focus on the most probable enemy attack method and AAs.
- Maneuver.
  - Clear and secure flanks (at least 500 meters) and the farside of the area to be cleared.
  - Provide security for the cleared area.

- Fire support. Ensure that the area-clearance team has a FIST coordinator. The FIST should be collocated with the support force OIC.
- Mobility/survivability. Establish minefield control points along the area to be cleared.
- CSS.
  - Ensure that the medical team consists of one or two ambulances and that it is located with the breach force.
  - Ensure that all personnel wear flak vests or IBASIC (Figure 11-1, page 11-6).
- C2.
  - Determine the area length, using clearly definable perimeter points.
  - Coordinate with adjacent units, the host nation, NGOs, PVOs, and SOF.

TASK ORGANIZATION

The battalion TF will focus a company team (minus) as the main effort to conduct area clearance.

Support Force

This force is comprised of two maneuver platoons and an OIC. The support force provides flank security, forward security, and protection for the breach force. It neutralizes hostile forces that are encountered by the company team. The support force secures the area 500 meters beyond the area to be cleared. METT-TC factors will affect the actual distance based on the threat and the weapon systems. The support force OIC establishes static security positions around the area until the clearance operation is complete. He also has control of fires and the responsibility to neutralize any hostile force.

Breach Force

The breach force is comprised of an engineer platoon that is organized into sweep teams, a medical team, and an EOD team (or one that is on call). The sweep team (squad-size) is organized as shown in Figure 11-3, page 11-10. The breach force's mission is to sweep and clear the area of mine and explosive threats.

METHODS AND TYPES

The breach force OIC determines the perimeter of the area to be cleared and ensures that it is marked. The OIC divides the area into sections to be cleared (Figure 11-9). The sections should be no larger than 40 meters wide and 100 meters long. This is an optimal-sized area for a sweep team to clear at one time. The OIC assigns squad-size sweep teams to each section.

The squads clear their assigned sections using the sweeping techniques discussed earlier in this chapter. As the sections are cleared, they are marked for safety and control purposes. This process is continued until the entire area is cleared. Progress is reported to the company team commander as required.

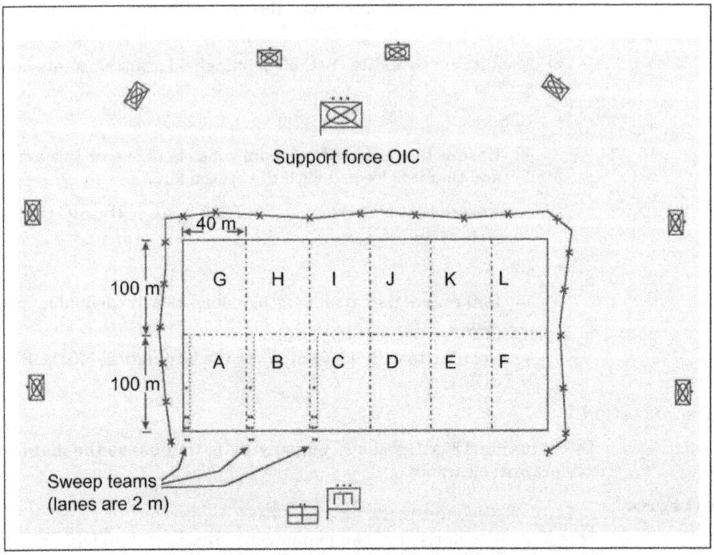

Figure 11-9. Area clearance site layout

IMPROVISED MINE THREAT

Mines are not always employed conventionally by military forces organic to the host nation or its enemies. In many cases, they are also employed by terrorists against allied forces or the host-nation populace. In these cases, the threat increases because of the improvised methods in which the mines were emplaced. In conventional emplacement of mines, a pattern emerges from the emplacing force's doctrine, and the threat can easily be reduced by using this knowledge. There is less pattern in the case of improvised mining methods, and this makes detection and removal very difficult.

Improvised mining has many different employment techniques. In most of the techniques shown below, a UXO can easily be employed in place of a mine:

- Coupling mines. Coupling is done by linking one mine to another, usually with detonating cord. When the initial mine is detonated, it detonates the linked mine. This technique is done to defeat countermine equipment.
- Boosting mines. Buried mines are stacked atop one another, and the farthest mine from the surface is fused. This reduces the probability of detection and increases the force of the blast.
- Sensitizing AT mines. On some nonmetallic AT mines, the pressure plate can be cracked and the spring removed or the mine's explosive can be cut into smaller blocks and employed as powerful AP mines.

The pressure plate can be removed from metallic AT mines and employed in the same manner. Alternatively, a pressure-fused AP mine can be placed on the top of an AT mine.

- Mixing training mines with live mines. Hostile forces can employ training mines at the start of a minefield and emplace live mines toward the end. The sweep element falsely believes that the minefield is phony and becomes complacent in its reduction activities. When this technique is used, live mines are painted to resemble training mines.
- Daisy-chaining mines. Command-detonated AP mines are commonly used in daisy chaining. Hostile forces link the mines with trip wires or detonating cord. When the initial mine is detonated, the other mines will detonate.

MINE LOCATIONS

Hostile forces normally place more than one mine in each mined area. Do not focus the detection effort solely on a horizontal mine threat, such as on the ground or in culverts. The mine threat is also vertical, such as in trees or attached to an overpass. Clearance efforts must accommodate the three-dimensional battlefield. Mines may be placed in—

- Frequently used roadways.
- Brush and other traffic obstructions placed on roadways.
- Bridge bypasses and fording sites.
- Road junctions.
- Obvious turnarounds, bypasses, culverts, ditches, and shoulders.
- Key logistic points (water, fuel, food).
- Debris along a route.

DISPOSITION OF MINES

The following actions should be taken when a suspected mine is found:

- Mark the suspected mine location; do not leave any mine unmarked.
- Search for electric wires and trip wires in the immediate area. Trace the wires in both directions to determine if items are attached to them. If there is nothing attached and the IPB does not state otherwise, cut loose trip wires and electric wires.

DANGER

Never cut taut trip wires. Alert the security element to search for an enemy that may be manning a command-detonated mine. Keep troops away from the mine until all the wires are traced and cut. Be alert for booby traps and ambush. If booby traps are found, use the clearance procedures outlined in Chapter 13.

- Probe the suspected mine location and uncover enough of the object to identify it. Other personnel should stay at least 30 meters away.

- — If the object is a mine, the prober withdraws and notifies the OIC. The OIC decides to bypass the mine, destroy it in place, remove it with a grapnel, or notify EOD for hand neutralization.
- — If the object is debris, get in a protected position and carefully remove the debris with a grapnel hook. Be alert for booby traps or AHDs wired to the debris.

MINE-REMOVAL TECHNIQUES

A mine can be bypassed, detonated in place, pulled out by a rope or a wire, or neutralized and removed by hand. The method used depends on the location of the mine, the type of the mine and the fuse, and the tactical situation. Methods of removal and actions on finding a mine should be addressed in the OPORD and rehearsed prior to executing the mission.

Trip-wire and tilt-rod mines can be detonated by throwing a grapnel, with a rope attached, past the trip wire or tilt rod and pulling the grapnel back to actuate the mine. Grapnels may be improvised from any available material, such as a bent drift pin or scrap material.

A hand-emplaced charge is the standard demolition material used to destroy a mine in place (see FM 5-250). A 1-pound block of explosive placed next to a mine is sufficient to detonate most mines. A charge can be placed next to each mine in a group, then the charges can be connected and fired simultaneously.

Rope or wire can be used to pull a mine out of its installed position. This is a safe method and only detonates mines that are equipped with AHDs. It also reduces noise and cratering. A tripod (Figure 10-21, page 10-23) makes it easier to pull a mine out of a hole on the first attempt. Use the following procedures to remove mines:

- Uncover only enough of the mine to expose a handle or a projection. Attach a 60-meter length of rope or wire to the mine or engage a grapnel. If there is no projection, engage a grapnel on the bottom side of the mine, opposite the direction of pull.

DANGER
Do not move the mine while uncovering it or attaching the rope because movement might detonate an AHD.

- Ensure that the covered area is not mined. Take cover and lie in a prone position at least 50 meters from the mine. Pull the rope to remove the mine from the hole.
- Wait at least 30 seconds before leaving cover and approaching the mine if the mine type is unknown. This guards against the possibility of a delay firing mechanism.
- Dispose of the mine according to the unit directive or SOP.

HAND NEUTRALIZATION

Appendix A discusses procedures for hand neutralization of US mines. Foreign mines and booby traps should only be neutralized by EOD personnel.

Mines are neutralized by hand, when—

- Units are conducting a covert breach.
- The mine is located on a bridge, building, or other facility required for use by friendly forces.
- Neutralization by other means is not possible.
- The mine can be positively neutralized by hand and is required for reuse.
- The mine type is unknown and recovery must be attempted for intelligence purposes.
- Chemical mines are located in areas where contamination would restrict the use of the area by friendly troops.

SAFETY

The following safety procedures should be observed during route and area clearance:

- Personnel should wear helmets and flak jackets to protect them against fragmentation. Sweep team members should wear IBASIC, if available.
- Vehicle floorboards should be sandbagged.
- Vehicles should be dispersed at 50-meter intervals. This ensures that a mine detonated by one vehicle will not cause casualties in other vehicles.
- One person at a time should be allowed at a suspected mine location.
- Personnel should assume that mines and explosive devices are equipped with AHDs until proven otherwise.
- Personnel should not run and should move only in previously cleared areas.
- Armored vehicles should have their hatches open to vent the pressure pulse from a mine detonation.
- Soldiers should wear ballistic and laser protective spectacles (BLPS) or lightly tinted, protective eyewear to reduce eye fatigue and improve their ability to recognize mine indicators.

REPORTS

Dissemination of information is the key to battlefield management. Units encountering minefields or explosive devices should follow a five-step process—stop, secure, mark, report, and avoid. Units must provide adequate information to their higher headquarters to ensure that follow-on elements are well informed. Information must include the known or suspected minefield location, types of mines (if known), the marking method, the time the minefield was encountered, and any additional information that may be of use to the clearing unit.

Division and maneuver brigades must establish a central control cell for mine clearance information. This cell receives and gathers all mine and explosive threat data within the unit's AO. Mine-contact reports are reported through maneuver command channels with a priority of *flash* or *immediate*. The information is then jointly controlled in the operations cell and the central mine control cell by the engineer staff officer, the G3/S3, and the Assistant Chief of Staff, G2 (Intelligence) (G2)/S2. The mine contact database is jointly maintained by the engineer and the G2/S2, who subsequently conduct pattern analysis and integrate it into intelligence and operational updates.

The central mine control cell performs the following actions:

- Maintains a current situation map and overlay that depict friendly and enemy mines and obstacles.
- Maintains and updates enemy obstacle (Figure 11-10, page 11-24) and route status (Figure 11-11, page 11-25) information.
- Receives and maintains minefield recording forms (US and foreign) within the unit's operational area (this includes host-nation minefield data, if available).
- Maintains a mine-contact database. (This could be a clearinghouse for future operations.)
- Processes, analyzes, and updates information; disseminates the information to subordinate commanders and staff.

Situation Report

Clearing units submit a situation report to higher headquarters if enemy activity is encountered or if an explosive device is discovered. This information should be tracked in the TOC and the CTCP. Information must be disseminated to subordinate units, especially CSS elements.

Progress Report

The clearing unit submits progress and completion reports until the clearance operation is complete. Progress reports must be timely and accurate. Report format and frequency are established in the OPORD before the clearance mission is executed.

ENEMY OBSTACLE REPORT
As of: ______

Obstacle No	Obstacle Location	Reporting Unit	Info DTG	Obstacle Description (Type of Mines/Wire)	Depth	DTG Marked	DTG Reduced	DTG Cleared	Lane/ Bypass Grids	Remarks

Figure 11-10. Sample enemy obstacle report

ROUTE STATUS REPORT
As of: ____________

Route Name	Responsible Unit	Status (Red, Amber, Green)	As Of	Start Point	End Point	Remarks
		○				
		○				
		○				
		○				
		○				
		○				
		○				
		○				
		○				
		○				
		○				
		○				
		○				

Figure 11-11. Sample route status report

Mine Incident Report

A mine incident includes any unplanned activity involving a mine, UXO, or a booby trap. It also includes near misses that could have resulted in damage or injury. The mine incident report (Figure 11-12) is a technical report that follows a serious incident report (SIR), and it should be submitted as soon as possible (local SOP will indicate time requirements).

MINE INCIDENT REPORT

DATE:

FROM: THRU: TO:

REFERENCE SIR #_____

A. Incident DTG	A1. dd/time/zone/mm/yy	
B. Incident location	B1. Map sheet/UTM/grid reference (8 digit)	Include a site sketch as an attachment.
	B2. Location (road, field, building)	
	B3. Emplacement (buried, surface-laid, off-route)	
C. Effects (to complement information already in the SIR)	C1. Casualties (rank, name, date of awareness training, time in the mission area, protection equipment used)	
	C2. Vehicle damage (number, type, extent of damage)	Include a photo if possible.
	C3. Collateral damage	
D. Device suspected	D1. Type of mine (AT, AP, make, model)	
	D2. Type of booby trap (pull, release, pressure)	
	D3. Type of UXO (dropped, thrown, projected)	
	D4. Unknown (detail, color, shape, size)	
E. Circumstances	E1. Activity at the time of the incident	
	E2. Degree of previous use of the route, area, location	
	E3. Date of previous clearance and proofing by engineers	
	E4. Where the route, area, or location is monitored	
F. Recommendations	F1. Recommendations to prevent reoccurrence	
G. Miscellaneous	G1. Any other pertinent data	

Figure 11-12. Sample mine incident report

PART THREE

Special Mining Operations

Part three provides tactical and technical information on special-mining operations, such as using booby traps and expedient devices. It also discusses mining in rivers, urban terrain, and unique environments. Restrictions and responsibilities are outlined in detail for the employment and the clearance of special mines and devices.

Chapter 12

Mining Operations in Special Environments

Mines are emplaced and encountered in all environments, some of which need special consideration to understand effective employment, detection, and/or removal.

STREAMBED AND RIVER MINING

EMPLOYMENT

Conventional AT mines are much more effective in water than on land because water transmits the shock effect better than air. Vehicle support members, tracks, and wheels are damaged by a mine blast. Small vehicles are overturned and almost completely destroyed. Because water amplifies and transmits shock waves, mines equipped with pressure-actuated fuses are subject to sympathetic detonation at greater distances in water than on land.

M15 and M19 AT mines can be used for streambed and river mining. The M21 AT mine should not be used because it is very difficult to arm and disarm underwater, and it can be easily functioned by drifting debris. To avoid sympathetic detonation, AT mines must be at least 14 meters apart in water that is less than 61 centimeters deep, and at least 25 meters apart in water that is deeper than 61 centimeters. The mined areas are chosen to take advantage of stream and adjacent area characteristics. Water depth within the minefield should not exceed 1 meter because it is difficult to work in deeper water, and pressure-actuated fuses are usually ineffective against waterborne vehicles.

Current velocity must be considered when emplacing mines in a streambed or a river. If the mines are placed deeper than 45 centimeters, they must be recovered by engineer divers:

- A lightweight diver has diving restrictions based on current velocity.
- A scuba diver is restricted to a maximum current velocity of 0.5 meter per second.

- A surface-supplied diver is restricted to a maximum current velocity of 1.3 meters per second.

Seasonal current velocity should also be considered if the minefield is to be in place for an extended period of time. Additional information on diving restrictions can be found in FMs 20-11 and 5-490.

Since sand in inland waters continuously moves downstream, it may be difficult to locate and remove mines planted on sandbars or downstream from sandbars. If the site has a muddy bottom, the mud should not be deeper than 46 centimeters and there must be a hard base underneath it. The enemy is unlikely to choose a fording point where vehicles mire easily. If underwater obstacles (gravel, rock, stumps) are bigger than the mine, the area cannot be easily mined. If such areas must be used, place the mines so that they are exposed to vehicle wheels or tracks. Armored vehicles usually enter and exit streams at points where the incline is less than 45 percent. After entering a stream, vehicles often travel upstream or downstream before exiting. Carefully examine riverbank formations and underwater obstacles to predict the trail a vehicle will use to ford the stream.

EMPLACEMENT

When emplacing mines in streams and rivers, always work in pairs. Prepare the mine on land near the emplacement site. Coat fuse threads and wells with silicone grease (a waterproof lubricant) or a heavy grease to minimize the chances of water leaking into the mine. Waterproof joints between the pressure plate and the mine case with silicone grease. As a rule of thumb, waterproofed mines are reliable up to 3 months when immersed without waterproof coverings. Secure the mine with outriggers to prevent drifting:

- Construct field-improvised outriggers with—
  - Two green limbs that are about 3 centimeters in diameter and 1 meter long. Green limbs are recommended because they are stronger and less likely to float than those which are dried out and dead. (Steel pickets, sign posts, fence rails, or similar items having the proper dimensions may also be used.)
  - Two pieces of clothesline, manila line, or similar material that are about 1 meter long.
- Fasten the limbs to the underside of the mine and secure them with the line (Figure 12-1).
- Approach the emplacement position from the downstream side. To prevent dragging the outrigger or contacting objects in the stream, carry the mine by grasping its sides, not by its carrying handle.
- Place the mine and the outrigger on the stream bottom. Stake down outriggers after they are emplaced to prevent drifting. If staking is impossible, place sandbags or large rocks on the outriggers for better anchorage.
- Arm the fuse.

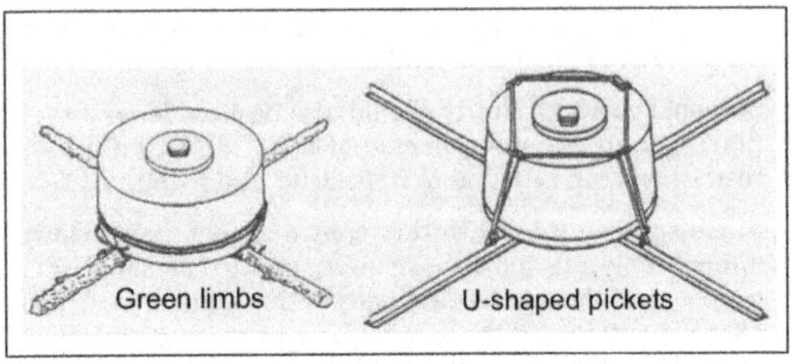

Figure 12-1. Outrigger techniques

Recovery

WARNING

Mines may have drifted downstream and/or been tampered with by enemy forces. Removal by any method other than explosive breaching (see TM 9-1375-213-12) is extremely hazardous and is not recommended. If the situation demands recovery by hand, proceed with utmost caution.

A two-person recovery team—

- Slowly proceeds 2 meters downstream from where the mine was emplaced and then carefully probes for the mine. If the mine was placed deeper than 45 centimeters, it must be recovered by engineer divers.
- Removes any foreign material from the top of the mine and disarms it.
- Carries the mine ashore and removes the fuse and the detonator.

Engineer divers normally emplace new mines or mines that have not been submersed in water; but if the mine and the fuse show no evidence of damage or deterioration, the mine can be resused. If the mine is reused, mark it to indicate that it has been immersed in water; for example, place the letter *W* on the pressure plate.

Recording

The minefield is recorded on DA Form 1355 (Figures 12-2a and 12-2b, pages 12-4 and 12-5).

Safety

In addition to normal safety measures, underwater mining requires evaluation of the tactical situation and application of special safety techniques. The turbidity, the velocity, and the depth of the water and the condition of the bottom require that laying-party personnel be able to swim well. Prolonged immersion of personnel, especially in cold temperatures, must be avoided. Sudden drop-offs, rocks, and other objects that are likely to cause personnel to lose their footing must be considered. Other safety measures include the following:

- Work in pairs.

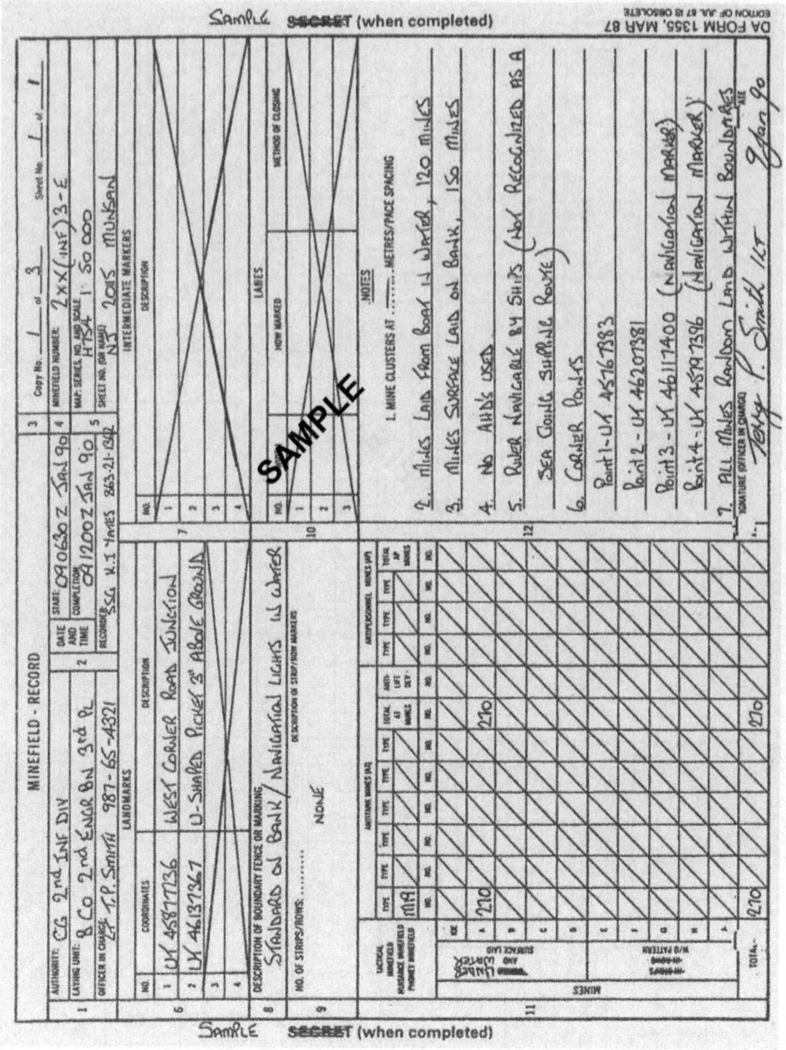
SAMPLE SECRET (when completed)

MINEFIELD - RECORD

1 AUTHORITY: CG 2nd INF DIV
LAYING UNIT: B Co 2nd ENGR BN 3rd PL
OFFICER IN CHARGE: 2LT T.P. SMITH 987-65-4321

2 DATE AND TIME START: 090630Z JAN 90
COMPLETION: 091200Z JAN 90
RECORDER: SSG K.I. YATES 863-21-1342

3 Copy No. 1 of 3 Sheet No. 1 of 1

4 MINEFIELD NUMBER: 2XX(INF)3-E
MAP: SERIES, NO. AND SCALE: H754 1:50 000

5 SHEET NO. (OR NAME): NJ 2015 MUNSAN

6 LANDMARKS

NO.	COORDINATES	DESCRIPTION
1	UK 45817736	WEST CORNER ROAD JUNCTION
2	UK 46137367	U-SHAPED PICKET 3' ABOVE GROUND
3		
4		

7 INTERMEDIATE MARKERS (NO. / DESCRIPTION)

8 DESCRIPTION OF BOUNDARY FENCE OR MARKING: STANDARD ON BANK/NAVIGATIONAL LIGHTS IN WATER

9 NO. OF STRIPS/ROWS: .... DESCRIPTION OF STRIP/ROW MARKERS: NONE

10 LANES (NO. / HOW MARKED / METHOD OF CLOSING)

11 MINES: TACTICAL MINEFIELD / NUISANCE MINEFIELD / PHONEY MINEFIELD; MINING UNDER WATER AND SURFACE LAID; IN-ROWS / IN-PATTERN W/O PATTERN
ANTITANK MINES (AT): TYPE M19, NO. 210; TOTAL AT MINES 210
ANTIPERSONNEL MINES (AP)
TOTAL: 210 ... 210

12 NOTES
1. MINE CLUSTERS AT ....METRES/PACE SPACING
2. MINES LAID FROM BOAT IN WATER, 120 MINES
3. MINES SURFACE LAID ON BANK, 150 MINES
4. NO AHD'S USED
5. RIVER NAVIGABLE BY SHIPS (NOT RECOGNIZED AS A SEA GOING SHIPPING ROUTE)
6. CORNER POINTS
Point 1 - UK 45767383
Point 2 - UK 46207381
Point 3 - UK 46117400 (NAVIGATIONAL MARKER)
Point 4 - UK 45797396 (NAVIGATIONAL MARKER)
7. ALL MINES RANDOM LAID WITHIN BOUNDARIES

SIGNATURE (OFFICER IN CHARGE): Terry P. Smith 2LT 9 Jan 90

DA FORM 1355, MAR 87 EDITION OF JUL 67 IS OBSOLETE

SAMPLE SECRET (when completed)

SAMPLE

Figure 12-2a. Sample DA Form 1355 (front side) for river mining

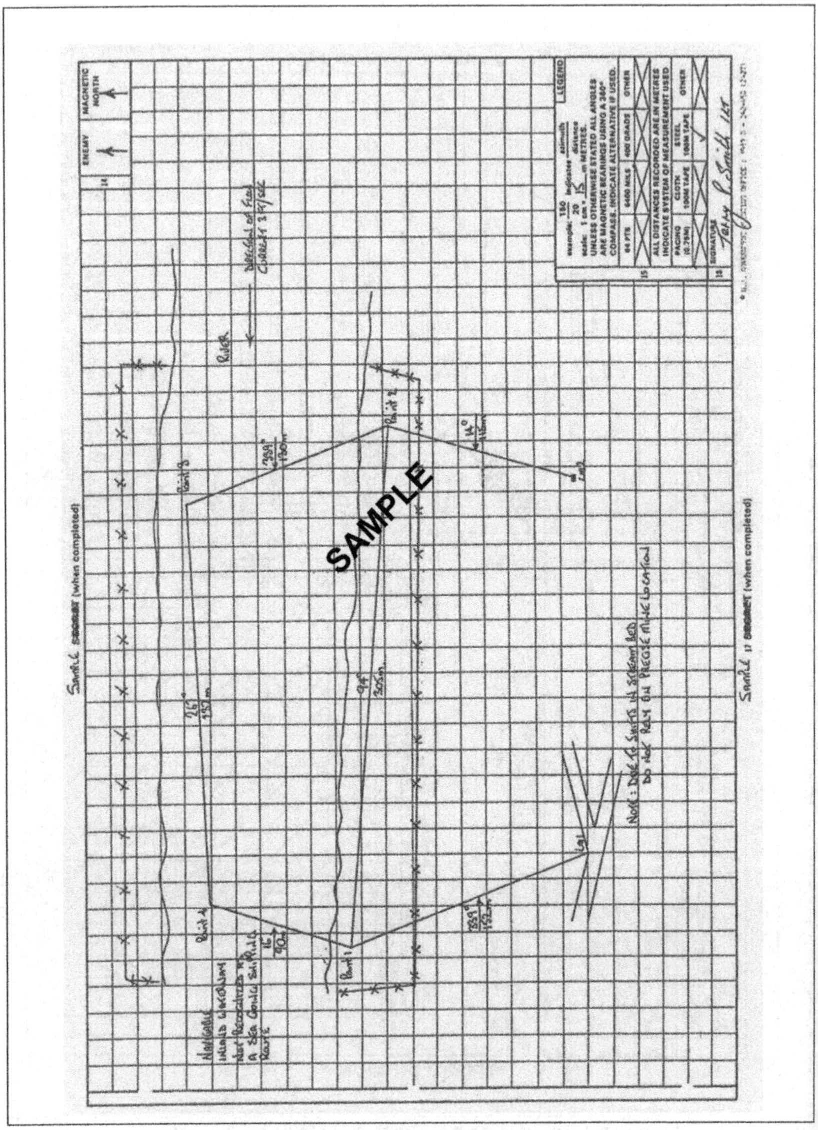

Figure 12-2b. Sample DA Form 1355 (inside) for river mining

- Emplace mines from upstream to downstream to prevent personnel and equipment from being swept into the mined area.
- Stay on the downstream side of the mine when arming the fuse.
- Place the mine as flat as possible on the bottom to prevent drifting. Use green saplings or other nonbuoyant material to construct outriggers.
- Do not arm the mine before it is laid.
- Carry the mine horizontally or edgewise to the current to reduce water resistance on the mine's pressure plate.

URBAN-TERRAIN MINING

Characteristics of urban areas (such as a high proportion of hard-surfaced roads) prohibit a simple transition from open- to urban-area mine employment techniques and doctrine. The advantages of abundant cover and concealment, maneuver restrictions, and observation already possessed by the defender of an urban area can be significantly enhanced by the proper use of mines. Terrain modified through the process of urbanization provides a unique battle environment.

The major characteristics of urban terrain which are likely to impact on mine warfare include the following:

- Multistoried buildings add a vertical dimension to the battle. Basements and floors become part of the battle scene. The vulnerability of combat vehicles increases because attack from above or below is likely.
- Fighting is done at close range, often face-to-face, and seldom exceeds 50 meters. Some weapons, particularly large-caliber weapons, are unsuitable at a short range.
- Sewers, subways, and tunnels provide covered and concealed passageways for the movement of troops on both sides. Detailed knowledge of the location and the status of these tunnels is needed to successfully wage an urban battle.
- Streets and parking lots are modified to withstand continuous use by vehicles. Major routes and lots are paved. A high density and complex pattern of streets provide numerous avenues of advance. Burying mines is extremely difficult. Most mines are surface-laid and camouflaged with rubble and debris to avoid detection.
- Movement by vehicle is difficult. Streets are littered by rubble and cratered if the city has been bombed or subjected to artillery attack. Bridges and overpasses are likely to be destroyed or blocked. Traffic flow is highly channelized.
- Extensive map and chart data are needed by the commander. For example, the commander should know the locations of telephone, electric, gas, water, and sewer connections; substations; and generating and pumping stations.

ANTIPERSONNEL MINES

US policy prohibits the use of non-self-destructing AP mines for all US forces except those on the Korean Peninsula. However, US forces can expect to encounter AP mines that are emplaced by other countries in support of MOBA. They are employed to block infantry approaches through or over underground passageways; open spaces; street, roof, and building obstacles; and dead spaces.

When AP mines are encountered **(Korea Only: or used)** in MOBA, mine locations are recorded on DA Form 1355 as shown in Figure 12-3)

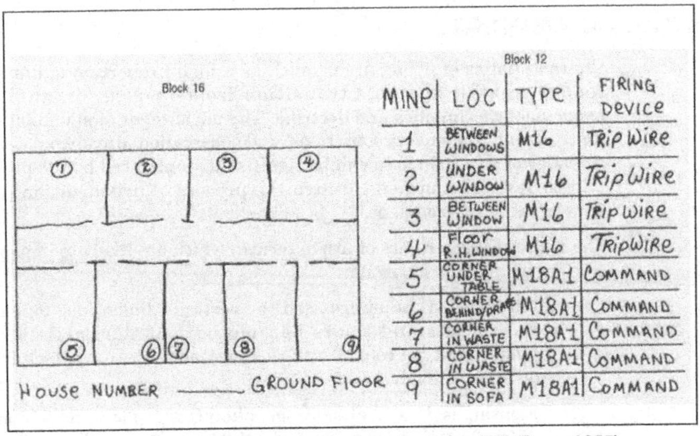

Figure 12-3. Building sketch and mine plan (DA Form 1355)

Underground Passageways

Subways, sewers, cellars, and utility tunnels provide protected movement routes for troops. In large cities where underground systems are numerous and complex, limited manpower may dictate that forces employ obstacles to block key passageways with wire and AP mines. (See Figure 12-4, page 12-8.)

Open Spaces

Open spaces include gaps between buildings, courtyards, residential yards, gardens, parks, and parking lots. In some cases, mines can be concealed in rubble or buried. However, the characteristics of most terrain surfaces, coupled with limited time and resources, dictate that mines be surface-laid. (See Figure 12-5, page 12-8.)

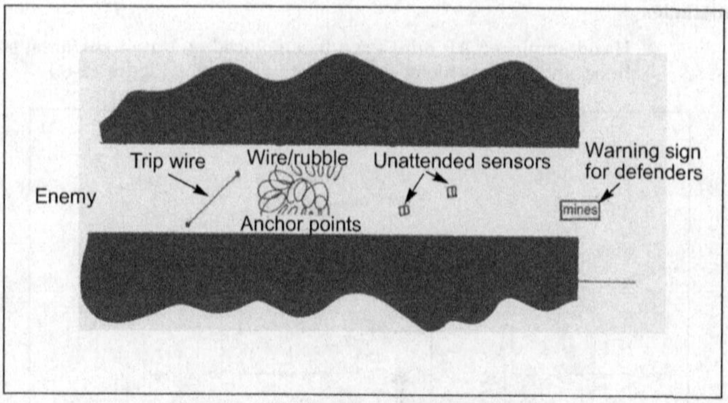

Figure 12-4. Underground passageway

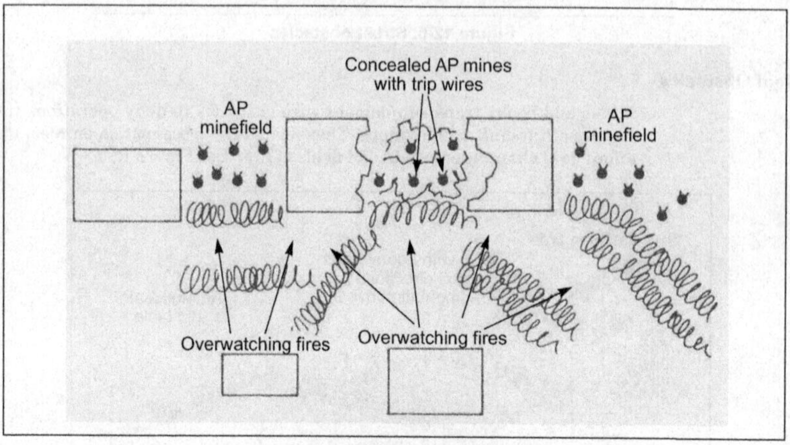

Figure 12-5. Open spaces

Street Obstacles

Hand-emplaced AP mines can be emplaced on street surfaces, on railroad lines, and in areas along shallow waterways. (See Figure 12-6.)

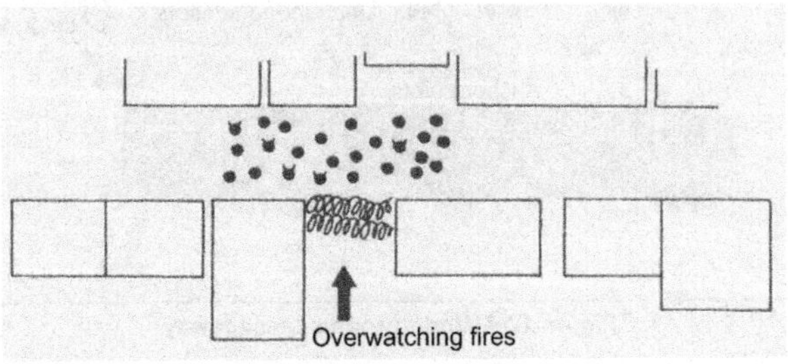

Figure 12-6. Street obstacles

Roof Obstacles

Mines and booby traps supplement wire obstacles to deny operations that require air assault onto rooftops. They also prevent occupation on roofs that afford good observation points and fields of fire. (See Figure 12-7.)

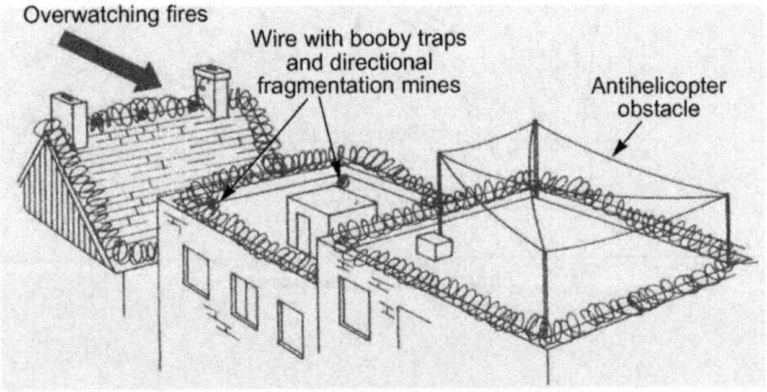

Figure 12-7. Roof obstacles

Building Obstacles

Building obstacles include areas within and adjacent to buildings. Forces can lay mines in conjunction with wire obstacles to deny infantry access to covered routes and weapon positions (Figure 12-8).

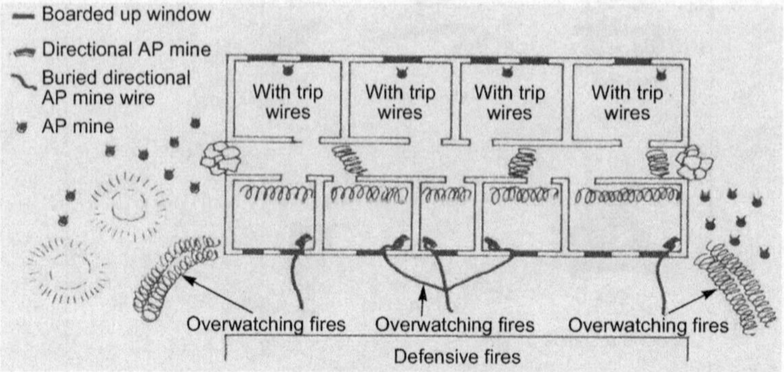

Figure 12-8. Building obstacles

Dead Spaces

Obstacles and mines can be emplaced to restrict infantry movement in areas that cannot be observed and in areas that are protected from direct fire.

Employment

The following AP mines are effective in urban terrain:

- **M14 (used by US forces in Korea only). Its small size makes it ideal for obscure places, such as stairs and cellars. It can be used in conjunction with metallic AP and AT mines to confuse and hinder breaching attempts. (See Figure 12-9.)**
- **M16 (used by US forces in Korea only). With trip-wire actuation, its lethal radius covers large areas such as rooftops, backyards, and cellars. An added advantage can be gained by attaching twine or wire to the release-pin ring to expediently rig the mine for command detonation. (See Figure 12-10.)**
- M18A1 (claymore). Numerous innovative applications of claymore munition deployment can be found for defensive warfare in urban areas (Figure 12-11, page 12-12). With remote firing, a series of claymore mines along a street establishes a highly effective ambush zone. Mines can also be employed on the sides of buildings, in abandoned vehicles, or in any other sturdy structure. Numerous opportunities exist for effectively sited, well-concealed mine employment above the terrain surface. Claymore munitions can be used to fill the dead space in the FPF of automatic weapons. They present a hazard when used in confined, built-up areas. Exercise caution when using them close to friendly forces because there is a danger of backblast.

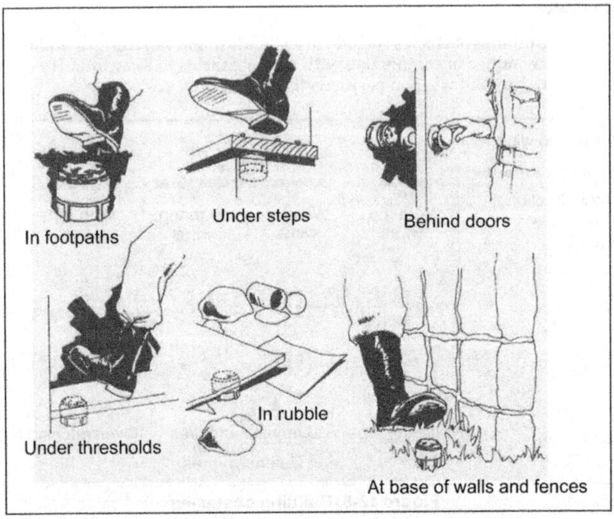

Figure 12-9. Probable M14 AP mine emplacement

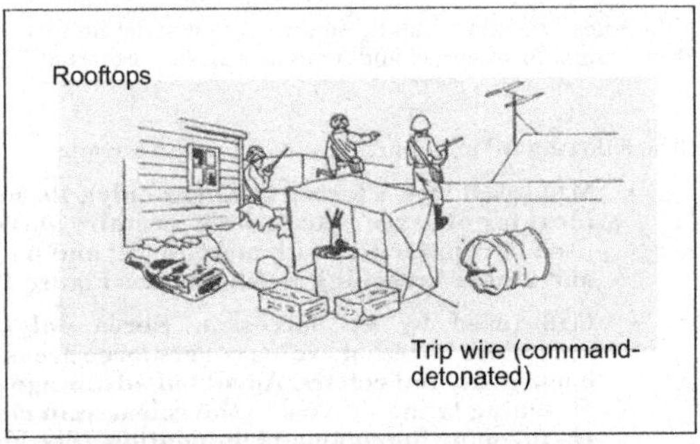

Figure 12-10. Probable M16 AP mine emplacement

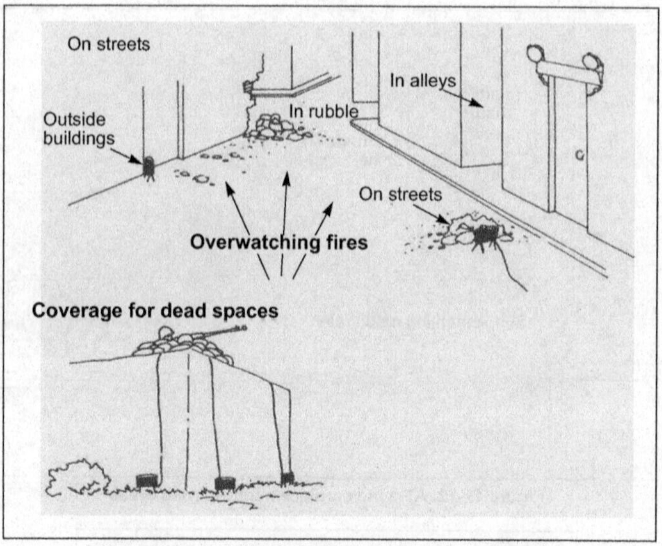

Figure 12-11. Probable M18A1 munition emplacement

CONVENTIONAL ANTITANK MINES

Enemy tanks, infantry fighting vehicles (IFVs), and direct-fire support weapons are restricted to streets, railroad lines, and, in some instances, waterways. (See Figure 12-12.) M15, M19, and M21 mines are used primarily in tactical and nuisance minefields; but they are occasionally used in protective minefields. They should be employed with other obstacles and covered by fire. Conventional AT mines emplaced in streets or alleys block routes of advance in narrow defiles. Concealment of large AT mines is accomplished by placing them in and around rubble and other obstacles. Extensive labor requirements generally prohibit burying mines in difficult terrain types.

In dispersed residential areas, obstacles are required to reduce the enemy's infantry mobility through and between houses and in open areas. They also prevent armored vehicles from moving between houses and along streets. AT minefield patterns should extend outward from the streets, incorporating open areas between buildings and streets to prevent easy bypass of street obstacles.

Significant labor and mine materials are required to deploy conventional mines between widely spaced buildings, in high-rise construction, and in industrial and transportation areas. Therefore, SCATMINEs should be seriously considered as viable alternatives. Some situations, such as the one shown in Figure 12-13, provide opportunities for the effective employment of mines in tactical and nuisance minefields.

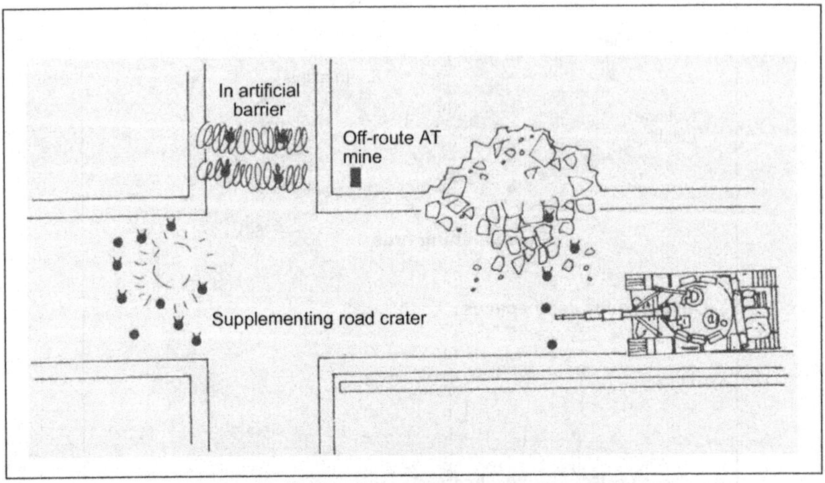

Figure 12-12. AT mine emplacement in urban areas

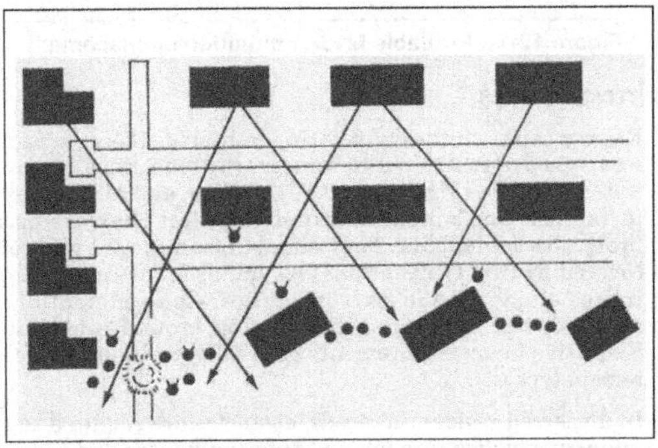

Figure 12-13. AT mine emplacement in industrial and transportation areas

SCATTERABLE MINES

Area-Denial Artillery Munitions and Remote Antiarmor Mines

In addition to the advantages (such as reducing required resources and emplacement time) applicable to all SCATMINE systems, ADAMs and RAAMs have two specific advantages. They are the most rapidly deployed SCATMINE systems, and preplanning artillery-delivered minefields increases the rate at which nuisance minefields can be emplaced. Secondly, these mines can be delivered under enemy fire. Employment of ADAMs and RAAMs is most effective when the enemy's intentions are known and their forces are committed to an avenue of advance. (See Figure 12-14, page 12-14.)

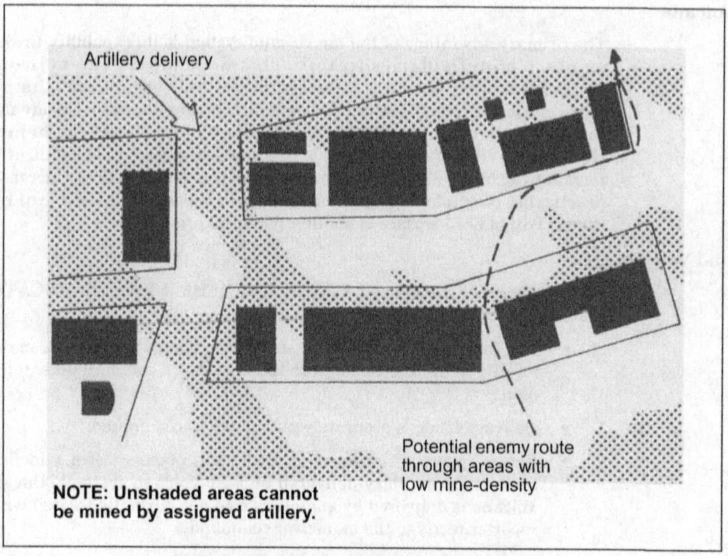

Figure 12-14. ADAM/RAAM employment

Using ADAMs/RAAMs in urban terrain involves five specific problem areas:

- Difficulty in precise minefield siting. Accurate siting is extremely critical due to the typically restrictive avenues of advance and may be futile due to the difficulty in adjusting artillery rounds in an environment that obscures observation. Further, buildings tend to create unmined shadow zones.
- Uncertainty of ADAM/RAAM survivability upon impact with a building or ground surfaces that are characteristic in urban areas.
- Likely availability of artillery firing units. ADAM/RAAM emplacement may not be a priority of the maneuver commander, because his supporting units may not have enough ADAM/RAAM munitions on hand. Assuming the availability of artillery assets for an ADAM/RAAM mission could prove disastrous for defending forces.
- High detectability of these mines on bare and lightly covered surfaces. This permits the enemy to seek out unmined passageways or pick through lightly seeded areas. If you use the doctrinal guidelines for emplacing artillery-delivered mines on top of the advancing enemy or immediately in front of them, the desired obstacle intent (disrupt, turn, fix, block) and enhanced fires are achieved.
- Difficulty in achieving a good, random pattern. Hard-surfaced areas cause mines to bounce and roll. Some mines (especially AT mines) will land on top of buildings and are ineffective.

Air Volcano

The primary advantage of the air Volcano system is its capability to site and emplace minefields accurately. This depends on the helicopter's maneuverability over the selected minefield terrain and the proper coordination between ground forces and aviation support. Disadvantages include vulnerability and the high replacement cost of the helicopter. However, in view of the system's operational concept, employment in urban terrain (which provides little exposure of the helicopter) actually increases the practicality of employing this system in urban areas. Mine survival rate on impact with a hard surface is another potential problem.

Ground Volcano

Three aspects of the ground Volcano distinguish it from other SCATMINE systems:

- The dispenser is organic to supporting combat engineers, making it readily available to support the maneuver commander's defensive plan.
- Delivery siting is accurately pinpointed to the ground.
- Better opportunities exist to record the presence of a minefield. In contrast to artillery-delivered and air Volcano systems, the ground Volcano is delivered by engineers who are normally located with and report directly to the maneuver commander.

Some primary factors may degrade ground Volcano deployment in urban terrain. The requirement to emplace minefields before an actual attack in order to reduce system vulnerability is the most significant factor. This makes the minefield detectable and provides more reaction time for the enemy to alter their scheme of maneuver. The delivery of mines depends on terrain trafficability. The prime mover and the launch vehicle must negotiate the terrain over which mines are to be dispensed.

Modular Pack Mine System

The MOPMS is ideally suited for employment in urban terrain (Figure 12-15, page 12-16). The module can be hidden from enemy view, and the mines can be dispensed after attackers are committed to a route of advance. Additionally, mines can be emplaced rapidly under enemy fire. In contrast to other SCATMINE systems, the commander controls when and where mines are dispensed and how they are detonated, regardless of the enemy situation.

Gator

When considered for employment in urban terrain, Gators encompass the same problems as artillery-delivered and air Volcano mine systems.

Deception Measures

Phony minefields can be established rapidly with negligible effort and cost. They have the distinct advantage of blocking the enemy but not friendly forces. Although it is difficult to fake a surface-laid minefield, expedients such as soup pans, seat cushions, and cardboard boxes have historically proven effective in delaying and channelizing attacking forces. These objects, as well

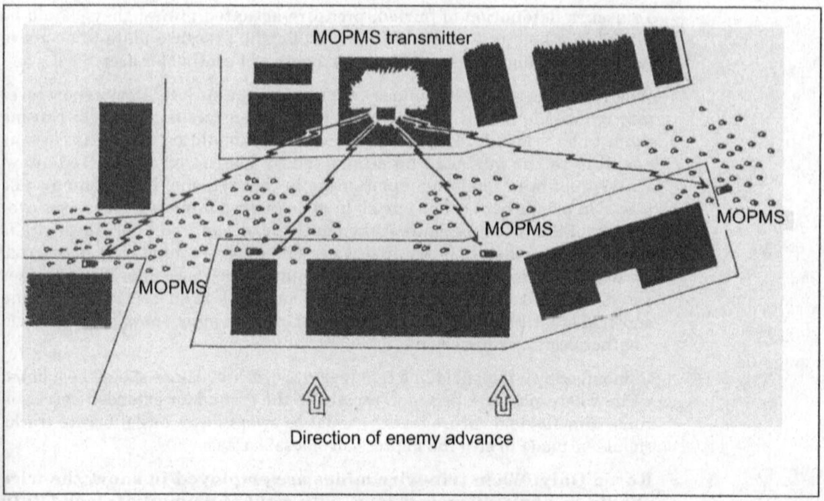

Figure 12-15. MOPMS employment

as other ones readily available in urban areas, can be used as phony minefields or used to cover real mines. A more realistic phony minefield could be created with inert or training mines.

Inadequate minefield camouflage in urban terrain is viewed as a critical constraint in deploying conventional mines and SCATMINEs. Smoke can be deployed from various dispensers, but it must be dense and accurately employed and released.

SPECIAL ENVIRONMENTS

COLD REGIONS

Mine employment in cold regions poses special problems—the principal one being emplacement. Mine burial is extremely difficult in frozen ground. The freezing water in soil causes it to have high strength and penetration resistance, so digging times are greatly increased if not impractical. However, there are several means to overcome this problem. In some cases, the minefield can be laid out before the soil freezes. To do this, dig holes for each individual mine and insert a plug into the hole to protect its shape and prevent it from being filled in. A wide variety of material can be used for plugs. Ideally, the plug should be economical, easy to remove, and rigid enough to maintain the depth and shape of the hole. Sandbags, plastic bags filled with sand or sawdust, or logs make excellent plugs. If the minefield cannot be prechambered, mechanical means can be used to dig holes. When available, civilian construction equipment (particularly large earth augers) can be used to drill holes for mine emplacement.

To assure detonation of buried, pressure-actuated mines, they should be placed in a hole that is shallow enough for the pressure plate to be above ground. Covering spoil should be a maximum of 1 centimeter deep.

When burial is impossible, mines are placed on the surface. Heavy snow cover may reduce the effectiveness of both buried and surface-laid mines by causing them to be bridged. Mines laid in deep snow should be placed as close as possible to the surface and supported by boards or compacted snow. Waterproof mines before emplacement in cold regions. Mines can also be placed in plastic bags before burial. In some cases, a layer of ice may form over the top of the pressure plate. Although the load required to break the ice is slightly higher than that required to activate the fuse, thin layers of external ice will have little effect on mine functioning. When possible, tilt-rod mines should be used in cold regions because they are less susceptible to ice and snow. Magnetic mines are not significantly affected by snow, although cold weather decreases battery life.

Camouflaging a minefield in a cold region is difficult. Mines should be painted white when snow is expected to remain on the ground for extended periods of time. Minefield signature tracks should be swept away, or deliberate tracks should be made to give the impression of a safe area.

Korea Only: When trip-wire mines are employed in snow, the wire should be about 10 meters long, with a slight amount of slack left in the wire. The trip wires should be supported approximately 46 centimeters above the ground to avoid degradation by snowfall.

JUNGLES

Fuses and explosive components deteriorate very rapidly in jungle climates. As a result, mines and mine material require more frequent and extensive maintenance and inspection. Waterproof mines employed in humid climates. The rapid growth of vegetation hinders maintenance recovery and removal. Dense vegetation may cause mines to become inoperable or windblown foliage can detonate them. FM 90-5 provides detailed information on jungle operations.

DESERTS

In desert climates, fuses and explosive components deteriorate slowly. The terrain and the situation determine how mines will be emplaced. Mine boards will normally be required to provide support in soft, shifting sand. Mines emplaced in the desert have a tendency to shift position, and the spacing between mines and rows should be increased to prevent sympathetic detonation. Blowing sand may expose buried mines or cover surface-laid mines. Sand may also cause mines to malfunction. It is important to realize the difficulty of accurately recording minefield locations in vast, open, desert areas void of recognizable terrain features. More mines are required for desert operations. Typically, desert minefields are much larger and have a lower density than those used in Europe or Korea. FM 90-3 provides detailed information on desert operations.

Chapter 13

Booby Traps and Expedient Devices

During war and OOTW, booby traps can be found anywhere at anytime. They can kill or incapacitate their unsuspecting victims. This chapter provides information on booby-trap employment concepts, detection techniques, marking and recording procedures, and removal guidelines.

This chapter also provides an overview of expedient devices and their employment considerations.

SECTION I. SETTING BOOBY TRAPS

US policy restricts the use of booby traps by US personnel. This does not preclude their use by other countries, so US forces may encounter them during operations.

The use of booby traps is limited only by the imagination of the force employing them. They—

- Are usually explosive in nature.
- Are actuated when an unsuspecting person disturbs an apparently harmless object or performs a presumably safe act.
- Are designed to kill or incapacitate.
- Cause unexpected, random casualties and damage.
- Create an attitude of uncertainty and suspicion in the enemy's mind, thereby, lowering his morale and inducing a degree of caution that restricts or slows his movement.

Many booby traps are constructed using military equipment and ammunition. Improvised traps are used during counterinsurgency missions in low-intensity conflicts.

The corps commander is the employment authority for booby traps. He can delegate this authority to the division commander. If authority is given to set booby traps, US personnel will adhere to the rules for international law applicable to armed conflict. There are several uses of booby traps that are prohibited. Remember, these restrictions are not observed by all countries; US personnel must still be cautious when approaching objects in areas where booby traps are supposedly prohibited.

International law prohibits the use of booby traps as follows:

- Booby traps and other devices are prohibited if they are attached to or associated with—

- — Internationally recognized protective emblems, signs, or signals.
- — Sick, wounded, or dead personnel.
- — Burial or cremation sites or graves.
- — Medical facilities, equipment, or supplies.
- — Children's toys or other portable objects or products that are designed for their feeding, health, hygiene, clothing, or education.
- — Food or drink.
- — Kitchen utensils or appliances except in military establishments, military locations, or supply depots.
- — Objects that are clearly religious in nature.
- — Historic monuments, works of art, or places of worship.
- — Animals or their carcasses.

- Booby traps are prohibited in cities, villages, and other areas that contain civilians if combat between ground forces is not taking place or does not appear to be imminent, unless—
  - — Booby traps are placed on or in the close vicinity of a military objective.
  - — Measures (guards, warning, or fences) are taken to protect civilians from booby-trap effects.

TACTICS

Booby traps are psychological weapons. They make the enemy cautious and slow it down. These actions, in turn, cause enemy casualties. Do not waste time attempting to set elaborate traps that are undetectable or impossible to disarm. Also, do not waste time developing difficult sites, because simple traps usually have the same chance of catching the enemy. Even if booby traps are detected and cleared, their aim is achieved.

The principles governing the use of booby traps and nuisance mines are identical, so consider using them in conjunction with one another. They have characteristics that make them suitable for use in different situations:

- Nuisance mines are quicker to lay and safer to use than booby traps, and they are normally used in outside locations where they can be buried.
- Booby traps are normally used in urban areas, structures, and places where mines are easily detected.

Booby traps and nuisance mines are particularly suited for defensive operations. They are used to—

- Slow the enemy's advance.
- Deny the enemy use of facilities and material.
- Warn of enemy approach.

- Deter the enemy from using ground not covered by direct fire.
- Plan defensive operations.

In offensive operations, booby traps and nuisance mines are employed on an opportunity basis during raids and patrols. Formal instruction is not usually issued by the staff.

Exercise caution when using bobby traps in offensive operations because they may hinder the operation. In advance and pursuit operations, they are primarily used by patrols and raiding parties. They slow down enemy follow-up actions and hinder the enemy's repair and maintenance teams after raids.

The following considerations pertain to defensive operations but may be relevant to offensive operations and must be considered when briefing troops:

- Booby trapping is rarely given a high priority and is usually peripheral to other engineer tasks.
- Nuisance mines are more cost-effective than booby traps, unless booby traps are used in situations that allow their full potential to be exploited. If it is easier, use nuisance mines instead of booby traps.

To maximize the effect of booby traps and nuisance mines, the staff provides engineer commanders with the following information:

- Purpose. Booby traps are time-consuming and dangerous to set. Do not waste time and effort setting traps that are unlikely to be actuated or that are not specifically designed to achieve the required aim. For example, if booby traps are being used against troops, small, simple traps designed to incapacitate will achieve this result just as well as complicated ones with large charges. If the aim is to destroy vehicles, use mines.
- Location. The precise location for booby traps and nuisance mines can only be determined by the setting unit. Areas must be delineated and recorded so that there is no threat to friendly forces in the event of reoccupation.
- Time setting starts and time available for setting. The time setting starts affects other engineer tasks, and the length of time available for setting governs the number of men required.
- Number of safe routes required. Safe routes are important during general withdrawals where authority has been given to booby-trap positions as they are evacuated. They also provide safe areas for the covering force to launch counterattacks.
- Likelihood of reoccupation. Even if the enemy has not detonated booby traps, they might have interfered with them. Therefore, do not set booby traps when areas are to be vacated to meet short-term tactical requirements or when reoccupation is expected soon.

Intelligence personnel provide information to assist the setting unit in maximizing the effect of booby traps. The nature and the type of traps required depend on the enemy unit. For example, while paying particular attention to dead space and defilade positions, use mines or widely dispersed traps (with large charges) against a mechanized enemy. Conversely, use small traps and AP mines (in places that afford cover) against an infantry enemy.

SITING

If the first obstacle or installation the enemy strikes is booby-trapped or nuisance-mined, he is delayed while he clears it. The enemy is further delayed by an increased degree of caution. His troops know that additional traps and mines can be encountered. Booby traps and nuisance mines are generally located—

- In and around buildings, installations, and field defenses.
- In and around road craters or any obstacle that must be cleared.
- In natural, covered resting places along routes.
- In likely assembly areas.
- In the vicinity of stocks of fuel, supplies, or materials.
- At focal points and bottlenecks in the road or rail systems (particularly the ones that cannot be bypassed).

The setting-party commander is responsible for the detailed siting and design of booby traps. Consider all the information about the enemy soldier and his operating procedures when selecting places and objects to trap. Also, consider the traps from the enemy's point of view and assess the courses open to the enemy when he encounters them. This can expose weaknesses in your initial plan and bring about changes to the proposed layout, or it can result in a different location being selected. In addition, determine the effort required by the enemy to bypass the traps. This shows whether the imposed delay justifies the effort required to set the booby traps in the selected location.

TYPES OF TRAPS

Booby traps are designed to—

- Be actuated by persons carrying out their normal duties.
- Take advantage of human nature.

The following booby traps can often be detected because they are designed to make the person do something:

- Bait. Usually consists of objects that arouse someone's interest, such as attractive or interesting items that have apparently been left behind or discarded during a rapid evacuation.
- Decoy. The most common decoy consists of two traps—one designed to be detected, the other designed to actuate when personnel deal with the first one. The first trap can be a dummy. A classic form of a decoy is to place booby traps or nuisance mines in locations from which the decoy mine can be removed.
- Bluff. A bluff is a hoax and usually consists of a dummy trap.

- Double bluff. A double bluff only appears to be a bluff. Personnel believe the trap is safe or can be disarmed. For example, a number of traps can be set that are disarmed when the detonating cord is removed from the charge. The double bluff is achieved by setting another trap that appears to be the same, but it actually explodes when the detonating cord is removed from the charge. Double bluffs rely on a reduced awareness and alertness caused by repetition.

COMPONENTS AND PRINCIPLES

There are two initiation methods for explosive booby traps—electric and nonelectric. Both methods can be constructed using many different types of FDs. FDs can be secured to the charge (direct connection) or located away from it (remote connection). They are actuated by one or more methods. It is impossible to describe every type of trap that may be encountered; however, most are constructed and operated by using components similar to those listed below:

- FD.
- Power source (battery, for example).
- Connection (usually detonating cord or electric wires).
- Blasting cap.
- Main charge.

Figure 13-1, page 13-6, shows how typical electric and nonelectric traps can be made.

ACTUATION METHODS

Many sophisticated booby-trap devices are now being manufactured that operate on vibration, sound, temperature change, and other methods. Current intelligence on the booby trap being used in the AO should be gathered so that countermeasures can be developed and practiced. Most FDs found in the combat zone are simple mechanisms designed to be actuated by pull, pressure, pressure release, or tension release (Figure 13-2, page 13-7).

METHODS OF CONNECTION

Procedures can be varied when it is safe to do so. For example, instead of connecting the FD to a charge already in position, preconnect trap components and then position the trap.

Small charges (up to 1 kilogram) are sufficient for AP traps, but larger quantities can be used to increase their effect. Shrapnel can be produced by packing stones, scrap metal, nails, or other material around the charge. AT traps require large charges (up to 6.75 kilograms for wheeled vehicles and 11.25 kilograms or more for tracked vehicles).

REMOTE

Follow the procedures listed below when assembling a remotely connected trap using an M142 FD (similar to the illustration in Figure 13-3, page 13-7):

- Design the trap and collect necessary materials.

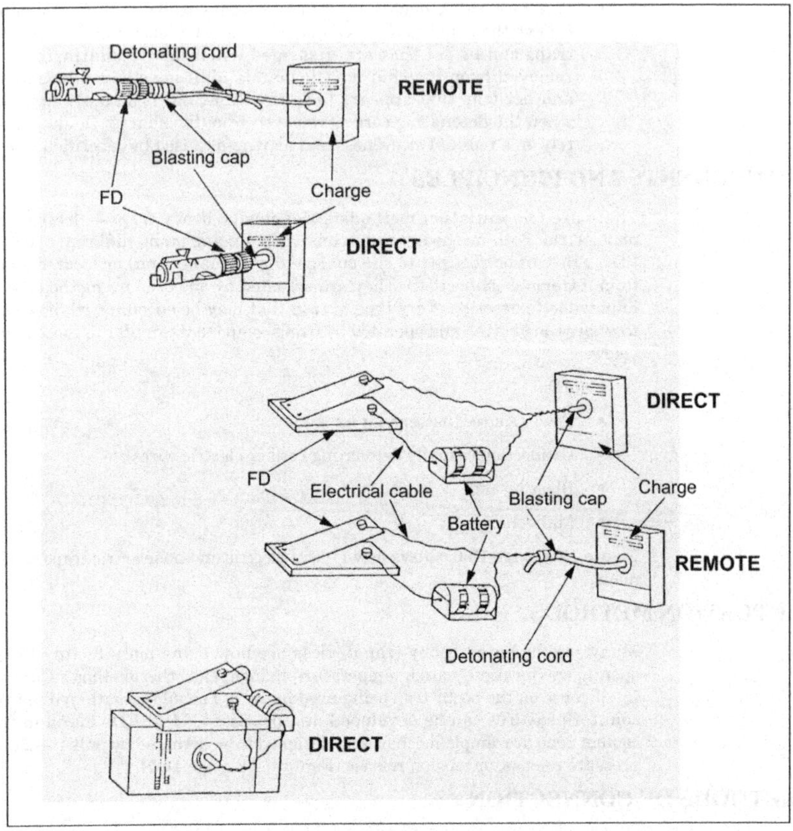

Figure 13-1. Typical electric and nonelectric booby traps

- Test the M142 FD.
- Lay the detonating cord from the charge location to the FD location.
- Position the charge.
- Connect the detonating cord to the charge.
- Prepare the coupler.
- Tape a length (46 centimeters, minimum) of detonating cord to the coupler's blasting end.
- Prepare and position the M142, set it to operate in the desired manner, and remove the round- or square-headed pin.

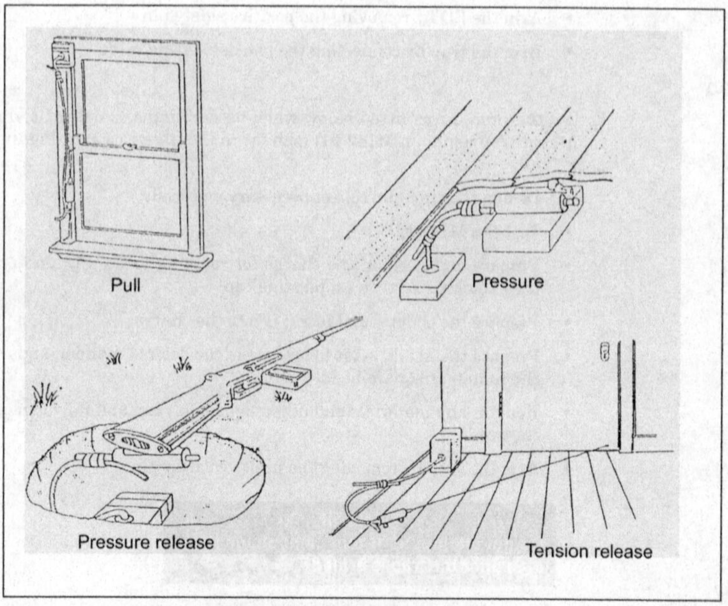

Figure 13-2. Methods of actuation

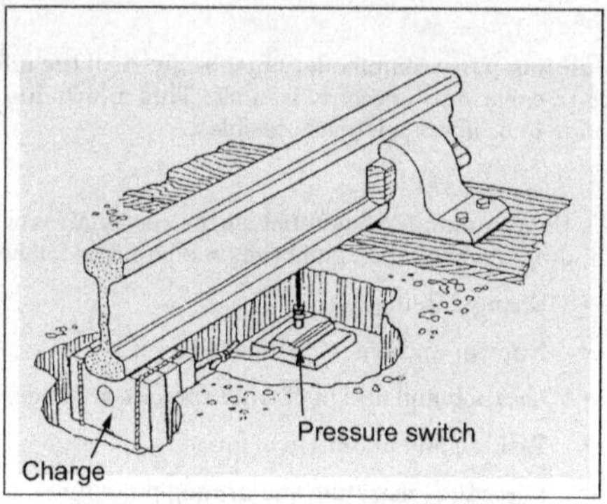

Figure 13-3. Remotely connected traps

- Ensure that the two detonating cords are not touching but can be conveniently connected when necessary.
- Remove any materials and other signs of laying, and fully camouflage the area.

- Arm the FD by removing the positive safety pin.
- Arm the trap by connecting the two detonating cords.

DIRECT

Follow the procedures listed below when assembling a nonelectric, directly connected trap using an M142 FD (similar to the illustration in Figure 13-1, page 13-6):

- Design the trap and collect necessary materials.
- Test the M142 FD.
- Prepare and position the charge for coupling to the FD, and prepare the explosive to receive a blasting cap.
- Prepare the coupler and insert it into the charge.
- Prepare the M142, set it to operate in the desired manner, and remove the round- or square-headed pin.
- Remove any materials and other signs of laying, and fully camouflage the area.
- Arm the trap by removing the positive safety pin.

DANGER
Do not attempt to further camouflage the area after the trap is armed.

PLANNING, SETTING, AND RECORDING

TIMELINESS

The setting-party commander must be given all the information and sufficient time to carry out a reconnaissance. This allows his plan and the soldier's briefing to be as complete as possible.

ORDERS AND BRIEFING

Once the commander has finished the reconnaissance, he makes a detailed plan, prepares the order, and briefs soldiers. The following points are covered:

- Enemy intelligence.
- Number and types of traps to be set.
- Location and design of traps (precise or general).
- Tasking and allocation of areas.
- Recording, marking, and arming procedures.
- Control measures.
- Time requirements.
- Material and equipment availability.
- Rehearsal and equipment preparation.

REHEARSAL

Whenever possible, rehearse laying procedures. Thoroughly exercise control and safety measures so that any flaws in the system will be discovered. Lay practice traps and carry out arming and recording procedures. If you cannot rehearse at the trap location, hold the rehearsal in a similar area.

ORGANIZATION AND PROCEDURE

Setting booby traps is a dangerous task and it must be carefully controlled. Exact drills cannot be prescribed, but the following procedures should be followed as closely as possible:

- Control point. The commander must establish a control point upon arrival at the trap area. The control point forms a headquarters and a material holding area. Also, safe routes start at the control point. If the rehearsal is conducted at another area, assemble all the troops at the control point before setting starts. At this time, relate the ground to the plan, confirm control measures, and check FDs and equipment for serviceability and adequacy.
- Control of parties. Each setting party consists of one or two men. Clearly define the area where each setting party will work (use tape, if necessary). Position traps so that accidental detonation will not cause friendly casualties. Strictly control the entry to areas where traps are being set. In most situations, only the commander, setting parties, and a recorder enter areas. Troops who are not assigned tasks should remain at the control point unless otherwise authorized by the commander.
- Marking. As with nuisance minefields, mark booby-trapped areas until they are evacuated. When possible, use standard booby-trap signs (Figure 13-4). When signs are not available, use temporary markings (any barrier or fence) to clearly indicate a booby-trapped area. Remove the temporary marking after the area is evacuated and before the enemy reconnoiters the area.

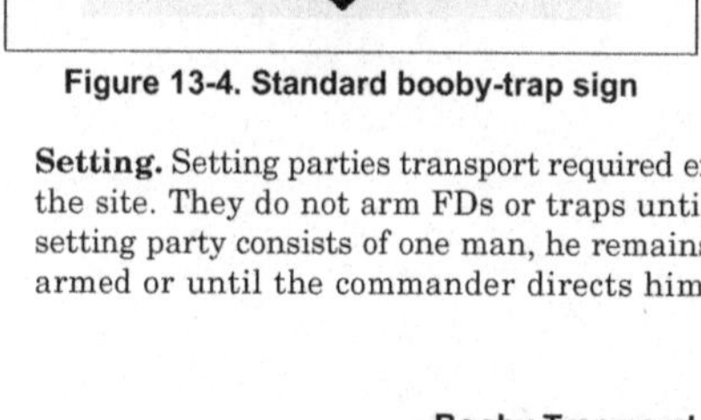

Figure 13-4. Standard booby-trap sign

- **Setting.** Setting parties transport required explosives and material to the site. They do not arm FDs or traps until ordered to do so. If the setting party consists of one man, he remains with the trap until it is armed or until the commander directs him to leave. In a two-man party, one man returns to the control point and reports the completion of the task, and the other man remains with the trap. A setting party enters and leaves its area by the same route. If the route is not clearly apparent, personnel lay tapes to guide the commander and the recorder to the area.

- Inspection and arming. The setting-party commander inspects traps for safety and camouflage before giving the order to arm them. He devises an arming plan that enables the setting parties to withdraw to the control point in a safe and logical manner. For example, in open areas, start arming at the furthest mine from the control point; in buildings, arm from the top down.

Reporting and Recording

Booby traps do not distinguish between friend and foe. Correct reporting and recording procedures must be observed to avoid friendly casualties. The procedures for recording booby traps are the same as for nuisance mines (see Chapter 8). When traps are set in areas occupied by the enemy, they are not recorded. In this case, it is sufficient to note their location and type in after-action reports.

Tactical Reports

Information about booby-trapped areas must be provided in stages from the time they are established until they are cleared. This procedure allows up-to-date information to be annotated on operational maps at all concerned units and headquarters (see Table 13-1).

Booby-Trap Records

The setting-party commander compiles records for all booby-trapped areas on DA Form 1355 (Figures 13-5a and 13-5b, pages 13-12 and 13-13). Strictly observe the instructions printed on the back of the form. Booby-trap records provide detailed information about the composition of an area, and they allow the area to be cleared quickly, with minimum casualty risk. They contain complete information on the number and types of traps set, as well as accurate and precise details on the location and design of individual traps. If a single record is not large enough to record all the traps set in a definable area, clearly note this fact and reference serial numbers of adjacent records. Submit the completed DA Form 1355, in triplicate, to the next higher headquarters.

Table 13-1. Tactical reports

Report	By Whom and When Initiated	To Whom Forwarded	Details to be Included
Intention to set (required)	Division commander, on deciding to set booby traps	• Corps HQ • Controlling HQ	• Purpose • General area to be booby-trapped (by grid coordinates) • Estimated number and types of traps • Estimated starting date and time • Estimated completion date and time
Siting (optional)	• Setting unit, on completion of reconnaissance • Division ordering setting, only if materially different from intention-to-set report	• Division ordering setting • Next higher HQ	• Boundaries of booby-trapped area (by grid coordinates) • Number and approximate positions of safe routes, if any • Estimated number and types of traps
Start of setting (required)	• Setting unit, when setting starts • Division ordering setting	• Division HQ ordering setting • Next higher HQ	• Time setting started • Estimated time of completion • Number and types of traps being set • Any change to the detail given in the siting report, and whether or not it was a result of orders
Completion (required)	• Setting unit, when setting is completed • Division HQ ordering setting	• Division HQ ordering setting • Next higher HQ	• A trace on the largest scale map available, showing the boundaries of the trapped area • Full details of the area and buildings trapped (with sketch maps) • Full details of traps set, including design and location • Full details of all safe routes • Total number and types of traps • Time and date of completion
Changes (required)	Division HQ ordering change	• Next higher HQ • Any other unit holding information on existing traps	Full details of change

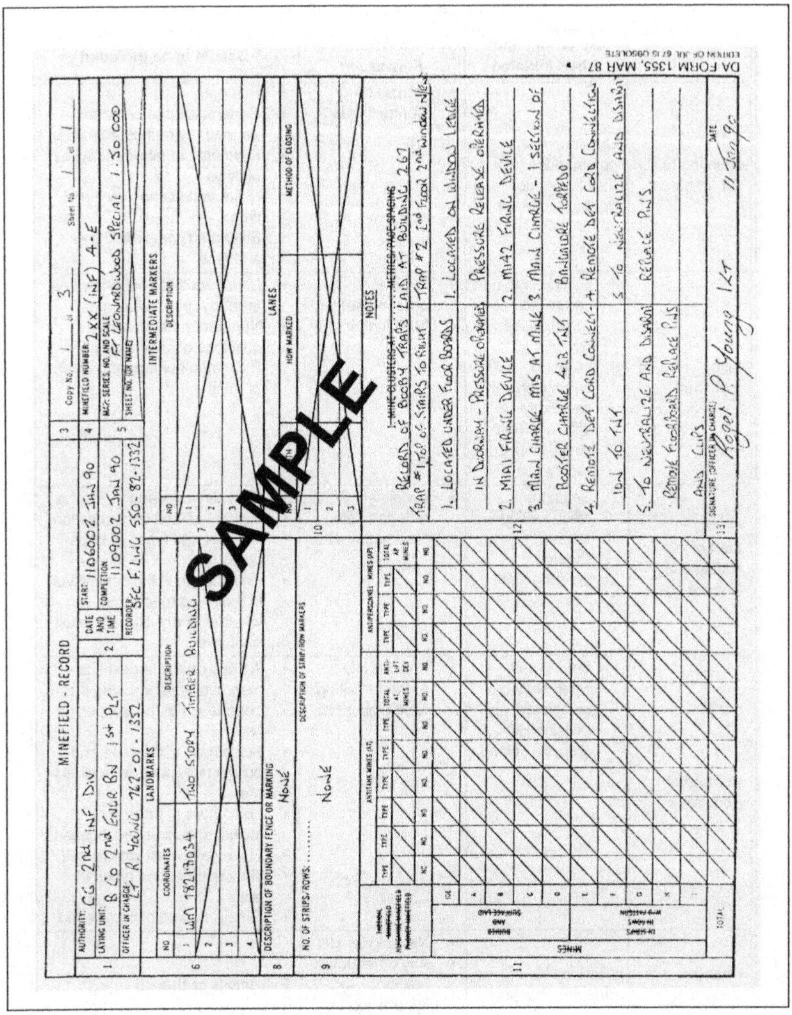
SAMPLE

MINEFIELD - RECORD

1 AUTHORITY: CG 2nd INF DIV
LAYING UNIT: B Co 2nd ENGR BN 1st PLT
OFFICER IN CHARGE: LT R. YOUNG 762-01-1352

2 DATE AND TIME — START: 1106002 JAN 90 — COMPLETION: 1109002 JAN 90
RECORDER: SFC F. LING 550-82-1332

3 Copy No. 1 of 3 Sheet No. 1 of 1

4 MINEFIELD NUMBER: 2XX (INF) A-E
MAP SERIES, NO. AND SCALE: FT LEONARDWOOD SPECIAL 1:50 000

5 SHEET NO. (OR NAME)

6 LANDMARKS
NO / COORDINATES / DESCRIPTION
1 WD 18217034 TWO STORY TIMBER BUILDING

7 INTERMEDIATE MARKERS — NO / DESCRIPTION

8 DESCRIPTION OF BOUNDARY FENCE OR MARKING: NONE

9 NO. OF STRIPS/ROWS: .... DESCRIPTION OF STRIP/ROW MARKERS: NONE

10 LANES — NO / WIDTH / HOW MARKED / METHOD OF CLOSING

11 MINES — ANTITANK MINES (AT) / ANTIPERSONNEL MINES (AP)

12 NOTES
1 MINE CLUSTERS AT .... METRES/PACE SPACING
RECORD OF BOOBY TRAPS LAID AT BUILDING 267
TRAP #1, TOP OF STAIRS TO RIGHT — TRAP #2 2nd FLOOR 2nd WINDOW N...
1. LOCATED UNDER FLOOR BOARDS IN DOORWAY - PRESSURE OPERATED
2. M1A1 FIRING DEVICE
3. MAIN CHARGE M15 AT MINE BOOSTER CHARGE 4LB TNT
4. REMOVE DET CORD CONNECT- ION TO TNT
5. TO NEUTRALIZE AND DISARM: REMOVE FLOORBOARDS, REPLACE PINS AND CLIPS
1. LOCATED ON WINDOW LEDGE PRESSURE RELEASE OPERATED
2. M142 FIRING DEVICE
3. MAIN CHARGE - 1 SECTION OF BANGALORE TORPEDO
4. REMOVE DET CORD CONNECTION
5. TO NEUTRALIZE AND DISARM: REPLACE PINS

13 SIGNATURE (OFFICER IN CHARGE): Roger P Young 1LT — DATE: 11 Jan 90

DA FORM 1355, MAR 87 — EDITION OF JUL 67 IS OBSOLETE

Figure 13-5a. Sample DA Form 1355 (front side) for a booby-trapped area

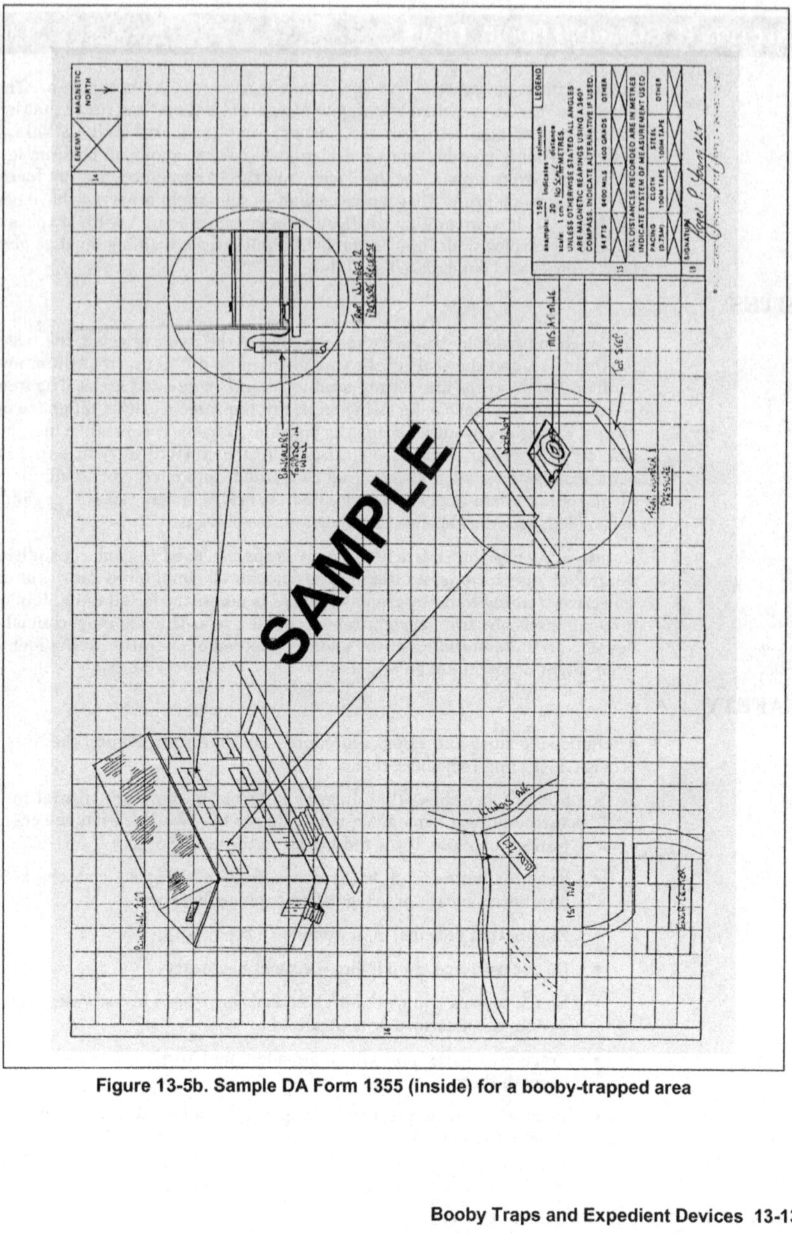

Figure 13-5b. Sample DA Form 1355 (inside) for a booby-trapped area

SECTION II. CLEARING BOOBY TRAPS

Soldiers must be aware of the threat that is presented by booby traps. They must also receive sufficient training to recognize locations and items that lend themselves to booby-trapping. Avoid overemphasis and strike a balance between what is possible and what is probable. For example, in medium-level operations, overemphasis of the booby-trap threat can be counterproductive by slowing momentum. This causes casualties that might otherwise have been avoided. In low-intensity conflicts where the use of booby traps and improvised explosive devices are probably widespread, training must be given high priority and emphasized at all levels.

SITES

Although many booby-trapped sites are similar, the items selected, the reason for their use, and the scale of the threat are quite different. In medium-level conflicts, booby traps are mainly used in recently contested areas. The items selected and the reasons for using traps are the same as those taught to our troops. Therefore, by anticipating the presence of traps, it is possible to isolate and bypass trapped areas. If this is not practicable, you can plan countermeasures, such as avoiding convenient and covered resting places along routes where mines can be located. At bridge or ferry sites that cannot be avoided, you must ensure that they are free of traps.

In low-intensity conflicts where booby traps are used to cause casualties, delays, or disruptions, no items or areas can be considered safe. Quality collective training in booby-trap awareness is necessary for all units. Rapidly disseminate booby-trap incident reports to all levels. This allows personnel to develop an understanding of the enemy's method of operation and a feel for what might or might not be targets.

SAFETY

The following rules and safety procedures can save your life and the lives of others—learn and remember them:

- Suspect any object that appears to be out of place or artificial in its surroundings. Remember, what you see may well be what the enemy wants you to see. If you did not put it there, do not pick it up.
- Examine mines and booby traps from all angles, and check for alternative means of firing before approaching them.
- Ensure that only one man works on a booby trap.
- Do not use force. Stop if force becomes necessary.
- Do not touch a trip wire until both ends have been investigated and all devices are disarmed or neutralized.
- Trace trip wires and check for additional traps located along and beneath them.
- Treat all parts of a trap with suspicion, because each part may be set to actuate the trap.

- Wait at least 30 seconds after pulling a booby trap or a mine. There might be a delay fuse.
- Mark all traps until they are cleared.
- Expect constant change in enemy techniques.
- Never attempt to clear booby traps by hand if pulling them or destroying them in place is possible and acceptable.

INDICATIONS

Successful detection depends on two things—being aware of what might be trapped and why, and being able to recognize the evidence of setting. The first requirement demands a well-developed sense of intuition; the second, a keen eye. Intuition, like mine sense, is gained through experience and an understanding of the enemy's techniques and habits. A keen eye is the result of training and practice in the recognition of things that might indicate the presence of a trap. The presence of booby traps or nuisance mines is indicated by—

- Disturbance of ground surface or scattered, loose soil.
- Wrappers, seals, loose shell caps, safety pins, nails, and pieces of wire or cord.
- Improvised methods of marking traps, such as piles of stones or marks on walls or trees.
- Evidence of camouflage, such as withered vegetation or signs of cutting.
- Breaks in the continuity of dust, paint, or vegetation.
- Trampled earth or vegetation; foot marks.
- Lumps or bulges under carpet or in furniture.

DETECTION

Detection methods depend on the nature of the environment. In open areas, methods used to detect mines can usually detect booby traps. Look for trip wires and other signs suggesting the presence of an actuating mechanism. In urban areas, mine detectors are probably of little use. You have to rely on manual search techniques and, if available, special equipment.

CLEARING METHODS

The method used to neutralize or disarm a trap depends on many things, including time constraints, personnel assets, and the type of trap. Remember, a trap cannot be considered safe until the blasting cap or the detonating cord has been removed from the charge. This is your first objective and is particularly important for electric traps, which may contain a collapsing circuit.

Use the safest method to neutralize a trap. For example, if the FD and the detonating cord are accessible, it is usually safer to cut the detonating cord. This method does not actuate the trap, but inserting pins in the FD might.

COMBAT CLEARANCE

Clearing booby traps and nuisance mines in AOs is done primarily by engineers. Therefore, engineer advice is important during the planning stages of any operation where booby traps are likely to be encountered. Intelligence regarding the possible presence and types of traps must be provided to engineer units as soon as possible. This allows the unit to take necessary action and provide relevant training. Clearance of booby traps cannot be undertaken as a secondary task, because engineer clearing teams might require protection that necessitates combined arms training. Before engineer planning can start, the staff provides commanders with the latest intelligence information and, if possible, the following information:

- Amount of clearance required.
- Acceptable damage.
- Time requirements.
- Availability of special equipment.
- Security requirements.

Intelligence information regarding the nature, type, and location of traps has a direct bearing on the number of clearing parties necessary and the degree of protection required. For example, in built-up areas where traps have to be cleared in buildings that offer protection from enemy fire, direct protection is usually provided by the normal combat situation. On the other hand, in open areas where clearing parties may be required to clear traps covered by direct enemy fire, protection arrangements must be more specific.

Engineer commanders must be aware of the time needed to clear various types of traps in differing terrain situations. Remember, increasing the number of clearance parties may not necessarily reduce the time required to clear traps. This is particularly true when traps are set close together or set deep along a narrow front that is the only available route.

Initially, clear areas of immediate tactical importance and traps that present a specific threat. For example, clear only the portion of a building required for observation and those traps presenting an immediate hazard. This enables clearing parties to concentrate on other areas of tactical importance.

Clearing traps by hand is the only way that damage can be avoided and security guaranteed. When it is vital to avoid equipment or structure damage, consider using available EOD assets. It is often necessary to balance the requirement to remain silent and avoid damage with the requirement to maintain momentum.

When traps are being cleared in direct-support combat situations, they are normally dealt with by using unit resources and locally manufactured or acquired aids. Specified equipment is rarely available. Equipment varies with the situation but usually consists of selected items from Table 13-2. In areas with a high incidence of booby traps, assemble and reserve special clearing kits.

Table 13-2. Clearing equipment

Item	Remarks
Unit Equipment	
Mine detector	
Mine probe/knife/bayonet	
Eye protection	
Body armor	Flak jacket, Kevlar helmet
Booby-trap signs	
Mine marking tape	
Mine markers	Chalk, felt-tipped pens
Electrician's adhesive tape	
Flashlight	
Tape measure	
Mine marking cones	
Trademan's tools	Saws, pliers, hammers, screwdrivers
Self-developing camera and film	
Locally Manufactured/Acquired	
Trip-wire feeler	
Grapnel and rope	
Hand mirrors	
Pins, wire, nails	For use as safety pins
Meat hooks	For use as block and tackle

CLEARANCE IN SECURE AREAS

POLICY AND PLANNING

Formal clearing procedures must be followed in secure areas. This type of clearance can be done by engineers or EOD personnel. It is subject to time constraints when traps or suspected traps are located in urgently required installations or facilities, such as supply dumps or telecommunication centers.

Commanders reconnoiter their areas of responsibility, make detailed plans, prepare orders, and brief their men. The following points are covered:

- Intelligence.
- How the operation is to be conducted (include acceptable damage and methods of clearing).
- Action on finding traps.
- Marking.
- Disarming procedures.
- Tasking and allocation of areas.
- Equipment available.
- Control measures.

- Time requirements.
- Rehearsal and equipment preparation.

Control Point

The operation commander establishes a control point upon arrival at the area to be cleared. The control point functions as the headquarters and the material holding area. It is also the point from which all clearance starts. Its suitability might not have escaped the enemy's notice, so it must be cleared before it can be used.

Control and Size of Parties

The size of clearing parties varies depending on the location being cleared. The following rules apply:

- Each party is controlled by an NCO.
- Only one party works in a particular subarea.
- The distance between parties is sufficient so that detonation in one area does not endanger persons in other areas.
- The NCOIC of each party is in visual, radio, or voice contact with every person in his party.
- Only one person works on a booby trap.

Marking

Booby traps and nuisance mines can be placed in diverse locations. It is impossible to standardize a method for marking areas, individual traps, or safe routes. Any form of prominent, permanent marking can be used.

Uncleared Areas

The perimeter marking of uncleared areas can take any clearly recognized form. Standard minefield or booby-trap signs suspended from a single-strand fence are recommended. The spacing of the signs is the same as standard minefield marking (see Chapter 2). As the area is cleared, the fence should be progressively moved.

Individual Traps

Because booby traps can have more than one means of actuation, do not attempt to place a cone or other marker over any part of a trap. Use sufficient signs to ensure that the trap can be detected and accurately located. In buildings, clearly mark rooms containing traps and, where possible, indicate the exact location of traps.

Internal Marking

The internal marking system depends on the area being cleared. One good system is to divide the total area into subareas, clear and mark safe lanes between the subareas, and then use the lanes as safe routes.

Clearing of Open Areas

Roads, road shoulders, bridges, obstacles, and other structures must be cleared in open areas. The main threat comes from nuisance mines, so regard each potential site as a nuisance minefield and use established minefield clearing procedures. After the site is cleared, adapt the drill to suit the situation.

Clearing of Buildings

Buildings are excellent locations for booby traps. Booby traps are concentrated inside the building, so carefully organize search and clearing procedures. In most cases, buildings are required for use, and excessive damage must be avoided.

A two-man clearing party can clear an average-size structure. No one else should enter the building until it is cleared. In very large buildings, two or more parties can be employed, if they work as far apart as possible and have clearly defined boundaries. As a general rule, two walls or two floors is the minimum distance between parties.

Exterior Reconnaissance and Entry

Before approaching a building, check the surrounding area for booby traps and nuisance mines. Carry out a reconnaissance to determine the point of entry and clear the way to it. When selecting the point of entry, consider the following:

- Doorways. Never consider doorways to be safe unless the door is fully open and the entrance is clear. If a house is built on a concrete slab, it is not likely to have a pressure FD in the floor.
- Windows. Windows are excellent locations for booby traps. Pay particular attention to the ground outside and the floor inside because they are classic sites for pressure FDs. Use the following procedures if access must be gained through a window:
  - Pull the window if it is unlocked and can be moved. If it is locked, use a small charge or a heavy object to break the glass.
  - Select a stationary window that cannot be opened versus a window that can be opened when both types are available, because the window that can be opened is more likely to contain a booby trap.
  - Deal with blinds and curtains in a manner similar to procedures used for windows.
  - Use a mirror to examine inside the window frame before entering the building.
- Mouseholes. If you decide not to enter the structure through a door or a window, use explosives to make a mousehole in the wall, roof, or floor. This offers a remote, safe method of creating an access point, but it can also detonate nearby traps. Exercise caution if entering through the roof of a two-story building, because it is far more difficult to clear booby traps when going down steps than it is when going up them.

Search Techniques

It is impossible to establish a search drill that is suitable for use in all buildings. It is essential, however, for each working party to develop its own drill and follow it. Domestic dwellings should be searched in the following order:

1. Floors and furniture.
2. Walls (including doors, windows, fireplaces, and cupboards).
3. Fittings (including light switches and pictures).
4. Ceilings.

Mark each area or item *SAFE* as it is cleared. This can be conveniently done by using chalk or a felt-tipped pen.

Use the following techniques and precautions when searching buildings. They can save your life—learn and remember them!

- Check both sides of a door before opening it. Do this by drilling a hole through the door and using a mirror to check the other side. You can also check or open doors by pulling or blowing the lock and hinges with a small charge.
- Examine floor coverings for signs of disturbance. The presence of FDs is often indicated by loose floorboards, bulges or tears in carpets, or loose tiles.
- Use a pulling cable to move furniture and to open cupboard doors and drawers.
- Check upholstered furniture and beds by remotely dropping a heavy object onto them.
- Treat every switch with suspicion, since electrical wiring provides a ready-made circuit for booby traps. To explode all the traps connected to the normal power supply, disconnect the power at the fuse board, turn all the switches on, and then reconnect the power. Repeat the procedure with the switch turned off in case the switch has been reversed. Remember, this procedure will not disclose traps that use a battery. Exercise caution when using switches, even if the power is disconnected.
- Leave doors, windows, cupboards, and drawers open after they have been cleared.
- Mark all routes, areas, and items that have been cleared.
- Check plumbing by remotely turning on all water taps and allowing the water to run for at least one minute.
- Check toilet tanks before flushing.
- Beware of light-sensitive devices in dark places (closets, cabinets, basements, attics, chimneys).
- Allow at least 30 seconds for an explosion after pulling anything, because there might be a delay fuse.

CLEARING INSTALLATIONS AND FACILITIES

Clearing by hand is necessary in installations and facilities (fuel dumps, ammunition dumps, electric substations) where an explosion could result in the loss of resources. In other situations, the item's importance or the resulting damage might not be obvious. For example, a small charge placed against the control valves of a dam or against the main cable entering a telephone exchange results in unforeseen damage that can take days to repair. Therefore, you should seek a specialist's advice, if possible, when clearing booby traps in industrial areas and unfamiliar locations.

CLEARING OBSTACLES

If an enemy has time to create obstacles, he also has time to set booby traps and lay nuisance mines. The obstacle itself is usually clear of traps to encourage a false sense of security and lead troops into more dangerous areas. Therefore, regard all obstacles as booby-trapped until proven otherwise. The simplest, safest way to deal with movable obstacles is to pull them. Before an obstacle can be pulled, you must first clear the area from which the pull will be made.

CLEARING SECURE AREAS

When clearing secure areas and time is not a major factor, use specialized clearance equipment as much as possible. The following equipment might be available for use:

- Cameras. Cameras have a wide range of applications. They can be used with different types of film, such as infrared and ultraviolet, to disclose evidence that is indiscernible to the naked eye. For example, infrared photography reveals differences in the heat emitted by objects and can often disclose recent digging and buried or concealed objects.
- Explosive detector dogs (EDDs). Although EDDs can detect minute quantities of explosives and the presence of trip wires, they are trained to detect the charge and not the FD. This extremely limits their usefulness in detecting booby traps. They also tend to become confused if the area contains explosive odors other than those emitting from booby traps.
- Electronic countermeasures. Electronic countermeasures can be used to explode electric booby traps and to prevent remotely controlled, improvised explosive devices from being detonated by radio.
- Robots. In their simplest form, robots can be used to detonate or neutralize booby traps. More sophisticated models can be remotely controlled to carry out simple tasks, such as videotaping or cutting wires.
- Body armor.
- Electric meters.
- X-ray equipment.

CLEARANCE METHODS

- Pulling. This method uses a grapnel and a rope to pull the trap. It is used when the resulting damage is acceptable. It is the safest method and is particularly applicable to traps set in open areas. Do not disturb any part of a booby trap when placing the grapnel and pulling the cable. Carefully select the site from where the pull is to be made because it might be mined or trapped. When a booby trap is pulled and does not explode, wait at least 30 seconds before approaching it in case delay devices have been used. Disposal of unexploded traps depends on their condition when inspected. The procedure for pulling booby traps is similar to that for pulling mines (see Chapter 11).
- Trip wires. Check the area for AP devices before proceeding. Place a grapnel hook as close as possible to the trip wire. Do not touch the trip wire until the pulling party is in a covered area.
- Pull and release. Pull away objects that conceal and operate pull and release mechanisms.
- Pressure mechanism. Pull pressure mechanisms from under objects that conceal and operate them. If this is impossible, blow them in place. In many cases, it might be easier to pull the charge rather than the FD. Take extreme care when attempting this, because additional mechanisms are often concealed in or under the main charge.
- Destroying in place. When destroying booby traps in place, explode a small charge near the booby trap's charge. Again, use this method only if damage from the explosion is acceptable. When it is impossible to place the explosive close enough to ensure detonation of the main charge, carefully place it alongside the mechanism. Do not assume the main charge is safe to handle just because the mechanism has been destroyed. Actuate pressure mechanisms by suspending one-half pound of explosive above the pressure plate.
- Clearing by hand. This method involves neutralizing, disarming, removing, and disposing of traps without causing damage. It is extremely hazardous and should only be used when pulling or destructing traps in place is impossible or unacceptable. Clearance should only be conducted by EOD personnel or experienced engineers. Carefully examine all aspects of the trap before deciding how to clear it.
- Explosive line charge. Using this device produces quick results when only a narrow path is required through a booby-trapped area. It gives clearance for the same distance to either side, only where it is in contact with the ground.
- Armor. This method is used where traps with small charges (designed as AP devices) are located in open areas. Armored vehicles track back and forth over the area. This shortens the clearing time with little risk of casualties.

- Fire. If traps are set in grass or dense vegetation, fire can be used to burn away camouflage material and expose traps. In most cases, sufficient heat is generated to burn or explode the traps. Unexploded traps are considered extremely sensitive and must not be cleared by hand.

IMPROVISED TRAPS

In low-intensity conflicts, there is a high incidence of improvised mines and booby traps. It is impossible to provide a complete list of FDs and improvised demolition material that can be used. However, most improvised traps are variations of those described below.

An electric FD requires a current to be passed between two contacts. The ways in which this can be achieved are limited only by the imagination. Examples of simple electric FDs that can be manufactured using household items or appliances are shown in Figure 13-6, page 13-24. Examples of improvised mechanical FDs are shown in Figures 13-7 and 13-8, page 13-25.

The simplest and most accurate method of incorporating delay is to use some form of clock or timer. However, any mechanism or chemical reaction that takes a measurable time to complete its function can be used. Examples of simple improvised delay devices are shown in Figures 13-9 and 13-10, page 13-26.

NONEXPLOSIVE TRAPS

Nonexplosive traps are typically used in tropical or rain-forest regions. Ideal construction materials abound, and concealment in surrounding vegetation is relatively easy. No prescribed procedures exist for clearing nonexplosive traps. Each trap must be cleared according to its nature.

PUNJI

The punji (Figure 13-11, page 13-27) is one of the simplest traps. It is normally made from locally available material, such as sharpened stakes, nails, or heavy-gauge wire. It is placed in concealed pits or in places that give cover from fire. The enemy normally smears the spikes or cutting edges with excrement or poison. After clearing a punji, sterilize cuts and abrasions immediately and see a medical officer as soon as possible.

CLOSING TRAP

The side-closing trap (Figure 13-12, page 13-27) consists of two wooden slats that are studded with spikes. The slats slide along a pair of guide rods controlled by heavy bands. When the prop holding the slats apart is dislodged, the slats spring together and implant the spikes into the portion of the body passing between them.

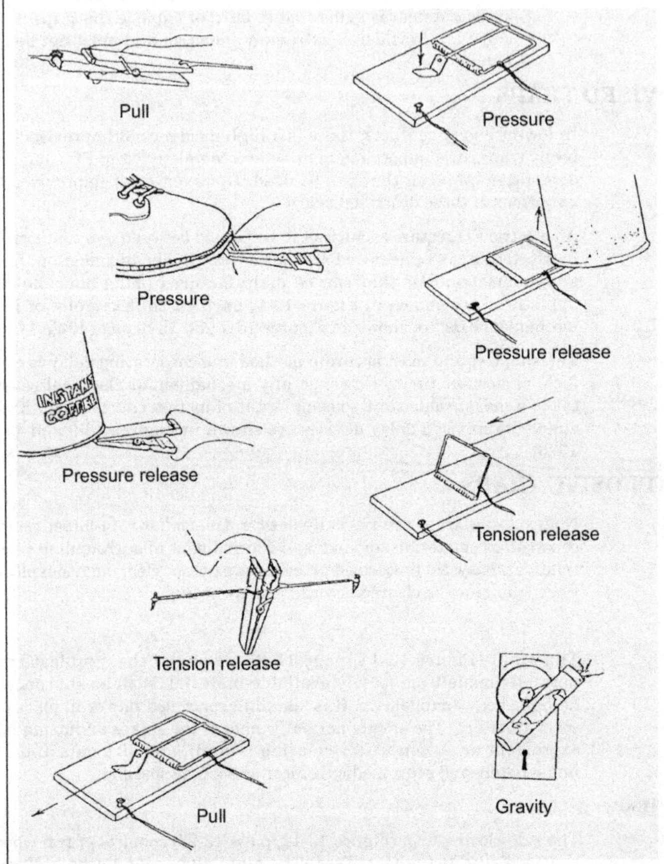

Figure 13-6. Improvised electrical FDs

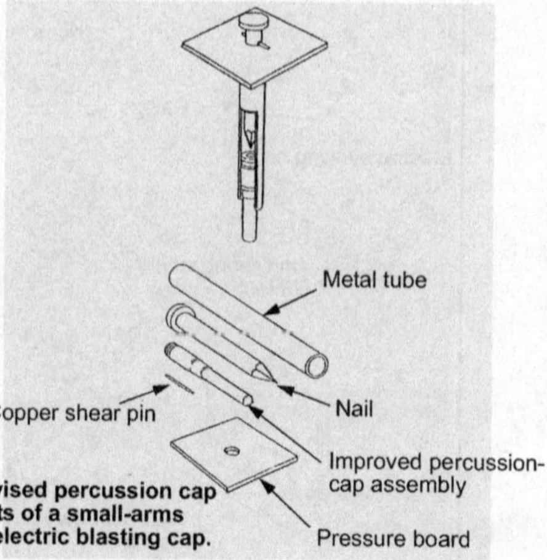

NOTE: An improvised percussion cap assembly consists of a small-arms round and a nonelectric blasting cap.

Figure 13-7. Improvised nonelectric FDs (shear-pin operated)

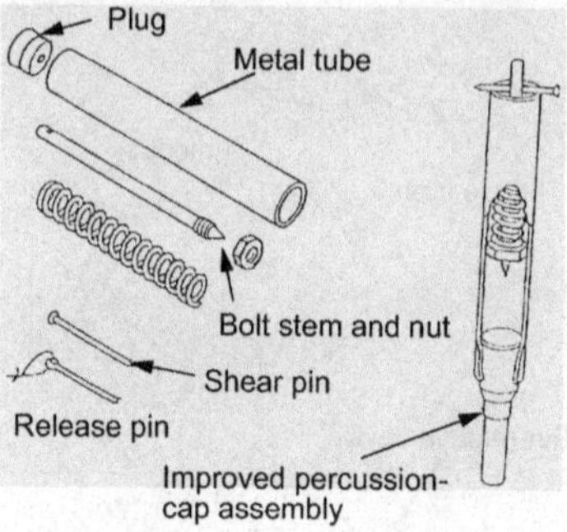

Figure 13-8. Improvised nonelectric FDs (spring-operated)

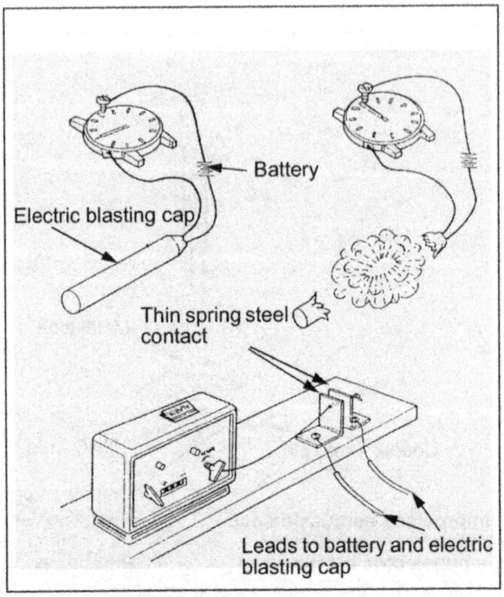

Figure 13-9. Improvised, electric delay devices

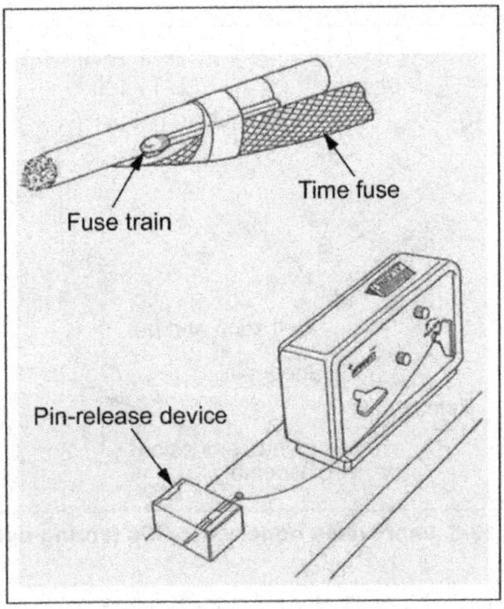

Figure 13-10. Improvised, nonelectric delay devices

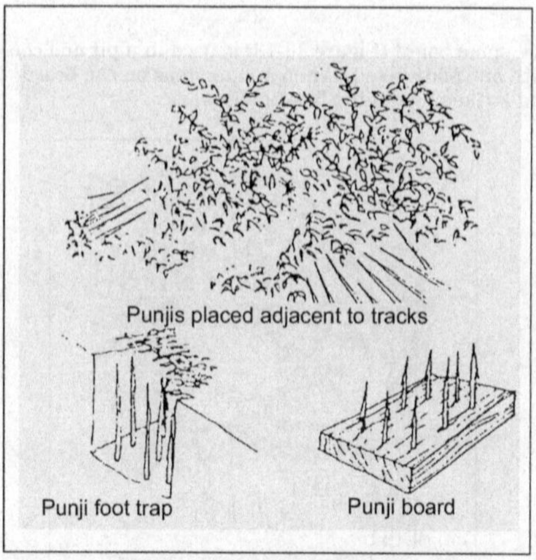

Figure 13-11. Typical punjis

Figure 13-12. Side-closing trap

SPIKE BOARD

The spike board (Figure 13-13) is used in a pit and consists of a treadle board with one end spiked. When a man steps on the board, the spiked end flies up and strikes him in the face or chest.

Figure 13-13. Spike board

VENUS FLYTRAP

The Venus flytrap consists of a rectangular framework with overlapping barbs emplaced in a pit. The one shown in Figure 13-14 is made from a metal container that is sunk into the ground until the top is flush. It is then covered with grass and/or leaves. The barbs inflict injury, especially when the victim attempts to withdraw his leg from the trap. These traps are typically located on tracks and along road edges.

Figure 13-14. Venus fly trap

SECTION III. EXPEDIENT DEVICES

Expedient devices are constructed in the field with locally available material. They are employed against vehicles or personnel in the same manner as other mine systems. Expedient devices—

- Supplement a unit's low supply of conventional mines.
- Hinder reconnaissance, clearance, and neutralization of minefields.
- Create enemy attitudes of uncertainty and suspicion to lower morale and slow movement).

AUTHORIZATION

Because expedient devices have nonstandard design and functioning, take special precautions to protect friendly forces. Consider neutralization, disarming requirements, and adequate marking procedures. The use of expedient devices is restricted under the Convention of Conventional Warfare. Expedient devices have the same international restrictions as booby traps. The corps commander is the employment authority for expedient devices. He can delegate this authority to the division commander. If authority is given to use expedient devices, US personnel will adhere to the rules for international law that are applicable to armed conflict.

EMPLOYMENT AND CONSTRUCTION TECHNIQUES

If issued mines are not readily available on the battlefield, expedient devices can be manufactured in the field. Construction varies based on available materials and the ingenuity of the personnel who are fabricating the devices. Expedient devices pose a potential safety hazard to friendly forces—those who are constructing them and those who may later encounter them. Construction should be performed by personnel who are familiar with the materials being used. Innovative designs should be checked and tested before arming and emplacing the devices.

As a minimum, test the fusing mechanism separately to ensure that it functions as designed. Improper fuse operation is the most common cause of malfunction. Also, test the fuse and the firing chain (base charge, blasting cap, and detonating cord) without the main charge to ensure proper operation. Emplace the device after satisfactory performance of the firing mechanism. First, emplace heavy items (such as artillery shells) that are used as the main charge, and then add the firing mechanism. Take care when moving or emplacing expedient devices because their nonstandard manufacture and potentially faulty construction make them highly sensitive to jars and shocks. Construct devices at the emplacement site whenever possible.

Expedient devices are prepared in the field using standard US FDs, detonators, and demolition materials. All devices discussed in this chapter can be made to function electrically or nonelectrically using modernized demolition initiators (MDIs). AP devices must be command-detonated.

HIGH-EXPLOSIVE, ARTILLERY-SHELL DEVICE

The HE, artillery-shell device (Figure 13-15) can be readily adapted to expedient mining. Remove the artillery fuse and replace it with a standard FD and a length of detonating cord or with an MDI blasting cap. If properly assembled, a destructor may also be used. If a destructor is not available, firmly pack the fuse well with composition C4 explosive and insert a length of knotted detonating cord or a blasting cap.

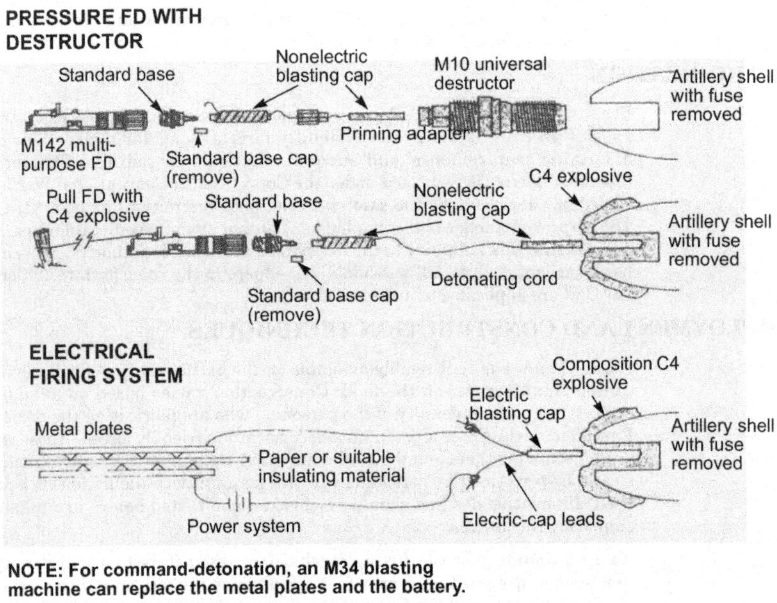

Figure 13-15. HE, artillery-shell device

The device can be activated by a variety of methods depending on the type of FD used. When MDI blasting caps are used, the device is command-detonated. The device can also be adapted to function electrically by adding an electric cap and a power source.

This device can be used as an AT or an AP device. When used as an AP device, it must be command-detonated.

NOTE: Use only serviceable US ammunition that has remained in the possession of US forces. Never use captured ammunition or UXO found on the battlefield. It may be armed, booby-trapped, or deteriorated.

PLATTER CHARGE

The platter charge (Figure 13-16) consists of a suitable container that is filled with uniformly packed explosive and placed behind a platter. The platter is metal (preferably round, but square is satisfactory) and weighs 1 to 3 kilograms. The explosive required is equal to the weight of the platter. The container may not be necessary if the explosive can be held firmly against the platter (tape can be used). The charge should be primed from the exact rear center, and the blasting cap should be secured with a small amount of C4 to ensure detonation.

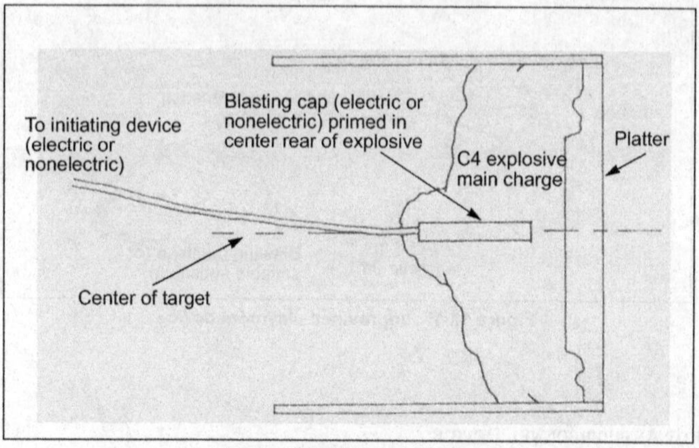

Figure 13-16. Platter charge

The charge should be aimed at the direct center of the target. The effective range (primarily a matter of aim) is approximately 35 meters for a small target. With practice, experienced personnel can hit a 55-gallon drum (a relatively small target) at 25 meters with about 90 percent accuracy.

The platter charge can be used as an AT or an AP device. When used as an AP device, it must be command-detonated.

IMPROVISED CLAYMORE

For the improvised claymore device (Figure 13-17, page 13-32), a layer of plastic explosive is attached to the convex side of a suitably dense, curved base (such as wood or metal). A hole must be made in the exact rear of the base. A blasting cap is placed in the hole to prime the device. Shrapnel is fixed to the explosive with a suitable retainer (cloth, tape, mesh screen).

The device must be command-detonated. Command detonation is best achieved with electrical priming or an MDI. A blasting device is attached to the electric cap via firing wires laid at least 50 meters from the device. Ensure that personnel have adequate cover when detonating the improvised claymore.

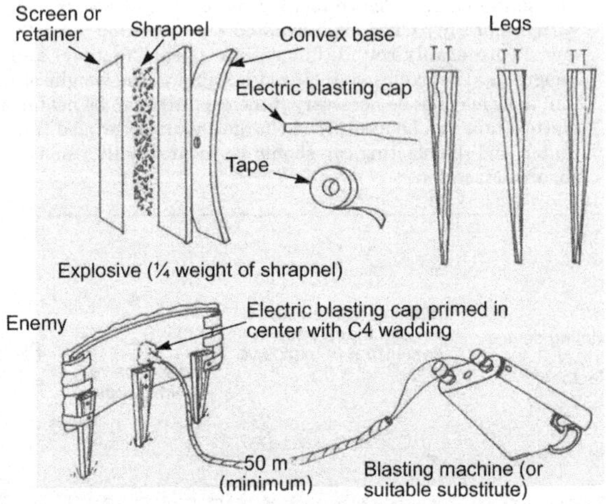

Figure 13-17. Improvised claymore device

GRAPESHOT ANTIPERSONNEL DEVICE

Place shrapnel in the bottom of a cylindrical container to make a grapeshot AP device (Figure 13-18). The shrapnel is tamped and held in place with a suitable separator (wadding). Explosive (approximately one-quarter the weight of the shrapnel) is packed to a uniform density behind the wadding. The device is primed in the center of the explosive with an electric cap or an MDI.

NOTE: The United Nations Convention of Certain Conventional Weapons mandates that all fragment munitions produce fragments that are visible by X ray (such as metal or rock).

This device must be command-detonated. The explosive propels the shrapnel outward from the container. The grapeshot is very effective against personnel targets.

BARBWIRE ANTIPERSONNEL DEVICE

The barbwire AP device (Figure 13-19) can be made directional by placing the wire against an embankment or a fixed object. This causes the force of the explosion to expel the barbwire fragments in the desired direction. One roll of standard barbwire is placed into position, and one block of C4 is placed in the center of the roll and primed. This device must be command-detonated.

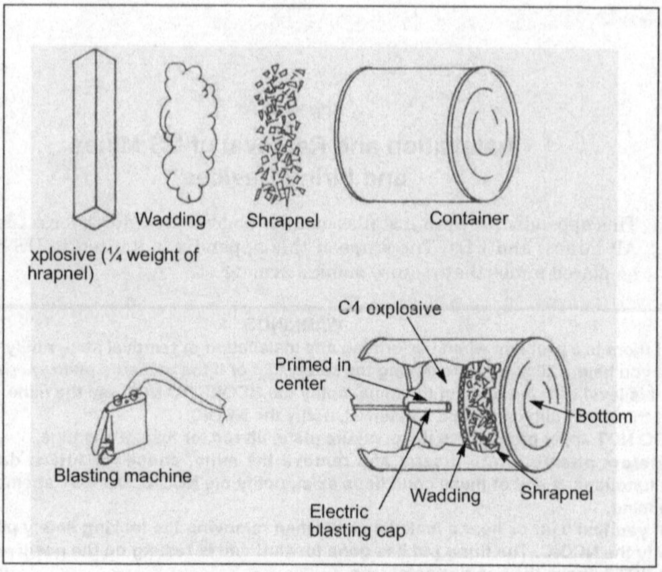

Figure 13-18. Grapeshot AP device

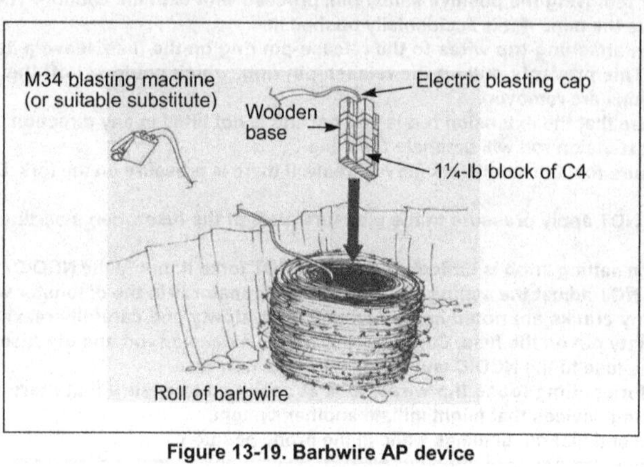

Figure 13-19. Barbwire AP device

Appendix A

Installation and Removal of US Mines and Firing Devices

This appendix provides installation and removal procedures for AT mines, AP mines, and FDs. The scope of this appendix is limited to US hand-emplaced mines that require manual arming.

WARNINGS

1. If there is a problem when performing any installation or removal step, notify the NCOIC.
2. If you hear a click when removing the safety clip or if the pressure plate snaps downward so it is level with the body of the mine, notify the NCOIC. DO NOT use the mine.
3. If the safety clip cannot be reinserted, notify the NCOIC.
4. DO NOT apply pressure to the pressure plate, tilt rod, or fuse at any time.
5. Before attempting to disarm and remove the mine, check for AHDs, damage, and malfunctions. If any of these conditions exist, notify the NCOIC. DO NOT attempt to disarm the mine.
6. If you feel a jar or hear a metallic click when removing the locking safety pin, stop and notify the NCOIC. The firing pin has gone forward and is resting on the positive safety pin. DO NOT remove the positive safety pin.
7. After removing the positive safety pin, proceed with extreme caution. The slider pin can detonate the mine if it is accidentally pushed in.
8. When attaching trip wires to the release-pin ring on the fuse, leave a little slack in the wires. This prevents pull on the release-pin ring, which could set off the mine when the safety pins are removed.
9. Ensure that the extension rod is vertical and is not tilted in any direction. A 20-degree tilt of the extension rod will detonate the mine.
10. Ensure that the safety fork moves freely. If there is pressure on the fork, DO NOT remove it.
11. DO NOT apply pressure to the pressure plate of the fuse when inserting it into the fuse well.
12. If the setting knob is difficult to turn, DO NOT force it; notify the NCOIC.
13. DO NOT adjust the setting knob while the detonator is in the detonator well.
14. If any cracks are noted in the plastic collar, slowly and carefully reassemble the stop and safety pin on the fuse. Carefully remove the extension rod and the fuse from the mine. Give the fuse to the NCOIC and replace it with a new fuse.
15. Before cutting loose trip wires, look at each end to ensure that there are no electric-producing devices that might initiate another system.
16. Arm and disarm all mines while in the prone position.

SECTION I. ANTIPERSONNEL MINES

AP mines can kill or incapacitate their victims. They can be fused by pressure, wire, or command detonation and contain a blast, bounding-fragmentation, or direct-fragmentation warhead.

M14

The M14 mine (Figure A-1) is a low-metallic, blast AP mine. It has a plastic body and an integral plastic fuse with a steel firing pin.

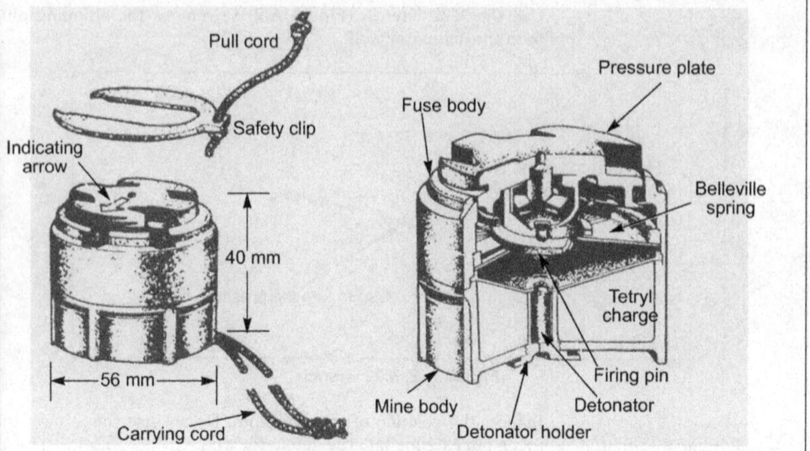

Figure A-1. M14 AP mine

CHARACTERISTICS

Main Charge	Diameter	Height	Weight	No Mines per Box	Weight per Box
Tetryl, 28.4 g	56 mm	40 mm	99.4 g (fused)	90	19.8 kg

- **Korea Only: The M14 is employed in tactical and nuisance minefields.**
- The M14's size allows for employment in large numbers, and rapid concealment is possible.
- The M14 is buried to prevent the target from knocking it over.

- The M14 is not designed to kill but to penetrate a boot or a foot.
- The M14 is difficult to detect because of plastic construction.
- The M14 requires an operating force of 11.5 to 13.5 kilograms to activate.

INSTALLATION

WARNING
Emplace and remove the mine while in the prone position.

- Inspect the mine.
  - Do not use the mine if it is dented, cracked, or damaged.
  - Use the M22 wrench (Figure A-2) to remove the shipping plug from the detonator well.

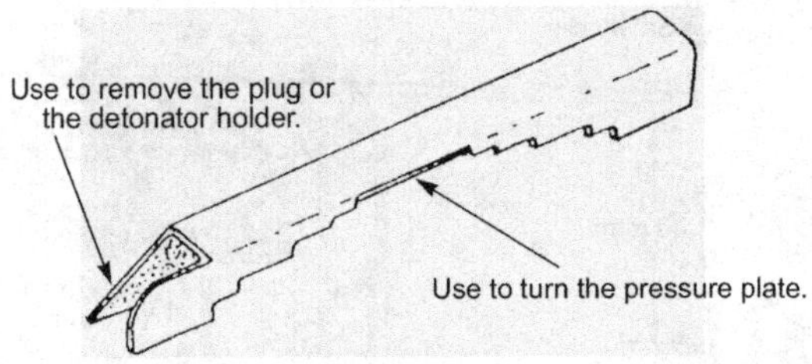

Figure A-2. M22 wrench

  - Inspect the position of the firing pin. Do not use the mine if the firing pin extends into the detonator well.
  - Inspect the detonator well for foreign material. If foreign material is present, remove it by carefully tapping the mine against the palm of your hand. If you cannot remove the debris, replace the shipping plug and do not use the mine.
- Test the pressure plate.
  - Ensure that the arrow is in the *SAFE* position.
  - Use the M22 wrench to turn the pressure plate from the *SAFE* position to the *ARMED* position (Figure A-3, page A-4).
  - Grasp the mine in one hand and remove the safety clip with the other hand. Listen for a click, indicating that the firing pin has dropped. Recheck the fuse well (Figure A-4, page A-4).
  - Replace the safety clip.

— Use the M22 wrench to turn the pressure plate back to the *SAFE* position.

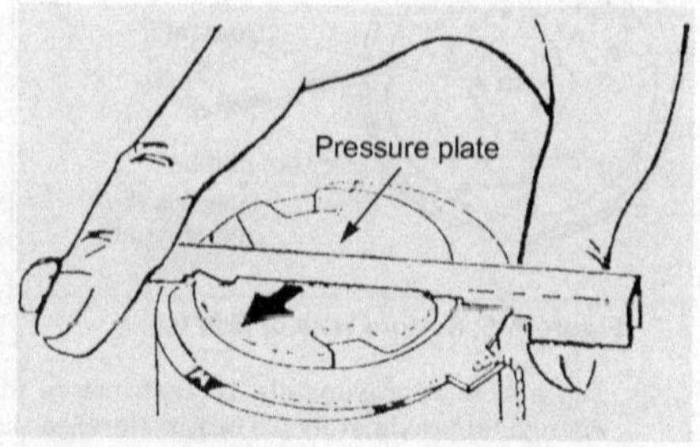

Figure A-3. M14 mine in *ARMED* position

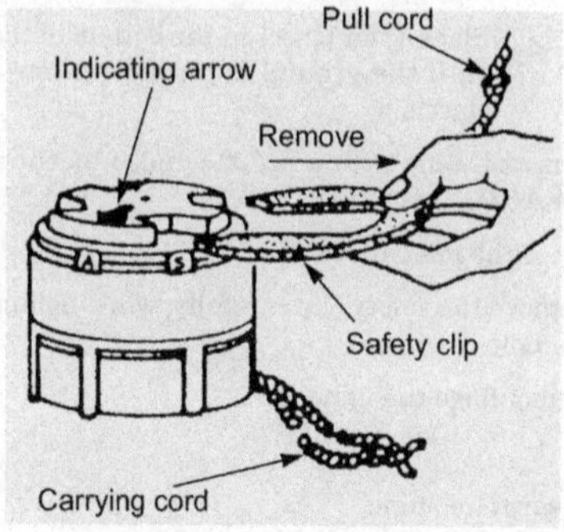

Figure A-4. Removal of safety clip

- Ensure that a metallic washer is attached (glued) to the bottom of the mine. If a washer is not attached, notify the NCOIC and do **not** use the mine. Apply silicone on the outside of the washer to prevent water damage, and screw the detonator into the well on the bottom of the mine (Figure A-5). **NOTE: Ensure that the gasket is tightly wedged between the detonator and the washer and that the washer is properly seated between the gasket and the body of the mine so that water cannot enter the mine.**
- Dig a hole to fit the mine.

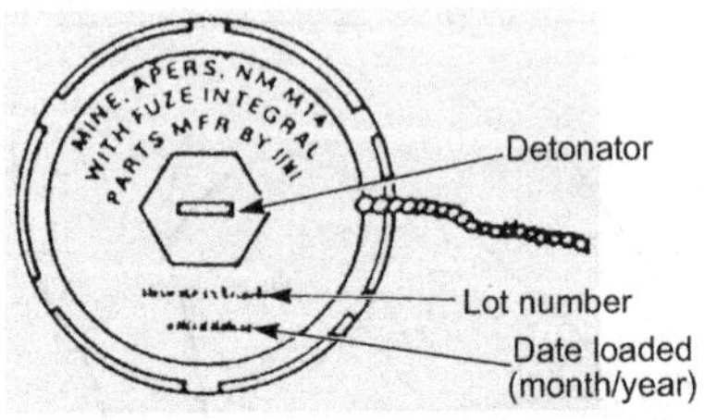

Figure A-5. Bottom view of M14 mine

- Dig a hole approximately 10 centimeters in diameter and deep enough (approximately 3.5 centimeters) so that the pressure plate extends above the ground.
- Inspect the ground at the bottom of the hole.
  - Ensure that the ground is hard enough to support the mine when pressure is applied to the pressure plate.
  - Place a flat object in the bottom of the hole for the mine to rest on if the ground is too soft. Allow additional depth for the objects.

- Arm the mine with an M22 wrench by turning the pressure plate to the *ARMED* position.
- Place the mine in the hole.
- Remove the safety clip carefully, while holding the mine body firmly in the hole.
- Camouflage the mine.

REMOVAL

- Disarm the mine.
  - Clear the soil away from the mine carefully.
  - Grasp the body of the mine firmly with one hand, and insert the safety clip with the other hand.
  - Use the M22 wrench to turn the pressure plate to the *SAFE* position.
- Remove the mine from the hole.
  - Turn the mine over, and carefully remove the detonator from the detonator well.
  - Screw the shipping plug into the detonator well.
  - Give the detonator to the NCOIC.

M16

The M16-series mine (Figures A-6, page A-6) is a bounding-fragmentation AP mine. Once actuated, the mine is propelled out of the ground (to a height of approximately 1.8 meters) and explodes. The mine consists of an M605 fuse, a propelling charge, and a projectile that are contained in a sheet-steel case. The fuse is screwed into the top of the case and extends through the center of the projectile to the bottom of the case, where the propelling charge is located. The remaining space inside the case is occupied by the projectile. Earlier versions of the mine are also available for issue. The principal difference between the old and new versions is the construction of the detonators and boosters.

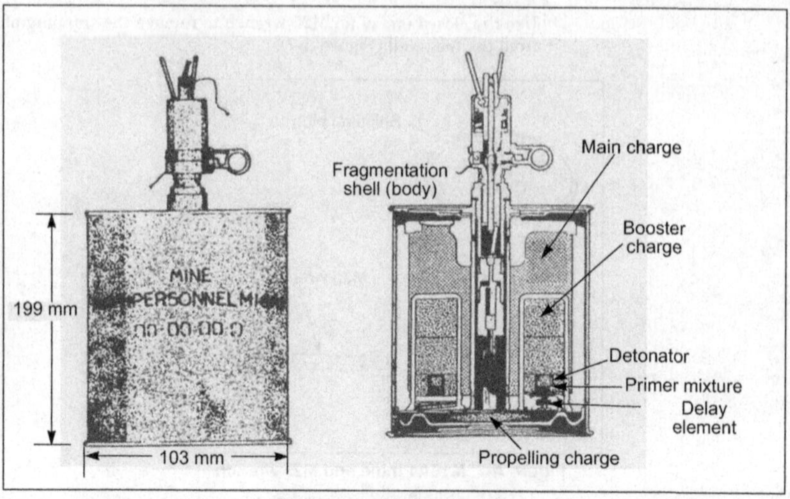

Figure A-6. M16A1 AP mine

CHARACTERISTICS

Main Charge	Diameter	Height	Weight	No Mines per Box	Weight per Box
TNT	103 mm	199 mm	3.5 kg	4	20.25 kg

- **Korea Only: The M16 is employed in protective, tactical, and nuisance minefields.**
- The M16 is used to defeat dismounted assaults and breaching operations.
- The M16 is pressure-actuated (3.6 to 9 kilograms) or pull-actuated (1.4 to 4.5 kilograms).

- The M16 and M16A1 have a casualty radius of 27 meters; the M16A2 has a casualty radius of 30 meters. The danger radius for friendly forces is 183 meters for all M16 mines.

INSTALLATION

WARNING
Emplace and remove the mine while in the prone position.

- Inspect the mine.
  - Do not use the mine if it is dented, cracked, or damaged.
  - Use the closed end of an M25 wrench to remove the shipping plug from the fuse well (Figure A-7).

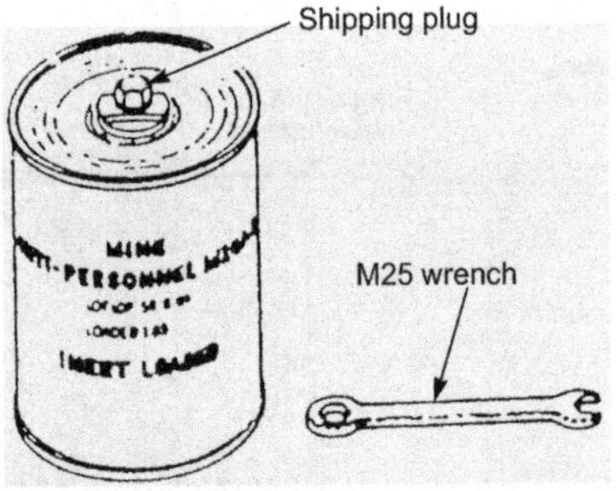

Figure A-7. M16A1 mine and M25 wrench

  - Inspect the fuse well for foreign material. If foreign material is present, turn the mine upside down and gently tap the bottom with your hand to dislodge the material. If you cannot remove the debris, replace the shipping plug and do not use the mine.
  - Examine the fuse assembly carefully for evidence of damage or missing safety pins. Ensure that the safety pins move freely in their holes and that the rubber gasket is around the fuse base (Figure A-8, page A-8).

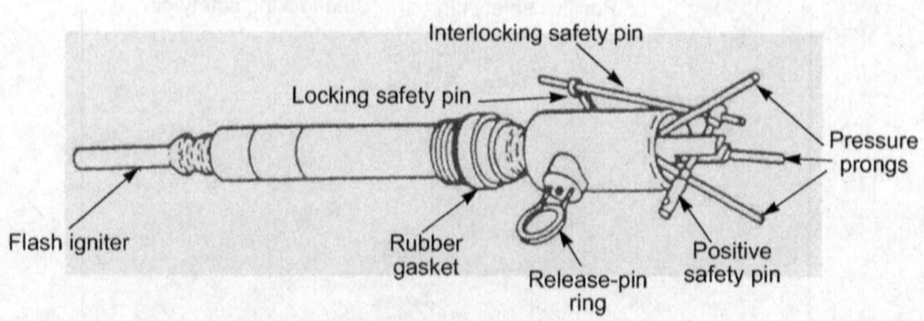

Figure A-8. M605 fuse

> **WARNING**
> **DO NOT use any fuse that was manufactured before 1957.**

- Fuse the mine.
  - Use the open end of the M25 wrench to tighten the bushing adapter on the fuse well.
  - Screw the fuse assembly into the fuse well by hand and tighten the assembly with the M25 wrench. Ensure that the rubber gasket is between the fuse body and the bushing adapter. **NOTE: For long-term use, smear a thin layer of silicone grease or similar lubricant on the fuse and threads.**
- Dig a hole to fit the mine.
  - Pressure installation. Dig the hole so that only the pressure-prong tips are above ground level.
  - Trip-wire installation. Dig the hole so that the release-pin ring is above ground level.
- Emplace the mine.
  - Place the mine in the hole and ensure that the safety pins remain in place.
  - Cover the mine with soil to the bottom of the release-pin ring.
  - Press the soil firmly around the sides of the mine.

Pressure Installation

- Remove the locking safety pin (Figure A-9). The interlocking safety pin will come free.

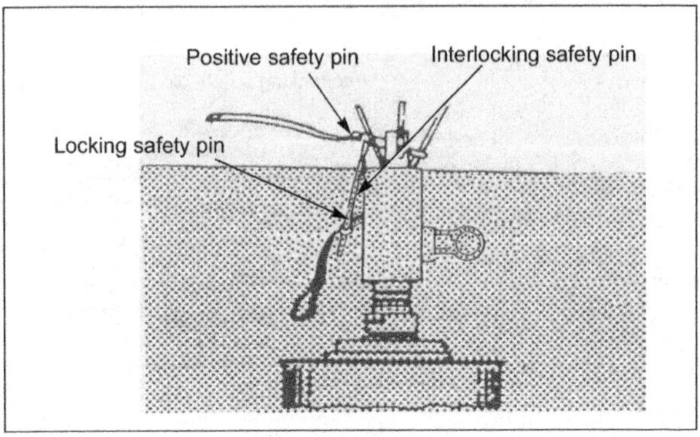

Figure A-9. Safety pins

- Arrange the pull cord on the positive safety pin so that it withdraws easily. Remove the metal collar.
- Finish covering the mine with soil until only the pressure prongs are above ground level.
- Camouflage the mine, place excess soil in sandbags, and remove sandbags from the area.
- Arm the mine by removing the positive safety pin (Figure A-9).
- Give the safety pins and the shipping plug to the NCOIC.

Trip-Wire Installation

- Install the trip wire (Figure A-10).

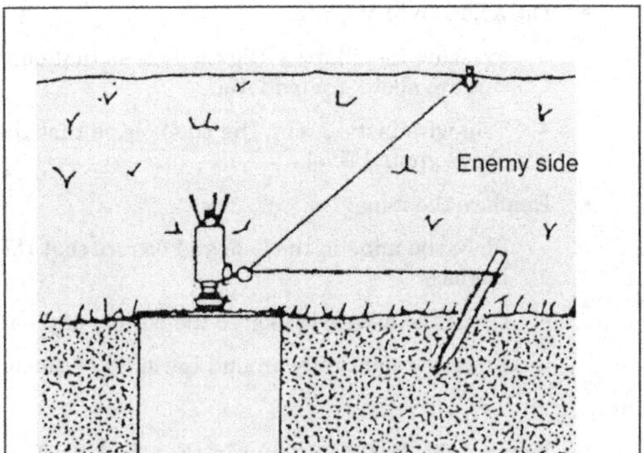

Figure A-10. Buried mine with a trip wire

- Cover the mine with soil and press it firmly around the sides of the mine. Leave the release-pin ring and the pressure prongs exposed.
- Tie off trip wires, approximately 10 meters from the mine. The wires should form a wide V, with the opening toward the enemy.
- Attach trip wires to the release-pin ring on the fuse.
- Remove the locking safety pin. The interlocking safety pin will come free.
- Arrange the pull cord on the positive safety pin so that it withdraws easily.
- Camouflage the mine, place excess soil in sandbags, and remove sandbags from the area.
- Arm the mine by removing the positive safety pin.
- Give the safety pins and the shipping plug to the NCOIC.

REMOVAL

> **WARNING**
> **Before attempting to disarm and remove the mine, ensure that the metal collar (Figure A-11) over the top of the striker on the M605 fuse is in place.**

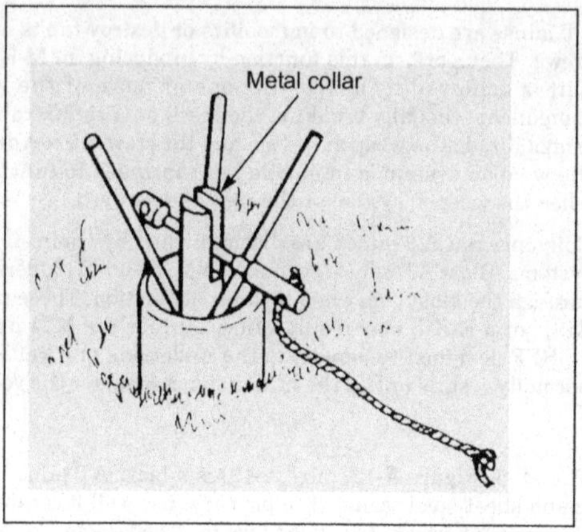

Figure A-11. Metal collar on an M605 fuse

- Disarm the mine.
  - Clear the soil carefully from the top of the fuse to the positive safety-pin hole. When using the M605 fuse, clear away all the soil from the fuse area.
  - Insert the positive safety pin through the positive safety-pin hole.
  - Insert the locking safety pin through the locking safety-pin hole.
  - Cut the slack trip wires that are attached to the release-pin ring.
- Check for AHDs.
  - Hold the mine body firmly in place with one hand.
  - Feel for AHDs with the other hand by digging around the sides of and underneath the mine.
- Remove the mine.
  - Remove the mine from the hole. Ensure that the safety pins remain in place.
  - Remove the M605 fuse with the M25 wrench.
  - Replace the shipping plug in the fuse well.

SECTION II. ANTITANK MINES

AT mines are designed to immobilize or destroy tanks and vehicles and their crews. They perform this function by producing an M-Kill or a K-Kill. An M-Kill is achieved by destroying one or more of the vehicle's vital drive components (usually breaking the track on a tank), causing the target to be immobilized. The weapon system and the crew are not destroyed in an M-Kill; the weapon system is immobile but continues to function. A K-Kill results when the weapon system or the crew is destroyed.

Conventional AT mines are distinguished by their effects and their fusing systems. Blast AT mines, such as the M15 and M19, derive their effectiveness through the blast generated by their detonation. These usually produce an M-Kill, but a K-Kill may result. Mines such as the M21 use a shaped charge or an SFF designed to penetrate the underside of a vehicle's armor. A K-Kill normally results unless the mine detonates under the vehicle's track.

M15

The M15 (Figure A-12, page A-12) is a blast AT mine that is contained in a round sheet-steel casing. The primary fuse well is located in the top center of the mine. There are two secondary fuse wells—one on the side and one on the bottom. The primary fuse well accepts the M603 pressure-actuated fuse. Standard FDs can be used in the secondary fuse wells with the M1 activator. The M624 tilt-rod-actuated fuse can also be used with this mine.

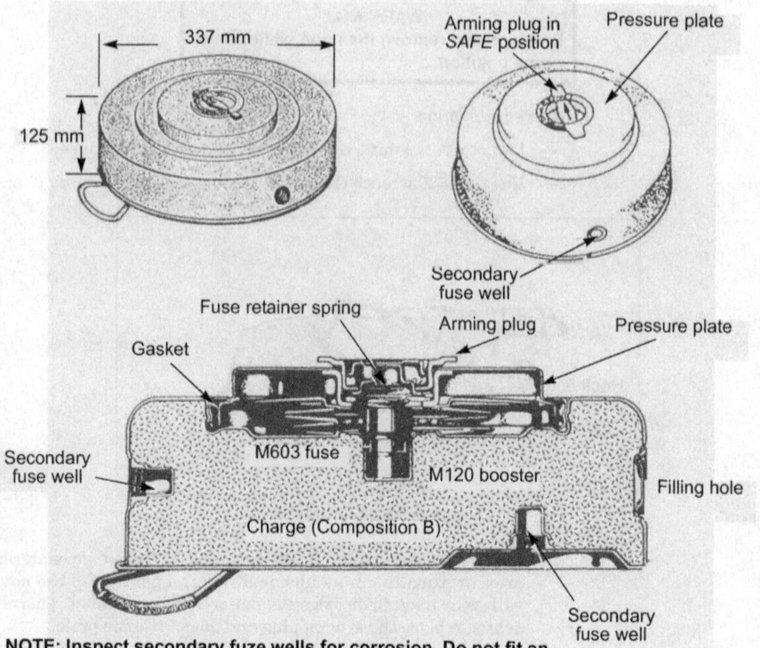

NOTE: Inspect secondary fuze wells for corrosion. Do not fit an M1 activator into a corroded fuze well. In training, return any mine with a corroded fuze well to the ASP as unserviceable.

Figure A-12. M15 AT mine

CHARACTERISTICS

Main Charge	Diameter	Height	Weight	No Mines per Box	Weight per Box
Comp B, 9.9 kg	337 mm	125 mm	13.5 kg	1	18 kg

- The M15 is employed in protective, tactical, and nuisance minefields.
- The M15 is surface-laid or buried.
- The M15 requires a force of 158 to 338 kilograms to detonate the M603 fuse and a force of 1.7 kilograms to deflect the tilt rod and detonate the M624 fuse.
- The M15 is designed to defeat heavy tanks.
- The M15 produces an M-Kill upon contact.

INSTALLATION USING THE M624 FUSE

WARNING
Emplace and remove the mine while in the prone position.

- Inspect the mine.
  - — Do not use the mine if it is dented, cracked, or damaged.
  - — Use the M20 wrench (Figure A-13) to remove the arming plug.

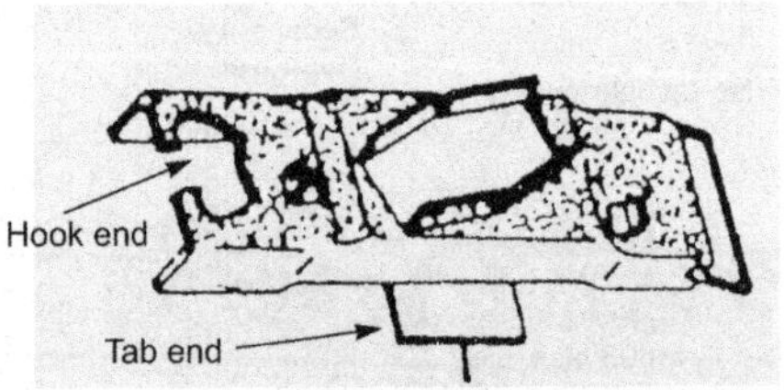

Figure A-13. M20 wrench

  - — Inspect the fuse well for foreign material. If foreign material is present, turn the mine upside down and gently tap the bottom with your hand to dislodge the material. If you cannot remove the debris, replace the arming plug and do not use the mine.
  - — Ensure that the booster retainer ring is seated in the fuse well. If the retainer ring is missing, replace the mine.
- Inspect the fuse.
  - — Remove the M624 fuse from the metal shipping container and inspect it for serviceability.
  - — Inspect the plastic collar of the fuse by looking down through the top of the pressure ring. If the safety pin is missing or improperly assembled, do not use the fuse (Figure A-14, page A-14).
  - — Do not use the fuse if the plastic collar appears to be cracked.
- Fuse the mine.
  - — Remove the M624 fuse from its fiber sleeve.
  - — Remove the end closure on the M624 fuse. **NOTE: For long-term emplacement, coat the fuse threads and gasket with silicone grease before removing the end closure (Figure A-15, page A-14).**

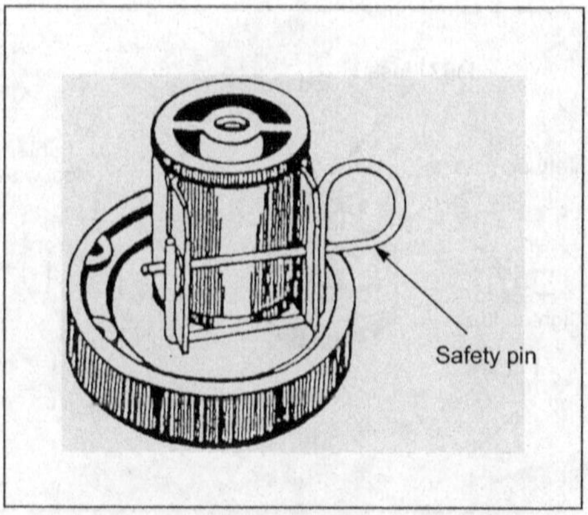

Figure A-14. Correct safety-pin configuration

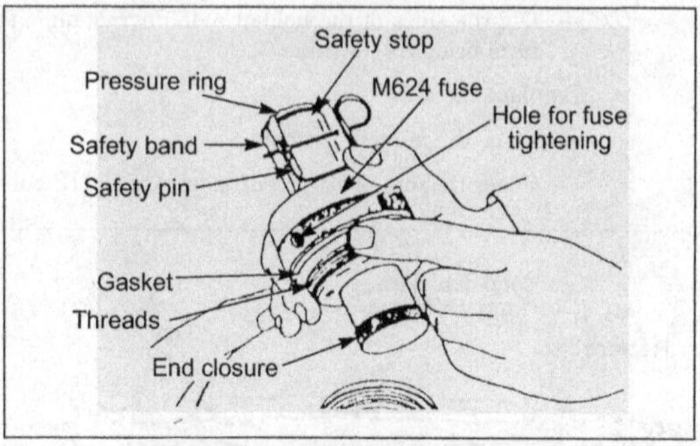

Figure A-15. Greasing the M624 fuse

— Screw the fuse hand-tight into the fuse well.

— Remove the extension rod from its packaging.

— Tighten the fuse by inserting the unthreaded end of one extension rod piece into the hole on the side of the fuse. Turn the fuse a quarter turn (Figure A-16).

— Remove the extension rod for further use after the fuse is secure.

NOTE: The M15 AT mine (with the M624 fuse) can be buried or surface-laid. If surface-laid, it must be staked in place.

- Dig a hole to fit the mine.

  — Dig a hole deep enough so that the top of the pressure plate will be at ground level.

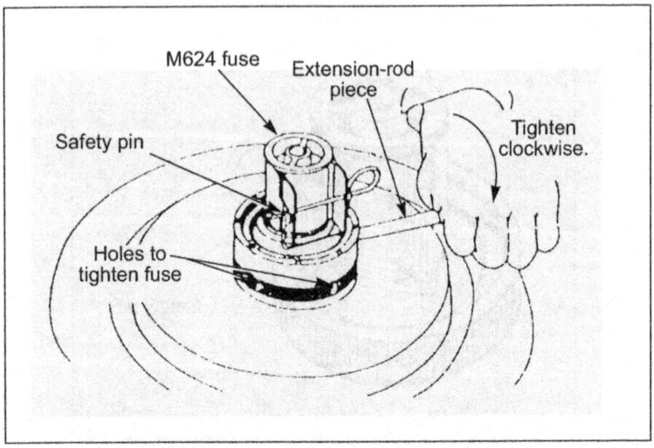

Figure A-16. Tightening the fuse with the extension rod

- — Dig the sides of the hole at a 45-degree angle to prevent vehicles from bridging the mine.
- Emplace the mine.
  - — Place the mine in the hole.
  - — Cover the mine with 2 centimeters of soil (Figure A-17).

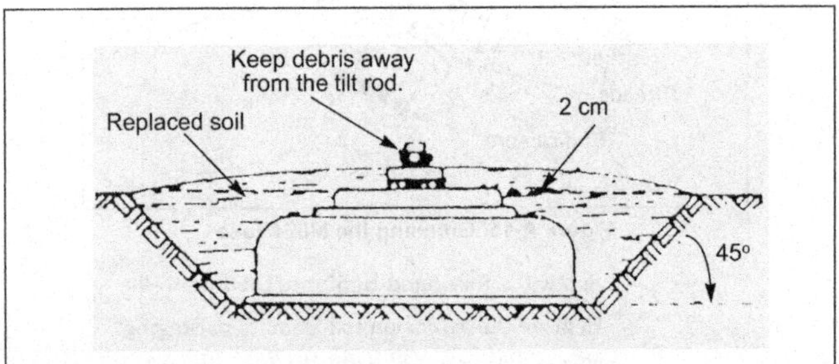

Figure A-17. M15 mine in the hole

NOTE: The M15 AT mine (with the M624 fuse) can be used in the tilt-rod or pressure role. In the tilt-rod role only, assemble all three pieces of the extension rod (Figure A-18, page A-16) and thread the extension rod into the threaded pressure ring of the fuse (Figure A-19, page A-16).

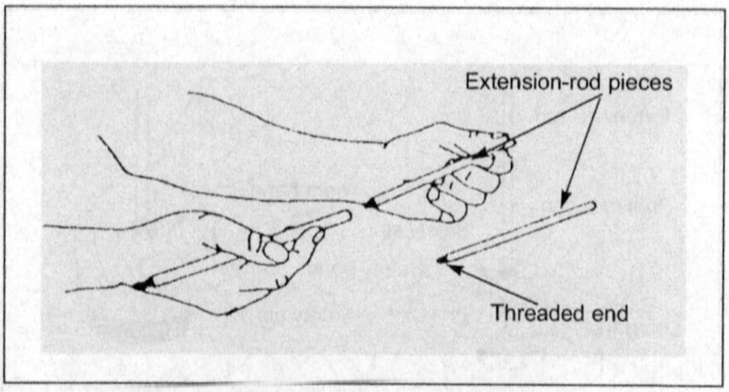

Figure A-18. Extension-rod assembly

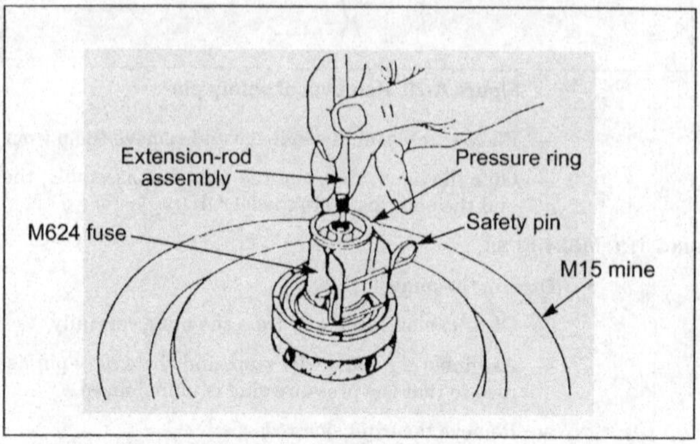

Figure A-19. Assembly of the extension rod into the fuse ring

- Arm the mine.
  - Use your right hand to raise the safety pin to the horizontal position, and grasp the safety band and safety stop with your left hand. Note the position of the thumb in Figure A-20.
  - Remove the safety pin with your right index finger, pulling it to the right.
  - Remove the safety stop carefully while holding the safety band in place.
  - Remove the safety band.
- Camouflage the mine.
  - Camouflage the mine with twigs, grass, or other material in the area. Place mines with extension rods in tall grass, if possible. Ensure that no pressure is applied to the tilt rod or the fuse.

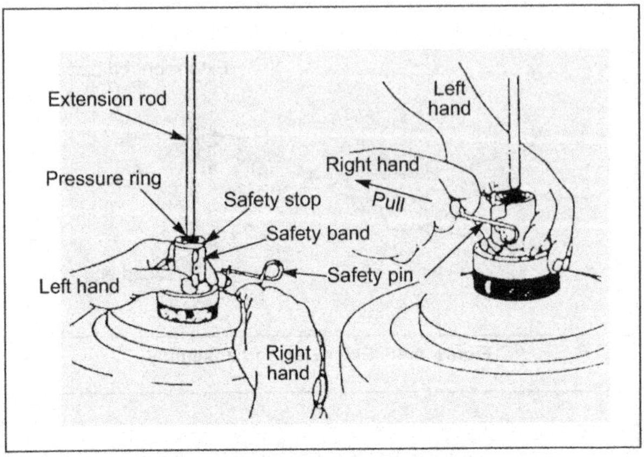

Figure A-20. Removal of safety pin

- Place excess soil in sandbags and remove them from the area.
- Give the band, the stop, the pull-ring assembly, the arming plug, and the end closure to the NCOIC.

Removal Using the M624 Fuse

- Disarm the mine.
  - Clear camouflage away from the mine carefully.
  - Assemble the band, the stop, and the safety-pin assembly on the fuse so that the pressure ring is immobilized.
  - Remove the extension rod.
- Check for AHDs.
  - Hold the mine firmly in place with one hand, without putting pressure on the fuse.
  - Feel for AHDs with the other hand by digging around the sides of and underneath the mine.
- Remove the mine.
  - Remove the mine from the hole.
  - Remove the fuse from the mine; use the extension rod if necessary.
  - Replace the end closure on the fuse.
  - Install the arming plug into the fuse well.

Installation Using the M603 Fuse

- Inspect the mine.
  - Do not use the mine if it is dented, cracked, or damaged.

— Use the M20 wrench to remove the arming plug from the mine.

— Inspect the fuse well for foreign material. If foreign material is present, turn the mine upside down and gently tap the bottom with your hand to dislodge the material. If you cannot remove the debris, replace the arming plug and do not use the mine.

— Ensure that the booster retainer ring is seated in the fuse well. If the retainer ring is missing, replace the mine.

- Perform a function check with the arming plug.

— Turn the setting knob to the *ARMED* position. Ensure that the shutter bar moves across the bottom of the arming plug (Figure A-21)

— Turn the setting knob to the *SAFE* position. Ensure that the shutter bar moves back across the bottom of the arming plug (Figure A-22). **NOTE: If the shutter bar does not go into the *SAFE* or *ARMED* position, notify the NCOIC.**

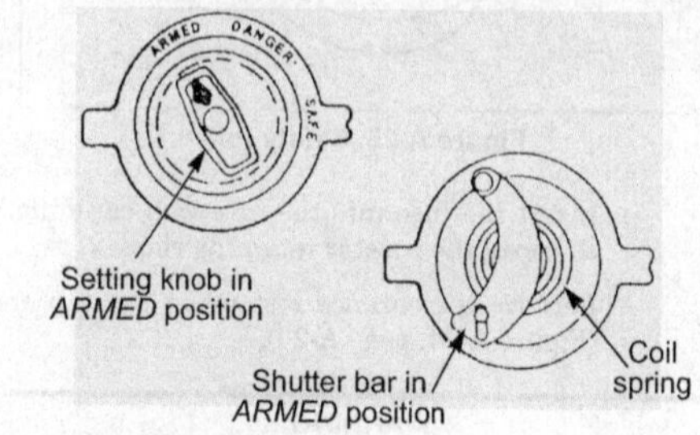

NOTE: A coil spring may not be present in older models.

Figure A-21. *ARMED* position

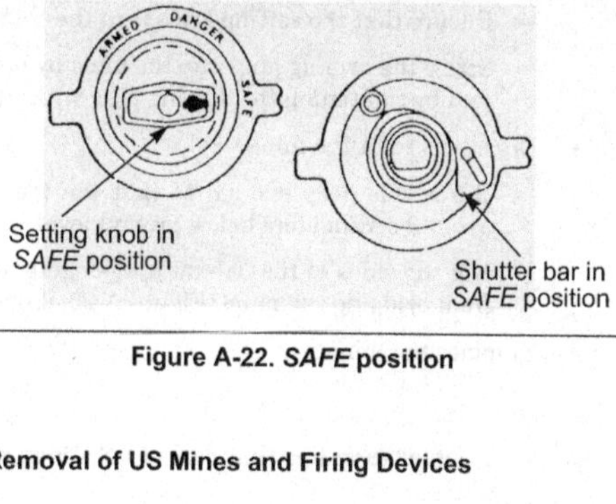

Figure A-22. *SAFE* position

- Fuse the mine.
  - Remove the M603 fuse from the metal shipping container and inspect it for serviceability. The green end of the detonator must show in the bottom of the fuse.
  - Remove the safety fork; use the hooked end of an M20 wrench if necessary (Figure A-23).

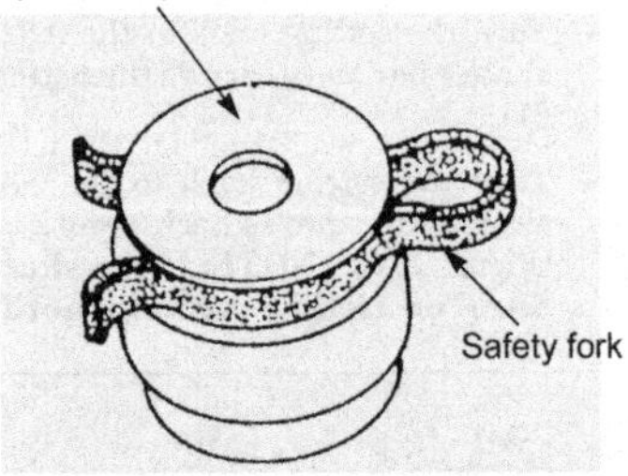

Figure A-23. Safety fork

  - Insert the fuse into the fuse well carefully until it seats securely on top of the booster retaining ring.
  - Perform a clearance test using the tab end of the M20 wrench (Figure A-24, page A-20).

WARNING
If the fuse pressure plate interferes with the tab end of the M20 wrench, investigate the cause and notify the NCOIC. DO NOT arm the mine.

NOTE: For long-term emplacement, smear a thin layer of silicone grease or similar lubricant on the arming plug, the threads, and the gasket.

  - Ensure that the setting knob is in the *SAFE* position.
  - Screw the arming plug into the mine by hand. Ensure a watertight seal by tightening the arming plug with the M20 wrench.
- Dig a hole to fit the mine.
  - Dig a hole deep enough so that the top of the pressure plate is about 3 centimeters below ground level.
  - Dig the sides of the hole at a 45-degree angle to prevent vehicles from bridging the mine (Figure A-25, page A-20).
- Emplace the mine.

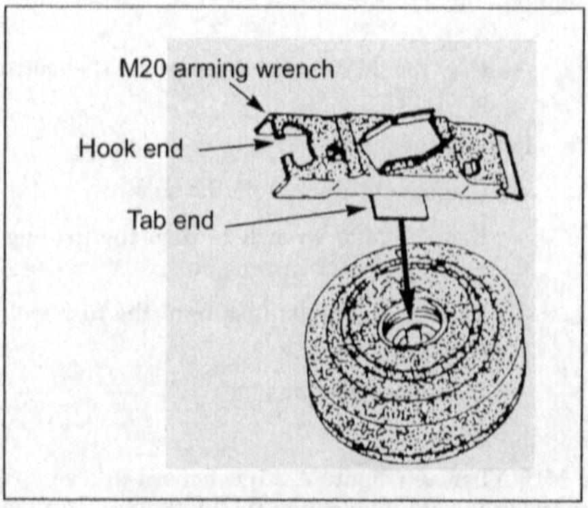

Figure A-24. Clearance test

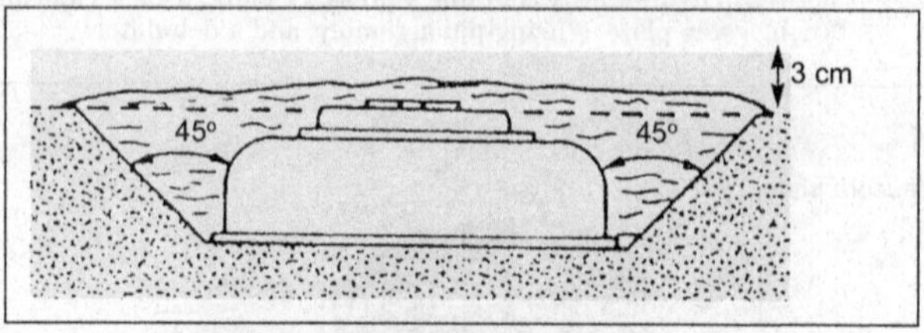

Figure A-25. M15 mine in the hole

- — Place the mine in the hole.
- — Cover the mine with soil until it is level with the top of the pressure plate.

- Use the M20 wrench to arm the mine by turning the setting knob from the *SAFE* position to the *ARMED* position.
- Camouflage the mine.
  - — Cover the mine with 3 to 5 centimeters of soil.
  - — Camouflage the mine, place excess soil in sandbags, and remove sandbags from the area.
  - — Give the safety clip to the NCOIC.

REMOVAL USING THE M603 FUSE

- Disarm the mine.
  - — Clear the soil from the top of the mine carefully.
  - — Hold the mine firmly in place with one hand, without putting pressure on the pressure plate.
  - — Feel for AHDs with the other hand by digging around the sides of and underneath the mine.

— Use the M20 wrench to turn the setting knob to the *SAFE* position.

- Remove the mine.

— Remove the mine from the hole.

— Use the M20 wrench to turn the arming plug counterclockwise, and remove the arming plug.

— Remove the M603 fuse from the fuse well and replace the safety fork.

— Install the arming plug.

M19

The M19 AT mine (Figure A-26) is housed in a square, plastic case and holds 9.45 kilograms of Composition B (HE charge). It consists of an M606 integral pressure fuse and two secondary fuse wells—one in the side and one in the bottom. The fuse body contains a pressure plate, a Belleville spring, a setting knob, a step plate, a firing-pin assembly, and a detonator.

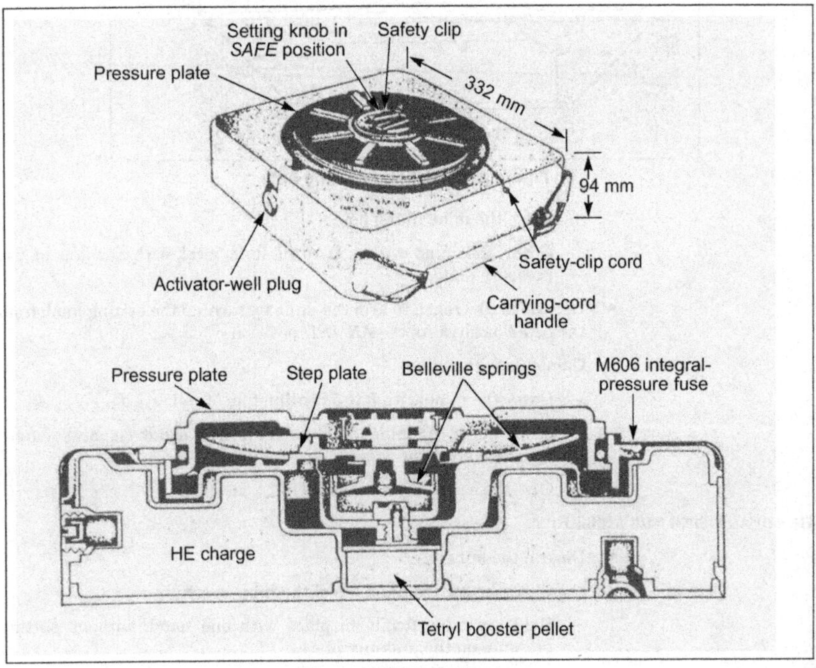

Figure A-26. M19 AT mine

CHARACTERISTICS

Main Charge	Diameter	Height	Weight	No Mines per Box	Weight per Box
Comp B, 9.45 kg	332 mm	94 mm	12.6 kg	2	32.5 kg

- The M19 is employed in tactical and nuisance minefields.
- The M19 is buried or surface-laid.
- The M19 requires a force of 157 to 225 kilograms to detonate.
- Standard FDs may be used with the M2 activator in any of the secondary fuse wells of the M19.

INSTALLATION

> **WARNING**
> **Emplace and remove the mine while in the prone position.**

- Inspect the mine.
  - Do not use the mine if it is dented, cracked, or damaged.
  - Remove the M606 fuse from the fuse well by turning it counterclockwise a quarter turn (Figure A-27).
  - Ensure that the rubber gasket is on the M606 fuse.

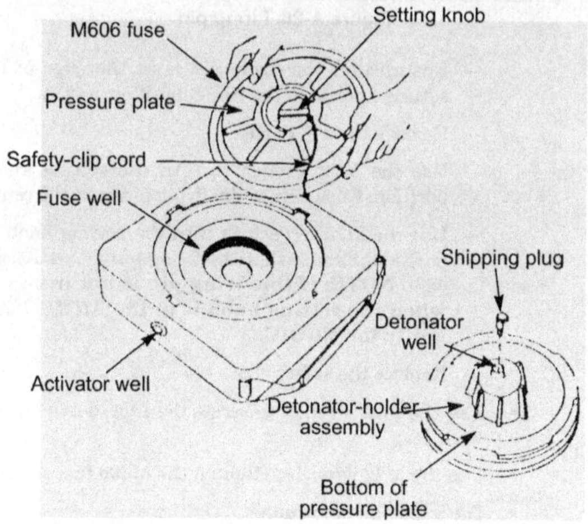

Figure A-27. Removal of the pressure plate

— Remove any foreign material from the fuse well.

— Ensure that the setting knob is in the *SAFE* position and that the safety clip is in place.

— Use the M22 wrench to remove the shipping plug from the detonator well.

— Inspect the detonator well for foreign material. If foreign material is present, gently tap the pressure plate with your hand to dislodge the material.

- Test the position of the firing pin (Figure A-28).

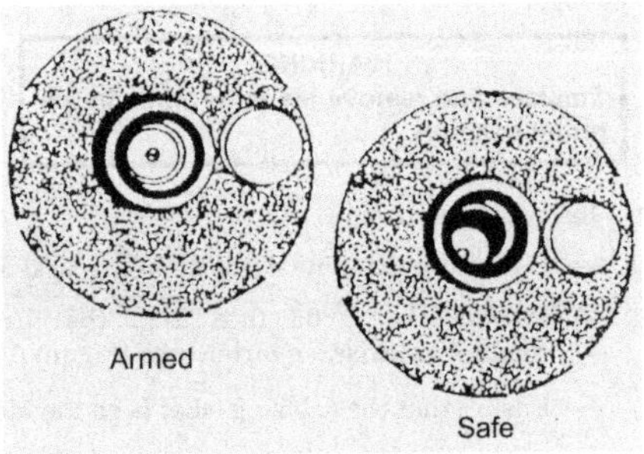

Figure A-28. Firing pin

— Ensure that the firing pin is at the edge of the well when the setting knob is in the *SAFE* position.

— Remove the safety clip.

— Use the M22 wrench to turn the setting knob to the *ARMED* position. Ensure that the firing pin is in the center of the well.

— Use the M22 wrench to turn the setting knob back to the *SAFE* position. Ensure that the firing pin moves back to the side of the well. **NOTE: If the firing pin is not in the correct position when the setting knob is in the *ARMED* or *SAFE* position, notify the NCOIC.**

— Replace the safety clip.

- Use the M22 wrench to screw the M50 detonator into the detonator well.
- Use the M22 wrench to tighten the M606 fuse into the fuse well.
- Dig a hole to fit the mine.

— Dig a hole deep enough so that the top of the pressure plate will be even with or slightly below ground level.

- Dig the sides of the hole at a 45-degree angle to prevent vehicles from bridging the mine.

- Emplace the mine.
  - Place the mine in the hole.
  - Cover the mine with soil until it is level with the top of the pressure plate.
- Arm the mine.
  - Remove the safety clip.
  - Use the M22 wrench to turn the setting knob from the *SAFE* position to the *ARMED* position.
- Camouflage the mine.
  - Cover the mine with 3 centimeters of soil.
  - Camouflage the mine, place excess soil in sandbags, and remove sandbags from the area.
  - Give the safety clip and the shipping plug to the NCOIC.

REMOVAL

- Disarm the mine.
  - Clear the soil from the top of the mine carefully.
  - Hold the mine firmly in place with one hand, without putting pressure on the pressure plate.
  - Feel for AHDs with the other hand by digging around the sides of and underneath the mine.
  - Use the M22 wrench to turn the setting knob to the *SAFE* position.
  - Replace the safety clip on the M606 fuse.
- Remove the mine.
  - Remove the mine from the hole.
  - Use the M22 wrench to remove the M606 fuse by turning it counterclockwise and lifting it out of the fuse well.
  - Use the M22 wrench to remove the detonator from the detonator well.
  - Replace the shipping plug in the detonator well.
  - Replace the pressure plate in the mine.

M21

The M21 AT mine (Figure A-29) utilizes a direct-energy warhead that is designed to produce a K-Kill. It is used in conjunction with the M607 fuse. The M21 produces a K-Kill against heavy tanks, unless the mine is activated under the track. It can be buried with a tilt rod, or it can be surface-laid with or without a tilt rod. If the M21 is surface-laid with a tilt rod, it must be staked to prevent it from being knocked over and causing the warhead to be directed away from the target. The M21 is not compatible with any mechanical mine-dispensing system.

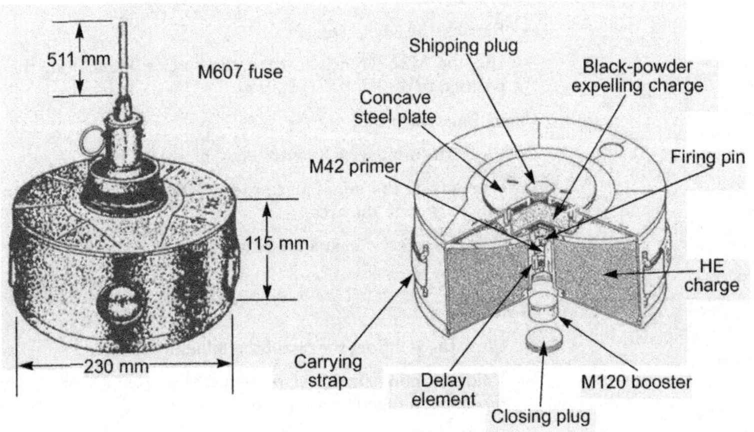

Figure A-29. M21 AT mine

CHARACTERISTICS

Main Charge	Diameter	Height	Weight	No Mines per Box	Weight per Box
Comp H6, 4.95 kg	230 mm	115 mm	7.8 kg	4	41 kg

- The M21 requires a minimum of 130.5 kilograms of pressure to detonate.
- The M21 can be used with a tilt-rod assembly, requiring 1.7 kilograms of pressure on the extension rod to cause a 20-degree deflection.
- The M21 is the only conventional US AT mine with a direct-energy warhead.

INSTALLATION

WARNING
Emplace and remove the mine while in the prone position.

- Inspect the mine.
  - Do not use the mine if it is dented, cracked, or damaged.
  - Ensure that the cotter pins on the fuse pull-ring assembly and the fuse-closure assembly are securely in place (Figure A-30).
  - Inspect the fuse to ensure that the neck portion behind the tile collar is not cracked.

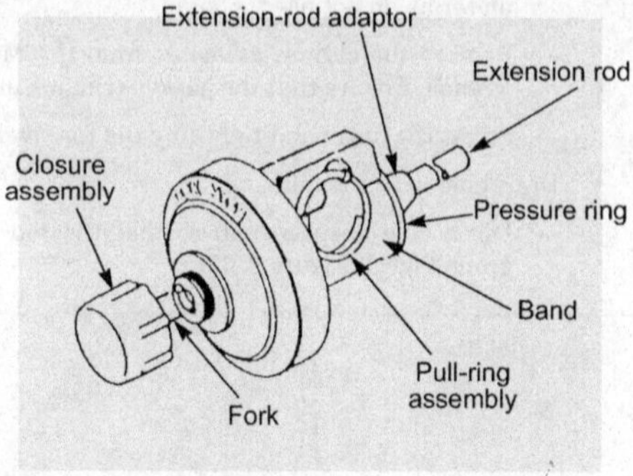

Figure A-30. M607 fuse

- Insert the booster.
  - Use the screwdriver end of the M26 wrench (Figure A-31) to remove the closing plug from the bottom of the mine.

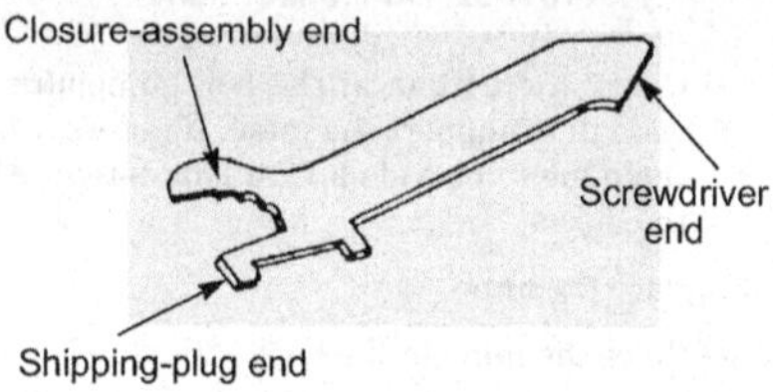

Figure A-31. M26 wrench

- Inspect the booster well for foreign material. If foreign material is present, gently tap the top of the mine with your hand to dislodge the material. If you cannot remove the debris, replace the closing plug and do not use the mine.
- Insert the M120 booster (with the washer side toward the fuse) into the booster well.
- Replace the closing plug with the M26 wrench.

- Fuse the mine.
  - Remove the shipping plug from the fuse well with the M26 wrench.
  - Inspect the fuse well for foreign material. If foreign material is present, gently shake the mine to dislodge the material. If black powder falls out of the fuse well or you cannot remove the foreign material, do not use the mine.
  - Remove the closure assembly from the M607 fuse with the M26 wrench. Ensure that the gasket remains in place on the fuse.
  - Screw the fuse hand-tight into the fuse well.
- Dig a hole to fit the mine.
  - Dig a hole deep enough so that the top of the mine will be at ground level (Figure A-32).

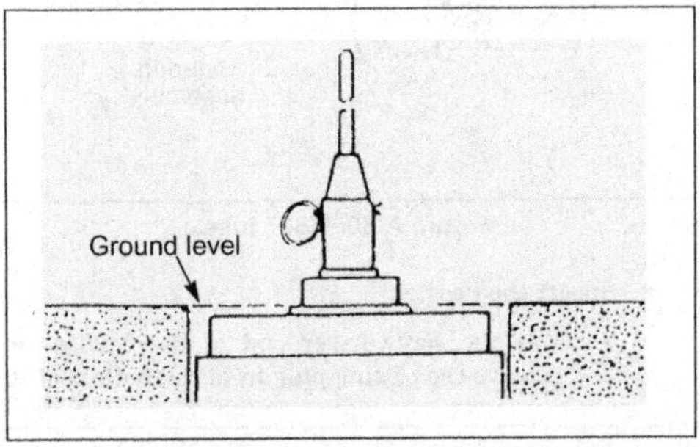

Figure A-32. Buried M21 mine

  - Check the bottom of the hole to ensure that the ground is solid enough to support the mine. If necessary, place a flat object under the mine to provide a firm foundation. Allow additional depth for the object.
- Emplace the mine.
  - Place the mine in the hole.
  - Cover the mine with soil until it is level with the top of the mine. Ensure that no soil falls around or under the plastic collar.

— Press the soil firmly around the sides of the mine.

- Assemble the extension rod. **NOTE: For pressure operation, do not use the extension-rod assembly.**

— Screw the extension rod onto the M607 fuse.

— Ensure that the extension rod is pointing straight up.

- Arm the mine.

— Squeeze the end of the cotter pin together on the pull ring.

— Remove the cotter pin by holding the fuse firmly in one hand and pulling the pull ring with the other hand.

— Remove the band and the stop slowly and carefully from the neck of the fuse (Figure A-33).

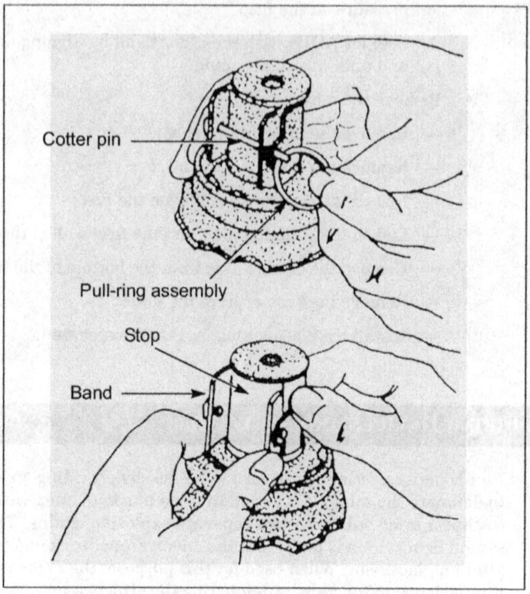

Figure A-33. Removing the band and the stop

- Camouflage the mine.

— Camouflage the mine with twigs, grass, or other material in the area. Place mines with extension rods in tall grass, if possible. Ensure that no pressure is applied to the tilt rod or the fuse.

— Place the excess soil in sandbags and remove them from the area.

- Give the band, the stop, the pull ring, the shipping plugs, and the closure assembly to the NCOIC.

Removal

- Disarm the mine.
  - Clear the camouflage away from the mine carefully.
  - Attach the band and the stop to the fuse.
  - Insert the cotter pin into the band and the stop. Spread the ends of the cotter pin.
  - Remove the extension rod.
- Check for AHDs.
  - Hold the mine firmly in place with one hand, without putting pressure on the fuse.
  - Feel for AHDs with the other hand by digging around the sides of and underneath the mine.
- Remove the mine.
  - Remove the mine from the hole.
  - Remove the fuse from the mine.
  - Install the closure assembly on the fuse.
  - Install the shipping plug into the fuse well of the mine.
  - Remove the closing plug from the bottom of the mine.
  - Remove the booster from the mine.
  - Install the closing plug into the booster well.

Section III. Firing Devices and Activators

An FD performs the function of a mine fuse by providing an alternative means to detonate the mine. It is normally used in conjunction with a standard fuse so that a mine will have two separate explosive chains. The purpose of the second firing chain is to prevent the enemy from disarming or removing mines after emplacement. When used for this purpose, the FD is called an AHD and it is designed to function by detonating the attached mine or another explosive charge nearby if unauthorized personnel attempt to remove or tamper with the mine. **NOTE: US forces will not employ AHDs on AP mines.** Both the M19 and the M15 have two secondary fuse wells for attaching an FD and an activator.

There are two standard US FDs—M5 pressure release and M142 multipurpose. They utilize a spring-loaded striker and a standard base and are designed to function in one or more of the following modes:

- Pressure.
- Pressure release.
- Tension.
- Tension release.

M5 PRESSURE-RELEASE FIRING DEVICE (MOUSETRAP)

The M5 FD (Figure A-34) is activated by the release of pressure. Lifting or removing a restraining weight releases the striker to fire the cap.

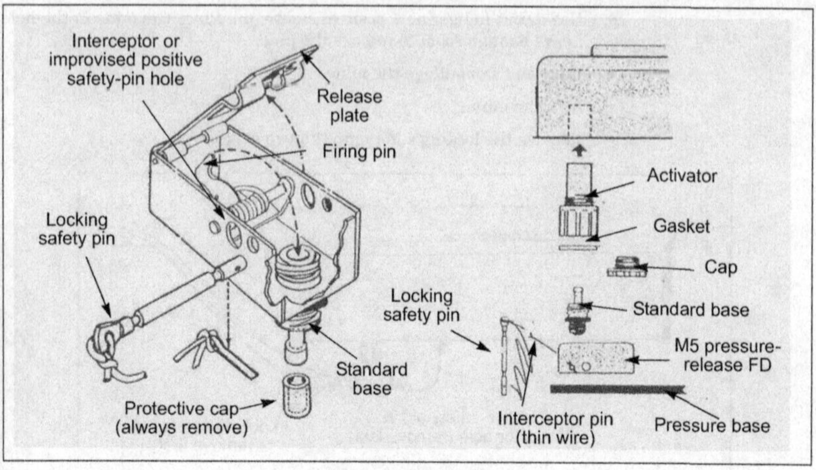

Figure A-34. M5 FD

CHARACTERISTICS

- Case: Metal.
- Color: Olive-drab.
- Length: 445 millimeters.
- Width: 239 millimeters.
- Height: 175 millimeters.
- Internal action: Mechanical with hinged striker release.
- Initiating action: Removal of restraining weight, 2.25 kilograms or more.
- Accessories: Pressure board.
- Safeties: Safety pin and hole for interceptor pin.
- Packaging: Four complete FDs and four plywood pressure boards are packaged in a paper carton, five cartons are packaged in a fiberboard box, and 10 fiberboard boxes are shipped in a wooden box.

INSTALLATION

- Dig a hole deep enough to bury the mine on a firm foundation. The pressure plate should be slightly above ground level.

> **WARNING**
> **Ensure that the mine and the FD are resting on a firm foundation before removing the pins.**

- Remove the protective cap from the standard base. Assemble the FD.
- Use a coat hanger as a positive safety pin. Place the mine in the hole. Leave enough room to remove the pins.
- Bury and camouflage the mine.
- Arm the mine.
- Remove the locking safety pin (Figure A-35).

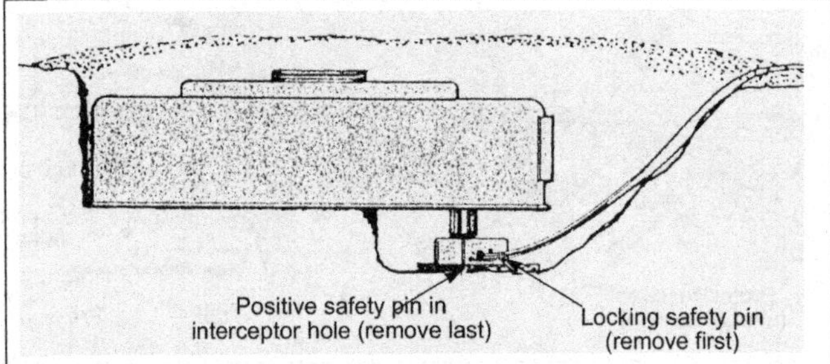

Figure A-35. Arming the M15

- Remove the positive safety pin (interceptor pin). **NOTE: If the positive safety pin is difficult to remove or if you hear a click when removing the locking safety pin, carefully remove the mine and replace the FD.**

REMOVAL

- Uncover the mine carefully and inspect it for tampering. Locate and carefully uncover the FD.

> **WARNING**
> **DO NOT release the pressure being applied to the device.**

- Insert the positive safety pin into the interceptor hole, and then insert the locking safety pin into the safety-pin hole.

- Disarm and remove the mine. Recover the FD.

M142 MULTIPURPOSE FIRING DEVICE

The M142 FD (Figure A-36) can be designed to function in the following modes:

- Pressure (11 kilograms or more).
- Pressure release (between 2 and 67 kilograms).
- Tension (3 kilograms or more).
- Tension release.

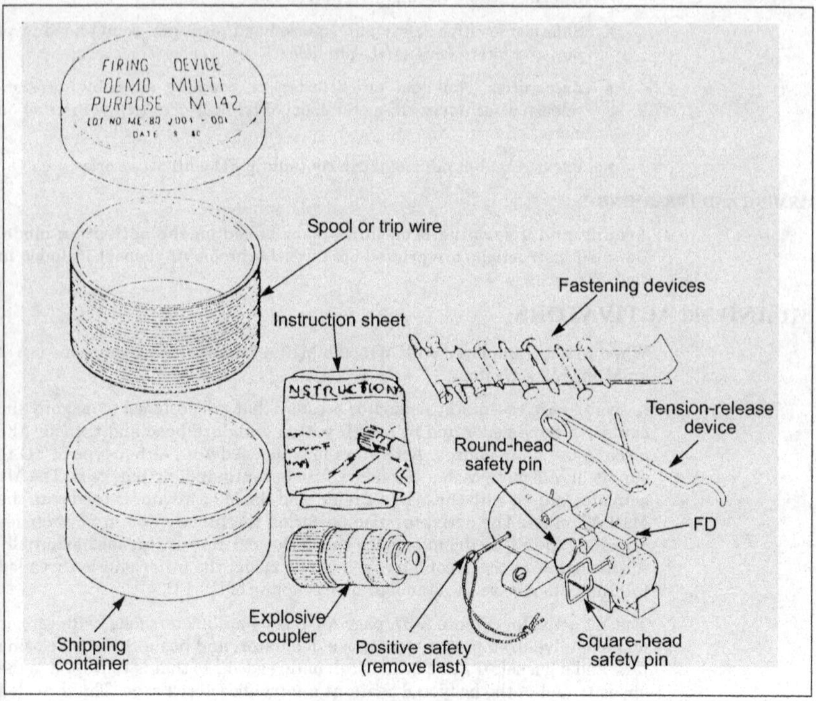

Figure A-36. M142 FD

Although primarily intended for booby-trap applications, the M142 can be readily adapted as an AHD. The M142 comes with a coupling device and a primer that accepts a standard nonelectric blasting cap. The initiating action sets off an explosive chain that passes from the FD and the primer to the blasting cap, and then via the detonating cord to the main charge. However, the coupling device with the primer will not initiate the detonating cord alone without a blasting cap attached, and it is not adaptable to any activator or secondary fuse well. When the M142 is used as an AHD, the coupling device is removed and an M1 or M2 standard base is used.

CHARACTERISTICS

- Case: Plastic.
- Color: Olive-drab.
- Diameter: 190.5 millimeters.
- Length: 571.5 millimeters.
- Internal action: Spring-driver striker.
- Safeties: Positive safety pin, square-head pivot pin, round-head pivot pin, and alternative safety-pin hole.
- Accessories: Nail and screw fasteners, coupling assembly, tension-release attachment, 15-meter spool of trip wire, and vinyl instruction sheet.
- Packaging: Round, metal can containing FD with accessories.

ARMING AND DISARMING

Arming and disarming procedures vary based on the activation mode. Detailed instructions are printed on a weatherproof, vinyl sheet included in each FD package.

M1 AND M2 ACTIVATORS

When FDs are employed with M15 and M19 AT mines, they require the use of an M1 or M2 activator.

Activators are essentially detonator boosters that are designed to magnify the explosive force generated by an FD with a standard base and transfer the force to the main charge. Activators may be used with either type of FD to supply an AT mine with a secondary fuse for antihandling purposes. The M1 activator is used with the M15 AT mine, and the M2 activator is used with the M19 AT mine. The activator also performs the function of an adapter for attaching the FD to the mine. One end of the activator is threaded externally for insertion in the secondary well of the mine; the other end is threaded internally to receive the standard base coupling of the FD.

The M1 activator (Figure A-37, page A-34) is 54 millimeters long (with cap), is made of olive-drab plastic, contains a detonator, and has a threaded closing plug and a gasket. It has a cylindrical, unthreaded cap that is cemented to the opposite end of the body and contains a tetryl booster charge. The threaded end, which screws into the mine, is 25 millimeters in diameter.

The M2 activator is similar to the M1 except that it contains an HE pellet, and its overall length, with cap, is 53 millimeters.

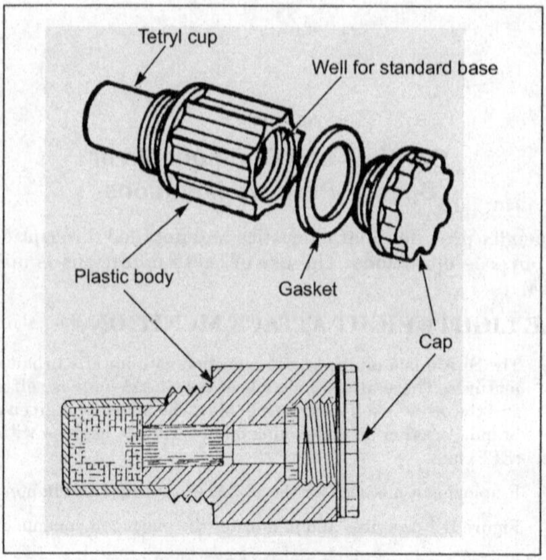

Figure A-37. M1 activator

Appendix B

Controls and Components of Special-Purpose Munitions

This appendix provides characteristics and detailed descriptions of US special-purpose munitions. The use of these munitions is outlined in Chapter 4.

SELECTABLE LIGHTWEIGHT ATTACK MUNITION

The SLAM is a multipurpose munition with antidisturbance and antitamper features. There are two models of the SLAM—one is self-neutralizing (M2) and the other is self-destructing (M4). The M2 is solid green and has no labels, brands, or other distinguishing marks. The M4 is green with a black warhead (EFP) face.

Employment methods for the SLAM are outlined in Chapter 4.

Figure B-1 describes and illustrates the major components of the SLAM.

Component	Description
Mounting holes (1)	The mounting holes are used to secure the carrying strap or the mounting wire to the SLAM when attaching the SLAM to trees and so forth.
Bore sights (2)	Two bore sights and an omega sight are located on the top of the SLAM and are used to aim the SLAM at targets.
Selector switch (3)	The selector switch is used to select operating modes and times. It has eight detent positions. The switch is against a stop (in the shipping position), which is the only switch position that allows the SLAM to fit in the reusable environmental protective pack. Turning clockwise, there are three positions for selecting the operating time (4, 10, and 24 hours). Setting any of these positions will select an internal sensor mode of operation, which is a magnetic sensor for mine mode and a passive infrared sensor for side-attack mode. These three positions will cause the SLAM to self-destruct (M4) or self-neutralize (M2) at the end of the selected operating time. Continuing clockwise, the last four positions select an internal timer, which sets the minutes until demolition. These positions are 15, 30, 45, and 60 minutes.
Activation-lever shear pin (4)	There is a shear pin mounted across the SLAM's lever slot. If the shear pin is sheared, thereby breaking the seal, the lever may have been pulled and the SLAM may be an electronic dud. If the shear pin is broken, it should only be used in the command-detonation mode.
Safety pin (7)	The safety pin slides from the body and starts the SLAM's timing. It is pried from its latch with the tip of the lever. Once the safety pin is pulled, it cannot be reinserted.

Figure B-1. SLAM components

Component	Description
Passive infrared sensor (8) and cover (9)	The SLAM is equipped with a passive infrared sensor that detects trucks and light armored vehicles by sensing the change in background temperature as vehicles cross in front of the SLAM. The sensor is directional and is aligned with the EFP. The sensor is active when the SLAM is operating with the selector switch set to 4, 10, or 24 hours and the sensor cover is removed to expose the infrared sensor (such as, during the side-attack mode). The SLAM will self-destruct (M4) or self-neutralize (M2) if the selected time expires before it is detonated by vehicle passage.
Blasting-cap well and plug (10)	The threaded plug seals the blasting-cap well. It is removed to mount a standard military blasting cap with a priming adapter.
Warhead (11)	The warhead is an EFP that is designed to defeat light armored vehicles. The EFP forms within the first 5 inches of flight and has an effective range of 25 feet.
Housing assembly (12)	The housing assembly contains the fusing, electronics, and S&A components. It also provides a structural interface for the warhead, the sights, the activation lever, the passive infrared sensor, the selector switch, and the safety pin.

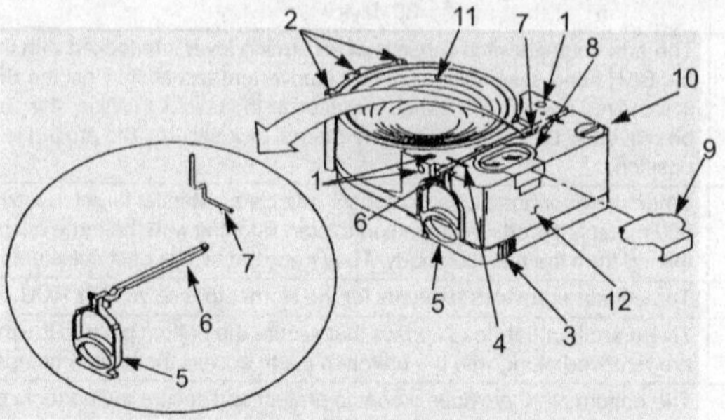

Figure B-1. SLAM components (continued)

M93 HORNET

The M93 Hornet is a lightweight (35 pounds) AT/antivehicular munition that one person can carry and employ. It is a one-time use, nonrecoverable munition that is capable of destroying vehicles using sound and motion as detection methods. The Hornet will automatically search, detect, recognize, and engage moving targets, using top attack at a maximum standoff distance of 100 meters. It is employed by units equipped with an M71 RCU. The RCU is a hand-held encoding unit that interfaces with the Hornet when the remote mode is selected at the time of employment. After encoding, the RCU can be used to arm the Hornet, reset SD times, and destruct the Hornet. Employment methods of the Hornet are outlined in Chapter 4.

Figure B-2 describes and illustrates the major components of the Hornet. Figure B-3, page B-4, describes and illustrates the controls and indicators of the Hornet.

Component	Description
Support legs (1)	Support legs are used to stabilize the Hornet when it is deployed.
Active battery-pack cover (2)	The active battery-pack cover provides a seal to protect and secure the active battery pack. The latch is lifted up to remove the cover, the active battery pack is installed, and the cover is then reinstalled and latched down. A line secures the battery-pack cover to the control panel of the munition.
SD switch (3)	The SD switch is a six-position rotary switch that is used to select the SD time and unlock the arm control switch. The SD switch is also used to unlock the arming lever. This is done by rotating the switch to the setting "U." A red lock element is extended 1/8 inch from the side of the munition when the SD switch is in the unlock position. The SD time is preset to Setting 1 when the Hornet is shipped. SD times are as follows: Setting — Time 1 — 4 hours 2 — 48 hours 3 — 5 days 4 — 15 days 5 — 30 days
Arm control switch (4)	The arm control switch consists of an arming lever interlocked with the SD switch and the S&H band assembly to prevent inadvertent actuation. Until the S&H band assembly is removed and the SD switch is placed in the unlock position, the arming lever cannot be moved to the arm position. An internal lock secures the arming lever in the arm position.
Microphones (5)	When the geophone seismic sensor detects a potential target, usually at ranges up to 600 meters, it alerts the munition to start listening with the three microphones that extend from the munition body. They track the two loudest noise sources that are heard.
Antenna (6)	The antenna provides a means for the Hornet to receive M71 RCU commands.
Capture screws (7)	These are four flat-head screws that secure the bottom plate to the munition body. They are removed along with the bottom plate to access the battery compartment.
Bottom plate (8)	The bottom plate provides a seal to protect and secure the battery compartment and connect the batteries once they are installed.
D-cell batteries (9)	The battery compartment houses four D-cell batteries. A drawing on the inside of each battery tube shows battery orientation.
Dowel pin (10)	The dowel pin ensures that the bottom plate is in the correct orientation to properly connect the batteries.

Figure B-2. Hornet components

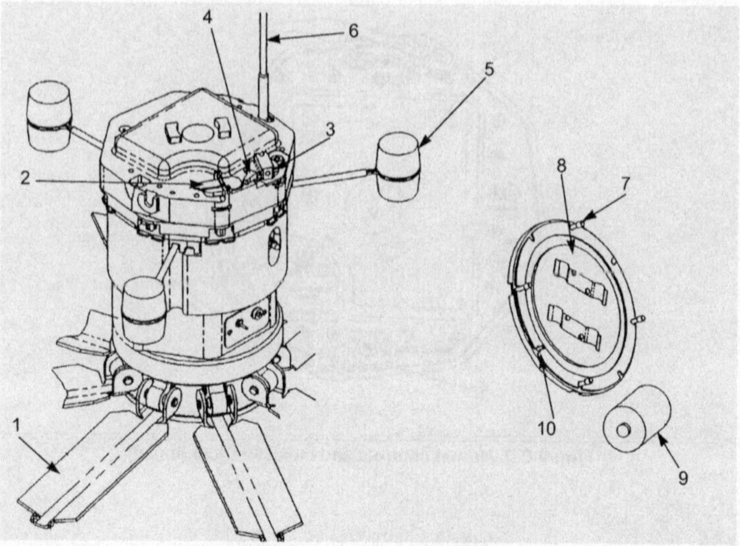

Figure B-2. Hornet components (continued)

Component	Description
Magnetic coupling device (MCD) (1)	This device is used as part of the RCU interface. The RCU interface consists of the MCD and keyed tabs. In the remote arming mode, the RCU is placed on top of the MCD and minefield code data is transferred to the munition. Upon successful encoding, the status light begins to flash.
Target switch (2)	The target switch is a toggle switch used to select the type of target engagement. This gives the operator the choice between detecting and destroying only heavy armored vehicles or all vehicles.
Manual select switch (3)	The manual select switch is a push-button switch, protected by a plastic cover that must be removed to access the switch. Successful activation of the switch will cause the status light to flash. This switch is used to allow the operator to employ the Hornet without the RCU.
Status light (4)	The status light is a visual indicator for the operator during the munition setup. It is a green light-emitting diode (LED) that indicates a self-test was successfully performed or an operating-mode selection was successfully selected.
SD switch (5)	See Figure B-2, page B-3.
Arming lever (6)	See Figure B-2.
Active battery-pack cover (7)	See Figure B-2.

Figure B-3. Hornet controls and indicators

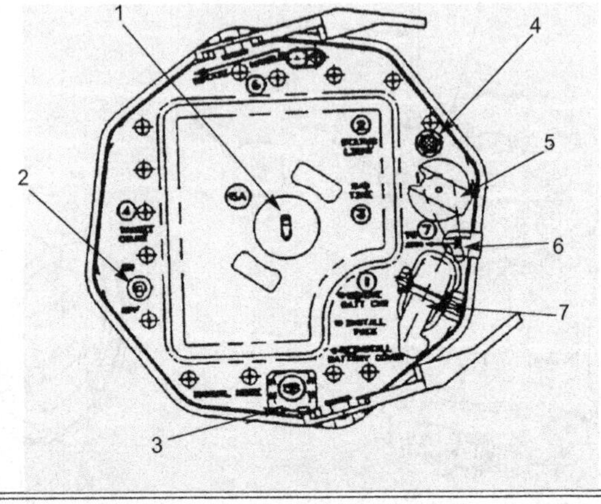

Figure B-3. Hornet controls and indicators (continued)

Appendix C

Threat Mine/Countermine Operations

This appendix is intended to complement the information presented in other manuals on threat obstacle tactics. It applies to most threat armies and their surrogates. Commanders should use this information to give added realism to unclassified training, although obstacle employment norms can change with METT-TC factors for a given AO. Therefore, preoperational training on templating, intelligence, reconnaissance, and reduction procedures must be based on the best information available before deployment.

Appendix G contains a compilation of countermine data.

MINE OPERATIONS

Threat formations contain considerable organic minefield emplacement capability. Threat rapid-mining capability presents a serious challenge to friendly maneuver.

To lay mines and place obstacles rapidly during offensive operations, threat armies form a special team from regimental and divisional assets. This team is called a *mobile obstacle detachment* (MOD). The MOD places AT mines on the most likely avenues for armored attacks or counterattacks. MODs are positioned on the flanks of a march formation for rapid deployment and are normally close to AT reserves. During the march, MODs reconnoiter avenues into the flanks and identify the most likely avenues for tank movement. At secured objectives, MODs reinforce existing obstacles and place new obstacles to assist in the defeat of counterattacks.

The combined arms commander orders the organization of MODs and determines their composition based on the combat situation and available troops. Engineer elements in a division MOD come from the divisional engineer battalion and normally consist of three armored tracked mine layers known as *GMZs* (Figure C-1, page C-2). This platoon-sized element has two or three trucks that carry mines for immediate resupply. For the regimental MOD, the regimental engineer company normally provides a platoon-sized unit equipped with two or three GMZs. The platoon travels in BTR-50/60s and has 600 AT mines.

The GMZ dispenses mines at a predetermined spacing of 5.5 meters. Mine-laying helicopters also support the MOD. The HIP and HIND-D helicopters carry two or three dispenser pods of AP or AT mines. Artillery-fired SCATMINEs can also support the MOD. Three GMZs can lay a 1,200-meter, three-row minefield, containing 624 mines, in 26 minutes. Doctrinally, this minefield would be broken into several minefields, each 200 to 300 meters long.

Threat armies use obstacles extensively throughout the depth of their defense, and their tactics are chosen well. Shallow obstacles are reduced quickly and easily. For example, a shallow, one-row minefield is essentially reduced by blowing one or two mines in the row. A threat rapidly emplaced minefield

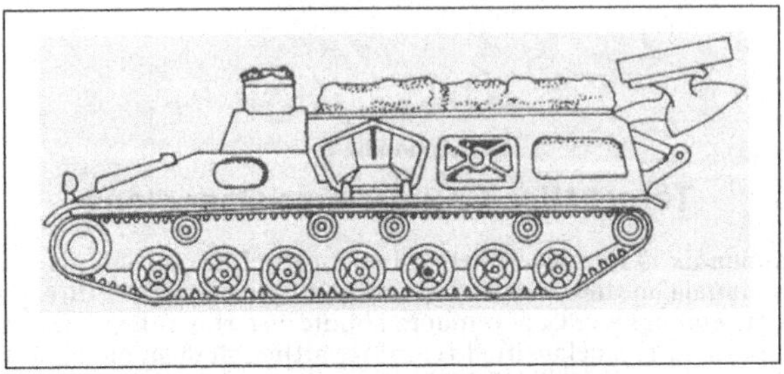

Figure C-1. GMZ armored tracked mine layer

consists of three or four 200- to 300-meter rows, spaced 20 to 40 meters apart, with mines spaced 4 to 6 meters apart. As a rule, the minefield covers the depth of a football field.

Table C-1 provides detailed information on standard threat AT and AP minefields. Terrain and tactical situations dictate the actual dimensions and distances of minefields.

Table C-1. Normal parameters for threat-style minefields

AT Minefields	
Front (situation-dependent)	200 to 300 meters
Depth	40 to 120 meters
Number of rows	3 or 4
Distance between rows	20 to 40 meters
Distance between mines	4 to 6 meters for antitrack mines; 9 to 12 meters for anithull mines
Outlay, normal	550 to 750 antitrack mines per kilometer; 300 to 400 antihull mines per kilometer
Outlay, increased effect	1,000+ antitrack mines per kilometer; 500+ antihull mines per kilometer
Probability of destruction	57% for antitrack mines (750 per kilometer); 85% for antihull mines (400 per kilometer)
AP Minefields	
Front (situation-dependent)	30 to 300 meters
Depth	10 to 150 meters
Number of rows	3 or 4
Distance between rows	5+ meters for blast mines; 25 to 50 meters for fragmentation mines
Distance between mines	1 meter for blast mines; 50 meters (or twice the lethal radius of fragmentation) for fragmentation mines
Outlay, normal	2,000 to 3,000 HE/blast mines per kilometer; 100 to 300 fragmentation mines per kilometer
Outlay, increased effect	2 to 3 times the normal outlay
Probability of destruction	15 to 20% for HE/blast mines (2,000 per kilometer); 10 to 15% for fragmentation mines (100 per kilometer)

Figure C-2 shows a standard rapidly emplaced minefield. The threat army typically uses such a minefield when they are in a hasty defense (offense is temporarily stalled).

Figures C-3 shows a standard antitrack minefield.

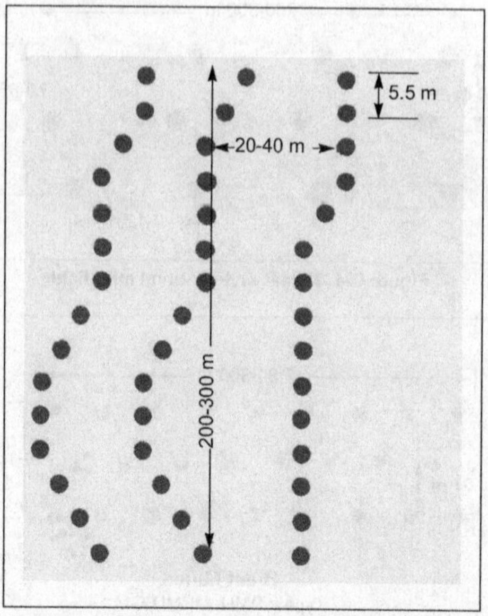

Figure C-2. Threat-style rapidly emplaced minefield

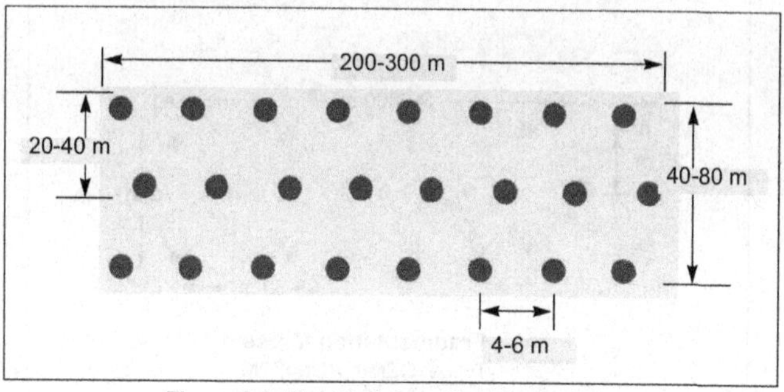

Figure C-3. Threat-style antitrack minefield

Figure C-4 shows a standard antihull minefield.

Figure C-5 shows a standard AP minefield.

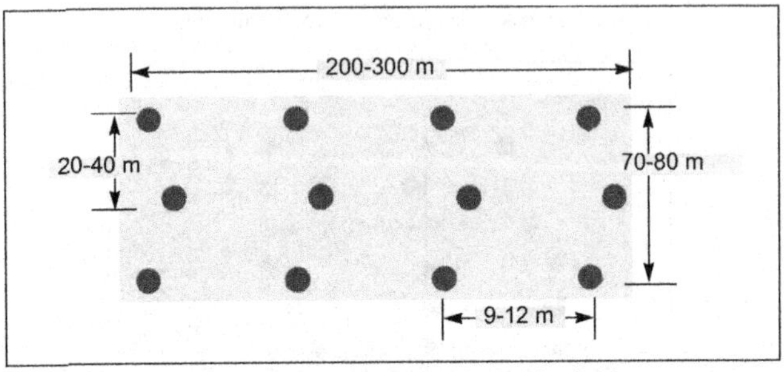

Figure C-4. Threat-style antihull minefield

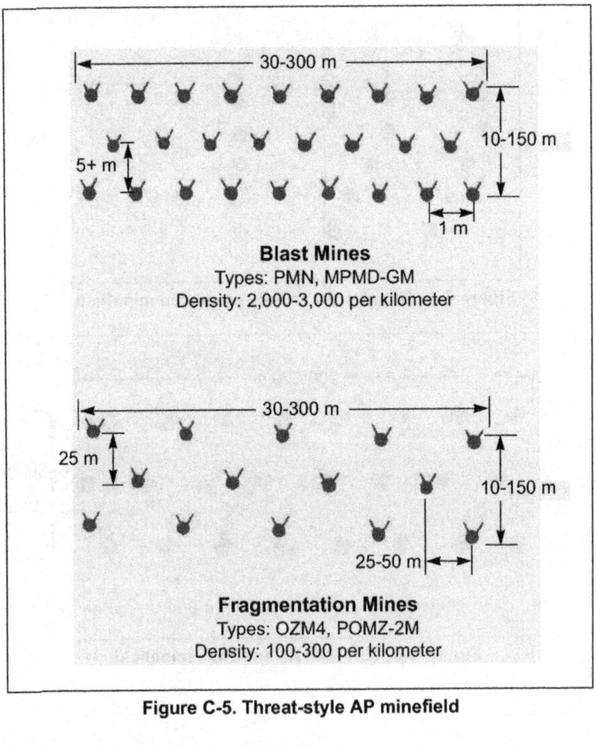

Figure C-5. Threat-style AP minefield

Threat armies also emplace mixed minefields. They are not the same as US mixed minefields. Threat armies normally emplace three rows of AT mines, then several rows of AP mines. AT and AP mines **are not** mixed in the same row.

Threat engineers use two fundamental drills to emplace mines:

- When emplacing armed mines, the drill uses a crew of five sappers. The first crew member (the senior man and operator) is in the mine-layer's seat and monitors the operation of the mine layer and the motion of the mines in the guide chute. He also sets the mine spacing and controls the actions of the GMZ. The second and third members take mines out of containers and place them in the intake chute at intervals between the guide tray's drive chain. The GMZ driver steers the vehicle along the indicated route at the established speed.
- When emplacing unarmed mines, two or three additional sappers are assigned to arm the mines. After emplacing the mines, one sapper trails the mine layer, marks emplaced mines with pennants, and partially camouflages the mines. The remaining sapper(s) then arm the mines.

Special precautions are taken when emplacing AP minefields. Threat doctrine only allows PMN mines to be surface-laid from mine layers. POMZ-2M mines are emplaced with the truck-and-tray technique. Extra effort is required to assemble, emplace, and deploy the trip wire and to camouflage the POMZ-2M mine.

Using three GMZs, a threat MOD can emplace 1,200 meters of a three-row AT, surface-laid minefield, containing 624 AT mines, in 26 minutes. This does not include the 12- to 15-minute reload and travel times. Travel and reload times increase during limited visibility.

Threat forces can also have ground-emplaced SCATMINE capability. One such system is the UMZ SCATMINE system (Figure C-6, page C-6). There are three UMZ truck-mounted SCATMINE systems in each combat regiment. The UMZ consists of six firing modules mounted on the back of a Zil-131 truck. Each module has 30 firing tubes, for a total of 180 firing tubes per system. Depending on the type of minefield desired, the UMZ can lay 180 to 11,520 mines without reloading. The UMZ can launch an AT or AP minefield 30 to 60 meters from the vehicle while the truck is driving 10 to 40 kph. It takes two men 1½ to 2 hours to reload the UMZ. One UMZ can lay a three-row minefield, 150 to 1,500 meters long, depending on the type of mine that is used.

UMZ vehicles are usually deployed together as a mobile obstacle/mine-laying detachment. The UMZ is used to lay minefields that protect subunit positions and flanks and the boundaries between subunits. UMZ-laid minefields also cover firing lines and gaps in combat formations. The UMZ can quickly close gaps in existing minefields and increase the density of mines on armor avenues of approach.

For hand-emplaced SCATMINEs, there is a man-portable SCATMINE dispenser. The PKM weighs 2.63 kilograms (without the mine canister) and consists of a single launch tube with a base mount, a blasting machine, and a reel of electric ignition wire. The operator loads a propelling charge and a mine canister into the launch tube and mounts the tube on the edge of a trench or firing parapet. He then aims the tube, connects the ignition wire to

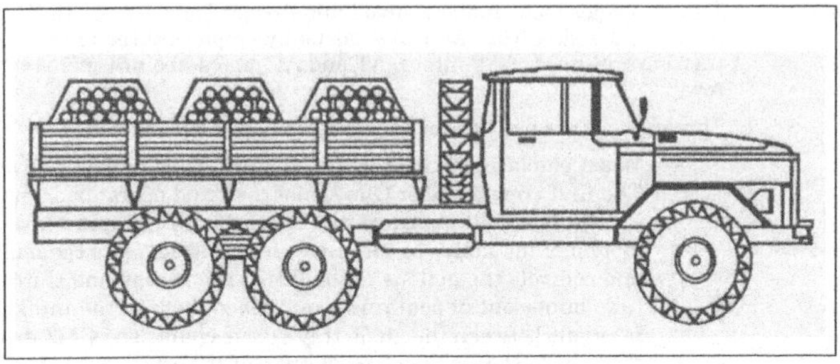

Figure C-6. UMZ SCATMINE system

the tube, and moves to a safe distance. At an initiating point, the operator connects the ignition wire to the blasting machine and initiates the system. The PKM propels the canister 30 to 100 meters, depending on the type of mine. It lays an AP minefield that is 10 by 20 meters (POM-1S mine canister), 10 by 40 meters (POM-2S mine canister), or 20 by 10 meters (PFM-1S mine canister). It takes a trained operator 5 minutes to set up the PKM and create a minefield. The PKM can also be used to launch the PTM-1S and PTM-3 AT mine canisters.

Threat forces use the PKM to lay minefields that protect subunit positions and flanks and the boundaries between subunits. PKM-laid minefields also cover firing lines and gaps in combat formations. The PKM can quickly close breaches in existing minefields and increase the density of mines on armor avenues of approach.

The type and complexity of an obstacle depends on the installing unit. Maneuver and artillery soldiers usually install simple single-system minefields that are protective in nature. Engineer soldiers install complex obstacles that can include AHDs. Engineer obstacle placement is usually equipment-intensive. Threat engineer effort generally concentrates on tactical obstacles unless maneuver soldiers are unable to employ the necessary protective obstacles. Threat units continue to improve the obstacles, supporting their positions by marking the friendly side of the obstacles, burying mines, and adding AHDs.

CHEMICAL MINES

Chemical land mines are AP mines with command- or target-detonated fuses, and they are filled with a persistent chemical (nerve or blister) agent. US policy prohibits their use by US personnel. However, this does not preclude their use by other countries, and US forces may encounter them during operations. When used, they are normally used in defense and retrograde operations. They are mixed with HE mines to form a HE chemical minefield. Chemical mines are normally encountered in tactical or nuisance minefields, and some countries use them in protective minefields. When an integrated HE chemical minefield is laid, it serves the following purposes:

- Chemical mines discourage the use of explosive, rapid mine-clearing devices because they create a chemical hazard in the area.

- HE mines reduce the speed of enemy forces crossing the minefield. Speed is further reduced by forcing the enemy to use protective clothing and masks.

Chemical mines will usually be added to existing HE minefields by laying additional strips of chemical mines in a random pattern or by adding HE chemical strips or rows to the front or rear of existing fields (Figure C-7).

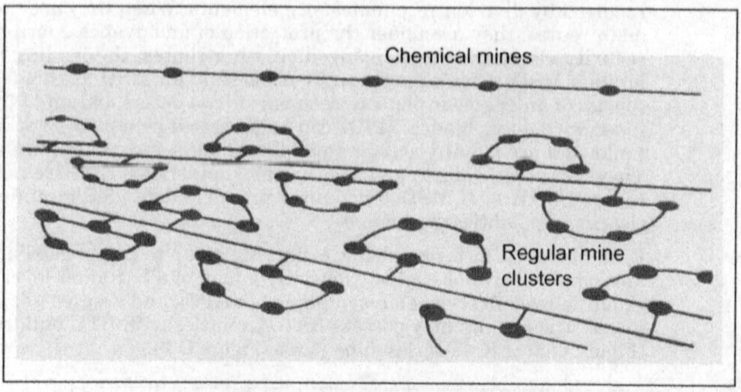

Figure C-7. Chemical-mine employment

No particular branch is responsible for clearing chemical mines. Planning chemical countermine operations is a brigade-level responsibility. When reducing chemical mines, consider prevailing and expected wind conditions. Commanders must ensure that friendly troops are protected when chemical agents are released. The release of chemical agents occurs as a result of enemy fire or friendly breaching attempts. Contact-actuated chemical mines are not likely to create a major downwind hazard because only single mines or small groups may be set off at one time.

COUNTERMINE OPERATIONS

In offensive operations, threat engineers clear lanes through obstacles when they cannot be bypassed. Although clearing obstacles applies to the march and the defense, the most critical performance of this task occurs during the attack. Engineers can be required to clear mines delivered by air, artillery, and rockets well ahead of NATO's forward edge. They must breach obstacles contained within NATO strongpoints. Threat forces must also clear their own minefields when making the transition from defense to offense. In the offense, threat forces breach or bypass remotely delivered minefields in their form-up areas or routes of movement to the attack line. They also breach obstacles along the forward edge of the battle area and deep within NATO defenses.

Although clearing passages through obstacles is a primary task for threat engineers, any maneuver element may encounter mines. Engineers may not be able to respond to every encounter, so maneuver troops are also required to breach through remotely emplaced obstacles.

ORGANIZATION

A movement support detachment (MSD) supports the movement of maneuver forces. It is task-organized from divisional or regimental engineer assets and can be platoon- to company-size. The MSD is equipped with route- and mine-clearing vehicles and devices. Depending on the mission (which comes directly from the combined arms commander or the chief of engineer services), an MSD is capable of filling craters, clearing minefields, preparing bypasses around major obstructions, and identifying NBC-contaminated areas.

The divisional engineer battalion can form two or three MSDs. During marches, MSDs travel in advance of the main body and clear obstructions reported by division reconnaissance elements. When they are deployed on main routes, they are under the protection of an advance guard or forward security element. When deployed on other routes, the leading regiments provide MSDs from organic engineer assets. An MSD at this level might consist of an engineer platoon, with one or two dozers and up to three tanks fitted with dozer blades. MSDs can be protected by a platoon of infantry or tanks and are usually accompanied by chemical-reconnaissance personnel. They can detect, mark, and breach hasty minefields that are not properly covered by fire. If MSDs encounter properly defended minefields, their clearing capabilities are limited.

Each battalion forms an obstacle-clearing group to create gaps in explosive and nonexplosive obstacles. Normally a part of a battalion-level MSD, the group follows first-echelon companies in APCs and creates gaps for those forces. These units may possess BAT-M vehicles with BTU bulldozer blades (Figure C-8) or KMT-series mine plows (Figure C-9).

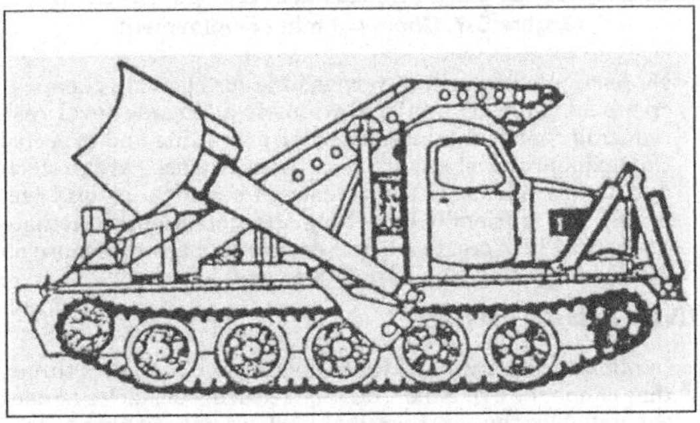

Figure C-8. BAT-M with BTU bulldozer blade

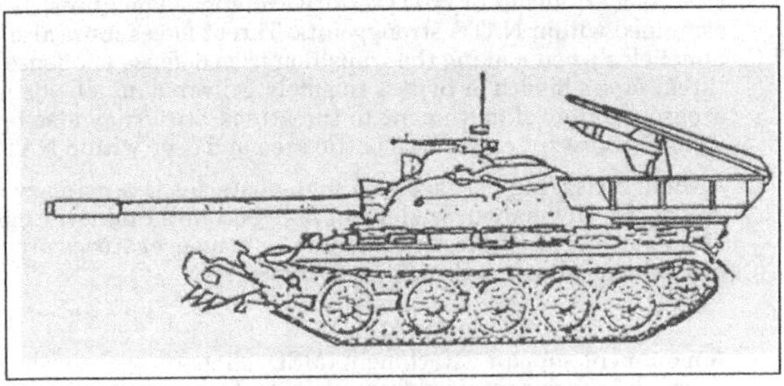

Figure C-9. KMT-4 plow

An obstacle-clearing detachment is created when more resources are needed to clear obstacles and debris. This usually occurs in urban environments and under conditions of massive destruction. An obstacle-clearing detachment is similar to an MSD, but its sole mission is to clear debris. Like an MSD, its composition depends on the mission scope, the mission objective, and the tempo of the offensive.

The divisional engineer battalion of the motorized rifle or tank division has a sapper company to clear obstacles. The company commander receives a mission to clear minefields. He then determines the exact location of the obstacle, ascertains the assets to devote to the task, and plans the methodology for success. Teams may be created to manually breach lanes using probes, IMP portable mine detectors (Figure C-10), and shovels. Larger tasks may necessitate the use of vehicle-mounted DIM mine detectors (Figure C-11), armored vehicle mine plows and/or rollers (Figure C-12, page C-10), and explosive line charges. When necessary or more practical, mines are explosively destroyed in place.

Figure C-10. IMP portable mine detector

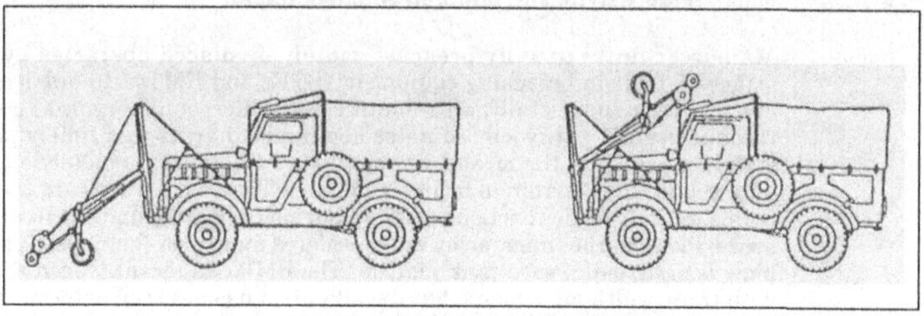

Figure C-11. DIM mine detector

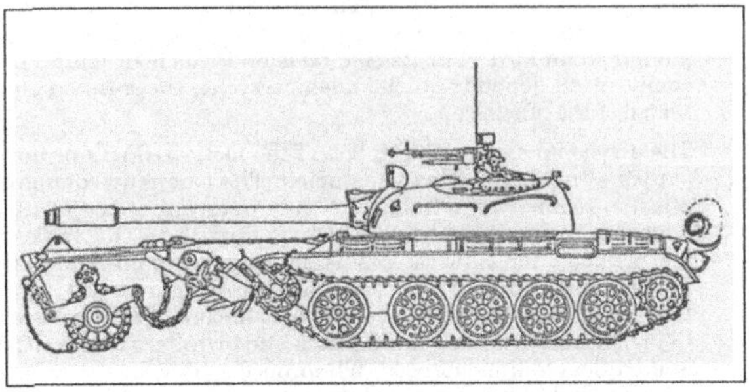

Figure C-12. KMT-5 plow-roller combination

The engineer company of the motorized rifle or tank regiment has breaching equipment such as KMT-series mine plows and rollers and BTU bulldozer blades located in its technical platoon. Because of limited assets in the technical platoon, coupled with the responsibility of forming its own MSD, the regiment can receive a sapper section from the divisional sapper company. An additional IMR armored engineer tractor (Figure C-13), BTR-50/60, and M1979 armored mine clearer (Figure C-14) and manual breaching equipment come with the sapper section.

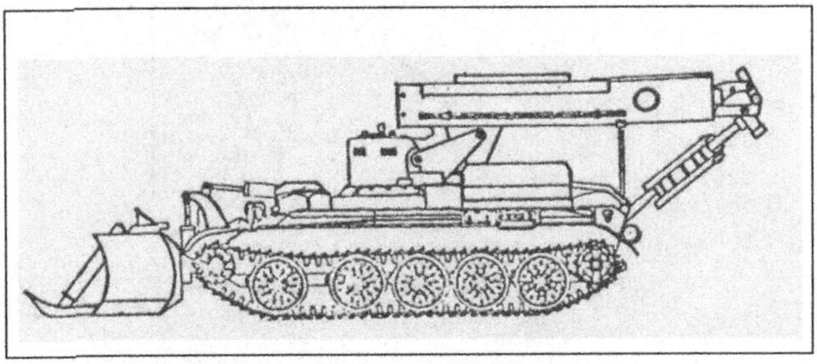

Figure C-13. IMR armored engineer tractor

Maneuver units usually breach remotely emplaced obstacles by using attached, built-in breaching equipment (BTUs and KMTs). In order to carry out this task successfully, all subunit commanders must organize constant reconnaissance, notify subordinates about mined areas in a timely manner, train personnel on the means and methods for handling remotely emplaced mines, and clear terrain in a timely manner. They must also train their own teams for independent actions when removing combat equipment from mined areas. Plows in the threat army are considered maneuver-force assets, and one plow is assigned to each tank platoon. The BMP has recently been equipped with track-width mine plows, but the allocation has not been determined.

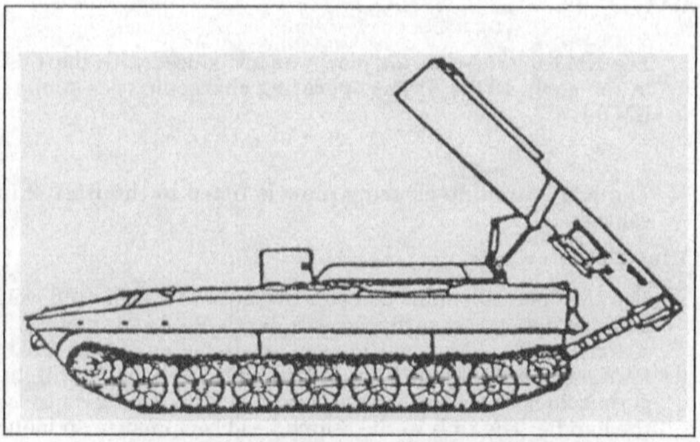

Figure C-14. M1979 armored mine clearer

EQUIPMENT

Several pieces of equipment are used by threat armies to detect and clear mines.

BAT-M Dozer

The BAT-M dozer (Figure C-8, page C-8) is a modified artillery tractor with a hydraulically operated bulldozer blade and crane. It is sometimes called a *roader* by Russians. The BAT-M dozer clears obstacles, fills craters, prepares bridge approaches, and performs other heavy pioneer tasks. It can also be configured for snowplowing.

The second generation BAT-M is the BAT-2. The BAT-2 is able to carry an 8-man engineer squad and operate in an NBC environment. It is replacing the BAT-M.

KMT-Series Plows and Rollers

KMT-4

The KMT-4 mine-clearing plow (Figure C-9, page C-8) was developed in the 1960s to fit on a T-545/55 tank. It actually consists of two plows (one mounted in front of each track), and each plow has five attached teeth. When the plow is lowered, the teeth dig into the ground and remove mines from the path of the tank. A plow is lighter than a roller and permits tanks to retain their cross-country mobility. The estimated clearing speed is 10 kph, and the depth of clearance is 10 centimeters.

Three plows are issued per tank company (one per platoon). However, these assets are normally held in the engineer company of a tank or MRR.

KMT-5

The KMT-5 mine-clearing plow-roller combination (Figure C-12, page C-10) consists of two plows and two rollers attached to the front of a tank hull. The plows or the rollers can be used, depending on terrain features, the type of soil, and the mine fuse. Plows and rollers cannot be used simultaneously. The rollers function against pressure-fused mines. The system can survive 5 to 6 kilograms of explosives, five or six times. The KMT-5 also includes a luminous lane-marking device for night operations.

KMT-6

The KMT-6 mine-clearing plow was introduced with the T-64 and T-72 tanks in the early 1970s. It has operating characteristics similar to those of the KMT-4.

KMT-10

The KMT-10 mine-clearing plow is fitted to the BMP-2 infantry combat vehicle.

IMP Portable Mine Detector

The IMP portable mine detector (Figure C-10, page C-9) weighs 7 kilograms and can detect mines buried to a depth of 45 centimeters. It has a tubular search head (one transmitting and two receiving coils encased in plastic) and a four-section handle. Power is furnished by four flashlight batteries that permit 20 hours of continuous operation. Two tuning controls are mounted on the handle. The coils in the search head compromise an induction bridge and are initially balanced for zero coupling. When the head passes over a metallic object, the induction bridge becomes unbalanced and produces an audible signal in the headset.

DIM Vehicle-Mounted Mine Detector

The DIM vehicle-mounted mine detector (Figure C-11, page C-9) is primarily used to clear roads during convoys and road marches. It sweeps at a speed of 10 kph with a 2.2-meter width. It can detect metallic mines at a depth of 25 centimeters. The brakes on the DIM automatically engage when a mine is detected. Cross-country use of the DIM is limited.

IMR Armored Engineer Tractor

The IMR armored engineer tractor (Figure C-13, page C-10) is mounted on a modified T-54/55 chassis. The turret is removed and a hydraulic crane, which can be fitted with either a grab or an excavator bucket, is emplaced. An adjustable, hydraulically operated blade is mounted on the front. The crane operator is provided with an armored cupola. The IMR can operate in an NBC environment.

M1979 Armored Mine Clearer

The M1979 armored mine clearer (Figure C-14, page C-11) is mounted on the chassis of an amphibious 122-millimeter, 2S1 self-propelled howitzer. It has a turret-like superstructure that contains three rockets on launch ramps. These, along with the upper part of the superstructure, are hydraulically elevated for firing. The rocket range is estimated at 200 to 400 meters. Each rocket is connected to 170 meters of mine-clearing hose via a towing line. The hose is folded and stowed in the uncovered base of the turret and connected to the vehicle with a cable. The cable allows the vehicle crew to reposition the hose after launching.

This chapter implements STANAG 2990.

Appendix D

Air Volcano

The air Volcano system provides a three-dimensional capability that allows units to emplace minefields in deep, close, and rear operations. It provides US forces with the capability to employ minefields rapidly under varied conditions. The air Volcano can be used to emplace tactical minefields; reinforce existing obstacles; close lanes, gaps, and defiles; protect flanks; and deny the enemy use of potential air-defense sites. Volcano minefields are ideal for flank protection of advancing forces and for operating in concert with air and ground cavalry units on flank-guard or screen missions.

COMPONENTS

The air Volcano system (Figure D-1) consists of an M87-series mine canister, an M139 dispenser, and vehicle-specific mounting hardware (UH-60 Blackhawks require a jettison kit).

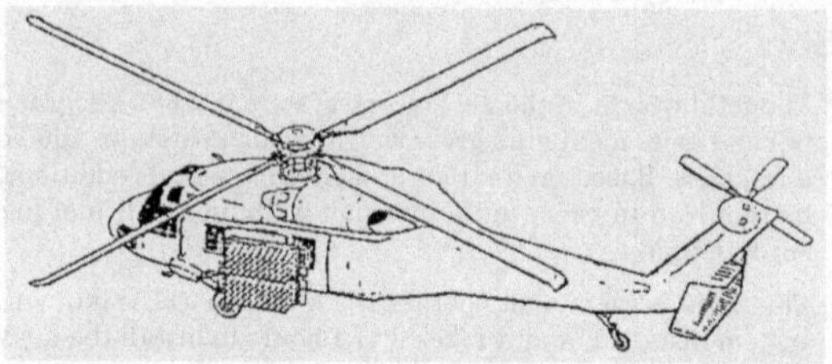

Figure D-1. Air Volcano system

M87-Series Mine Canister

The M87-series mine canister is the same canister used for the ground Volcano system:

- M87. Prepackaged with five AT mines, one AP mine, and a propulsion device inside a tube housing.
- M87A1. Prepackaged with six AT mines and a propulsion device inside a tube housing.

The mixture of mines is fixed and cannot be altered in the field. The mines in each canister are electrically connected with a web that functions as a lateral dispersion device as the mines exit the canister. Spring fingers mounted on each mine prevent the mine from coming to rest on its edge. AT mines have a delay-arm time of 2 minutes 30 seconds; AP mines have a delay-arm time of 4 minutes. All canisters are capable of dispensing mines with 4-hour, 48-hour, or 15-day SD times. SD times are selected prior to dispensing and do not require a change or modification in the base M87-series mine canister.

M139 Dispenser

The M139 dispenser consists of an electronic DCU and four launcher racks; each rack holds 40 M87-series mine canisters. The racks provide the structural strength and the mechanical support required for launch and provide the electrical interface between the mine canisters and the DCU. Mounting hardware for the UH-60A Blackhawk includes a jettison subassembly to propel the Volcano racks and canisters away from the aircraft in the event of an emergency.

The operator uses the DCU to electrically control the dispensing operation from within the aircraft. The DCU provides controls for the arming sequence and the delivery speed selection, and it sets mine SD times. The DCU allows the operator to start and stop mine dispensing at anytime. A counter on the DCU indicates the number of canisters remaining on each side of the aircraft.

Mines are dispensed from their canisters by an explosive propelling charge, 35 to 70 meters from the aircraft's line of flight. The aircraft flies at a minimum altitude of 1.5 meters, at speeds of 20 to 120 knots. It can deliver up to 960 mines per sortie.

LIMITATIONS

The total weight of the air Volcano system is 2,886 kilograms. An aircraft will be close to its maximum gross weight when it contains the Volcano system and a full crew. Based on weather and environmental conditions, the aircraft may be required to execute the mission without a full fuel load, thus reducing enroute time.

The flight crew cannot operate the M60D machine gun with the air Volcano system installed, and it takes 3 to 4 hours to install the air Volcano system on a UH-60A Blackhawk.

EMPLOYMENT

The air Volcano is the fastest method for emplacing large tactical minefields. When employed by combat aviation elements in support of maneuver units, close coordination between aviation and ground units ensures that mines are emplaced accurately and quickly. Although placement is not as precise as it is with ground systems, air Volcano minefields can be placed accurately enough to avoid the dangers inherent in minefields emplaced by artillery or jet aircraft.

Air Volcano minefields can be emplaced in friendly and enemy territories. They should not be planned in areas of enemy observation and fire, because the helicopter is extremely vulnerable while flying at the steady altitude, the speed, and the path required to emplace the minefield. The air Volcano is the best form of a situational obstacle because of its short emplacement time. Its employment varies depending on the type of operation.

DEEP OPERATIONS

Employment

The air Volcano is employed in deep operations to—

- Disrupt enemy formations along key AAs and choke points.
- Fix enemy formations in EAs to enhance target acquisition by attack helicopters, CAS, artillery, or a combination of weapon platforms.
- Provide area denial of possible enemy artillery and ADA sites that will affect future friendly schemes of maneuver.

AH-64 security is essential to air Volcano missions because the target area must not be in the enemy formation's direct line of sight. It may require two suppression of enemy air defense (SEAD) missions to get the air Volcano across the FLOT to the target location. This removes the security aircraft from the battle. All air Volcano missions require planning, designation, and control during the execution of air corridors (routes).

Aviation Configuration

The air Volcano can be used in deep operations, but it has distinct limitations. The aircraft can maintain a speed of 80 to 90 knots (UH-60A) for 90 to 120 kilometers, depending on the wind and the temperature (see Table D-1, page D-4).

The aircraft cannot employ the door guns for self-defense with the Volcano system mounted. The attack aircraft averages 100 to 120 knots while targeting enemy formations that are 150 to 200 kilometers deep. In order for the air Volcano to go deep, the attack aircraft must slow down or special security/escort aircraft must be assigned for protection. The security/escort can be AH-64s or OH-58D Kiowa Warriors (KWs). KWs have many advantages—they are not a primary killing platform, and they are very good at quickly securing an area. The air Volcano requires one or two security aircraft for protection, and there should also be a backup aircraft for the mission.

Fire-Support Coordination

The division main (light force) or brigade main (mechanized force) FSE is responsible for coordinating and executing fires in support of air Volcano missions. Upon completion of the target meeting, the FSE, the assistant division/brigade engineer, and the G3/S3 air representative will coordinate to ensure that the air-coordination/tasking order will support the mission and the planned SEAD fires.

If the mission is a deep aviation attack, the aviation element FSE is responsible for coordinating through the forward command post to the division/brigade main FSE. If the forward command post has *jumped forward* or has not deployed, the aviation element FSE will coordinate directly with the division/brigade main FSE.

Table D-1. Air Volcano capabilities and limitations

Pressure Altitude (ft)	Temperature (°C)	Maximum Weight (lb)[1]	Cruise Speed (knots/kph)	Fuel Burn Rate (lb per hr)	Endurance (hr + min)[2]	Maximum Range (nm/km)[3]	Planning Combat Radius (nm/km)[4]
0	+20	22,000	80/148	965	1 + 57	154/285	60/111
			90/166	1,040	1 + 48	160/296	
			100/182	1,145	1 + 35	158/292	
	+30	22,000	80/148	975	1 + 56	156/289	59/109
			90/166	1,045	1 + 48	160/296	
			100/182	1,160	1 + 34	156/289	
	+40	21,600	80/148	995	1 + 53	150/277	58/107
			90/166	1,055	1 + 45	156/289	
			100/182[7]	1,175	1 + 33	154/285	
2,000	+20	22,000	80/148	960	1 + 58	157/290	60/111
			90/166	1,060	1 + 45	157/290	
			100/182	1,155	1 + 35	158/292	
	+30	21,000[5]	80/148	1,000	1 + 28	117/217	45/83
			90/166	1,045	1 + 23	124/229	
			100/182[7]	1,175	1 + 12	120/222	
	+40	20,000[6]	80/148	1,005	1 + 21	28/52	7/13
			90/166[7]	1,075	1 + 19	29/53	
			100/182[7]	1,150	1 + 16	27/49	

NOTE: Based on UH-60A with an aircraft torque factor of 97.5.

[1]Maximum weight for deep operations only; no interdiction capability.

[2]Endurance includes 20-minute fuel reserve.

[3]Does not compensate for winds. This is the maximum range line-of-site distance.

[4]The general planning figures used for time-on-target missions; includes compensation for 10-knot winds, makeup for lost time, and a 5-minute on-station time.

[5]1,800 pounds of fuel at takeoff

[6]700 pounds of fuel at takeoff

[7]Operation at this air speed may be limited to 30 minutes due to target limits.

LIMITATIONS:

- There are no modifications to mount the M60D machine gun on the air Volcano aircraft.
- Crews cannot use the M60D with the system mounted on an aircraft.
- Crews cannot fly the aircraft with full tanks above +30°C and 2,000-ft pressure altitude.
- Aircraft may require a rolling takeoff and landing, depending on ambient weather conditions.

CLOSE OPERATIONS

Employment

The air Volcano is employed in close operations to fix enemy formations in EAs, turn advancing formations into desirable terrain that supports the friendly scheme of maneuver, disrupt formations enough to slow the enemy down, and block key AAs. Multiple missions may be required, depending on the intended effect. The air Volcano can be used to reseed existing minefields or to close lanes and gaps. The target area must be clear of friendly forces before an air Volcano mission is executed.

Use of the air Volcano in close operations should be a primary planning consideration. It can quickly reach the outer edge of the forward operating base where AAs need a minefield obstacle. The threat level will be lower, and the station time will increase.

Aviation Configuration

Two air Volcano aircraft should be used (one primary, one backup). The requirement for security aircraft depends on METT-TC factors, but security should be used whenever possible.

Fire-Support Coordination

The forward command post FSE coordinates and executes fires in support of air Volcano missions. The FSE, the engineer liaison officer, and the G3/S3 representative coordinate to ensure that the air coordination/tasking order supports the mission and the planned SEAD fires. The division/brigade main will be available to support the forward command post as necessary.

The brigade/TF FSE is responsible for coordinating through the forward command post to the division/brigade main FSE. If the forward command post has jumped, the brigade/TF FSE coordinates directly with the division/brigade main FSE.

REAR OPERATIONS

Employment

The primary purposes of the air Volcano in rear areas is to protect key terrain from possible airborne/air-assault forces and to fix/disrupt enemy forces long enough to allow the tactical combat force or ready-reserve force time to react and meet the changing enemy situation.

The least preferred employment method is to deliver tactical minefields to brigade and corps support areas. This employment tactic is normally used when all other available assets have been exhausted. The flexibility of the air Volcano system makes it ideal for employment against a mounted Level III threat in the rear. The target area should be out of the direct view/fire of the threat and on a choke point that allows cover for the reacting forces.

Aviation Configuration

The air Volcano aircraft could be employed individually or with security/escort aircraft. The use of OH-58D KWs as security aircraft allows units to develop the situation and helps place minefields in the proper location to assist inbound attack aircraft or fires. If the air Volcano aircraft is not provided security aircraft, it is recommended that ground forces provide covering fires.

Fire-Support Coordination

The division/brigade rear FSE coordinates and executes fires in support of air Volcano missions. The FSE, the engineer liaison officer, and the G3/S3 representative coordinate to ensure that the air coordination/tasking order supports the mission and the planned SEAD fires. The division/brigade main will be available to support the division/brigade rear as necessary.

The headquarters element that controls the rear area coordinates with the division/brigade rear FSE. The division/brigade rear FSE coordinates with the division/brigade FSE for fire support and air assets.

MINEFIELD EFFECTS

Turn

A turn minefield manipulates enemy maneuver in a desired direction. It forces or entices enemy formations to move in a different direction rather than breach the obstacle. This means the bypass must be easily identified. Turn minefields are extremely lethal, with approximately 80 percent probability of mine encounter. The typical width is 557 by 320 meters for air Volcano. Figure D-2 shows two turn minefields combined to create a turn-effect obstacle group. It takes 160 canisters (800 AT/160 AP mines) to emplace one turn minefield. One air Volcano aircraft can lay one turn minefield (see Table D-2).

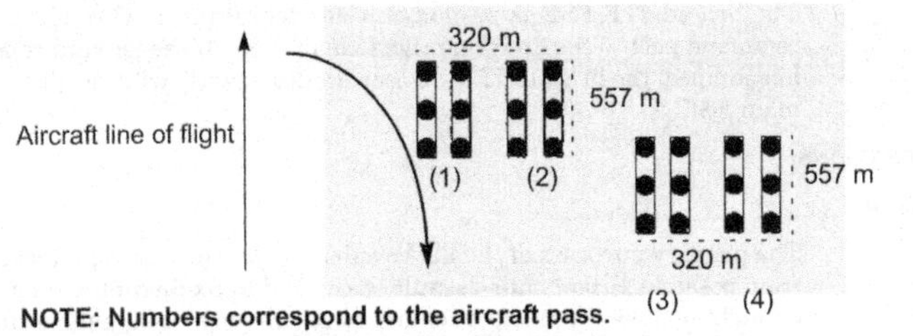

Figure D-2. Turn obstacle

Table D-2. Air Volcano minefield data

Type of Minefield	Depth (m)	Frontage of Minefield (m)	Number of Strips	Canisters per Strip	Total Canisters	Minefields per Aircraft
Disrupt	120	278	1	40 (20 each side)	40	4
Fix	120	278	1	40 (20 each side)	40	4
Turn	320	557	1	80 (40 each side)	160	1
Block	320	557	1	80 (40 each side)	160	1

Block

A block minefield (Figure D-3) is designed to stop an enemy advance along a specific AA or allow it to advance at an extremely high cost. Block minefields are obstacles with intensive integrated fires. They should be employed in a complex obstacle scheme with road craters or bridge demolitions enhancing the effectiveness of the minefield. One air Volcano aircraft can lay a 557- by 320-meter block minefield, using all 160 canisters (800 AT/160 AP mines). The probability of mine encounter is more than 80 percent. One Volcano aircraft can lay one block minefield, making two passes side by side (see Table D-2).

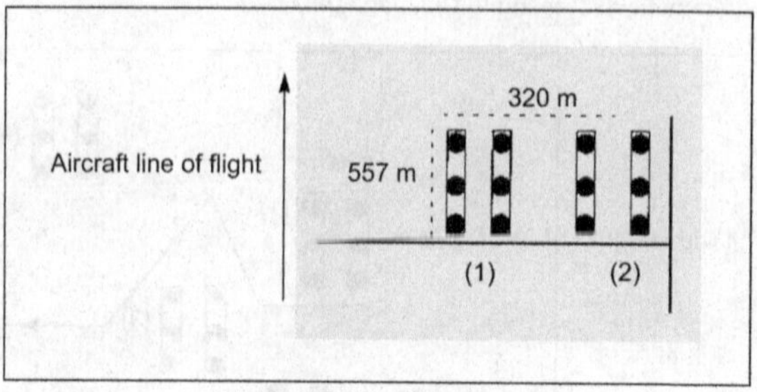

Figure D-3. Block obstacle

Disrupt

A disrupt minefield (Figure D-4) fractures and breaks up enemy formations. It causes premature commitment of reduction assets, interrupts C^2, and alters timing. A disrupt minefield is not resource- or time-intensive. The probability of mine encounter is approximately 50 percent, and the typical width is 278 by 120 meters. It takes 40 canisters (200 AT/40 AP mines) to emplace one disrupt minefield. One air Volcano aircraft can lay four disrupt minefields (see Table D-2).

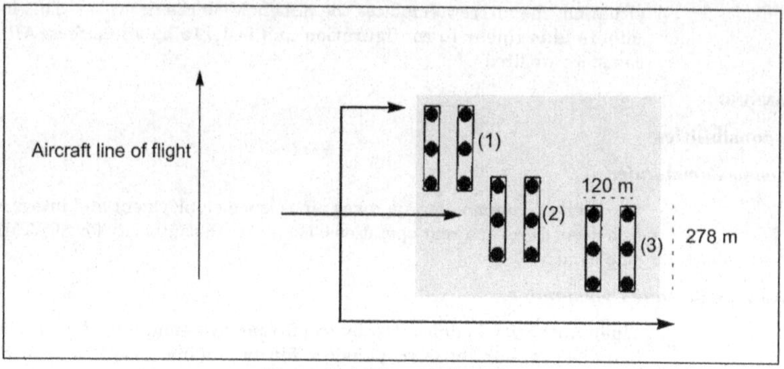

Figure D-4. Disrupt obstacle

Fix

A fix minefield (Figure D-5) slows enemy formations within a specified EA. It gives friendly forces time to acquire, target, and destroy enemy formations. A fix minefield is employed in depth and causes enemy formations to react and breach repeatedly. The air Volcano fix minefield does not look impenetrable to enemy formations. The probability of mine encounter is 50 percent, and the typical size is 278 by 120 meters. It takes 40 canisters (200 AT/40 AP mines) to emplace one fix minefield. One air Volcano aircraft can lay four fix minefields (see Table D-2, page D-6).

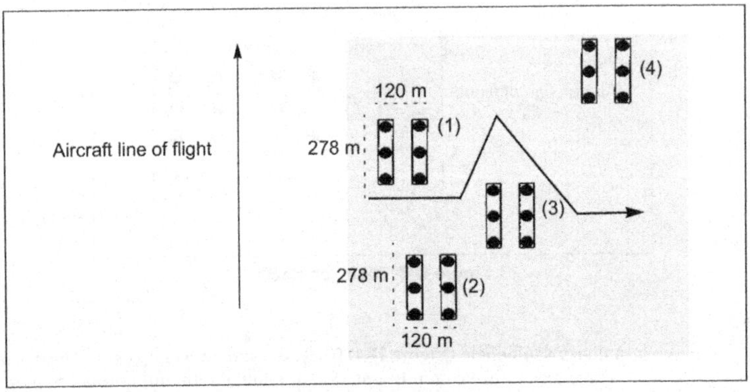

Figure D-5. Fix obstacle

Linear

If the threat situation allows, the aircrew makes a pass to confirm the minefield end points and the suitability of the terrain. In a high-threat situation, the aircrew emplaces the minefield on the first pass. This hasty minefield is linear in configuration and is 1,115 by 75 meters. All 160 canisters are fired.

PLANNING

Responsibilities

Division Commander

The division commander approves air Volcano employment and integration into deep, close, and rear operations. He is also the authority for SCATMINE employment.

Maneuver Brigade Commander

When authority is delegated by the division commander, the maneuver brigade commander is responsible for employing air Volcano in close operations and in supporting follow-on missions. He is responsible for approving target nominations to be submitted to the division. The maneuver brigade commander receives the air Volcano aircraft and its crew in OPCON status.

Aviation Brigade Commander

The aviation brigade commander is responsible for integrating the air Volcano into deep aviation attacks and for shaping EAs in the division AO. He executes the air Volcano missions in deep, close, and rear operations. The aviation brigade commander submits target nominations to the division targeting cell through the aviation brigade engineer and provides support for transporting and loading Volcano mines.

Division Engineer

The division engineer is responsible for target nominations that support division missions or objectives in deep operations. He—

- Submits the nominations to the division targeting cell.
- Determines minefield characteristics.
- Incorporates the Volcano minefield, DTG of SD times, and safety zones into the division obstacle plan.
- Estimates the requirements for mine canisters and Class IV/V supplies.
- Determines the intent of the Volcano minefield as it is integrated into the division obstacle plan.
- Disseminates the SCATMINWARN to adjacent and subordinate units before the minefield is laid and one hour before the SD sequence of the minefield is initiated.

Brigade Engineer

The brigade engineer is responsible for target nominations that support objectives within his respective fight. He—

- Submits target nominations to the assistant division engineer.
- Synchronizes air Volcano missions into brigade operations.
- Assists the brigade FSE in planning SEAD, CAS, and Apache escort.
- Provides logistics estimates to the brigade S4 for coordination of ammunition requirements.
- Posts the operations map with the minefield's location, aviation graphics, the safety zone, and the DTG of SD times.
- Disseminates the SCATMINWARN to adjacent and subordinate units before the minefield is laid and one hour before SD sequence of the minefield is initiated.

Deputy Fire-Support Coordinator

The deputy FSCOORD recommends commanding general/chief of staff approval for target nominations developed by the division targeting cell. He is responsible for submitting division-approved air Volcano target nominations to the G3 air for inclusion in the air coordination order. He coordinates SEAD, CAS, and intelligence and electronic warfare for division-directed air Volcano missions. The deputy FSCOORD is also responsible for—

- Targeting (intelligence and asset coordination).
- The air tasking order.
- The G3 air coordination order.
- G2 collection and assessment.

G3 Air

The G3 air synchronizes coordination and deconfliction of division air space for air missions, SEAD, and CAS. When necessary, he submits the air coordination order to higher headquarters with the division-approved target list.

Emplacing Unit/UH-60 Company

The emplacing unit or the designated UH-60 company sets up and loads the air Volcano system in conjunction with the forward-area refuel point. The Volcano system is loaded on the UH-60 at the designated point. The emplacing unit/UH-60 company is also responsible for—

- Maintaining the unit basic load at the forward-area refuel point or ATP.
- Preparing the scatterable minefield report and record, forwarding it to the authorizing commander or the aviation brigade engineer, and verifying that the assistant division/brigade engineer receives the entire report.
- Coordinating air routes and corridors.
- Requesting SEAD and security aircraft.
- Planning the air-mission coordination meeting to refine or develop the aviation scheme of maneuver.
- Ensuring that pilots attend the unit's rehearsal.
- Posting aviation graphics on the current operations overlay.

Process

All target nominations are submitted through the division targeting cell. In the brigade, all target nominations go to the assistant division engineer for submission to the board. Nominations should be submitted 96 hours prior to the execution time (see Table D-3).

- The targeting cell validates the targets based on the SITEMP and recommends approval through the commanding general/chief of staff.
- Upon approval, the deputy FSCOORD turns the target numbers and the air requirements over to the G3 air, who adds the targets to the air coordination order.
- The deputy FSCOORD begins to plan SEAD/CAS requirements for division missions.
- The G3 issues the division WO so that units can be prepared to execute specific air Volcano missions.

Table D-3. Planning process (H-hour sequence)

H-96+00. Submit target nominations/receive approval. Submit CAS requests for enemy ADA sites along ingress and egress routes.

H-72+00. Receive mission. Division WO issued.
- Conduct air-mission coordination meeting.
- Conduct S2 update on enemy situation and pass down to executing unit.
- Submit additions or changes to the CAS request to FSE for enemy ADA sites.
- Post the air Volcano mission in the air coordination order.
- Disseminate the SCATMINWARN.

H-48+00. Mount the system. The system takes 3 to 4 hours with a crew of three on a hard surface. The time is doubled in a field environment and requires three or four personnel to assist in loading.

H-24+00. Conduct detailed mission analysis and planning. Issue WO to attack assets, including—
- NLT time for mission planning and upload completion.
- Time and location of air-mission brief.
- Units attending.
- Minefield location and type.
- Establishment of C^2 relationship.
- Face-to-face coordination with attack assets.
- Liaison officers planning multiple routes to minefield.
- Liaison officers coordinating with engineers, ADA, and FSEs for support. G4/S4 must be notified of reload plans and any other Class III/V requirements.
- Intelligence update.

H-20+00. Volcano integrated into deep attack or air-assault matrix by the aviation brigade.

H-8+00. Conditions-check matrix approved by the commanding general or designated authority, including—
- Weather.
- Threat suppression along route.
- Attack support.
- FLOT crossing coordination completion.
- FLOT initial crossing point.
- SEAD fire coordination.
- Mines available.
- Integration into execution matrix.
- Crews updated on ADA threat.
- Backup system and aircraft available.
- CAS coordination.
- Rehearsal completion with attack assets.
- Intelligence update.
- Close operation mission, including—
  - — Location of friendly forces.
  - — Type of marking and when it will be in place.

H-6+00. Conduct air-mission brief, including—
- S2 intelligence update.
- ADA threat update.
- Rehearsal held after brief.
- Brigade engineer and brigade fire-support officer being present.

H-4+00. Latest time for link-up of aviation assets. Fly to forward-area refuel point or tactical assembly area for mine load-up. Remember that—
- Loading canisters takes 1 to 3 hours with a crew of four to eight. Time varies greatly based on whether or not the canisters are in shipping containers.
- The loading time in a field environment is decreased if one pad is dedicated to air Volcano aircraft loading and arming.

H-0+30. Cross FLOT (estimated; exact time is based on route), to include—
- SEAD fired.
- CAS sorties.

H-hour. Minefield deployed, to include—
- Aircraft reports to the supporting brigade engineer if an infantry brigade mission.
- Aircraft reports to the aviation brigade engineer if a division mission.
- Engineer disseminates the scatterable minefield report and record to appropriate units.

- The division issues a fragmentary order (FRAGO) to order the execution of the air Volcano mission.
- The planning staff gives the aviation brigade a date and a time for the air Volcano system to be uploaded and prepared for employment.
- Air Volcano minefields are integrated into the scheme of maneuver as a directed, situational, or reserve obstacle as stated in FM 90-7. This includes integration into the COA that is synchronized during war gaming and included in the OPORD as part of the rehearsal.

Logistics

Air Volcano munitions are transported and handled in the same manner as Class V mines and explosives. The only significant difference in handling is whether air Volcano operations are conducted from the tactical assembly area or the forward operating base.

Echelon-above-corps transportation assets will transfer the air Volcano to corps storage areas (ammunition). Based on forecasts by the division ammunition officer, the corps support area will push air Volcanos to ASPs or ATPs in the division area (tactical assembly area). ASPs and ATPs in or near the tactical assembly area are normally operated by a direct-support ammunition company attached to the corps support group (forward). ATPs in brigade support areas are normally operated by forward support battalions. Mines are moved from the corps support area to ASPs/ATPs by transportation units from the corps support group (rear). Further throughput of mines from ASPs to ATPs is supported by transportation assets from the corps support group (forward). Based on the division ammunition officer's forecast and the availability of transportation, the corps support area will attempt to transfer bulk mines to the ASP/ATP that is best positioned to support requirements. ATPs issue mines to using units while ASPs handle bulk ammunition. ASPs can do emergency issue to using units, but this is usually done only for units using large quantities of bulk ammunition and having their own ammunition transportation support available (division or corps artillery).

Preparation and Coordination

If an air Volcano mission has been approved for a unit, an aviation brigade liaison officer coordinates with the S3 air, the engineer, and the air defense officer to outline air Volcano requirements.

Logistical Requirements

The engineer planner calculates Class IV/V supplies and requests them from the Assistant Chief of Staff, G4 (Logistics)(G4)/S4. The engineer and the assistant division aviation officer coordinate the location (forward-area refuel point, tactical assembly area, forward operating base) of the ATP where the UH-60 will be loaded and fueled. The ALO must provide the amount of air time available and the fuel required, and he must be prepared to discuss emplacement times based on Volcano locations. The air-mission brief will facilitate dissemination of this information.

Concept of the Operation

The scheme of maneuver, fires, and engineer operation must be outlined. The scheme of engineer operations outlines exact grid coordinates, the obstacle intent, and the minefield composition and size. The air Volcano can be emplaced under enemy contact, but additional control measures must be used to protect the aircraft and the crew. If the air Volcano is triggered by enemy action, the DST must be briefed during the air-mission brief. The DST must outline NAIs, including the type of sensors and triggers (long-range surveillance device [LRSD], SOF, UAV), TAIs, decision points, and execution criteria.

Control Points and Markers

The initial point, the approach marker, and minefield markers are designated. Approach and minefield markers must be visible from the air and be distinctly different from one another:

- The initial point is an easily identifiable terrain feature used for coordinating the entry point of the UH-60 into the sector.
- The approach marker allows the aircraft to set the altitude, the speed, and the final orientation to the minefield. The approach marker can also be a terrain feature.
- Minefield markers establish the limits of the desired minefield. Deep area minefields do not require minefield markers.

Terrain Analysis

A combined (aviation, maneuver, fire-support, and engineer planners) terrain analysis should be conducted using Terrabase or a similar product. Terrabase enables planners to analyze the effects of the terrain in a three-dimensional format. AAs, terrain references, TAIs, NAIs, line-of-sight profiles, and minefield locations can be confirmed during this analysis.

Air-Mission Brief

The air-mission brief is the most critical planning and coordination meeting at the execution level, and it must occur no later than H-6 in the planning process. During the air-mission brief—

- C^2 is established.
- Updates and changes to the situation are exchanged between the brigade or TF engineer, the fire-support officer, and the air-mission commander.
- The engineer and the air-mission commander use four control techniques (discussed later in this appendix) to ensure mission success. The primary technique for emplacement and the responsibilities for each control technique are outlined.
- Radio frequencies; points of contact; code words; identification, friend or foe, modes; and challenge and passwords are exchanged and disseminated.

- The forward-area refuel point location, the security aircraft, flight routes, and lethal/nonlethal SEAD are identified (if already approved).

Mounting

The system must be mounted no later than H-48 in the planning process.

Execution

Siting

The key to proper emplacement is the location of the minefield in relation to existing terrain features. If appropriate, the minefield should tie into an existing terrain feature to prevent easy bypass or fording. Using Terrabase enhances siting and emplacement procedures in deep operations. Reconnaissance must be conducted to verify the location.

Movement

Loading. The launcher rack functions as the carrier and launcher platform for 40 mine canisters. The rack has 40 keyholes for mine canisters, a green latch that latches the mine canister to the rack, and a red latch that arms the mine canisters. The rack has two electrical receptacles—one for the connector and one for the launcher rack cable from the DCU. While looking at the canister side of the rack, rows are 1 through 4 from bottom to top and columns are 1 through 10 from left to right.

Arming.

- Due to the weight of the Volcano mine system, a large, open area that is clear of obstacles must be selected. The site should have a hard surface if possible. If a hard surface is unavailable, inspect the ground to ensure that it is firm enough to support the weight of the aircraft. Perforated steel planking or two pieces of 1-inch plywood (4 by 4 feet) may be used as a field-expedient surface in soft areas.
- Concentrations of nonessential personnel or frequently traveled vehicular routes should not be within 1,000 meters of the site. This distance is based on the total weight of explosives and the safe fragmentation distance found in FM 5-250. When using the M88 training canisters, the minimum distance is 30 meters.
- Two Underwriters Laboratories, Incorporated (UL)-listed 10BC fire extinguishers and a grounding rod (minimum safety equipment) must be available at the arming point. This equipment is provided by personnel who deliver the mines.
- The number of personnel allowed access to the site should be held to a minimum. All personnel involved in the arming will receive a safety brief that includes—
  - Ammunition handling and inspection procedures.
  - Loading procedures.
  - Emergency procedures and rendezvous points.
- Emergency procedures.

— Fire. In the event of a fire away from the mines, attempt to contain or extinguish the fire by any available means. If the fire is near the mines or in them, clear the area to a minimum distance of 1,000 meters and notify fire-fighting personnel immediately. When training with M88 canisters, clear the area to a minimum distance of 30 meters.

— Accidental discharge. Immediately clear the area to a distance of 640 meters and notify EOD. The mines arm approximately 2½ minutes after firing. When training with M88 canisters, terminate arming until the problem can be identified and corrected.

— Failure to fire. Remove the canister from the aircraft and place it in the dud pit. Notify EOD immediately. When training with M88 canisters, remove the canister from the aircraft, separate it from the other canisters, repack it, and return it to the ASP.

- Site layout (Figure D-6).

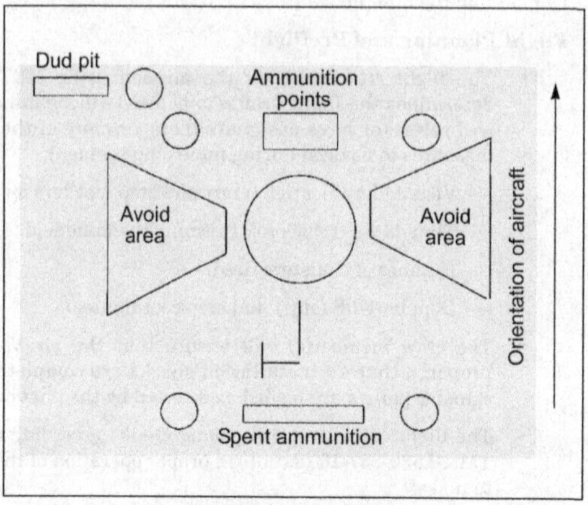

Figure D-6. Site layout

— Berming of the site is not required for a tactical arming point.

— The following rules apply when the site is located next to a refuel point:

> A minimum of 1,000 meters must exist between arming points and refuel points when the total quantity of explosives is less than 600 kilograms. For quantities greater than 600 kilograms, refer to FM 5-250.

NOTE: Each M87 canister contains 3.4 kilograms of explosives; a full load (160 canisters) contains 550 kilograms of explosives.

> The refuel point for armed aircraft must be located at least 375 meters from other aircraft refueling points.

> Parked, armed aircraft must be at least 36 meters from other armed aircraft to prevent the detonation of explosives on adjacent aircraft. This distance will not prevent damage to adjacent aircraft; a 130-meter distance is required to prevent damage by fragments and to ensure that the aircraft remains operational.

— A dud pit (bermed when possible) for damaged or misfired ammunition should be established beyond the ammunition points.

— Arming points should be laid out as shown in Figure D-6.

Dearming. After the mission is complete, the aircraft returns to the arming point for dearming. Spent canisters should be discarded at least 30 meters from the aircraft, at the 4- and 8-o'clock positions. Live canisters should be returned to ASPs for future use or repackaging. Canisters that misfire should be placed in the dud pit.

Flight Planning and Preflight.

- The flight crew analyzes the mission using METT-TC factors and determines the flight profile to be used during mine emplacement. It will select (or have designated) one or more of the following control measures to be used during mine emplacement:
  - — Visual identification (start and stop markers on the ground).
  - — Time-lapse (tables to determine the minefield length).
  - — Number of canisters fired.
  - — Doppler/GPS (start and stop coordinates).
- The crew member(s) will ensure that the air Volcano is installed properly, that all installation checks are completed, and that mine canister pallets are loaded as directed by the pilot or the SOP.
- The flight crew conducts ground checks according to the checklist in TM 1-1520-237-10 to confirm proper operation of the air Volcano prior to takeoff.

Before Arrival at the Target Area.

- During the equipment check, the crew chief turns on the DCU power-control switch, verifies that no malfunctions were indicated during the initial built-in test, and turns off the DCU power-control switch.
- After completion of *run-up* with the aircraft at flight idle, the crew chief turns on the DCU power-control switch.
- Before arrival at the release point, the pilot will make the following checks (listed on the Volcano card [a sample is shown in Figure D-7]):
  - — Verify that the DCU is on.
  - — Verify that the mine SD time is properly set.

VOLCANO CARD			
N	Emplacement Date:		
	Unit	Engr	ASP
	Call Sign		
	Frequency		
	Location		
	Remarks		

Not to scale

	1st pass	2nd pass	3rd pass	4th pass
Initial Point	A	D	G	J
SD Time				
Ground Speed	knots kph	knots kph	knots kph	knots kph
No Tubes				
Altitude				
Track				
Start Time	B	E	H	K
Stop Time	C	F	I	L

Figure D-7. Sample Volcano card

— Announce the ground speed in knots or kph, as required. (The pilot and the crew chief will acknowledge.)

— Announce the number of canisters the crew chief will count down. If deployment is 40 canisters per run (20 per side) and there are two runs, the crew chief will count down from 80 to 60 on the first run and 60 to 40 on the second run. (The crew chief will acknowledge.)

— Announce the altitude for employment. (The pilot will acknowledge.)

— Announce the course for the delivery track. (The pilot will acknowledge.)

— Announce the delivery time based on the setting of the ground speed. (The crew chief will acknowledge.)

- Before arrival at the initial point, the crew chief will—
  - — Ensure that the DCU fire-circuit switch's safety pin and streamer are removed.
  - — Ensure that the DCU fire-circuit switch is enabled.
  - — Place the interface control-panel arming switch to the *ARM* position. He will verify that the jettison advisory light indicates *armed* and that no fault codes are displayed on the DCU.

At the Target Area.

- The pilot simultaneously announces "mark," presses the go-around switch, and starts timing the run when he is over the minefield start point.
- The pilot maintains a ground speed of 5 kph/3 knots and an altitude of 3 meters during the mine-dispensing pass. The pilot is responsible for flying the aircraft within the prescribed limitations.
- The crew chief announces the mine-canister count, as the canisters are dispensed, by counting down in 10s. He then announces the last three canisters. For example, the pilot announces a canister count of 60, the crew chief calls out "80, 70, 3, 2, 1, mark."
- The pilot terminates mine dispensing when the grid location is reached.

After Mission Completion.

The crew chief will—

- Place the interface control-panel arming switch to the *SAFE* position and verify that the armed advisory capsule is extinguished.
- Place the DCU fire-circuit switch to the *OFF* position. Install the safety pin and the streamer.
- Prepare and submit a SCATMINWARN.

EMPLACEMENT

The air Volcano is fast and flexible, but it is difficult to accurately dispense mines within the confines of the minefield marking. The desired obstacle-effect norms for the air Volcano require extensive planning, preparation, coordination, and positive control during emplacement. The critical aspect of the air Volcano is getting the right amount of mines in the specified location and in the desired density.

The detailed coordination focuses on positive control. Positive control of an air Volcano mission requires a redundancy of control techniques to minimize errors in minefield size and location. These control techniques must compensate for poor visibility, wind speed and direction, and navigational errors.

The following control techniques are used by the engineer and the air-mission commander to ensure that Volcano minefields match specific obstacle-effect norms. Units rely on these techniques to accomplish the mission, and they are part of the Volcano air-mission brief:

- Visual identification. Focuses on the visual identification of minefield emplacement. As part of the preparation for a Volcano minefield, an engineer element erects airfield panel markers to mark start and end points. This provides a visual signal for the engineer and the air-mission commander to start and stop firing Volcano canisters. The pilot depresses the launch switch over the first marker to start firing and depresses it again over the second marker to stop firing. This control technique is good for open terrain with adequate visibility and little canopy coverage.

- Time lapse. Focuses on when to stop firing Volcano canisters. The UH-60's air speed and the type of minefield being laid determines the amount of time it takes to lay a minefield. The air Volcano has six air speed settings—20, 30, 40, 55, 80, and 120 knots. Table D-4 shows the time required to lay minefields and the full load time.

Table D-4. Air Volcano dispensing times based on air speed

Knots	Disrupt and Fix Minefields	Turn and Block Minefields	160 Canisters per Load
20	27 seconds	54 seconds	108 seconds
30	18 seconds	36 seconds	72 seconds
40[1]	13 seconds	27 seconds	54 seconds
55	9 seconds[2]	18 seconds	39 seconds
80	6 seconds[2]	13 seconds[2]	27 seconds
120	4 seconds[2]	9 seconds[2]	18 seconds
Width of minefield (meters)	278.8	557.5	1,115
No passes per minefield	1	2[3]	1
No canisters per pass	40	80	160

[1]Recommended air speed

[2]Recommended only if absolutely necessary

[3]Blackhawks in pairs can lay turn and block minefields in one pass, firing 80 canisters each.

The following example is provided to show how Table D-4 is used:

Example: The mission is to install an air Volcano disrupt minefield. The UH-60 is traveling at 40 knots (this is entered on the DCU), and the pilot initiates (depresses the launch switch) at the identification of the Volcano start marker or the grid location on the ground. The pilot depresses the launch switch a second time after 13 seconds have elapsed.

- Number of canisters fired. Focuses on when to stop firing Volcano canisters. The number of Volcano canisters dispensed also determines when firing is terminated. There is a digital readout on the DCU (for the left and right side) that shows the number of canisters remaining. The pilot stops firing when the required number of canisters have been fired (see Table D-4, page D-19).

Using the disrupt minefield example above, the UH-60 starts the mission with a full load (80 canisters on each side of the aircraft). The pilot initiates (depresses the launch switch) at the identification of the Volcano start marker or the grid location on the ground. The pilot depresses the launch switch a second time after 20 canisters have been expended on each side. The DCU counts down from the total number of canisters. When the DCU reads 60 right/60 left, the pilot depresses the switch to end the firing process. Ideally, the timing of delivery and the number of canisters fired are done simultaneously. As the crew chief counts down the timer, the pilot and the crew chief monitor the number of canisters remaining on the DCU digital readout.

- Doppler/GPS. Focuses on when to start and stop firing Volcano canisters using the UH-60's Doppler/GPS guidance and navigation set. This set provides the present position or destination in latitude and longitude (degrees and minutes) or grid coordinates. As part of the preparation for the Volcano minefield, exact grid coordinates are needed to determine the approach points and the limits of the minefield. These coordinates are provided to the air-mission commander during the air-mission brief. The pilot enters the grid coordinates into the Doppler/GPS on the primary and the backup aircraft. During execution, the air-mission commander monitors the Doppler/GPS and determines the time to target, when to initiate firing, and when to terminate firing.

Outside Friendly Territory

Reconnaissance of the proposed site for the air Volcano minefield will be conducted before mines are emplaced. This could include—

- LRSD.
- Apache gun tapes.
- UAV overflights.
- Imagery.

Key terrain or landmarks are used to identify start and end points for aviation assets. It is unlikely that military marking will be employed based on the proximity to enemy forces and the probability of early detection if man-made markers are present. Fencing the minefield is not required until the area has been secured by friendly forces.

Within Friendly Territory

Within friendly territory, air Volcano minefields should be fenced and marked with NATO identification signs to protect friendly forces.

Fencing is installed before the air Volcano minefield is delivered, and it is located 100 meters from the centerline of the minefield and 100 meters from the start and end points (Figure D-8).

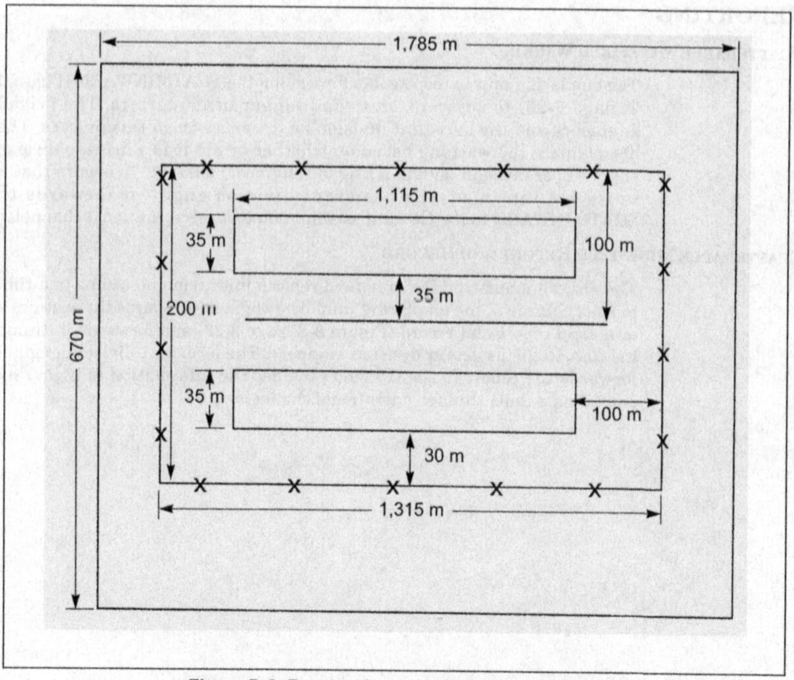

Figure D-8. Fencing for an air Volcano minefield

Start and end points should be marked with man-made devices such as VS-17 panels. During limited visibility, start and end points should be marked with infrared or heat-producing sources. Key terrain features and landmarks should still be used to identify start and end points.

Fencing the minefield is not viable when the minefield duration is short or civilians on the battlefield are an issue. In this case, CA, public affairs, and PSYOP personnel should be involved in letting friendly personnel know the minefield location. This could include—

- Leaflet drops.
- CA teams disseminating information.
- Host-nation support.
- PSYOP.

REPORTING

SCATTERABLE MINEFIELD WARNING

The emplacing unit is responsible for issuing the SCATMINWARN (Figure 8-7, page 8-23) to adjacent units and higher headquarters. The brigade engineers and the assistant division engineer assist in this process. They disseminate the warning based on whether or not it is a brigade (brigade engineer) or division (aviation brigade engineer) mission. To ensure that all units are informed, the assistant division engineer forwards the SCATMINWARN to the G3 for dissemination through operational channels.

SCATTERABLE MINEFIELD REPORT AND RECORD

The aircraft emplacing the minefield reports initiation and completion times to the engineer of the emplacing unit. The engineer prepares the scatterable minefield report and record (Figure 8-8, page 8-23) and forwards it through his unit to the assistant division engineer. The assistant division engineer forwards the report to the G3 who provides the information to higher and subordinate units through operational channels.

Appendix E

Safety and Training

Mine training is inherently dangerous, in part, because several different types of mines and fuse systems are used throughout the world. Detailed safety instructions for each type of mine are provided throughout this manual. This appendix merely points out the safety aspects of live-mine training that are common to all types of mines.

Conduct mine training as if the mines were live. This is the only way soldiers form a habit of correctly and safely handling mines and gain a true appreciation of the requirements and the time it takes to perform an actual mine-warfare mission. Live-mine training gives soldiers the confidence they need to handle mines and their components. Accidents can usually be traced to ignorance, negligence, deliberate mishandling, overconfidence, mechanical failure, or fright. The first four can be overcome by training and proper supervision. Mechanical failure rarely happens; but if it does, it can be controlled by training and proper supervision. The last item, fright, is mastered through well-controlled, live-mine training.

STORAGE

There are three types of mines used in mine training:

- Inert. Does not contain explosives.
- Practice. Contains an LE charge or a smoke-producing component to simulate detonation.
- HE. Involves actual mines used in combat

Conventional mines are painted to enhance concealment, retard rusting of exposed metal parts, and help identify the type of mine and filler (HE, LE, or chemical agent). Older manufactured mines are painted according to the Five-Element Marking System; newer mines use the Standard Ammunition Color-Coding System (see Table E-1, page E-2).

NOTE: Mines that are color-coded and marked according to the old system have been on hand for several years. Ensure that all ammunition, whether color-coded according to the old or new system, is properly and fully identified.

Always handle mines with care. The explosive elements in fuses, primers, detonators, and boosters are particularly sensitive to mechanical shock, friction, static electricity, and high temperatures. Boxes and crates containing mines should not be dropped, dragged, tumbled, walked on, or struck. Do not smoke within 50 meters of a mine or its components.

Table E-1. Mine color-coding system

Type of Ammunition	Five-Element Marking System (Old)	Standard Ammunition Color-Coding System (New)*
Persistent casualty chemical agent	Gray with green markings and two green bands	Gray with green markings and two 12-mm green bands
Nerve agents	Gray with green markings and two or three green bands	Gray with green markings and three 12-mm green bands
Incendiary	Gray with violet markings and one violet band	Light red with black markings and one yellow band
HE	Olive drab with yellow markings	Olive drab with yellow markings
Practice mines	Blue with white markings	Blue with white markings
Inert mines	Black with the word *INERT* in white	Blue with the word *INERT* in white
*Chemical ammunition containing an HE has one 6-mm yellow band in addition to the other markings.		

When it is necessary to leave mines in the open—

- Set them on dunnage at least 5 centimeters above the ground.
- Place a waterproof cover (such as canvas) over them, and leave enough space for air circulation.
- Dig drainage trenches around stacks of mines to prevent water from collecting under them.
- Protect mines and their components against moisture by waterproofing them with grease coatings, tar paper, or tarpaulins.

Additional maintenance procedures are as follows:

- Do not open mine boxes in a magazine, at an ammunition dump, or within 30 meters of an explosive store. Use copper or wooden safety tools, if available, to unpack and repack mines.
- Do not fuse mines within 30 meters of an explosive or ammunition holding area. Mines can be fused at the mine dump.
- Use specifics authorized by the US Army Materiel Command and applicable TMs to disassemble mines and their components.
- Remove safety pins, safety forks (clips), and other safety devices as the last step when arming the mine; and replace them before the mine is moved again. These devices prevent accidental initiation of the mine while it is being handled.
- Place tape over open fuse cavities and secondary fuse wells. Ensure that they are clear of obstruction and free of foreign matter before attempting to install the fuse, the detonator, or the FD.
- Take steps to prevent moisture or water from accumulating around the mine and subsequently freezing if the temperature fluctuates around freezing. Mines usually function satisfactorily at temperatures between 40 and 160°F. Most mines are not appreciably affected by temperature changes, but mines can become neutralized by ice formations (see Chapter 12).

- Observe proper procedures when recovering mines. Ensure that components do not show evidence of damage or deterioration.
- Ensure that practice or inert mines or their components are not present when live mines or their components are being used.
- Do not mix inert mines with live mines.
- Do not display live mines or their components in museums, demonstrations, models, or similar layouts. Only inert equipment can be used for displays.
- Handle explosive materials with appropriate care. The explosive elements in primers, blasting caps, and fuses are particularly sensitive to shock and high temperatures.
- Assemble activators, standard bases, and FDs before installing them. Do not carry them in the pockets of your clothing.
- Do not point FDs at anyone.
- Camouflage the mine before removing the positive safety pin when possible.

NOTE: Additional storage and safety precautions are outlined in TM 9-1300-206.

LIVE-MINE TRAINING

NOTE: No live-mine training is authorized with M14 mines. Units outside Korea will not use live M16A1 mines in tactical or protective minefield training.

Live-mine training is conducted by preparing, laying, arming, neutralizing, and disarming live mines (with live fuses and components) in a training environment.

Supervisors must adhere to the following safety considerations when conducting live-mine training:

- Only personnel who are qualified and certified according to the local range SOP are allowed to supervise activities or training in which live mines or their components are used.
- Minimum personnel requirements to conduct live-mine training are—
  - — Range officer (OIC).
  - — Range safety officer (RSO).
  - — One NCO supervisor for each arming bay.
  - — Mine-explosive breakdown NCO.
  - — One medic per four arming bays.
  - — Guards, as required by the range SOP.
- Sound organization is a must before live-mine training can begin. The OIC and supervising NCOs conduct a demonstration/briefing to ensure that the practice runs smoothly.

- The training officer must foresee hazards that can occur through personnel nervousness or material failure. The commander should conduct a risk assessment according to AR 385-10.
- The OIC takes his place at the control point or post. Once he is satisfied that all safety regulations have been observed, he orders the first detail to start training.
- Soldiers are trained on inert and practice mines before arming live mines, according to the guidelines established by the Standards in Training Commission.
- Fuses are not inserted into mines until ordered by the OIC.
- An NCO supervisor must be present when soldiers arm live mines. He ensures that soldiers adhere to the proper procedures and regulations.
- Only one soldier arms a mine at any given time.
- Personnel disarm one mine before arming the next one.
- Personnel never arm an M16 AP mine in the trip-wire mode during live-mine training.
- Personnel never remove the positive safety pin from the M16 AP mine during live-mine training.
- Instructors inspect fuses and mines for serviceability before starting practice.
- Instructors inspect mines and their components for damage and excessive wear after each student has gone through the station. Replace the mine and the fuse if damage or wear is found.
- All personnel wear a helmet (with the serviceable chin strap fastened) and body armor when arming and disarming mines.
- Ear protection is not permitted in the arming bays. The student must be able to hear the supervisor and certain distinct noises (such as a firing pin dropping).
- Instructors post guards at all entrances to the range. The guards communicate with the RSO by radio, wire, voice, or signal. No one enters the range without permission from the RSO.
- Instructors keep mine records and inventory sheets. They maintain accountability of all mines and fuses, before and after each exercise.
- Instructors draw and return supplies; check equipment for issue; and ensure that live mines are safe, serviceable, and unarmed. They ensure that the requirements contained in AR 385-63, range regulations, and SOPs are observed and that no one does anything to prejudice safety.
- Instructors clearly mark the word *LIVE* on all live mines and their components that are used for live-mine training. Live mines are maintained separately from practice and inert mines.

- Live AHDs are not used with live mines during training, but they can be used with practice and inert mines.
- Arming and disarming are conducted in the prone position.
- Waiting personnel are located in a bunker, behind a suitable barricade, or at a safe distance from live-mine training.
- Supervisors ensure that live-mine training is not rushed. There are no shortcuts. Supervisors must allow soldiers ample time to arm and disarm mines. Most soldiers are already in a high state of stress from dealing with live munitions, and rushing them only serves to heighten their stress level.

LIVE-MINE DEMONSTRATIONS

Live-mine demonstrations show mine characteristics and capabilities using M16 and M18 AP mines and M15, M19, and M21 AT mines. The appropriate authority must authorize the demonstration, and firing personnel must be fully conversant with all safety and technical aspects pertaining to live-mine firing.

An OIC and an RSO are appointed for each activity involving live-mine firing. The amount of explosive contained in the mine cannot exceed the maximum amount allowed for the range, and one mine is fired at a time.

Upon arriving at the range, the instructor and his assistants establish areas according to the following rules (signs are posted for large demonstrations):

- Firing point. Sited outside the danger area and near the OIC to facilitate coordination, commentaries, and firing.
- Spectator area. Sited outside the danger area and within earshot of the commentator. It is large enough to provide a good view of the explosion.
- Supply area. Any suitable area away from spectators.
- Explosive area. Sited away from supplies and spectators.
- Mine area. Mines are set out in full view of the OIC and spectators. Individual mines are at least 25 meters apart.
- Target area. Targets are positioned and inspected by spectators before the blasting cap is inserted into the mine.

M16 ANTIPERSONNEL MINE

- Safety distance. 300 meters.
- Firing procedures.
  - Roll out 300 meters of firing cable and attach it to a stake or picket in the ground (leave at least 1 meter of free end). Test the firing cable for continuity.
  - Place the mine in the ground (dig in level with the surface). Remove the shipping plug.
  - Test a blasting cap (under a sandbag) with the demolition test set.

— Attach the ends of the blasting-cap leads to the ends of the electric cable and insulate the joints with tape. Place the blasting cap into the fuse well (see Figure E-1, page E-6).

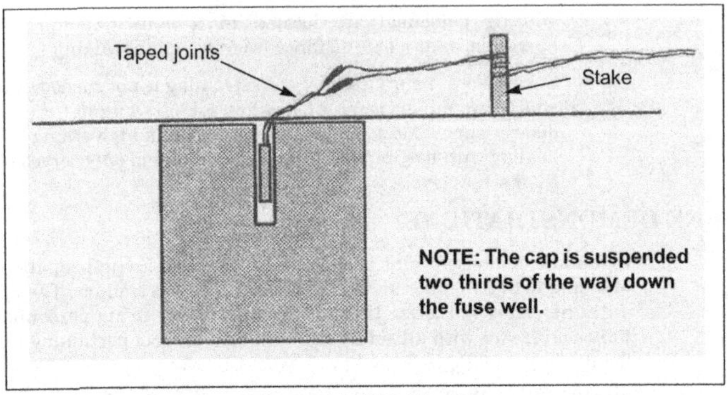

Figure E-1. M16 AP mine

- Suggested target. A circle of tar paper, 6 meters in diameter, supported by 1.8-meter pickets. Spectators can later view shrapnel effects.

NOTE: The procedure detailed here dispenses with the M605 igniter. The mine cannot be detonated by pull or pressure. The expulsion charge and millisecond delay fuses are still operated, and the mine bounds out of its casing (which remains in the ground) before exploding in the air. Although the normal firing delay is removed, it does not detract from the demonstration. The blasting cap is suspended two-thirds of the way down the fuse well to initiate the expelling charge and delay elements.

- Misfires. In the event of a misfire, the RSO disposes of the mine by placing a block of C4 as close to the mine as possible, without touching it. He destroys the mine by normal nonelectric means.

M18A1 ANTIPERSONNEL MUNITION

- Safety distance. 300 meters.
- Firing procedures.
  - — Roll out 300 meters of firing cable and attach it to a stake or picket in the ground (leave at least 1 meter of free end). Test the firing cable for continuity.
  - — Place the mine on the ground and ensure that the front of the mine faces away from the firing point. Remove the shipping plug.
  - — Test an electric blasting cap (under a sandbag) with the demolition test set.

— Attach the ends of the blasting cap leads to the ends of the electric cable and insulate the joints with tape. Place the blasting cap into the detonator well (see Figure E-2).

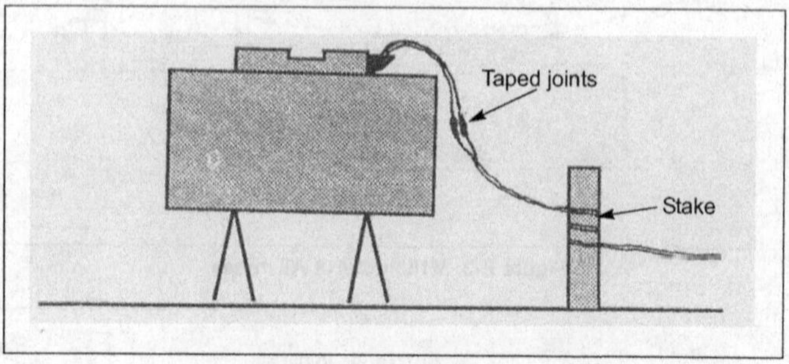

Figure E-2. M18A1 AP mine

- Suggested target. Several E-type silhouette targets, 15 to 100 meters from the mine.

NOTE: The procedure detailed here applies only to demonstration firings. Standard accessories are used on all other occasions. The mine explodes instantaneously and clearly illustrates the sound of an M18A1 explosion.

- Misfires. In the event of a misfire, the RSO disposes of the mine by placing a block of C4 as close to the mine as possible, without touching it. He destroys the mine by normal nonelectric means.

M15, M19, AND M21 ANTITANK MINES

- Safety distance. 1,000 meters.
- Firing procedures.

  — Roll out 1,000 meters of firing cable and attach it to a stake or picket in the ground (leave at least 1 meter of free end). Test the firing cable for continuity.

  — Place the mine in the ground and leave the top exposed. A target is used only when the mine can be placed without disturbing the target. A derelict vehicle is a suitable target.

  — Place a block of C4 on top of M15s and M19s (see Figure E-3, page E-8).

  — Remove the shipping plug from the booster well of the M21 and pack the well with C4. Insert an electric blasting cap into the C4 (see Figure E-4, page E-8).

NOTE: Do not remove safety devices. Keep arming dials in the *SAFE* position. The mine explodes instantaneously and clearly demonstrates the blast/shaped-charge effect.

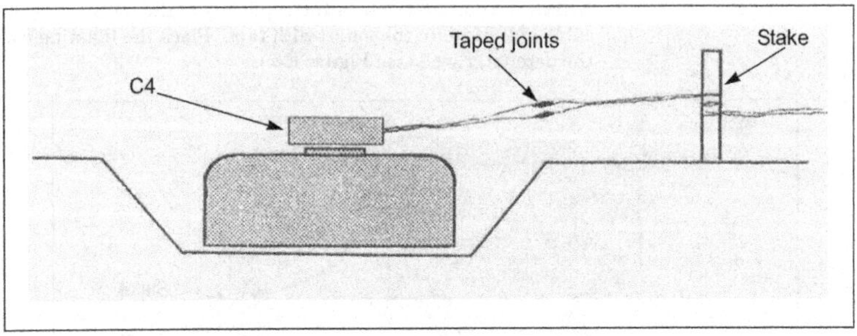

Figure E-3. M15 and M19 AT mines

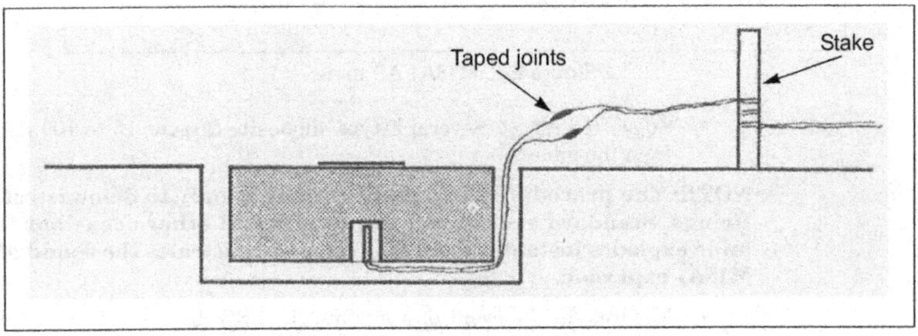

Figure E-4. M21 AT mine

- Misfires. In the event of a misfire, the RSO disposes of the mine by placing a block of C4 as close to the mine as possible, without touching it. He destroys the mine by normal nonelectric means.

RISK ASSESSMENT FOR LIVE-MINE DEMONSTRATIONS

The following risk assessment is provided as a guideline for live-mine demonstrations using an M16 AP mine. It must be carefully reviewed before conducting a demonstration. Live-mine demonstrations can be conducted in a safe manner. The risk of injury to personnel is significantly minimized if you adhere to established procedures.

During the demonstration, mines are not armed with standard fuses. They are activated by electric blasting caps placed inside the fuse wells.

A demonstration shows the effectiveness of an M16 AP mine. Spectators do not handle the mines or explosives. To show the effectiveness of an M16 mine, a sheet of paper is placed in a semicircle around the mine. Spectators remain in bunkers or at a safe distance while mines are primed with electric blasting caps and detonated. After the mines have been detonated and the RSO has cleared the area, spectators are allowed to view the results of the detonated mine. Misfires are handled by the RSO.

Figure E-5 is a risk assessment prepared by the Department of Transportation.

QUALITATIVE RISK ASSESSMENT

Qualitative risk-assessment techniques are used to place a value on the level of risks created by hazards in an operation. The principal qualitative technique is the risk-assessment code (RAC) described in MILSTD-882C. This method was established as a common way to set priorities for Department of Defense (DOD)-wide hazard-abatement programs and uses an RAC matrix format to combine the concepts of frequency and severity into a single, numerical code. It is very useful in comparing different risks (such as those from different programs) or even differences (such as health versus safety risks).

RACs are implemented for the Army in Army Regulation (AR) 385-10, which describes the two qualities of hazard severity and hazard probability as follows:

- Category I—CATASTROPHIC. May cause death or loss of a facility. In this case, loss does not mean a period of interrupted service; it means destruction of the facility or operation.
- Category II—CRITICAL. May cause severe injury, severe occupational illness, or major property damage.
- Category III—MARGINAL. May cause minor injury, minor occupational illness, or minor property damage.
- Category IV—NEGLIGIBLE. Probably would not affect personnel safety or health, but is in violation of specific standards.

Mishap probability is the probability that a hazard will result in a mishap, based on an assessment of factors such as location, exposure in terms of cycles or hours of operation, and affected population. The expression combines the idea of the probability of an event and the exposure to the event. These probabilities are expressed as letters conforming to the following system:

- Subcategory A—Likely to occur immediately.
- Subcategory B—Probably will occur in time.
- Subcategory C—May occur in time.
- Subcategory D—Remote.
- Subcategory E—Improbable.

The two qualities are combined to yield an RAC by using the following table:

Risk-Assessment Code Table

Hazard Severity	Mishap Probability				
	A	B	C	D	E
I	1	1	2	3	4
II	1	2	3	4	5
III	2	3	4	5	5
IV	—	—	—	—	—

In using the RAC system, it is important to note that the IA and IIA classifications are termed "imminent danger." Though their RAC codes of 1 are the same as that of the IB entry, their immediacy makes them more critical. The codes are useful in assessing an operation as it begins, but they must be updated as the operation continues, the facility ages, and so forth to account for degrading condition or performance.

For risk managers, there are some important organizational/management considerations to RAC codes. AR 40-10 also contains an RAC system, but due to a difference in definitions, the health RAC code may be a lower number, indicating a higher degree of risk. This is important to managers who are comparing health risks to others risks under AR 385-10; the health issue would always win if no compensation or consideration was factored into the codes. From the managerial standpoint, it must be remembered that RACs are judgmental and not necessarily held to be the same by different managers or evaluators. When differences in perception occur, the differences are likely based on either the understanding of the operation's behavior or the criteria for selecting the probability and severity. It is wise to listen to the basis of others' RAC choices and attempt to develop a common understanding.

Figure E-5. Excerpt from Risk-Assessment Techniques Manual, prepared by the Department of Transportation's Transportation Safety Institute, August 1986

Qualitative risk-assessment techniques are used to prepare estimates of risk levels using performance data, when available, to improve the accuracy of risk estimates used in risk-acceptance decision making. These assessments are numeric values representing the safety risk of an Army activity, system operation, or comparable endeavor, based on actuarial or derived numeric data. Though RACs are numerical, they are derived from judgments and are not demonstrable in records of performance. If it is desirable that performance be measured, it is necessary that quantified estimates of risk levels be established, that risk levels be predictive so that future performance has a base of comparison, and that risk levels be assigned numeric values.

Figure E-5. Excerpt from Risk-Assessment Techniques Manual, prepared by the Department of Transportation's Transportation Safety Institute, August 1986 (continued)

RISK ASSESSMENT FOR LIVE-MINE TRAINING

The United States Army Engineer Center, Department of Instruction, obtained information for the following risk assessment from the Collective Training Branch, Department of Training and Doctrine, and from the Engineering Branch for Engineer Officer Basic Course demolitions training. Hazards are identified and analyzed on preliminary hazard-analysis work sheets (see Figures E-6 through E-15, pages E-11 through E-21). Risk-assessment codes are assigned to each hazard based on the severity and the probability of occurrence.

References used in the risk-analysis process include this manual and the following publications:

- DA Pamphlet (Pam) 350-38.
- AR 385-10.
- AR 385-16.
- AR 385-63.
- Soldier Training Publication (STP) 5-12B1-SM.
- Training Circular (TC) 25-8.
- TM 9-1345-203-12.
- TM 43-0001-36.

Operation Name: Arming the M15 AT Mine
Date Prepared: 11 October 1990
Sheet Number: 1 of 1

Hazards	System Effects	Causal Factors	RAC	Actions	Controlled RAC	Standards/Comments
Damaged mine used	None	Lack of training/improper training Improper supervision Incomplete or no inspection	IV B (-)	Do not use mine. Contact QASAS to investigate. Turn mine in for investigation/ disposal.	IV D (-)	Proper training is conducted. Soldiers are proficient on inert mines first. One instructor is assigned to one soldier. All training is conducted in the prone position. Mine training is done after identified by a unit's METL and only with select personnel. Training is conducted in the proper environment. Soldiers are in the proper protective gear. No trip wires or AHDs are used with an M15 mine. Using the cap to perform the fuse depth check is not authorized. Use the wrench.
Too much force on extension rod/ pressure ring (M624 fuse)	Mine detonates	Soldier bends/pushes on rod or fuse	I B (1)	Do not remove the safety stop and band if the extension rod is used. **OR** Remove the extension rod before removing safety stop and band.	I D (3)	
Too much pressure on pressure plate (M603 fuse)	Mine detonates	Soldier applies excessive pressure	I D (3)	See comments.	I D (3)	
Fuse improperly seated in the fuse well (M603 fuse)	Mine detonates	Excessive debris/corrosion in well Lack of training/improper training Improper supervision Incomplete or no inspection	I C (2)	Perform depth check with wrench.	I D (3)	
Improper handling of fuse (M603 fuse)	Fuse detonates	Pressure applied to fuse with safety fork removed	II D (4)	Handle fuse properly.	II D (4)	

Figure E-6. Preliminary hazard-analysis work sheet (arming M15)

Operation Name: Disarming the M15 AT Mine
Date Prepared: 11 October 1990
Sheet Number: 1 of 1

Hazards	System Effects	Causal Factors	RAC	Actions	Controlled RAC	Standards/Comments
Camouflage removed	Mine detonates	Too much pressure on pressure plate Damage to the mine	I D (3)	Do not apply pressure to pressure plate. See comments.	I D (3)	See comments for arming an M15 AT mine (Figure E-8, page E-13). See risk assessment for command-detonating a mine in place with explosives (Figure E-16, page E-21).
Unable to turn the mine to SAFE or reinstall safety ring, safety stop, or safety pin	Mine detonates	Too much resistance (M603 fuse) is applied to turning indicator. Safety band (M624), safety stop, or safety pin has become damaged. Foreign material has entered the fuse well.	I C (2)	Perform controlled detonation in place.	None	
Fuse removed/ safety fork installed	Fuse detonates	Pressure is placed on the pressure plate of fuse. Fuse is not removed.	II D (4)	Follow proper procedures.	II D (4)	

Figure E-7. Preliminary hazard-analysis work sheet (disarming M15)

Operation Name: Arming the M16-Series AP Mine
Date Prepared: 11 October 1990
Sheet Number: 1 of 1

Hazards	System Effects	Causal Factors	RAC	Actions	Controlled RAC	Standards/Comments
Damaged mine used	None	Lack of training/improper training Improper supervision Incomplete or no inspection	IV C (-)	Do not use mine. Contact QASAS to investigate. Turn in the mine for investigation/ disposal.	None	Proper training is conducted. Soldiers are proficient on inert mines first. One instructor is assigned to one soldier.
Faulty fuse used (M605)	Mine detonates	Safety pins missing Prongs bent Fuse head does not turn freely Pins installed incorrectly	I B (1)	Do not use fuse. Contact QASAS to investigate. Turn mine in for investigation/ disposal. Conduct quality control check before training.	I D (3)	All training is conducted in the prone position. Mine training is done after the need is identified by a unit's METL and then only with select personnel. Training is conducted in the proper environment. Soldiers are in the proper protective gear.
Pressure applied to prongs after the positive safety pin is removed	Mine detonates	Pressure is applied inadvertently	I B (1)	Do not remove the positive safety pin at any time.	I D (3)	No trip wires or live AHDs are used with an M16 mine.
Fuse improperly armed	Mine detonates	Individual pushes in on release pin while or after positive safety pin is removed Wrong sequence for removing pins Locking safety pin is removed, click is heard, soldier continues	I B (1)	Do not remove the positive safety pin at any time.	I D (3)	In the event of damage or loss of any safety pins, stop training with this particular mine and destroy it.

Figure E-8. Preliminary hazard-analysis work sheet (arming M16)

Operation Name: Disarming the M16-Series AP Mine
Date Prepared: 11 October 1990
Sheet Number: 1 of 1

Hazards	System Effects	Causal Factors	RAC	Actions	Controlled RAC	Standards/Comments
Camouflage removed	Mine detonates	Too much pressure on M605 fuse pressure prongs	I C (2)	Provide close supervision. Do not rush the procedure. Do not camouflage mine.	I D (3)	Do not fully arm this mine; leave the positive safety pin in the fuse at all times. See comments for arming an M16 AP mine (Figure E-10, page E-15). See risk assessment for command-detonating a mine in place with explosives (Figure E-16, page E-21).
Replace safety pins	Mine detonates	Pins not replaced Pull on release-pin ring Pressure on fuse pressure prongs Pins replaced in wrong order	I C (2)	Provide close supervision. Do not rush the procedure.	I D (3)	
M605 fuse removed	Mine detonates	Safety pins not properly in place, thus allowing pressure to be applied to the fuse, pressure prongs, or release pin	I C (2)	Provide close supervision. Assure that pins are secure and replaced correctly.	I D (3)	

Figure E-9. Preliminary hazard-analysis work sheet (disarming M16)

Operation Name: Arming the M19 AT Mine
Date Prepared: 11 October 1990
Sheet Number: 1 of 1

Hazards	System Effects	Causal Factors	RAC	Actions	Controlled RAC	Standards/Comments
Damaged mine used	None	Lack of training/improper training Improper supervision Incomplete or no inspection	IV B (-)	Do not use mine. Contact QASAS to investigate.	IV B (-)	Proper training is conducted. Soldiers are proficient on inert mines first. One instructor is assigned to one soldier.
Improper handling of detonator	Detonator detonates	Soldier handles detonator improperly Lack of training/improper training Improper supervision Incomplete or no inspection	I D (4)	Handle detonator properly.	III D (5)	All training is conducted in the prone position. Mine training is done after identified by a unit's METL and only with select personnel.
Too much pressure on pressure plate	Mine detonates	Soldier applies too much pressure	I D (3)	See comments.	I D (3)	Training is conducted in the proper environment.
Weakened Belleville spring	Mine detonates at pressure lower than designed pressure	Belleville spring weakening	I C (2)	Inform soldiers of less pressure required to detonate.	I D (3)	Soldiers are in the proper protective gear. No trip wires or AHDs are used with a live M19 mine.

Figure E-10. Preliminary hazard-analysis work sheet (arming M19)

Operation Name: Disarming the M19 AT Mine
Date Prepared: 11 October 1990
Sheet Number: 1 of 1

Hazards	System Effects	Causal Factors	RAC	Actions	Controlled RAC	Standards/Comments
Camouflage removed	Mine detonates	Too much pressure on pressure plate Damaged or malfunctions	I D (3)	Do not camouflage mine.	I D (3)	See comments for arming an M19 AT mine (Figure E-12, page E-17). See risk assessment for command-detonating a mine in place with explosives (Figure E-16, page E-21).
Unable to turn the mine to SAFE	Mine detonates	Too much resistance turning the dial indicator Foreign material in the fuse well	I C (2)	Perform controlled detonation in place.	I D (3)	
Fuse assembly removed and detonator replaced with shipping plug	Detonator detonates	Dropped fuse assembly with detonator Dropped the detonator	II C (3)	Handle fuse assembly and detonator properly.	II D (4)	

Figure E-11. Preliminary hazard-analysis work sheet (disarming M19)

Operation Name: Arming the M21 AT Mine
Date Prepared: 11 October 1990
Sheet Number: 1 of 1

Hazards	System Effects	Causal Factors	RAC	Actions	Controlled RAC	Standards/Comments
Damaged mine used	None	Lack of training/improper training Improper supervision Incomplete or no inspection	IV B (-)	Do not use mine. Contact QASAS to investigate. Turn mine in for investigation/ disposal.	IV B (-)	Proper training is conducted. Soldiers are proficient on inert mines first. One instructor is assigned to one soldier. All training is conducted in the prone position. Mine training is done after identified by a unit's METL and only with select personnel. Training is conducted in the proper environment. Soldiers are in the proper protective gear. No trip wires or AHDs are used with a live M21 mine.
Too much force on extension rod/ pressure ring (M607 fuse)	Mine detonates	Soldier bends/pushes on rod or fuse	I B (1)	Do not remove the safety stop and band if the extension rod is used. **OR** Remove the extension rod before removing the safety stop and band.	I D (3)	
Fuse improperly seated in the fuse well	Mine detonates	Excessive debris/corrosion in well Lack of training/improper training Improper supervision Incomplete or no inspection	I C (2)	See comments.	I D (3)	
Improper handling of fuse	Fuse detonates	Pressure applied to fuse while mine is armed	II D (4)	Handle fuse properly.	None	

Figure E-12. Preliminary hazard-analysis work sheet (arming M21)

Operation Name: Disarming the M21 AT Mine
Date Prepared: 11 October 1990
Sheet Number: 1 of 1

Hazards	System Effects	Causal Factors	RAC	Actions	Controlled RAC	Standards/Comments
Camouflage removed	Mine detonates	Too much pressure Tilted extension rod Damaged or malfunctions	I B (1)	Do not use extension rod. Ensure close supervision.	I D (3)	See comment for arming an M21 AT mine (Figure E-14, page E-19).
Band, stop, and cotter pin replaced	Mine detonates	Too much pressure Tilted extension rod Damaged stop, band, or cotter pin	I B (1)	Do not use extension rod. Ensure close supervision.	I C (2)	Stop training with this particular mine and destroy it in the event of damage or loss of stop, band, or cotter pin.
Extension rod and/or fuse removed	Mine detonates	Too much pressure Tilted extension rod	I C (2)	Do not use extension rod. Ensure proper placement of band, stop, and cotter pin.	I D (3)	See risk assessment for command-detonating a mine in place with explosives (Figure E-16, page E-21).
Booster removed	Booster detonates	Soldier improperly handles booster	II D (4)	Handle booster properly.	None	

Figure E-13. Preliminary hazard-analysis work sheet (disarming M21)

Operation Name: Preparing charges and priming explosives to be used for mine demolition (detonated in place)
Date Prepared: 11 October 1990
Sheet Number: 1 of 1

Hazards	System Effects	Causal Factors	RAC	Actions	Controlled RAC	Standards/Comments
Defective wire or faulty test of wire	No detonation	Wire will not carry electrical charge to detonate explosive	IV D (-)	Do not use wire (replace it).	None	After testing, ensure that wire ends are twisted together. Only one demolitions person and one safety person will be at the mine when the blasting cap is inserted into the demolition. RSO is responsible for clearing misfires. RSO keeps a *misfire kit* under his control.
Electric blasting cap check done incorrectly or not at all	Blasting cap detonates	Cap improperly handled Cap not put under sandbag	III C (4)	Handle cap properly. Place cap under sandbag.	IV D (-)	
Blasting cap attached incorrectly to firing wire	Blasting cap detonates	Cap/firing wire improperly handled End of firing wire not shunted	III C (4)	Ensure that firing wire is shunted. Ensure proper handling of blasting cap.	IV D (-)	
Blasting cap inserted incorrectly/too forcefully into 1-pound demolition charge	Blasting cap and/or demolition charge detonates	Too much pressure on blasting cap Cap improperly handled	I D (3)	Ensure proper handling of blasting cap.	I D (3)	
Charge does not detonate	No detonation	Faulty blasting cap Faulty blasting machine Faulty firing wire	IV C (-)	OIC declares a misfire. Follow RSO procedures for a misfire.	IV C (-)	

Figure E-14. Preliminary hazard-analysis work sheet (command detonation)

Operation Name: Mine arming/disarming peripheral factors (not all-inclusive)
Date Prepared: 11 October 1990
Sheet Number: 1 of 2

Hazards	System Effects	Causal Factors	RAC	Actions	Controlled RAC	Standards/Comments
Good weather	No direct	Temperature between 45 and 70°F Clear or partly cloudy Wind speed less than 5 mph	IV A (-)	None	None	This list is not conclusive and will depend on the particulars of the unit, the training, and the range facilities. Additional peripheral factors a unit commander may wish to consider are— • Level of proficiency. • Time of event (day or night). • Availability and extent of emergency-response assets. • Train up (rehearsals and dry runs). • Terrain. • Location of instructor in relation to soldier during the training.
Minimal weather	No direct	Temperature between 32 and 44°F or 71 and 80°F Drizzle Wind speed between 5 and 15 mph	III A (2)	Ensure close supervision. Do not rush training to get in shelter. Postpone or cancel training.	III C (4)	
Bad weather	Wet, miserable soldiers Wet munitions	Temperature less than 32°F or greater than 80°F Moderate to heavy rain Snow or ice Wind speed greater than 15 mph	I C (2)	Postpone or cancel training.	I D (3)	
Lack of references	The sequential arming or disarming not done correctly, causing the fuse and/or mine to detonate	Current doctrine not available Questions or confirmation of techniques not quantified	I B (1)	Stop training if unsure of a task. Have proper references available.	I D (3)	

Figure E-15. Preliminary hazard-analysis work sheet (peripheral factors)

Operation Name: Mine arming/disarming peripheral factors (not all-inclusive)
Date Prepared: 11 October 1990
Sheet Number: 2 of 2

Hazards	System Effects	Causal Factors	RAC	Actions	Controlled RAC	Standards/Comments
Lack of support materials or components	Improper procedures or practices introduced, causing the fuse and/or the mine to detonate	Not enough or incorrect fuses, wrenches, and/or washers used	I C (2)	Ensure that proper support material is available. Stop training if the proper material is not available.	I D (3)	
Condition of soldiers	Improper procedures or practices introduced, causing the fuse and/or the mine to detonate	Long amount of training with little or no sleep Strenuous training conducted before or during live-mine training	I D (3)	Ensure that soldiers have adequate sleep. Allow for breaks during training. Stop training if soldiers appear to be heavily fatigued.	None	
Protective clothing	Improper or lacking protective clothing, increasing the severity of an accident if it does occur	Improper or lacking protective clothing	I C (2)	All soldiers participating in mine training will have Kevlar helmets, flak vests, and boots.	None	

Figure E-15. Preliminary hazard-analysis work sheet (peripheral factors) (continued)

Appendix F

Mine Awareness

Mine awareness should actually be entitled mine/UXO awareness. If only mines are emphasized, ordnance (bomblets, submunitions) may be overlooked, and it has equal if not greater killing potential. The main objective of mine awareness is to save lives, so it is important to ensure that soldiers are well-informed and thoroughly trained. This appendix outlines the tasks needed for soldiers and units to survive in a mined/UXO environment.

Mine awareness should be emphasized at all levels of command, and it involves soldier and leader skills. Soldier skills are a mix of individual and collective tasks that are required for an element to maintain its combat effectiveness in and around a mined environment. It is important to note that a soldier's basic mine-awareness skills are critical to his and the unit's survival. Leader skills involve planning missions, assessing situations, and tracking/disseminating mine information. A unit must be proficient in all mine-awareness skills to effectively operate in a mined environment.

SOLDIER

Soldier skills involve individual and collective tasks that are required for basic survival in a mined/UXO environment. They include minefield indicators, probing techniques, mine-detector operation, extraction drills, survival rules, casualty treatment, and evacuation drills.

Visual Indicators

Mine/UXO indicators are part of all combat operations. Understanding and recognizing mine indicators could determine whether or not a soldier becomes a casualty. The following may indicate the presence of mines/UXO:

- Trip wires.
- Signs of road repair (such as new fill or paving, road patches, ditching, culvert work).
- Signs placed on trees, posts, or stakes. Threat forces mark their minefields to protect their own forces.
- Dead animals.
- Damaged vehicles.
- Disturbances in previous tire tracks or tracks that stop unexplainably.
- Wires leading away from the side of the road. They may be firing wires that are partially buried.

- Odd features in the ground or patterns that are not present in nature. Plant growth may wilt or change color, rain may wash away some of the cover, the cover may sink or crack around the edges, or the material covering the mines may look like mounds of dirt.
- Civilians. They may know where mines or booby traps are located in the residential area. Civilians staying away from certain places or out of certain buildings are good indications of the presence of mines or booby traps. Question civilians to determine the exact locations.
- Pieces of wood or other debris on a road. They may be indicative of pressure or pressure-release FDs. These devices may be on the surface or partially buried.
- Patterns of objects that could be used as a sighting line. The enemy can use mines that are fired by command, so road shoulders and areas close to the objects should be searched.

PROBING

Probing is very time-consuming and is used primarily for clearing operations, self-extraction, and covert breaching operations. Detection of mines by visual or electronic methods should be confirmed by probing. Use the following procedures and techniques when probing for mines:

- Roll up your sleeves and remove your jewelry to increase sensitivity. Wear a Kevlar helmet, with the chin strap buckled, and a protective fragmentation vest.
- Stay close to the ground and move in a prone position to reduce the effects of an accidental blast. When moving into a prone position, the prober should—
  - Squat down without touching his knees to the ground.
  - Scan forward up to 2 meters and to the sides up to 3 meters for mine indicators.
  - Probe the area around his feet and as far forward as possible.
  - Kneel on the ground after the area is found to be clear and continue probing forward until he is in a prone position.
- Use sight and touch to detect trip wires, fuses, and pressure prongs.
- Use a slender, nonmetallic object as a probe.
- Probe every 5 centimeters across a 1-meter front.
- Gently push the probe into the ground at an angle that is less than 45 degrees.

DANGER
Use extreme caution when probing. If the probe is pushed straight down, its tip may detonate a pressure fuse.

- Apply just enough pressure on the probe to sink it slowly into the ground.

- If the probe encounters resistance and does not go into the ground freely, carefully pick the soil away with the tip of the probe and remove the loose dirt by hand. Care must be taken to prevent functioning the mine.
- When a solid object is touched, stop probing and use two fingers from each hand to carefully remove the surrounding soil and identify the object.
- If the object is a mine, remove enough soil to show the mine type and mark its location. Do not attempt to remove or disarm the mine. Use explosives to destroy detected mines in place or use a grappling hook and rope to cause mines to self-detonate. Metal grappling hooks should not be used on magnetic-fused mines.

Probing is extremely stressful and tedious. The senior leader must set a limit to the time a prober is actually probing in the minefield. To determine a reasonable time, the leader must consider METT-TC factors, weather conditions, the threat level, the unit's stress level, and the prober's fatigue level and state of mind. As a rule, 20 to 30 minutes is the maximum amount of time that an individual can probe effectively.

AN/PSS-12 Metallic Mine Detector

The AN/PSS-12 mine detector (Figure F-1, page F-4) is a man-portable metallic mine-detection system that is used to detect AT and AP land mines. Its search head contains two concentric coils—the transmitting (outer) coil and the receiving (inner) coil. During operation, the transmitting coil emits a magnetic field symmetrically above and below the search head. When this field encounters metal objects, it induces currents in these objects. When the induced currents reach the receiving coil, an auditory signal is provided to the operator. The detector is able to detect large metal mines from a distance of several feet, but can detect low-metal mines from only a few inches.

WARNING
It is important to understand that the mine detector is only effective when there is a sufficient amount of metal in the mine.

Unpacking

The system is stored and transported in a single carrying case.

- Open the pressure-relief valve in the carrying case.
- Release the latches on the carrying case and open the top.
- Remove the bag that contains system components.
- Unzip the bag and ensure that all components are present (Figure F-2, page F-4).
- Remove the following items from the bag carefully:
  - — Telescopic pole and search-head assembly with cable and plug.
  - — Electronic unit.

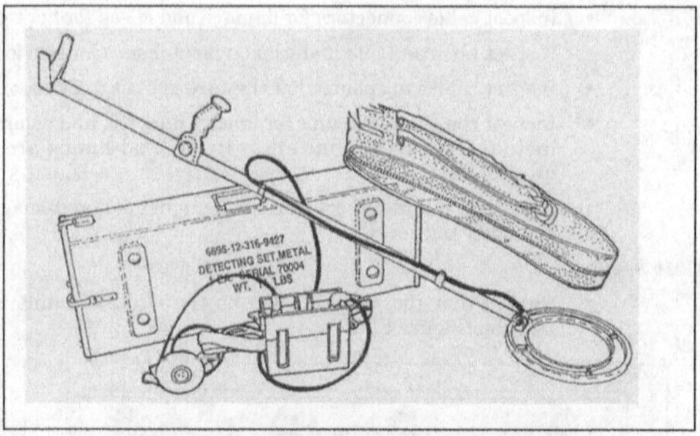

Figure F-1. AN/PSS-12 metallic mine detector

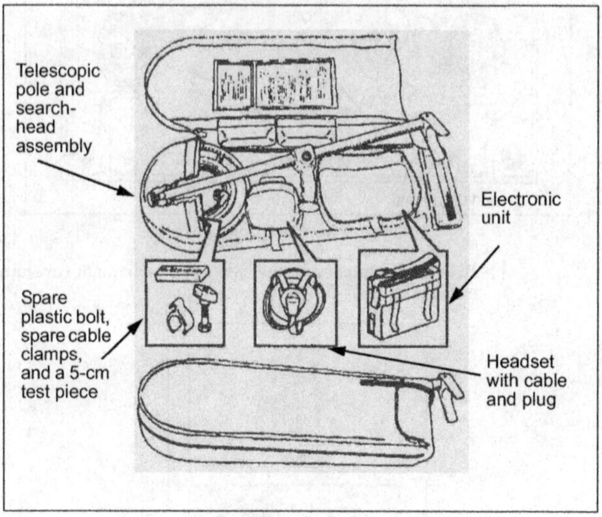

Figure F-2. AN/PSS-12 packed components

- — Headset with cable and plug.
- Ensure that the bag contains the following spare parts and test items:
  - — Spare plastic bolt.
  - — Spare cable clamps.
  - — 5-centimeter test piece.
- Inspect the search head for cracks or damage.

- Inspect cable connectors for damage and check for bent pins.
- Inspect hook and pile material on headphones for serviceability.
- Inspect cables to ensure that they are not cut, broken, or frayed.
- Inspect the electronic unit for cracks, damage, and completeness. This includes ensuring that all switches and knobs are present and functional.
- Ensure that telescopic pole is not bent, dented, or damaged and can be extended and retracted.

Electronic-Unit Setup

- Ensure that the power switch on the electronic unit is in the *OFF* position (Figure F-3).

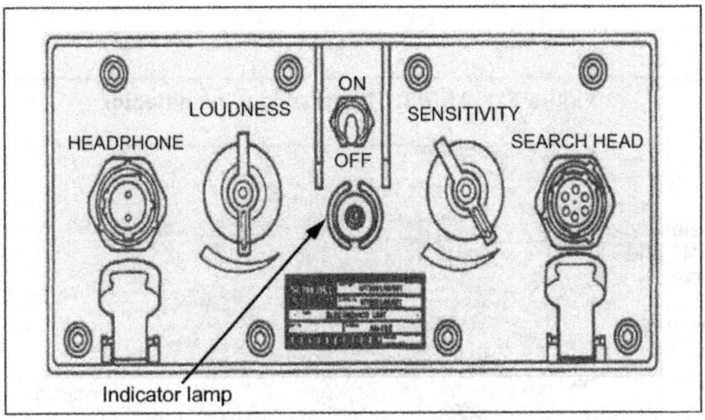

Figure F-3. Electronic unit

- Release the latches on the battery-compartment cover and remove the cover (Figure F-4).

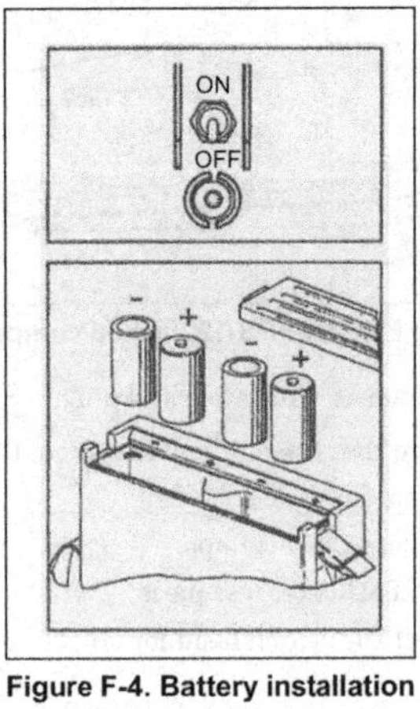

Figure F-4. Battery installation

- Insert batteries according to markings.

> **WARNING**
> **Ensure that the battery cover is completely closed and the latches are in the proper position. This prevents the inadvertent opening of the battery compartment during operation. Failure to do this could result in injury, damage to the equipment, and/or improper operation.**

- Reinstall and latch the battery cover. Ensure that the latches are in the proper closed position.
- Press in the catch (located just below the arm support) to unlock the telescopic pole from the transport position. Turn the outer tube until the catch snaps into the guide groove. Pull the telescopic pole out to a suitable length. Ensure that the telescopic pole is locked in one of three fixed positions.
- Connect the headphones and the detector-head cables to the electronic unit. Fit the protective caps of the cable connectors to the corresponding rubber caps on the electronic unit.
- Put on the headphones. Place the straps of the headphones over the forehead and on top of the head. The hook and pile material goes behind the head at the nape of the neck.
- Connect the detector-head cable to the lower (plastic) portion of the telescopic pole.
  - — Attach the lower cable clamp 5 centimeters above the wing nut.
  - — Attach the second cable clamp 5 centimeters down from the junction of the lower and upper (aluminum) portion of the telescopic pole.
  - — Attach the third cable clamp between the other two cable clamps. Do not attach the cable to this middle clamp. This clamp will become the sensitivity marker clamp.

> **WARNING**
> **The cable should not be attached to the aluminum portion of the pole, because it causes the sensitivity of the mine detector to fluctuate.**

- Adjust the position of the handle by loosening the knurl nut. Retighten the nut when the handle is properly positioned.
- Adjust the position of the search head so that it will be parallel to the ground.
- Attach the electronic unit to the operator's load-bearing equipment (LBE) belt using the belt clips. Right-handed soldiers should put the electronic unit on the left hip. Left-handed soldiers should put the electronic unit on the right hip. Put the safety strap over the shoulder.

Electronic-Unit Controls and Indicators

The AN/PSS-12 electronic unit has the following controls and indicators (Figure F-3, page F-5):

- On/off switch. This switch is the power control for the AN/PSS-12. It must be in the ON position when detection is being conducted.
- Loudness control. This is the volume control for the audible alarm. It should be adjusted to a comfortable listening level during use.
- Sensitivity control. The sensitivity control is used to adjust the detection characteristics of the mine-detector head. Details on how to set the sensitivity are provided below.
- Indicator lamp. This lamp indicates low battery voltage or a system malfunction. It provides a short flash when the unit is turned on and flashes continuously when the batteries are low or there is a malfunction.
- Audible signals.
  - The system provides an audible signal through the headset when the search head is over or very near a metal object. The signal is a continuous tone that at first resembles a growl. The tone increases to a higher pitch as the search head is moved closer to a metal source (such as a mine) and decreases in pitch when moved away.
  - The AN/PSS-12 has a check tone that is provided every 1 to 2 seconds. The tone resembles a clicking sound, and its purpose is to continuously inform the operator that the system is functioning satisfactorily. If the check tone disappears or its frequency decreases, discontinue searching and determine the malfunction.
  - A flashing indicator lamp and a continuous tone (the same tone heard with a target return, except it does not abate when moved away) on the headphone indicates low battery voltage. If the unit's indicator lamp flashes, change the batteries and readjust the unit.

Operation

Once the system is assembled, it is ready for adjustments prior to operation. Remove rings, watches, and other jewelry before adjusting or using the system.

Power Up/Sensitivity Adjustments

- Turn the sensitivity and loudness knobs completely counterclockwise.
- Look at the indicator lamp on the electronic unit and turn the on/off switch to the ON position. The lamp should give a short flash. If it does not, ensure that the batteries are correctly installed or insert new batteries. If the lamp flashes continuously, the battery voltage may be low.
- Hold the search head approximately 0.6 meter above the ground, and turn the sensitivity knob clockwise until a continuous tone is heard.

While this is being done, adjust the loudness control to a comfortable listening level.

- Turn the sensitivity knob slightly counterclockwise until the tone ceases. The check tone (a clicking sound) should still be heard every 1 to 2 seconds. Readjust the loudness control if necessary. If the check tone disappears or its frequency decreases, discontinue searching and determine the malfunction.
- Set the sensitivity in either of the following two ways:
  - The preferred method is to use the most difficult to detect type of mine (made safe without changing the metal content) that is expected to be encountered.
  - In the absence an actual mine, use the 5-centimeter test piece that comes with the AN/PSS-12 packed components.
- Bury the mine (if you have a mine) at the deepest depth that threat mines are expected to be found. If you do not have a mine, bury the 5-centimeter test piece vertically so that the metal portion is at the deepest depth which threat mines are expected to be found. In the absence of more specific information, bury the test piece at a depth of 5 centimeters). This places the top of the test piece flush with the ground's surface.
- Place the detector head lightly on the ground directly above the mine/test piece and adjust the sensitivity knob so that the mine is detected and you have an easily heard signal from the headset. Now, move the detector away from the mine/test piece until the signal ceases to be heard.
- Float the detector head lightly on the ground at a 0.3-meter-per-second movement rate across the mine/test piece. With the detector head in motion, check to see if the signal can again be easily heard. If not, increase the sensitivity until the signal can be easily heard.

NOTE: The higher the sensitivity is set (between just being able to detect the mine/test piece and where the signal from the soil first becomes audible) the less the chance of missing a mine, but the more the chance of a false alarm rate.

WARNING

In some conditions, you may not be able to detect a target mine/test piece at the deepest depth the threat mine is expected to be encountered. Do not use the AN/PSS-12 to detect mines in those conditions.

NOTE: Adjusting the sensitivity with the mine/test piece buried in soil that is similar to the soil where the detector will be used is very important. Otherwise, the sensitivity setting is simply a guess, because different soil types and moisture content influence the sensitivity of the detector.

Sensitivity Level

The AN/PSS-12's sensitivity setting drifts over time. During use, soldiers should check the detector's sensitivity every 1 to 2 meters of forward advance in the mine lane.

- With the sensitivity properly set using a buried mine/test piece, the detector is moved to an adjacent place on the ground over a clear area.
- To check the detector, a metal object (bayonet or other metal object) is slid down the lower (plastic) portion of the telescopic pole until the operator hears the same auditory signal that was emitted when the search head was placed over the buried mine/test piece.
- At this point on the lower plastic shaft, the middle plastic cable clamp (the middle clamp previously attached without the cable) is positioned. This clamp now becomes the sensitivity marker clamp.
- At approximately every 1 to 2 meters of forward advance in the mine lane (or more often if desired), the same metal object (bayonet) is moved down the shaft to the sensitivity marker clamp and the operator listens to determine if the same auditory signal is emitted. When the same auditory signal is not heard, the sensitivity knob is adjusted up or down until the same auditory signal is replicated and maintained.

NOTE: It is important that the operator maintain a consistent body posture and angle between the telescopic pole and the search head each time a sensitivity check is made.

DANGER
Low batteries may reduce detector performance well before the indicator light comes on. If you need to frequently adjust the sensitivity knob to maintain a constant sensitivity setting or if the indicator light comes on, discard all batteries and replace them with new ones. Failure to do this could result in injury or death.

DANGER
Discontinue searching and readjust the sensitivity if the check tone disappears or its frequency decreases. Failure to do this could result in injury or death.

Search Methods

- Sweep across the lane, keeping the entire search head in light contact with the ground. Sweeping speed should be approximately 0.3 meter per second. Each sweep must be a minimum of 1.5 meters and must overlap the edge of the lane by at least one-half the width of the search head. Each sweep should advance forward at no more than 6 inches per sweep. The sweep and advance must be maintained as described, otherwise gaps will be left between sweep paths or at the edge of the lane and a low-metal mine can be missed.

- Perform a sensitivity check at approximately every 1 to 2 meters of forward advance in the mine lane (or more often if desired) and make adjustments as necessary. It is important that the operator maintain a consistent body posture and angle between the telescopic pole and the search head each time a sensitivity check is made.
- Pat the search head lightly on the ground (each pat advancing no more than one-half the width of the search head) in low vegetation or rocky or uneven ground where smooth sweeping cannot be performed. Ensure that the wing nut is tight, so that the position between the head and telescopic pole is fixed.

NOTE: Many conditions of vegetation may not allow getting the search head on the ground. Do not push through vegetation to get the search head on the ground. Different environments will require slightly different techniques. If you cannot get the search head on or very near the surface of the ground, the detector will not detect low-metal mines.

NOTE: The AN-PSS/12 detector's performance is reduced when the cable between the electronic unit and the top cable clamp is permitted to hang unrestrained during sweeps. The soldier can control this problem by holding the cable in his free hand.

WARNING
Always be aware of the potential for booby traps. The search head should never be swept where its path cannot be visually cleared. If trip wires are a threat, other techniques to detect trip wires should be used before sweeping.

Alarm Investigation/Mine Identification

When an auditory signal is heard, the soldier must investigate further to determine if the signal is a reliable indication of a mine.

- End the sweep procedure and begin the investigation procedure at the first auditory indication from the detector of metal in the ground. The purpose of the investigation is to determine if the signal is repeatable and, therefore, a likely indication of a mine. If so, the investigation continues with the purpose of gaining more information concerning the size, type (high-metal or low-metal mine), and specific location of the signal source.

NOTE: In some cases, the detector may provide a weak signal that cannot be repeated during the initial investigation process. Do not give up on the reliability of the signal until the sensitivity of the detector has been checked to ensure that it has not drifted to the low side.

NOTE: A footprint is defined as the entire area on the ground where a mine or metallic source causes the detector to generate an auditory signal. Small footprints, often as small as 10 to 15 centimeters in diameter, will indicate low-metal mines. Footprints as large as 60 to 122 centimeters in diameter may indicate high-metal mines.

- Note the location on the ground (at the center of the search head) where the signal is first heard. Then attempt to develop a set of points on the ground that identifies the source footprint.
  - Develop this footprint by first moving the search head away from the signal source until no signal is heard and then sliding the search head toward the signal from several clock positions.
  - Note the specific location of the center of the search head at each location where the auditory signal begins. Once the signal is detected, the detector is no longer slid toward the potential source to avoid coming closer than necessary to a potential mine.
  - Repeat this process and mark or note the locations on the ground until the size and shape of the footprint can be understood. Typically, the pattern will resemble a semicircle with the 6 o'clock position nearest the operator.

WARNING
The large footprints of high-metal mines may mask signals from low-metal mines within the footprint. Always assume that there are low-metal mines within the large footprint area.

- Identify the center of very large footprints (if important). The airborne technique may be quicker than the method previously discussed to identify the perimeter.
  - The search head must be adjusted so that it can be held above the ground with the head parallel to the surface.
  - The detector can then be manipulated above the source until the signal can be heard at only a single point. Move the search head in a crossing pattern, which should produce a smaller and smaller signal area as the pattern is repeated at gradually increasing heights. Near the surface, the signal can be heard over a broad lateral area, but as the search head is raised higher off the ground, this area becomes progressively smaller. As the search head is raised higher off the ground, the area where the signal can still be heard is reduced to a point.
  - The center of the mine or metal source should be directly below this point. Do not ignore the potential for low-metal mines to be hidden in this large footprint on the ground.

WARNING
Always be aware of the potential for booby traps and UXO. The search head should never be swept where its path cannot be visually cleared. If trip wires are a threat, other techniques to detect trip wires should be used before sweeping.

Other Operational Actions

- Keep mine detectors at least 2 meters apart during setting, adjustment, and operational phases to prevent interference.
- Change the batteries immediately if the indicator lamp flashes, a constant audible tone is heard, or the check tone (clicking sound) stops and readjust the sensitivity. The search sensitivity is affected long before the lamp begins flashing, and the unit will be unusable until fresh batteries are installed.
- Ensure that only the lower part of the telescopic pole is used when the equipment is operated by a soldier in the prone position.
- Turn the unit off after completing the search operations.

Disassembly and Packing

- Ensure that the on/off switch on the electronic unit is in the *OFF* position.
- Disconnect the headphones from the electronic unit.
- Detach the cable connection on the electronic unit for the magnetic search head, and replace the protective caps on the plug and socket.
- Release the electronic unit's battery-cover latches, and remove the battery cover.
- Remove the batteries, and ensure that none of the battery cases have ruptured; if they have, notify your supervisor. Reinstall the battery cover and latch it.
- Remove the cable clips from the telescopic pole.
- Collapse the telescopic pole to its travel length, and turn its outer tube until it is locked by the catch. Loosen the plastic restraining bolt, and fold in the magnetic search head.
- Pack the components in the carry bag as shown in Figure F-2, page F-4. For long-term storage, do not put batteries in the carry bag. Close and zip the carry bag.
- Place the carry bag in the metal transport case, and latch the case. Close the pressure-relief valve.

As in probing, the senior leader must set a limit to the time an individual can use the mine detector. The time limit is determined by METT-TC factors, weather conditions, the threat level, the unit's stress level, and the individual's fatigue level and state of mind. As a rule, 20 to 30 minutes is the maximum amount of time an individual can use the detector effectively.

EVACUATION DRILLS

A well-developed, well-rehearsed evacuation drill is necessary to extract an individual or a unit from a mined area. Units must develop evacuation drills for dismounted and mounted operations. Each type of operation should include two drills—one using a mine detector (mounted extraction) and one without using a mine detector (dismounted extraction).

Mounted Extraction

- The convoy commander halts the convoy and reports to higher headquarters.
- No vehicles move and no troops dismount unless directed to do so.
- Elements provide 360-degree security from vehicles.
- Troops thrown from vehicles should not move. Personnel are extracted by using dismounted evacuation procedures if electronic detectors are not available.
- The senior leader, if engineers are not available, assesses the situation and directs vehicles to back up along the entry-route tracks. If an immediate threat exists, occupants of damaged vehicles evacuate out the rear of the vehicle and along the vehicle-entry tracks. If no immediate threat exists, occupants of damaged vehicles remain in the vehicle until it is extracted.
- Engineers, if available, sweep the area and provide a cleared path for movement. They—
  - Clear a lane that is wide enough for the towing vehicle.
  - Use all available tow cables to increase the distance before towing if an M88 is unavailable. Remember, an M88 has a wider track base than other tracked vehicles.
  - Ensure that all towing-shackle sets are complete and mounted.
  - Ensure that the towing vehicle has tow cables on the front and the rear if possible.
  - Ensure that rear cables are attached to the lower mounts. This allows the crew to recover the vehicle without touching the ground.
  - Pull the vehicle out at least two-vehicle lengths before switching to a tow bar. When towing a vehicle after a mine strike, the chance of fire is greater because of possible damage to the vehicle.
  - Provide first aid and conduct casualty evacuation or have medics provide treatment and medical evacuation.
  - Guide vehicles through the safe area.
  - Mark, record, and report the threat.
- Continue the mission.

Dismounted Extraction

- All personnel freeze and crouch into a low-silhouette position. Be cautious when making this movement to ensure that you do not detonate another mine. If a protective mask is worn on your hip, do not allow it to come in contact with the ground because contact may detonate a mine. Do not help casualties because you could also become a casualty.
- The leader designates a security element and a soldier to assist in casualty evacuation.
- Soldiers extract along the path they entered. If possible, they step in the same places as before; if impossible, they probe their way out.
- The security element, consisting of individuals who are not in the minefield, sets up security for the unit.
- The soldier extracting the casualty—
  - — Probes a 1-meter-wide path to the casualty and marks the cleared path with foot powder or marking tape as it is probed.
  - — Probes around the casualty to clear the area.
  - — Performs first aid.
  - — Carries the casualty out of the minefield along the cleared path. (Stretcher parties do not enter the area unless a 2-meter-wide path has been cleared to the casualty.)
- The unit marks the threat and assembles back at the rally point.
- The unit reports the incident when it is 50 to 100 meters away from the minefield. If soldiers are in the minefield and radio transmission is required, move the transmitter at least 300 meters from the minefield. This prevents accidental mine detonation from the radio signal.
- The unit provides first aid and conducts casualty evacuation or has medics provide treatment and medical evacuation.
- The unit continues the mission.

Extraction from Scatterable Mines

- The individual who discovers the mine initiates the alarm according to the unit SOP.
- Unit personnel at the command post receive the alarm and alert others.
- The unit TOC requests counterbattery fire if the mines are artillery-delivered.
- Vehicle commanders check the immediate area. They do not dismount. Inspect the vehicle for mines and/or trip wires. Note and record the location of all mines found on or around the vehicle.
- Personnel leave any vehicle that is touching or is blocked by AT mines in place until the remainder of the unit is out of the minefield.

- Unit leaders identify unmovable vehicles and designate one or more lanes for remaining personnel and vehicles to use when exiting the minefield.
- Leaders identify a clearance team to extract remaining vehicles and personnel. The clearance team—
  - — Uses visual means to locate mines and marks vehicle lanes at least 4 meters wide. Personnel mark lanes according to the tactical situation and the threat in the area; however, they mark them so that team personnel can reenter the minefield and recover equipment and vehicles.
  - — Destroys or removes all mines in lanes, using a grapnel hook or other means as necessary, in the sequence directed by the team leader. Personnel detonate unmovable mines to reduce personnel injury and equipment damage.
- Vehicle commanders direct personnel to ground-guide vehicles out of the minefield. Ground guides—
  - — Ensure that individual elements move only when directed by the chain of command.
  - — Place equipment that is not in contact with a mine or a trip wire onto vehicles.
  - — Direct vehicles to the designated exit lane or, if safer, allow vehicles to exit the minefield on their own.
- Clearance team personnel, aided by unit personnel, remove equipment and vehicles remaining after initial extraction from the minefield. They—
  - — Reenter the minefield using the same exit routes.
  - — Detonate the minimum number of mines necessary to remove vehicles and equipment from the minefield.
  - — Avoid touching mines. Personnel should take all possible precautions to ensure that mines are not jarred.
  - — Place sandbags near mines, if possible, to minimize damage to vehicles and equipment.
  - — Remove mines from equipment by using a line or other remote means. Ensure that the entire team is far enough from the mine to avoid casualties if it explodes.
  - — Place explosive charges to minimize vehicle damage when detonating mines on the ground.
- Clearance team personnel—
  - — Clear sufficient mines to allow for mission accomplishment if the position cannot be evacuated.
  - — Clear and mark communication lanes between positions.
- Continue the mission.

Survival Rules

Many of our allies have extensive experience in mine operations. Among them is Canada. They have produced several manuals and videotapes on mine awareness and have developed the following *survival rules*. They are very practical and can be applied to our soldiers as well.

- *If you did not drop it, do not pick it up.*
- *All terrain and structures are potentially mined or booby-trapped.*
- *Beware of areas associated with basic human needs. They could be mined or booby-trapped.*
- *Immediately report all confirmed or suspected mines.*
- *Leave mine disposal to the EOD personnel and combat engineers.*
- *Avoid touching or removing foreign objects, no matter how attractive. They could be mined or booby-trapped.*
- *Avoid verges because they could be mined. Stay on the traveled road.*
- *Mark and avoid UXOs if possible. Consider them unstable.*
- *Develop and rehearse effective evacuation drills.*

A convoy provides better protection against mine and UXO threat than a single vehicle. In convoy movement, some rules of thumb should be followed:

- The lead vehicle should be one of the heaviest (2½-ton, 5-ton) vehicles in the unit and be hardened against a mine threat.
- A high-mobility, multipurpose, wheeled vehicle (HMMWV) or a HEMMT should not lead a convoy unless absolutely necessary. These vehicles are extremely vulnerable to mine and UXO threat and are difficult to harden without commercial products.

Casualty Treatment and Evacuation

Casualty treatment and evacuation should be a part of every mission order. Incorporate the following elements into the unit SOP, and ensure that soldiers understand them. Emphasize that rushing to help a mine victim can lead to the rescuer becoming a casualty.

- Reassure the casualty.
- Do not panic and create another casualty.
- Call higher headquarters at the earliest opportunity, and request a specialist engineer and medical help as soon as possible.
- Extract yourself and mark the path as you go.
- Reenter along the marked path.
- Clear a path to the casualty.
- Administer first aid.
- Recover the casualty.

- Mark the area after it is clear, record it on the map, and report to higher headquarters.

Medical personnel and combat lifesavers should anticipate and train for the following injuries:

- Blast injuries with fragments embedded.
- Burns.
- Fractures.
- Traumatic amputations.
- Blunt trauma.
- Psychological anxiety reaction.

LEADER

Leader skills in effective planning, risk management, and mine-data tracking/dissemination can greatly enhance force protection and reduce casualties in a mined environment. Unit leaders must train themselves and their subordinates on mine-awareness tasks.

RISK MANAGEMENT

Risk management reduces the frequency of mine and UXO strikes and diminishes the physical effects when they do occur. It is used to identify hazards, define risks, identify methods for control, and identify responsibilities for implementation. The risk-management process enables commanders and staffs to define acceptable risk levels and implement controls until risks are commensurate with the mission. Risk management is a simple, five-step process:

- Identify hazards.
- Assess the risk of each hazard.
- Make risk decisions and develop controls.
- Implement controls.
- Supervise.

Identify Hazards

This is often the most difficult part of risk management. A mine hazard is the condition that results from the interactions of a mine, a catalyst (such as activation from a soldier or vehicle), and a common spatial relationship. These hazards are defined in terms of mine types (AT, AP, fuses, metal content, AHD), how a soldier might encounter mines (dismounted or mounted and the type of vehicle), and the locations where encounters would be most likely. Mines are seldom employed in isolation, so complete hazard definition includes the complex obstacle and the covering fires.

Assess the Risk of Each Hazard

This requires determining the probability of a mine strike and the effects of the strike. An effective risk assessment is critical for evaluating the combat effectiveness of a unit in a mined environment. Risk-assessment criteria is developed by using Table F-1. A sample risk assessment is shown in Table F-2, page F-19.

Table F-1. Risk-assessment criteria

			Hazard Probability				
			Frequent (A)	**Probable (B)**	**Occasional (C)**	**Remote (D)**	**Improbable (E)**
	Single Item		Likely to occur frequently	Will occur several times during life of item	Likely to occur sometime during life of item	Unlikely, but possible to occur during life of item	So unlikoly, can assume occurrence will not take place
	Fleet/Inventory		Occurs continuously	Will occur frequently	Will occur several times	Unlikely but reasonably expected to occur	Unlikely to occur, but possible
Hazard Severity	CATASTROPHIC (death or system loss)	I	High	High	High	High	Medium
Hazard Severity	CRITICAL (severe injury or occupational illness or major damage)	II	High	High	High	Medium	Low
Hazard Severity	MARGINAL (minor injury or occupational illness or minor damage)	III	High	Medium	Medium	Low	Low
Hazard Severity	NEGLIGIBLE (less than minor injury, illness, or damage)	IV	Medium	Low	Low	Low	Low

Make Risk Decisions and Develop Controls

This step requires decision makers to identify actions that can reduce the probability and/or severity to acceptable levels. This may be accomplished by taking actions to reduce the probability of a mine strike or by providing more protection to the soldier or materiel to reduce severity of a mine strike. Often, it is a combination of the two.

Implement Controls

Leaders must apply the identified controls to reduce the probability and severity of a mine strike.

Table F-2. Sample risk assessment

DIRECTIONS: Circle the appropriate number in each section and total them. Appropriate directions are at the end of the assessment block.

TRAINING (Circle One)
1. Experience working in mined areas
2. No experience working in mined areas, but trained
3. No experience or training working in or around mined areas

TYPE OF AREA (Circle One)
1. Area known by friendly forces to be clear of mines
2. Old confrontation line or suspected mined area
3. Area known to be mined

LIGHT AND WEATHER (Circle One)
1. Daylight with clear weather
2. Daylight with poor weather
3. Darkness

MISSION (Circle One)
1. One that your soldiers have done before
2. One that your subordinate leaders have done before
3. An unfamiliar type of mission

ON OR OFF ROAD (Circle One)
1. Approved division/corps route
2. Paved road
3. Unpaved road
4. Cross-country

SLEEP (Circle One)
1. 6 hours of sleep in last 24
2. 4 hours of sleep in last 24
3. 2 hours of sleep in last 24
4. Awake for more than 24 hours.

TYPE OF VEHICLES IN CONVOY (Circle One)
1. Armored
2. Mixed, armored vehicles leading
3. Wheeled
4. On foot

AVAILABILITY OF MINE INFORMATION (Circle One)
1. Updated mined-area graphics in each vehicle
2. Updated mined-area graphics in the lead vehicle or a reliable, knowledgeable guide is available
3. No upgraded mined-area graphics are available

GROUND COVER (Circle One)
1. Dry, hard ground with short grass
2. Dry, hard ground with long grass or vegetation
3. Wet, muddy ground or snow less than 10 centimeters deep
4. Snow more than 10 centimeters deep

ROAD USE (Circle One)
1. Heavy tracked vehicles or trucks recently used the road
2. Light wheeled vehicles recently used the road
3. No traffic observed on the road; some tire marks
4. No traffic observed on the road; no tire marks

10-16 Continue the mission. Keep following training and common-sense rules that apply to working around mines.
17-24 Continue the mission. Remember to stress safety and mine awareness when you brief soldiers. Ensure that leaders maintain positive control of personnel.
25-35 Consider postponing the mission until better conditions are attained. If you must continue the mission, constantly stress mine awareness and safety. Ask higher headquarters for engineer support to accomplish the mission. Conduct mine-awareness training.

Supervise

This step ensures that controls are implemented and that a measure of quality control exists to ensure a quantified level of clearance.

The key to using risk management successfully is to employ it at each echelon—from the commander, through the tactical planner, to the soldiers executing the mission. Each level identifies hazards, eliminates or reduces hazards as feasible, and communicates the residual hazards to the next lower echelon. As such, each echelon works as a filter to control unacceptable risks.

- Training provides soldiers with an understanding of equipment limitations and plays a critical role in the risk-management process. Capabilities and limitations of Army systems are taken into consideration during the development of doctrine and TTP.
- Risk management at the tactical planning level requires a thorough knowledge and awareness of the hazards and potential controls that can be employed. The planning process requires a methodical and disciplined technique to identify the hazards and develop appropriate controls for operating in a mined environment. The controls for countermine operations, discussed in Chapters 9 and 11 and in FM 3-34.2, provide a framework for risk-managing hazards associated with mines.
- The execution level is the culminating point of risk management. It is where soldiers and leaders employ the systems provided to accomplish the mission. The amount of residual hazards remaining after the filtering process from echelons above may well determine success. The individual soldier is the last element to control any residual hazards.

Optimizing the components of risk management at the tactical planning level is more challenging as emerging technologically dependent systems bring more variables into the mission. While tactical intelligence is the key element in identifying mine-related hazards, technical knowledge is the key element in assessing the risks associated with mine hazards. This knowledge assimilates the tactical intelligence with the capabilities of the unit's equipment, the performance of threat mines, and the protection provided to our soldiers by their vehicles or personal protective equipment. The staff engineer, using his engineer C^2 system to risk-manage each COA, provides the maneuver commander and his staff with information on risks and potential controls early on in the planning process. Each subsequent commander must perform the same analysis and incorporate the mine threat into risk management.

RECORDING AND MINE-DATA TRACKING

Obtaining and disseminating information are the keys to battlefield management. Units encountering minefields or explosive devices should follow a five-step process—stop, secure, mark, report, and avoid. Units must provide adequate information to their higher headquarters to ensure that follow-on elements are well informed. Information must include known or suspected minefield locations, types of mines (if known), the marking method, the time encountered, and any additional information that may be of use to the clearing unit.

Division and maneuver brigade engineer planning cells must establish a central control cell for mine-clearance information. The central control cell—

- Maintains a current situation map and overlay that depicts friendly and enemy mines and obstacles.
- Maintains and updates information on minefield tracking and route status.
- Receives and maintains minefield recording forms within the unit's AO (includes host-nation minefield data if available).
- Maintains a database of mine information.
- Processes, analyzes, updates, and disseminates the information to subordinate commanders and staff.

MINE-INCIDENT REPORT

A mine incident includes any unplanned activity involving a mine, UXO, or booby trap. It also includes near misses that could have resulted in potential damage or injury. The mine-incident report is a technical report that follows a serious-incident report. The report should be submitted as soon as possible (the local SOP will indicate time requirements). A sample mine-incident report is shown in Figure 11-12, page 11-26.

TRAINING

Modern combat is complex, lethal, and demanding. Soldiers must be capable of performing their missions in any type of battlefield environment. Current doctrine and TTP provide soldiers with guidelines to accomplish their tasks and quality equipment provides the means. The common thread that connects doctrine, tactics, and equipment is quality training. To fight and win, units must train their soldiers to execute all wartime missions successfully. They must use every training opportunity to improve soldier, leader, and unit task performance. Without quality training, no amount of world-class equipment can make the soldier effective or make him survive in a wartime environment. Even the best doctrine in the world is worthless unless soldiers receive effective training. This is especially true with mine-awareness training.

Soldiers must be trained to think mine awareness as well as perform mine-related actions. Decisions, actions, and reactions must become automatic to every soldier. This requires that all soldiers receive mine-awareness training early in their careers. It must begin at early entry training with basic individual tasks and continue through advanced unit training with collective tasks. Soldiers who survive mine threats can survive mine warfare, but it requires continual training.

INDIVIDUAL TRAINING

Basic Mine-Awareness Training

This is the most important phase for preparing soldiers to survive in a mine environment. Individual soldiers must be trained to a level that meets the environment they will face on the modern battlefield. Basic mine-awareness tasks (mine detection, survival rules, minefield indicators, and procedures for self-extraction from a minefield) must begin early in the soldier's career.

With an estimated 80 to 100 million mines deployed worldwide, countermine considerations must become second nature to mission planners. From CSS operations in the rear area to close combat operations in the main battle area, there is no place on the battlefield that is safe from mines. This is especially true with the threat of SCATMINEs that are deployed from aircraft and indirect-fire sources. The need for initial countermine training will get more critical as the modern battlefield becomes more lethal.

Annual Mine-Awareness Training

Units exercising military skills at the National Training Center, Joint Reserve Training Center, or the Combat Maneuver Training Center can reinforce and evaluate their awareness training. Soldiers should be fully trained on the individual and collective tasks that are required to support their unit's mine-related missions before deploying to the combat training centers. An effective method to meet this challenge is a well-planned annual training program that trains soldiers on individual and collective mine-awareness tasks. The plan should address the requirements for initial collective-task training and sustainment training.

Predeployment Mine-Awareness Training

With an ever-changing global environment, the probability of projecting US forces to various geographical locations is very real. It is extremely critical that units be situationally aware. Each flash point in the world contains its own type of threat. Many of these threats can be neutralized or reduced through quality predeployment training. Quality predeployment mine-awareness training enhances the ability of the unit to perform its mission and increases the confidence of soldiers.

Units must train intensively when alerted for a deployment. Predeployment training is the time to polish and refine training and to focus on theater-specific operations. Predeployment training is intended to augment individual and collective training, not replace it.

In-Theater Mine-Awareness Training

Training does not end with predeployment training. Units must train in the country where they deploy. They need to plan and conduct routine sustainment training on individual and collective mine-awareness tasks. This is an essential ingredient of force-protection operations. In-theater mine-awareness training reinforces the soldier's existing skills and places mine survival in the forefront of each soldier's mind.

Leader Training

Officers and NCOs must receive mine-awareness training in basic and career courses. Commissioned and noncommissioned leaders must be trained on the use and the employment of the equipment that they will use in the field. Officer basic courses and officer and NCO advanced courses must include access to modeling/simulation training. Using modeling/simulation enables the leader to practice decision-making, employment, and sustainment operations in a simulated mine environment. This develops the situational awareness required for timely, accurate decisions.

UNIT TRAINING

The Army Training and Evaluation Program remains the most effective measure of individual and collective training effectiveness short of actual combat. It provides leaders with the opportunity to verify the effectiveness of virtual and actual training without endangering soldiers with mine effects. Units must include mine-threat scenarios in their home-station training exercises. Basic missions include minefield detection, reduction, marking, proofing, and recording. Commanders are required to assess their unit's state of proficiency in mine-awareness tasks on a routine basis.

Mine awareness is a critical, perishable skill. It must be trained effectively and sustained on a continuous basis. If a unit properly trains its soldiers on mine awareness, it will maintain its force, boost its soldiers' confidence in their abilities, and accomplish its mission more effectively. Mine awareness is not a liability; it is an investment in a unit's future.

Appendix G

Countermine Data

This appendix provides a compilation of countermine information to assist commanders in the decision-making process for countermine operations.

BREACHING ASSETS VERSUS THREAT OBSTACLES

Tables G-1 and G-2, pages G-2 through G-8, provide general guidance on the effect of breaching assets versus threat obstacles.

FOREIGN MINE DATA

Tables G-3 through G-7, pages G-9 through G-12, provide foreign mine characteristics.

FOREIGN MINEFIELD EMPLACEMENT DATA

Table G-8, page G-13, provides foreign minefield emplacement data. Information is presented in the probability of occurrence based on historical data.

FOREIGN MINE DELIVERY SYSTEMS

Table G-9, pages G-14 and G-15, contains characteristics of foreign mine delivery systems.

NOTE: Blank areas in the above tables indicate that the information is not applicable or that it was unavailable when this manual was published.

Table G-1. Mounted breaching assets versus threat obstacles

Asset	Track-Width AT Mine	Full-Width AT Mine	Side-Atk AT Mine	Cmd-Det Blast Mine	AT Ditch and Other Gaps	Wire Obstacles	Remarks
MCB with dog-bone assembly	Is more than 90 percent effective to 12 inches deep at speeds less than 10 kph (soil dependent)	Do not use against magnetic-influence mines; is more than 99 percent effective against tilt-rod mines	Do not use	Do not use	May be effective at filling in some shallow gaps or trenches	Do not use	The MCB provides almost 100 percent clearance of simple pressure-fused AT mines when used with the MICLIC. The MCB has minimal impact on the M1's maneuverability and speed when not plowing. The main gun should be traversed to the side to prevent damage in the event of a mine detonation when plowing. The reduction site should be relatively flat and free of large rocks and other obstructions. Plowing operations should begin about 100 m short of the suspected minefield and continue 100 m past its suspected end (typically 2 to 2 1/2 minutes to cover 300 to 400 m overall). Plowing operations should be performed in a straight line to keep the vehicle out of uncleared ground. Fifteen percent of pressure-fused AT-mine encounters across the MCB's width can be expected to detonate under the skid shoe, probably disabling the blade. The MCB is vulnerable to coupled mines and may be stopped by log-hurdle obstacles laid within a minefield as part of a complex obstacle. A second plow should not be used to clear the center of the lane because this will simply push mines into the area cleared by the first plow.

Table G-1. Mounted breaching assets versus threat obstacles (continued)

Asset	Track-Width AT Mine	Full-Width AT Mine	Side-Atk AT Mine	Cmd-Det Blast Mine	AT Ditch and Other Gaps	Wire Obstacles	Remarks
MCB with IDA	See above	Is effective against magnetic-influence mines; is more than 99 percent effective against tilt-rod mines	Do not use	Do not use	See above	Do not use	These AT mines are fitted with magnetic-influence fuses and AHDs may disable the blade. Dismounts should expect to encounter some pressure-fused AP mines in the breached lane that may be missed by the blade. The MCB clears a 180 inch-wide lane but leaves a 26 inch-wide area uncleared of simple pressure-fused mines and a 54 inch-wide area uncleared of double-impulse mines (under the skid shoes) in the middle of the breached lane. The berms and furrows left by the MCB aid in immediately marking the breached lane. Live mines remain in both berms and may roll back into the cleared lane. Over time, these may vibrate free and roll back into the breached lane.
MCR (including Panther) with dog-bone assembly	Is more than 90 percent effective to 12 inches deep at speeds less than 5 to 15 kph (soil dependent) Has reduced effectiveness against some (*mostly Italian*) blast-hardened mines Do not use against double-impulse mines	Do not use against magnetic-influence mines; is more than 99 percent effective against tilt-rod mines	Do not use	Do not use	Do not use	Do not use	Normally used to detect the leading edge of a conventionally laid tactical minefield and to proof lanes/routes cleared by other means. Use caution when crossing on an AVLB. May become stuck or ineffective in muddy conditions. The MCR is vulnerable to coupled mines. The MCR is designed to withstand up to two detonations per side. The MCR clears a 160 inch-wide lane but leaves a 72 inch-wide area uncleared in the middle of the breached lane. **NOTE: The M113 also has 72 inches between tracks and must use extreme caution when following the MCR.**

Table G-1. Mounted breaching assets versus threat obstacles (continued)

Asset	Track-Width AT Mine	Full-Width AT Mine	Side-Atk AT Mine	Cmd-Det Blast Mine	AT Ditch and Other Gaps	Wire Obstacles	Remarks
MCR with IDA	See above	Is effective against magnetic-influence mines; is more than 99 percent effective against tilt-rod mines	Do not use	Do not use	Do not use	Do not use	Multiple rollers are required to provide the overlap necessary for proofing the route's entire width when the MCR is used for proofing during route-clearance operations. Vehicle speed is limited to 5 to 15 kph. The main gun should be traversed to the side to prevent damage in the event of a mine detonation when rolling a suspected mined area. The IDA may be damaged by detonated mines.

Table G-2. Dismounted breaching assets versus threat obstacles

Asset	Pressure-Fused AP Mine	Trip-Wire/Break-Wire-Fused AP Mine	Influence-Fused AP Mine	Cmd-Det Blast Mine	Wire Obstacles	Remarks
APOBS	Effectively clears 95 percent, including blast-resistant types	Effectively cuts 99 percent of trip wires in its path	Is ineffective	Is ineffective	Clears a footpath	Provides a 0.6- x 45-m footpath from a 25-m standoff Can be deployed by two men in 2 minutes Consists of two manpack units weighing approximately 55 lbs each
Bangalore torpedo	Effectively clears 85 percent of simple pressure-fused mines; is less effective against blast-hardened mines	Effectively cuts 99 percent of trip wires in its path	Is ineffective	Is ineffective	Clears a footpath	Provides a 0.6- x 15-m footpath Can be deployed by a squad in 5 minutes Consists of ten 1.5-m-long sections weighing 13 lbs each Is ineffective against AT mines
Grapnel hook and WLGH	Is ineffective	Effectively clears 99 percent of trip wires/break wires in a single pass	Is ineffective; however, it may snag the wires between the control unit and its mines	Is ineffective; however, it may snag the wires between the observer and his mines	Is somewhat effective at clearing trip-wire-fused mines integrated with wire obstacles	Has a range of 80 to 100 m when launched Can be thrown 15 to 20 m when hand thrown Must consider the lethal range/radius (up to 90 m in some cases) of fragmenting AP mines that may be encountered, and select a covered position or move back an adequate distance before pulling the grapnel Helps increase confidence by grappling an area several times Is hampered by thick vegetation/overhead obstructions which may reduce the range/ effectiveness of the WLGH

Table G-2. Dismounted breaching assets versus threat obstacles (continued)

Asset	Pressure-Fused AP Mine	Trip-Wire/Break-Wire-Fused AP Mine	Influence-Fused AP Mine	Cmd-Det Blast Mine	Wire Obstacles	Remarks
Probe	Is effective at locating all buried mines; however, extreme caution should be used when probing for small mines (less than 3-inch dia) and sensitive mines (more than 10 lbs actuating force)	Is ineffective; the soldier must rely on clearance by other means and visual detection	Do not use	Ensure that observer is neutralized before approaching	Is ineffective	
Wire cutter	Is ineffective	Do not cut a slack wire until both ends have been checked if break-wire-fused mines may be present; do not cut taught trip wires until both ends have been checked for tension-release FDs	Do not use	Ensure that observer is neutralized before approaching	Is effective; sound can be minimized by wrapping the wire with a cloth before cutting it	
Gunfire	Is effective against surface-laid munitions	Is effective against surface-laid munitions	Do not use	Is effective against surface-laid munitions	Is ineffective	Is used as a last resort Is a relatively slow method for clearing visually detected ordnance Ensures protection of weapons operator Works better with larger caliber weapons (up to 25 mm), incendiary ammo works best May be the only option open to artillery, MP, and transportation units for dealing with SCATMINEs in the absence of engineers and reduction assets

Table G-2. Dismounted breaching assets versus threat obstacles (continued)

Asset	Pressure-Fused AP Mine	Trip-Wire/Break-Wire-Fused AP Mine	Influence-Fused AP Mine	Cmd-Det Blast Mine	Wire Obstacles	Remarks
MiniFlail	Is effective against simple pressure-fused AP mines Is less effective against blast-hardened AP mines	Is effective	Is ineffective	Is armored against fragmenting threats Is vulnerable to command-detonated blast mines	Do not use	Do not use if the presence of AT mines is suspected, because AT mines can destroy the MiniFlail. Can clear at a rate of about 1,500 square meters per hour
AN/PSS-12	Is very effective at finding metallic mines Is less effective against low-metal mines	Cannot detect monofilament fishing line used for trip wires; it is undetectable by the AN/PSS-12. Visual detection and other neutralization means should be used first in any area suspected of having trip wires.	Do not use	Ensure that observer is neutralized before approaching	Is ineffective	Is less effective against low-metal mines and in areas contaminated with metal fragments (artillery, mortar, small arms) or areas that have heavily mineralized soils

Table G-2. Dismounted breaching assets versus threat obstacles (continued)

Asset	Pressure-Fused AP Mine	Trip-Wire/Break Wire-Fused AP Mine	Influence-Fused AP Mine	Cmd-Det Blast Mine	Wire Obstacles	Remarks
Demolitions	Place at least a 1-pound charge within 6 inches of simple pressure-fused mines Ensure that the charge is placed within an inch of blast-hardened mines	Place at least a 1-lb charge within 6 inches of the mine after the mine at the end of a trip wire has been located Can use elevated charges if necessary against the claymore and stake-type mines	Do not use	Ensure that observer is neutralized before approaching Can use elevated charges if necessary against claymore-type mines	Is effective when used to make an improvised bangalore torpedo	
Blast overshoes	May be effective against very small (up to 1 oz of explosive) mines Is ineffective against mines with more than 1-oz explosive weight	Can reduce casualty severity against fragmenting mines	Can reduce casualty severity against fragmenting mines	Ensure that observer is neutralized before approaching Can reduce casualty severity	Is ineffective	
IBASIC	Will protect personnel near the individual who detonates a pressure-fused AP mine	Can reduce casualty severity against fragmenting mines	Can reduce casualty severity against fragmenting mines	Ensure that observer is neutralized before approaching Can reduce casualty severity	Is ineffective	

Table G-3. Foreign track-width AT mines

Origin	Mechanically Planted	Scatterable	Metallic	Nonmetallic	Simple Pressure	Double Impulse	Blast Hardened	AHD
Russia	TM-46/57/62	PTM-1	TM-46/57/62M	TM-62B/D/P	TM-46/57/62		TM-46/57	TMN-46
China	Type 72	Type 84B	Type 72	Type 72	Type 84B	Type 72	Type 72	
Yugo	TMM-1		TMM-1	TMA-1/2/3/4/5	TMA-1/2/5		TMA-3/4	TMM-1, TMA-1/2/3/4/5
Czech	PT-MI-K, PT-MI-BA III		PT-MI-K, PT-MI-D	PT-MI-BA II/III	PT-MI-BA III, PT-MI-D		PT-MI-K, PT-MI-BA II	PT-MI-K, PT-MI-D
Italy	VS-1.6/2.2, TC-.6/6, SB-81	VS-1.6, SB-81		VS-1.6/2.2, TC-.6/6, SB-81			VS-1.6/2.2, TC-3.6/6, SB-81	VS-1.6/2.2
France	M-51, M-52			M-51, M-52	M-51, M-52			M-51
UK	L9 Barmine, MK-7		MK-7	L9 Barmine	L9 Barmine, MK-7	L9 Barmine, MK-7		L9 Barmine
Ger	PM-60, DM-11		DM-21	PM-60, DM-11	PM-60, DM-11, DM-21			PM-60, DM-11, DM-21

Table G-4. Foreign full-width AT mines

Origin	Mechanically Planted	Scatterable	Tilt Rod	Magnetic Influence	Seismic Influence	Coupled	AHD
Russia	TM-62/72/89	PTM-3	TMK-2	TM-62/72/89	TM-62/72/89		TM-62/72/89
China			Type 84C	Type 84A			Type 84A
Yugo	TMRP-6	KB-2	TMRP-6	TMRP-6, KB-2			TMRP-6, KB-2
Czech			PT-MI-P				
France	HPD series	Lance	M-51, M-52, Model 1954	HPD series		Model 1953	HPD series, M-51, Lance
Hungary	UKA-63		UKA-63				UKA-63
Italy	VS-HCT/HCT-2, SB-MV	VS-SATM1, VS-HCT		VS-HCT/HCT-2, SB-MV, VS-SATM1		Fuse option	VS-HCT/HCT-2, SB-MV, VS-SATM1
Sweden	FFV-028			FFV-028		Fuse 15	FFV-028
UK	L9 Barmine	HB-876	L9 Barmine, MK-7	L9 Barmine, HB-876			L9 Barmine, MK-7, HB-876
Germany	DM-31	AT-2, MIFF		DM-31, AT-2, MIFF			DM-31, AT-2, MIFF
Romania						MC-71	

Table G-5. Foreign side-attack AT mines

Origin	Command	Break Wire	Influence	AHD
Russia	MON-100/200/500, TM-83	LMG	TM-73/83	
Yugo		TMRP-6, COBRA		
France	MIACAH F1	MIACAH F1	MIACAH F1	
UK	L14	L14	AJAX	AJAX
Germany			DM-12	
Sweden		FFV-016, AT-4		

Table G-6. Foreign pressure-fused AP mines

Origin	Scatterable	Low Metal	Simple Pressure	Small Diameter	Low-Operating Threshold	Pronged Fuse	Blast Hardened	AHD
Russia	PFM-1		PMN, PMD-6		PMN, PMD-6, PFM-1	OZM series	PMN-2	
China	GLD-111	Type 72A, GLD-111	Type 72A, Type 58, GLD-111	GLD-111	Type-72, Type 58, GLD-111			Type 72A/B
Yugo		PMA-1/2/3	PMA-1, PMD-1		PMA-1/3, PMD-1	PROM-1/2, PMR-3	PMA-2	
Czech			PP-MI-D, PP-MI-BA		PP-MI-D	PP-MI-SR, PP-MI-SR II		
Italy	TS-50, VS-50, SB-33, VS-MK2	TS-50, SB-33, VS-MK2				Valmara 59/69	TS-50, VS-50, SB-33, VS-MK2	VS-50, SB-33, VS-MK2
France		M-59		M-59	M-59		M-59	
UK	Ranger		Ranger	Ranger				
Germany		DM-11	PPM-2		DM-11		DM-11	

Table G-7. Foreign trip-wire/break-wire-fused AP mines

Origin	Scatterable	Trip Wire	Break Wire
Russia	POM-2	POMZ-2, OZM series, MON series	OZM series, MON series
China		Type 58, Type 59, Type 66, Type 69	
Yugo		PROM-1/2/KD, PMR-2/3/4, MRUD	PROM-1/2/KD, PMR-2/3/4, MRUD
Czech		PP-MI-SR, PP-MI-SR II, PP-MI-SK	
Italy	VS-SAPFM3	Valmara 59/69, BM-85, P-25, P-40, VS-DAFM1	
France		M1951/1955, MAPED F1	M1951/1955
UK		PAD	
Germany		DM-31, SM-70	

Table G-8. Foreign emplaced minefields

	Russia	Iraq	North Korea	Bosnia
Manually				
Antitrack AT	1	1	1	1
Antihull AT	2	3	3	2
Side-attack AT	3	4	3	4
Blast AP	1	1	1	1
Directional AP	1	4		1
Bounding AP	1	1	2	1
Stake AP	1	4	1	1
Mechanically				
Antitrack AT	1	1	1	4
Antihull AT	2	3		1
Vehicle Scattered				
Antitrack AT	1			
Antihull AT	1			
Blast AP				
Directional AP				
Bounding AP				
Simple Frag	1			
Artillery/MLRS				
Antitrack AT	1	4		
Antihull AT		1		1
Blast AP				
Directional AP				
Bounding AP				
Stake AP	1			
Helicopter/Fixed-Wing Aircraft				
Antitrack AT	1	1		
Antihull AT	1			
Blast AP	1	1		
Directional AP				
Bounding AP				
Stake AP	1			
NOTE: Scored from 1 (most common) to 4 (very rare).				

Table G-9. Foreign mine delivery systems

Origin	System	Type[1]	Range	Mine Payload	Density[2]	Remarks
Russia	BM-21	R	20 km	PTM-1/3, POM-2	1.5	1,000 x 500 m per battery
	BM -22	R	35 km	PTM 1/3, POM-2	1.5	
	PKPI	F, H	NA	PTM 1/3, PFM-1, POM-2	1.5	Lays relatively narrow strips
	KSMU	F	NA	PTM 1/3, PFM-1, POM-2	1.5	
	UMZ	V	NA	PTM 1/3, POM-2	1.5	
	GMZ/PMZ-4	MP	NA	Varies	0.4 to 1.0	3 systems can emplace 1,200 m with 624 mines in 26 minutes
	VMR-1/2	H	NA	Varies	0.4 to 1.0	3 Mi-8s can lay 1,200 to 3,000 m with 1,200 mines
China	Type 83	R	15 km	Type 84	0.5	900 x 800 m per battery
Yugo	M87	R	50 km	KB-2	0.5	900 x 500 m per battery
Italy	FIROS 25	R	22 km	VS-HCL, VS-Mk, VS-SAPFM 3	0.5	1,000 x 500 m per battery
	FIROS 30	R	35 km	VS-HCL, VS-Mk, VS-SAPFM 3	0.5	
	DAT	H	NA	Varies	1.2	Lays relatively narrow strips of up to 320 AT mines
	Istrice	V	NA	SATM1, VS-Mk 2 EL, VS-SAPFM 3	Max 1.6	Typically 360 x 140 m
	ST-AT/V	MP	NA	Varies	0.3 to 1.5	240 mines per hour
	Grillo 90	MD	NA	SATM1, VS-Mk 2 EL	0.5	5 AT or 15 AP mines per dispenser
Sweden	FFV-5821	MP	NA	FFV-028	0.4 to 1.4	300 buried mines per hour or 500 surface-laid mines per hour
France	Minotaur	V	NA	Lance	0.7	1,200 x 600 m with 800 mines
	EBG	V	NA	Lance	1.5	60 x 600 m with 40 mines
	155-mm How	A	18 km	Lance		8 mines per round
	Matenin	MP	NA	HPD series	0.4 to 1.4	500 mines per hour
UK	JP-233	F	NA	HB 876		430 mines per Tornado aircraft
	Barmine layer	MP	NA	L9 Barmine	0.4 to 0.9	Up to 700 mines per hour
	Ranger	V	NA	L10		1,296 mines per dispenser

Table G-9. Foreign mine delivery systems (continued)

Origin	System	Type[1]	Range	Mine Payload	Density[2]	Remarks
Germany	Skorpion	V, H	NA	AT-2	0.4	1,500 x 50 m with 600 mines
	MW-1	F	NA	MIFF, MUSPA	0.5	55 to 500 m wide and 200 to 2,500 m long with 892 AT mines
	LARS	R	14 km	AT-2	0.5	1,836 AT mines per launcher, 1,000 x 500 m per battery
	MARS	R	30 km	AT-2	0.3	1,000 x 500 m with 336 AT mines per launcher
	FFV-5821	MP	NA	DM-31	0.4 to 1.4	300 buried mines per hour or 500 surface-laid mines per hour

[1] A = artillery, H = helicopter, F = fixed-wing aircraft, R = rocket, V = vehicle dispensed, MD = man-portable dispenser, MP = mechanical planter

[2] AT mines per meter of front

Appendix H

Metric Conversion Chart

This appendix complies with current Army directives which state that the metric system will be incorporated into all new publications. Table H-1 is a conversion chart.

Table H-1. Metric conversion chart

US Units	Multiplied By	Metric Units
Feet	0.3048	Meters
Inches	2.54	Centimeters
Inches	0.0254	Meters
Inches	25.4001	Millimeters
Miles	1.6093	Kilometers
Pounds	453.59	Grams
Pounds	0.4536	Kilograms
Metric Units	**Multiplied By**	**US Units**
Centimeters	0.3937	Inches
Grams	0.03527	Ounces
Kilograms	2.2046	Pounds
Kilometers	0.62137	Miles
Meters	3.2808	Feet
Meters	39.37	Inches
Meters	1.0936	Yards
Millimeters	0.03937	Inches

Glossary

1LT	first lieutenant
1SG	first sergeant
2LT	second lieutenant
AA	avenue of approach
ACE	armored combat earthmover
ACR	armored cavalry regiment
AD	armor division
ADA	air-defense artillery
ADAM	area-denial artillery munition
ADE	assistant division engineer
AHD	antihandling device
AI	area of interest
ALO	air liaison officer
ammo	ammunition
AO	area of operations
AP	antipersonnel
APB	antipersonnel blast
APC	armored personnel carrier
APDS	armor-piercing discarding sabot
APF	antipersonnel fragmentation
APL	antipersonnel land mine
APOBS	Antipersonnel Obstacle Breaching System
AR	Army regulation
ASP	ammunition supply point

ASTAMIDS	Airborne Standoff Minefield Detection System
AT	antitank
atk	attack
ATP	ammunition transfer point
attn	attention
AVLB	armored vehicle-launched bridge
AVLM	armored vehicle-launched MICLIC
BAI	battlefield air interdiction
bde	brigade
BICC	battlefield information control center
BLPS	ballistic and laser protective spectacles
bn	battalion
BOS	Battlefield Operating System
BP	battle position
BRDEC	US Army Belvoir Research, Development, and Engineering Center
C	Celsius
C^2	command and control
CA	civil affairs
cal	caliber
CAS	close air support
cdr	commander
CEV	combat engineer vehicle
CG	commanding general
CHS	combat health support
CL	centerline
cm	centimeter(s)
cmd	command
CMMC	corps material-management center

co	company
COA	course of action
COB	civilian on the battlefield
comp	composition
CS	combat support
CSR	controlled supply rate
CSS	combat service support
CTCP	combat trains command post
cu	cubic
Czech	Czechoslovakia
d	day
DA	Department of the Army
DAO	division ammunition officer
DCU	dispenser control unit
det	detonated
det	detonating
dia	diameter
div	division
DMMC	division material-management center
DOD	Department of Defense
DODIC	Department of Defense identification code
DST	decision support template
DTG	date-time group
DZ	drop zone
EA	engagement area
ea	each
EBA	engineer battlefield assessment
EDD	explosive detector dog

EFP	explosive-formed penetrating
EID	end item code
EM	enlisted member
engr	engineer
ENS	Explosive Neutralization System
EOD	explosive ordnance disposal
ESMB	explosive standoff minefield breacher
F	Fahrenheit
FCP	forward command post
FCS	fire control system
FD	firing device
FECS	Field-Expedient Countermine System
FIST	fire-support team
FLOT	forward line of own troops
FM	field manual
FM	frequency-modulated
FMFM	Fleet Marine Force manual
FPF	final protective fires
frag	fragmentation
FRAGO	fragmentary order
FSB	forward support battalion
FSC	federal stock classification
FSCL	fire-support coordination line
FSCOORD	fire-support coordinator
FSE	fire-support element
ft	foot, feet
ft	fort
FWF	former warring faction

g	gram(s)
G2	Assistant Chief of Staff, G2 (Intelligence)
G3	Assistant Chief of Staff, G3 (Operations and Plans)
G4	Assistant Chief of Staff, G4 (Logistics)
GDP	general defense plan
Ger	Germany
GPS	global-positioning system
H	hour of execution
HE	high-explosive
HEI-T	high-explosive incendiary tracer
HEMMS	hand-emplaced minefield marking set
HEMTT	heavy expanded mobility tactical truck
HEXJAM	hand-emplaced, expandable jammer
HMMWV	high-mobility, multipurpose, wheeled vehicle
HQ	headquarters
hr	hour(s)
HTF	how to fight
hvy	heavy
IBASIC	Improved Body Armor System, Individual Countermine
ID	infantry division
IDA	improved dog-bone assembly
IFV	infantry fighting vehicle
ILLUM	illumination
IMU	inertial measurement unit
in	inch(es)
inf	infantry
IOE	irregular outer edge
IP	improved plow

IPB	intelligence preparation of the battlefield
IR	infrared
ITV	improved TOW vehicle
IVMMD	interim vehicle-mounted mine detector
JAAT	joint air-attack team
Jan	January
Jul	July
K-Kill	catastrophic kill
kg	kilogram(s)
km	kilometer(s)
kph	kilometers per hour
KW	Kiowa Warrior
lb	pound(s)
LBE	load-bearing equipment
LE	low explosive
LED	light-emitting diode
LIN	line item number
LM	landmark
LOC	lines of communication
LOGPAC	logistics package
LRA	local reproduction authorized
LRP	logistics release point
LRSD	long-range surveillance device
LZ	landing zone
m	month
m	meter(s)
M-Kill	mobility kill
M-S	Miznay-Schardin

mag	magnetic
Mar	March
MBA	main battle area
MC	mobility corridor
MCAP	mine-clearing/armor-protection kit
MCB	mine-clearing blade
MCD	magnetic coupling device
MCO	Marine Corps order
MCOO	modified combined obstacle overlay
MCR	mine-clearing roller
MCRP	Marine Corps reference publication
MDI	modernized demolition initiation
MDV	mine-detection vehicle
Met+VE	meteorological data/velocity error
METL	mission-essential task list
METT-TC	mission, enemy, terrain, troops, time available, and civilian considerations
MHE	material handling equipment
MICLIC	mine-clearing line charge
MIDAP	minefield-detection algorithm and processor
MILSTD	military standard
min	minute(s)
min	minimum
MLRS	Multiple-Launched Rocket System
mm	millimeter(s)
MOBA	military operations in built-up areas
MOD	mobile obstacle detachment
MOPMS	Modular Pack Mine System
MOPP	mission-oriented protective posture

MOOTW	military operations other than war
MOUT	military operations on urbanized terrain
MP	military police
MRB	motorized rifle battalion
MRC	motorized rifle company
MRR	motorized rifle regiment
MSD	movement support detachment
MSR	main supply route
NA	not applicable
NAI	named area of interest
NATO	North Atlantic Treaty Organization
NAVSEA	Naval Sea Systems Command
NBC	nuclear, biological, chemical
NCO	noncommissioned officer
NCOIC	noncommissioned officer in charge
NE	northeast
NGO	nongovernment organization
NLT	no later than
nm	nautical mile
no	number
NSN	national stock number
OBSTINTEL	obstacle intelligence
Oct	October
OIC	officer in charge
OP	observation post
OPCON	operational control
OPLAN	operation plan
OPORD	operation order

P	pace(s)
pam	pamphlet
PIR	priority intelligence requirement
PIRS	passive infrared sensor
PL	platoon leader
PL	phase line
PLS	palletized load system
plt	platoon
PSG	platoon sergeant
psi	pounds per square inch
PSYOP	psychological operations
pt	point
PVO	private volunteer organization
QASAS	quality assurance ammunition specialist
R&S	reconnaissance and surveillance
RAAM	remote antiarmor mine
RAC	risk-assessment code
RCU	remote-control unit
RDX	cyclonite
recon	reconnaissance
rep	representative
RF	radio frequency
ROKUS	Republic of Korea, United States
RP	reference point
RSO	range safety officer
RTO	radio-telephone operator
S&A	safing and arming

S&H	safety and handling
S&T	supply and transport
S2	Intelligence Officer (US Army)
S3	Operations and Training Officer (US Army)
S4	Supply Officer (US Army)
SAW	squad automatic weapon
SCATMINE	scatterable mine
SCATMINWARN	scatterable minefield warning
sct	scout
SD	self-destruct
SE	southeast
SEAD	suppression of enemy air defense
sec	second(s)
SEE	small emplacement excavator
SFC	sergeant first class
SFC	sergeant first class
SFF	self-forging fragmentation
SIR	serious incident report
SITEMP	situation template
SLAM	selectable lightweight attack munition
SOEO	scheme of engineer operations
SOF	special-operations forces
SOP	standard operating procedure
SOSR	suppress, obscure, secure, and reduce
SPECS	Special Protective Eyewear Cylindrical System
SPOTREP	spot report
SSG	staff sergeant
SSN	social security number

STANAG	International Standardization Agreement
STP	soldier training publication
T/MDV	towing/mine-detection vehicle
TACC	tactical air control center
TAHQ	theater army headquarters
TAI	target area of interest
TC	track commander
TC	training circular
TCP	traffic-control post
TCP	tactical command post
TF	task force
tm	team
TM	technical manual
TMD	tactical-munition dispenser
TNT	trinitrotoluene
TO	theater of operations
TO	Air Force technical order
TOC	tactical operations center
TOW	tube-launched, optically tracked, wire-guided missile
TRADOC	United States Army Training and Doctrine Command
TRP	target reference point
TTP	tactics, techniques, and procedures
UAV	unmanned aerial vehicle
UK	United Kingdom
UL	Underwriters Laboratories, Incorporated
US	United States
USAF	United States Air Force
USN	United States Navy

UTM	Universal Transverse Mercator
UXO	unexploded ordnance
vs	versus
W	width
w/	with
WLGH	weapon-launched grapnel hook
WO	warning order
WP	white phosphorus
XO	executive officer
y	year
Yugo	Yugoslavia

References

SOURCES USED

These are the sources quoted or paraphrased in this publication.

Joint and Multiservice Publications

AR 385-63. *Policies and Procedures for Firing Ammunition for Training, Target Practice and Combat (MCO P3570.1A)*. 15 October 1983.

FM 20-11. *Military Diving (NAVSEA 0910-LP-708-8000)*. 20 January 1999.

FM 90-3. *Desert Operations (FMFM 7-27)*. 24 August 1993.

FM 90-13. *River-Crossing Operations (MCWP 3-17.1)*. 26 January 1998.

FM 101-5-1. *Operational Terms and Graphics (MCRP 5-2A)*. 30 September 1997.

STANAG 2036. *Land Mine Laying, Marking, Recording and Reporting Procedures*. Edition 5. 12 February 1987.

STANAG 2889. *Marking of Hazardous Areas and Routes Through Them*. Edition 3. 26 March 1984.

STANAG 2990. *Principles and Procedures for the Employment in Land Warfare of Scatterable Mines With a Limited Laid Life - ATP-50*. Edition 1. 15 September 1988.

TM 5-6665-298-10. *Operator's Manual AN/PSS-12 Mine Detecting Set (TM 5-6665-298-10)*. 28 February 1995.

TM 9-1345-203-12&P. *Operator's and Organizational Maintenance Manual (Including Repair Parts and Special Tools List) for Land Mines (TO 11A8-3-1)*. 14 January 1977.

TM 9-2590-509-10. *Operator's Manual Mine Clearing Blade for M1, IPM1, or M1A1 Abrams Tank (NSN 2590-01-230-8862) (TM 2590-10/1)*. 12 April 1990.

Army Publications

AR 40-10. *Health Hazard Assessment Program in Support of the Army Materiel Acquisition Decision Process*. 1 October 1991.

AR 385-10. *The Army Safety Program*. 23 May 1988.

AR 385-16. *System Safety Engineering and Management*. 3 May 1990.

DA Pam 350-38. *Standards in Weapons Training*. 3 July 1997.

FM 1-113. *Utility and Cargo Helicopter Operations*. 12 September 1997.

FM 3-34.2 (formerly FM 90-13-1). *Combined-Arms Breaching Operations*. 31 August 2000.

FM 5-7-30. *Brigade Engineer and Engineer Company Combat Operations (Airborne, Air Assault, Light)*. 28 December 1994.

FM 5-10. *Combat Engineer Platoon*. 3 October 1995.

FM 5-34. *Engineer Field Data*. 30 August 1999.

FM 5-71-2. *Armored Task-Force Engineer Combat Operations*. 28 June 1996.

FM 5-71-3. *Brigade Engineer Combat Operations (Armored)*. 3 October 1995.

FM 5-100. *Engineer Operations*. 27 February 1996.

FM 5-170. *Engineer Reconnaissance*. 5 May 1998.

FM 5-250. *Explosives and Demolitions*. 30 July 1998.

FM 90-5 (HTF). *Jungle Operations (How to Fight)*. 16 August 1982.

FM 90-7. *Combined Arms Obstacle Integration*. 29 September 1994.

FM 90-10 (HTF). *Military Operations on Urbanized Terrain (MOUT) (How to Fight)*. 15 August 1979.

FM 90-10-1. *An Infantryman's Guide to Combat in Built-up Areas*. 12 May 1993.

FM 101-5. *Staff Organization and Operations*. 31 May 1997.

MILSTD 882C (Revision). *System Safety Program Requirements*. 19 January 1993.

STP 5-12B1-SM. *Soldier's Manual: MOS 12B, Combat Engineer, Skill Level 1*. 14 February 2001.

TC 25-8. *Training Ranges*. 25 February 1992.

TC 31-34. *Demining Operations*. 24 September 1997.

TM 1-1520-237-10. *Operator's Manual for UH-60A Helicopters, UH-60L Helicopters, and EH-60A Helicopters*. 31 October 1996.

TM 9-1095-208-10-1. *Operator's Manual for Dispenser, Mine: M139 (NSN 1095-01-235-3139) (EIC: 3V8) with Mounting Kits for 5-Ton Vehicle (1095-01-252-2818) (EIC: 3V9) and M548A1 Vehicle (1095-01-331-6755) (Ground Volcano)*. 10 July 1992.

TM 9-1300-206. *Ammunition and Explosives Standards*. 30 August 1973.

TM 9-1345-203-12. *Operator's and Unit Maintenance Manual for Land Mines*. 30 October 1995.

TM 9-1345-209-10. *Operator's Manual for Modular Pack Mine System (MOPMS) Consisting of Dispenser and Mine, Ground: M131 (NSN 1345-01-160-8909) Control, Remote, Land Mine System: M71 (1290-01-161-3662) and Dispenser and Mine, Ground Training: M136 (6920-01-162-9380)*. 31 March 1992.

TM 9-1375-213-12. *Operator's and Unit Maintenance Manual (Including Repair Parts and Special Tools List): Demolition Materials*. 30 March 1973.

TM 9-1375-215-14&P. *Operator's, Unit, Direct Support and General Support Maintenance Manual (Including Repair Parts and Special Tools List) for Demolition Kit, Mine Clearing Line Charge (MICLIC)*. 31 January 1992.

TM 43-0001-36. *Army Ammunition Data Sheets for Land Mines (FSC 1345)*. 1 September 1994.

DOCUMENTS NEEDED

These documents must be available to the intended users of this publication.

DA Form 1355. *Minefield Record*. March 1987.

DA Form 1355-1-R. *Hasty Protective Row Minefield Record (LRA)*. September 2001.

DA Form 2028. *Recommended Changes to Publications and Blank Forms*. February 1974.

READINGS RECOMMENDED

These readings contain relevant supplemental information.

FM 5-490. *Engineer Diving Operations.* 31 March 1992.

FM 6-20-40. *Tactics, Techniques, and Procedures for Fire Support for Brigade Operations (Heavy).* 5 January 1990.

FM 6-20-50. *Tactics, Techniques, and Procedures for Fire Support for Brigade Operations (Light).* 5 January 1990.

TM 5-2590-214-10. *Operator's Manual for Roller Kit, Mine (NSN 2590-01-134-3724) (EIC: AKZ) and Mounting Kit, Mine Clearing Roller (2590-01-235-5458) (EIC: AKZ).* 30 March 1993.

TM 9-1095-209-10. *Operator's Manual for Mine Dispenser: M138 (Flipper).* 29 March 1990.

Index

A

Airborne Standoff Minefield Detection System, 10-3
air-mission brief, D-13
anchor point, 2-12
antihandling devices, 2-9, 6-31, 6-37
  M142 multipurpose FD, 1-7
  M5 pressure-release FD, 1-7
antipersonnel mines, 1-5
  characteristics, 5-3
  installation, A-3, A-7
  M14, 5-3, A-2
  M16, A-6
  removal, A-5, A-10
  sensing, types of, 1-5
  warheads, types of, 1-6
Antipersonnel Obstacle Breaching System (APOBS), 10-12
antitamper feature, 4-6
antitank mines, 1-4
  characteristics, 5-2
  M15, A-11, E-7
  M19, A-21, E-7
  M21, A-24, E-7
AP SCATMINEs, 3-1
  characteristics, 3-2
APOBS. *See* Antipersonnel Obstacle Breaching System.
area clearance, 9-1, 9-7, 11-15
area-denial artillery munitions, 3-11
area-disruption obstacle, 4-10
arming point, D-16
arming time, 3-6, 4-1
armored vehicle-launched MICLIC (AVLM), 10-7
artillery-shell AT device, 13-30
assault force, 11-7
AT SCATMINEs, 3-3
  characteristics, 3-4
AVLM. *See* armored vehicle-launched MICLIC.

B

bangalore torpedo, 10-13
barbwire AP device, 13-32
Battlefield Operating System, 2-34
bearing board, 5-6
bearing plate, 5-6
beginning-of-strip marker, 7-2
blasting machine, 3-27
block minefields, 2-13, 6-31, D-6
  MOPMS, 3-29
  row, 6-31
  Volcano, 3-24
booby traps, 13-1
  actuation methods, 13-5
booster charge, 1-1
breach force, 11-8, 11-18
breaching, 9-1, 9-2
break wires. *See* trip wires.
breaking a trip wire, 1-2
bypasses, 10-27
bypassing, 9-2

C

camouflage, 5-6
catastrophic kill (K-Kill), 1-4
centurion, 4-7
chemical land mines, C-6
claymore device, improvised, 13-31
clearing, 9-6
  buildings, 13-19
  equipment, 10-7
  methods, 13-22
  obstacles, 13-21
  open areas, 13-18
  secure areas, 13-17, 13-21
  sequence, 13-15
clusters, 7-4
combat clearance, 13-16
combat engineer vehicle with full-width mine rake, 10-21
command-detonation, 4-4
control measures
  site layout, 6-16, 6-17
control of parties, 13-9
control point, 13-9

I

conventional mines, 5-1
countermine data, G-1
countermine operations, C-6

D

DA Form 1355, 8-3, 12-3
DA Form 1355-1-R, 8-3, 8-17
DCU. *See* dispenser control unit.
decision and execution, 2-29
  obstacle-execution matrix, 2-29
  scheme-of-obstacle overlay, 2-29
demining, 9-2, 9-7
density, 3-8
detection, 10-1, 13-15
  Airborne Standoff Minefield Detection System (ASTAMIDS), 10-3
  AN/PSS-12 mine detector, 10-3
  electronic, 10-3
  mine rollers, 10-6
  physical (probing), 10-2
  visual, 10-1
detonator, 1-1
digging team, 6-16
dispenser control unit (DCU), 3-21
dispensing marker, 3-19
disrupt minefields, 2-10, D-7
  MOPMS, 3-29
  row, 6-28
  Volcano, 3-24
dud pit, D-16

E

emplacement
  ADAM/RAAM, 3-13
  Gator, 3-16
  hasty protective row minefield, 6-34
  Hornet, 4-14
  MOPMS, 3-28
  Volcano, 3-23
  water, 12-2
emplacement authority, 3-9
employment
  ADAM/RAAM, 3-12
  Gator, 3-15
  MOPMS, 3-28
  Volcano, 3-23
employment authority, 6-34
end-of-strip marker, 7-4
engineer divers, 12-1, 12-3
ENS. *See* Explosive Neutralization System.
ESMB. *See* explosive standoff minefield breacher.
expedient devices, 13-29
Explosive Neutralization System (ENS), 10-11
explosive standoff minefield breacher (ESMB), 10-11
extraction
  dismounted, F-14
  mounted, F-13

F

FECS. *See* Field-Expedient Countermine System.
fencing, 3-33, D-20, D-21
Field-Expedient Countermine System (FECS), 10-22
fire-support coordination line, 3-15
fire-support plan, 2-34
firing mechanism, 1-1, 1-2
fix minefields, 2-11, D-7
  MOPMS, 3-30
  row, 6-28
  Volcano, 3-24
force organization, 9-4
fragment hazard zone, 3-31, 3-32
fuse types, 1-2

G

gaps, 7-8
Gator, 3-14
gauntlet obstacle, 4-12
grapeshot AP device, 13-32
grapnel hook, 10-22, 10-23
  hand-thrown, 10-24
  weapon-launched, 10-24
Grizzly, 10-20

H

hand neutralization, 11-21
hand-emplaced explosives, 10-22
hand-emplaced mine marking system (HEMMS), 10-36
hand-emplaced mines, 5-1
haul capacity, Class IV/V, 2-45
HEMMS. *See* hand-emplaced mine marking system.
Hornet, 4-6

I

IBASIC. *See* Improved Body Armor System, Individual Countermine.
igniter, 1-1
Improved Body Armor System, Individual Countermine (IBASIC), 10-19
improved dog-bone assembly, 10-15, 10-17
improvised mining, 11-19
individual training, F-21
interdiction weapons, 4-14
interim vehicle-mounted mine detector (IVMMD), 10-5
intermediate markers, 8-10
IOE. *See* irregular outer edge.
irregular outer edge (IOE), 2-9, 6-2, 6-20, 7-3, 7-6
  baseline, 7-4
  short rows, 6-20
  short strips, 7-6
IVMMD. *See* interim vehicle-mounted mine detector.

L

landmarks, 6-3, 7-9, 8-9
lane marking, 10-27
  NATO standard, 10-36
lane widths, 9-4, 9-6
lane-closure team, 7-15
lanes, 7-7, 9-6
  closing, 7-15
  reducing, 9-4
laying a minefield, 6-20
laying a row minefield, 6-18, 6-23
  drill, 6-18, 6-20
laying party, 6-15, 7-11
layout
  siting, 6-16
lethality, 3-5, 3-7
lethality and density, 3-7
life cycle, 3-6
linear obstacle effort, 2-10, 2-39
live-mine firing demonstrations, E-5
live-mine training, E-3
logistical calculations, 6-3, 7-9

M

M1 and M2 activators, A-33
M139 dispenser, D-2
M14 AT mine, 5-3
M142 firing device, A-32
M15 AT mine, 5-1
M16 AP mine, 5-3, 5-4, E-5
M18A1 AP munition, E-6
M18A1 Claymore munition, 4-2
M19 AT mine, 5-1, 5-2
M21 AT mine, 5-1, 5-2
M5 firing device, A-30
M58A4 mine-clearing line charge, 10-7
M60 Panther, 10-18
M603 fuse, 5-2
M606 integral fuse, 5-2
M624 fuse, 5-2
M71 remote-control unit, 3-26
M87 mine canister, D-1
M87A1 mine canister, D-1
M93 Hornet, 4-6, B-1
magnetic-influenced mine, 4-3
manual breaching, 10-22
manual obstacle reduction, 10-22
markers, D-13
  entrance, 10-25
  entrance funnel, 10-26
  exit, 10-26
  far recognition, 10-26
  final-approach, 10-26
  handrail, 10-25
marking, 3-29, 3-30, 9-7, 10-24, 13-9, 13-18
marking devices, 10-34
marking of minefields and obstacle groups, 2-52
marking party, 6-15, 7-11
marking procedures, 6-25
MICLIC. *See* mine-clearing line charge.
mine
  clusters, 6-2, 7-1
  components, 1-2
  emplacement, 7-13
  incident, 11-26
  removal, 6-36, 11-21
  rows, 6-1, 6-16, 6-34
  spacing, 6-17
  strips, 7-1
  visual indicators, F-1
mine detector, AN/PSS-12, 10-3, F-3
mine dump, 2-43, 6-15
mine roller, 10-6
mine training, E-1
mine-awareness training, F-21
mine-clearing blade, 10-14
mine-clearing line charge (MICLIC), 10-7

mine-clearing roller, 10-14, 10-16
mine-clearing/armor-protection kit, 10-22
mine-dump party, 6-15, 7-11
minefield
density, 3-14
design, 2-5
gaps, 7-8
handover, 7-19
inspection and maintenance, 2-55
lanes, 6-3, 7-7
marking, 2-49
recording, 6-15
reduction, 10-7
reporting, 6-15
siting, 2-38, 6-15
turnover, 2-52
minefield marking set, 10-36
minefield packages, 2-40, 2-42, 2-43
minefield report and record, 2-53
minefield turnover report, 2-53
minefield variables, 2-7
minefields
block, 6-31
disrupt and fix row, 6-28
hasty protective row, 6-33
nuisance, 7-17
standard pattern, 7-1
turn, 6-29
minefields, types of, 2-1
mine-incident report, F-21
mines
antipersonnel
M14, 12-10
M16, 5-3, 5-4, 12-10
M18A1 (claymore), 12-10
M605 fuse, 5-4
prong-activated, 5-5
trip-wire-activated, 5-5
antitank
M15, 5-1, 12-1
M603 fuse, 5-2
M624 fuse, 5-2
M19, 5-1, 5-2, 12-1
M606 integral fuse, 5-2
M21, 5-1, 5-2, 12-1
sympathetic detonation, 5-8
conventional, 5-1
hand-emplaced, 5-1
mines tally sheet, 7-14
mine-strip packages, 2-43, 2-47
MiniFlail, 10-19
mission analysis, 2-19
mobile obstacle detachment, C-1
mobility kill (M-Kill), 1-4
Modular Pack Mine System (MOPMS), 3-26, 12-15
employment, 3-28
MOPMS. *See* Modular Pack Mine System.
mousetrap. *See* M5 firing device.

N

NATO markers, 10-36
NATO standard marking, 10-36
neutralization, 9-1
nuisance minefields, 7-17

O

obscuration, 9-4
obstacle control, 2-14
obstacle control measures, 2-14
belts, 2-16
groups, 2-17
restrictions, 2-18
zones, 2-15
obstacle effects, 2-5
obstacle emplacement authority, 2-14
obstacle intelligence, 9-2
obstacle material
Class IV, 2-39
Class V, 2-39
obstacle numbering system, 8-8
obstacle planning, 2-19
obstacle siting, 2-37
obstacle-turnover work sheet, 2-54
outriggers, 12-2
overlay symbols, 8-25

P

Panther, 10-18
phony minefields, 12-15
planning considerations, 11-17
planning factors, 2-20
planning process (air Volcano), D-11
platoon-size sweep team, 11-8
platter charge, 13-30
plowing, 10-16
pressure plates, 5-4
probability of encounter, 2-8
probability of kill, 2-9
probes, 10-22
probing, 10-2, F-2

progress report, 11-23
prongs, 5-4
proofing, 9-2, 9-7, 10-24
protective obstacles, 2-5

R

RAAM. *See* remote antiarmor mine.
radio-frequency jamming devices, 4-6
ranges of common weapons, 2-23
reconnaissance, 9-2
recording and mine-data tracking, F-20
recording party, 7-11
reduction, 9-1, 9-4, 10-1
regular strips, 7-5
reinforce a conventional minefield, 4-8
remote antiarmor mine (RAAM), 3-11, 3-12
remote control unit, 3-26
reporting and recording, 13-10
reporting procedures, 6-25
reports
- change, 8-3
- completion, 8-2
- initiation, 8-1
- intention, 8-1
- spot, 10-7
- strip feeder, 6-15, 6-22
- transfer, 8-2

resource factors, 2-10, 2-39
responsibilities (air Volcano), D-8
risk assessment, E-8, F-18
risk management, F-17
river mining, 12-1
route clearance, 9-1, 11-1
route clearance, methods of, 11-11
row mining, 6-1

S

safe standoff distance, 4-9
- Hornet, 4-10
- MOPMS, 4-9
- Volcano, 4-10

safe-separation countdown, 3-6
safety considerations, E-3
safety procedures, 13-14
safety tapes, 7-17
safety zone, 3-26, 3-30, 3-32
sample risk assessment, F-18
scatterable minefield reinforcement, 4-10
scatterable minefield report and record, 8-23, D-22
scatterable minefield warning, 3-10, 8-23, 8-24, D-22
scatterable mines, 3-1
- ADAM, 12-13
- capabilities, 3-5
- Gator, 12-15
- RAAM, 12-13
- Volcano
  - air, 12-15
  - ground, 12-15

scatterable mines, extraction from, F-14
selectable lightweight attack munition (SLAM), 4-3, B-1
- M2, 4-3
- M4, 4-3

self-destruct times
- SLAM, 4-4
- Volcano, 3-21
- windows, 3-7

self-forging fragmentation, 3-5
self-neutralization (SLAM), 4-4
sensing, types of, 1-5
setting party, 13-9
side attack (SLAM), 4-4
site layout, 3-24, 6-19, 6-34
siting, 2-37
siting party, 6-16, 7-10
siting-and-recording party, 6-18
situation report, 11-23
skim technique, 10-14
skip zone, 10-11
SLAM. *See* selectable lightweight attack munition.
special environments
- cold regions, 12-16
- desert, 12-17
- jungle, 12-17

squad drill, 6-24
STANAG 2036, 6-1, 6-34, 10-36
STANAG 2889, 10-36
standard AP minefield, C-4
streambed mining, 12-1
supply operations, 2-39
support force, 11-7, 11-18
suppression, 9-4
sweep operations, 11-13
sweep team
- squad-size, 11-10

sympathetic detonation, 5-8, 12-1

T

tactical minefields, 2-32
tactical munition dispensers, 3-15
tactical obstacles, 2-5
tactical-obstacle effects, 2-6, 3-7
task organization, 11-16
technical inspections, 2-55, 7-18
Terrabase, D-13
terrain analysis, D-13
theater air-tasking order, 3-15, 3-16
threat
  antitrack minefield, C-3
  equipment, C-11
  mine operations, C-1
  minefield parameters, C-2
  mixed minefields, C-5
  movement support detachment, C-7
tilt-rod fuses, 5-6
timed-demolition, 4-4
top attack, 4-6
traffic control, 9-6
traffic-control posts, 10-26
trip wires, 1-2, 1-6, 5-4, 6-2, 7-6
tripod, 10-23
turn minefields, 2-12, D-6
  MOPMS, 3-29
  row, 6-29
  Volcano, 3-24
turning points, 7-6
turnover, 2-52

U

uncrating mines, 2-42
urban-area mine employment, 12-6

V

Volcano, 3-21, 3-23
  air, 3-23, D-1
  emplacement, 3-23
  employment, 3-23
  fencing, 3-33
  ground, 3-23
  M87, 3-21
  M87A1, 3-21
  minefields, 3-24
  multiple-delivery mine system, 3-21

W

warheads, types of, 1-5
  blast AT, 1-5
  self-forging fragmentation, 1-5

Field Manual
No. 3-19.1

Headquarters
Department of the Army
Washington, DC, 22 March 2001

Military Police Operations

Contents

Page

PREFACE iv

Chapter 1 INTRODUCTION 1-1
Overview 1-1
Operational Framework 1-3
Battlefield Organization 1-4
Types of Military Police Units 1-7
Joint, Multinational, and Interagency Operations 1-7

Chapter 2 BATTLE COMMAND 2-1
Overview 2-1
Battlefield Visualization 2-1
Commander's Intent 2-2
Concept of Operations 2-2
Command and Control Relationships 2-2
Support Relationships 2-3
Staff Relationships 2-3

Chapter 3 THE THREAT 3-1
Overview 3-1
Rear-Area and Sustainment Operations 3-1
Reception, Staging, Onward Movement, and Integration Operations 3-2
Threat Levels 3-3
Threat Priorities 3-4
Threat Location 3-5
Countering the Threat 3-7

Chapter 4 MILITARY POLICE FUNCTIONS 4-1
Overview 4-1
Maneuver and Mobility Support 4-2
Area Security 4-6
Internment and Resettlement 4-9
Law and Order 4-10
Police Intelligence Operations 4-13

Distribution Restriction: Approved for public release; distribution is unlimited.

*This publication supersedes FM 19-1, 23 May 1988.

		Page
Chapter 5	**MILITARY POLICE SUPPORT TO ECHELONS ABOVE CORPS**	5-1
	Overview	5-1
	Military Police Support	5-2
Chapter 6	**MILITARY POLICE SUPPORT TO CORPS**	6-1
	Overview	6-1
	The Military Police Brigade (Combat Support)	6-1
	Command and Control	6-3
	Employment	6-3
Chapter 7	**MILITARY POLICE SUPPORT TO DIVISION**	7-1
	Overview	7-1
	Division Military Police Company	7-2
	Command and Control	7-2
	Military Police Employment	7-3
Chapter 8	**SEPARATE BRIGADES AND INITIAL/INTERIM BRIGADE COMBAT TEAMS**	8-1
	Overview	8-1
	Support to Separate Brigades	8-1
	Support to the Initial/Interim Brigade Combat Team	8-2
Chapter 9	**THE UNITED STATES ARMY CRIMINAL INVESTIGATION COMMAND**	9-1
	Overview	9-1
	Command and Control	9-2
	Wartime Support	9-2
	Organization	9-4
	Employment	9-8
Chapter 10	**OFFENSIVE AND DEFENSIVE OPERATIONS**	10-1
	Overview	10-1
	Support to Offensive Operations	10-1
	Support to Defensive Operations	10-3
Chapter 11	**STABILITY AND SUPPORT OPERATIONS**	11-1
	Overview	11-1
	Force Suitability	11-2
	Stability and Support	11-3
	Organizations and Capabilities	11-14
Chapter 12	**FORCE PROTECTION**	12-1
	Overview	12-1
	Support to Force Protection	12-1
Appendix A	**METRIC CONVERSION CHART**	A-1
Appendix B	**DOCTRINE AND TRAINING LITERATURE**	B-1
	Doctrine	B-1
	Tactics	B-2
	Techniques	B-2
	Procedures	B-2

Page

Appendix C BATTLEFIELD WORKLOAD ANALYSIS C-1

Appendix D AUTOMATED INFORMATION SYSTEMS IN SUPPORT OF BATTLE COMMAND D-1
Army Information Systems D-1
Military Police Automated Systems D-4

Appendix E POLICE INFORMATION ASSESSMENT PROCESS E-1
Implementation E-1
Sample Scenario E-6

Glossary Glossary-1

Bibliography Bibliography-1

Index Index-1

Preface

The Military Police (MP) Corps supports the commander across the full spectrum of military operations. This manual is the foundation for all MP doctrine as it relates to this support. It communicates to all levels of leadership and staffs how the MP provide a flexible and lethal force capable of operating across this full spectrum. As the keystone manual, it identifies what the MP train on and how their forces are organized and equipped in support of all Army echelons. Additionally, this manual provides guidance that can be used by United States (US) Army service schools, students, sister services, and federal agencies.

This manual is based on the purpose, organization, responsibilities, and goals of the US Army as set forth in Field Manuals (FMs) 100-1 and 3-0, as well as corps, division, and brigade manuals. Additionally, this manual is fully compatible with current joint, multinational, and interagency doctrine.

Appendix A contains a metric conversion chart.

The proponent of this publication is HQ TRADOC. Send comments and recommendations on Department of the Army (DA) Form 2028 directly to Commandant, US Army Military Police School (USAMPS), ATTN: ATSJ-MP-TD, 401 Engineer Loop, Suite 2060, Fort Leonard Wood, Missouri 65473-8926.

Unless this publication states otherwise, masculine nouns or pronouns do not refer exclusively to men.

Chapter 1

Introduction

The MP Corps has a strong history evolving over the past five decades. We, as a corps, continue to transform our organizations and doctrine as we have in the past to support the Army in the active defense of the 1970s, the AirLand battle of the 1980s, and now the force-projection doctrine of the 1990s. Our five MP functions clearly articulate the diverse role the MP play across the full spectrum of military operations. We cannot bask in our successes, nor reflect or celebrate. Our charter is to continue our legacy of stellar performance and strive to perfect it.

BG Donald J. Ryder

When the Army developed the Active Defense strategy in 1976, the US was facing the Cold War scenario of central Europe. Military strategy and doctrine were related to a single, focused threat that revolved around the countries in the Warsaw Pact. We were an outnumbered and technically inferior force facing an armor-dominated European battlefield. The MP Corps supported the Active Defense strategy by tailoring its forces to meet the threat. In 1982, when the AirLand Battle strategy was developed, US forces were still outnumbered, but were no longer technically inferior. Still threat-based and focused on a central European conflict, the AirLand Battle strategy used a relatively fixed framework suited to the echeloned attack of soviet-style forces. It delineated and clarified the levels of war; emphasized closed, concerted operations of airpower and ground forces; balanced the offense and the defense; and highlighted the synchronization of close, deep, and rear operations. MP doctrine kept pace with the Army's AirLand Battle strategy by supporting the battlefield commander through four basic missions—battlefield circulation and control, area security (AS), enemy prisoner of war (EPW), and law and order (L&O).

OVERVIEW

1-1. In October 1983, MP capabilities in the AirLand Battle strategy were tested during operation Urgent Fury in Grenada. The MP performed missions that ranged from assisting the infantry in building-clearance operations to assisting Caribbean peacekeeping forces in restoring L&O. These actions secured the MP's place in the combat-support (CS) role, demonstrating the professional knowledge and flexibility necessary for rapid transition from combat to CS to peacetime missions. The changing battlefield conditions of operation Urgent Fury set the stage for the demand of MP units today.

1-2. Evolving simultaneously with the changing definition of the modern battlefield, MP performance in Operations Hawkeye, Just Cause, and Desert Shield/Storm galvanized their ability to perform at any point along the operational continuum. With the publication of FM 100-5 in 1993, the Army adopted the doctrine of full-dimensional operations, relying on the art of battle command to apply those principles and to shift the focus from AirLand Battle to force-projection doctrine. This new doctrine was based on recent combat experience in a multipolar world with new technological advances. Already trained and expected to perform in this new strategy, MP support was already in place and fully operational. The MP continued to perform their basic battlefield missions and to refine their capabilities while supporting the battlefield commander as he deployed to contingency operations throughout the world.

1-3. In 1996, the MP Corps went through a doctrinal review process to determine if it was properly articulating its multiple performance capabilities in support of US forces deployed worldwide (see Appendix B). The review process identified the need to restructure and expand the EPW mission to include handling US military prisoners and all dislocated civilians. This new emphasis transformed the EPW mission into the internment and resettlement (I/R) function. The review process also identified the need to shift from missions to functions. In the past, the four battlefield missions adequately described MP capabilities in a mature theater against a predictable, echeloned threat. However, that landscape is no longer valid. Accordingly, the four MP battlefield missions have become the following five MP functions:

- Maneuver and mobility support (MMS).
- AS.
- L&O.
- I/R.
- Police intelligence operations (PIO).

1-4. These new MP functions are shaped by the following factors:

- The application of stability and support operations where the integration of joint, multinational, and interagency capabilities are common occurrence.
- The lack of traditional linear battlefields, requiring theater commanders in chief (CINCs) to request forces that meet a specific function to accomplish operational requirements.
- The impact of asymmetric threats (such as drug traffickers and terrorist factions) and the effects of man-made and natural disasters.
- The impact of advances in information and communication technologies and specifically in understanding the increased vulnerabilities presented by these technologies.

1-5. Articulating MP capabilities along functional lines benefits the MP and the Army echelon commander as well as the combatant commander. Since there is a multinational, interagency, and sister-service overlap of security services, the importance of including MP leaders and staffs early in the operational planning process cannot be overemphasized. This means before units are designated, unit boundaries are drawn, and unit missions are assigned. Early involvement ensures the proper development of common security responsibilities, communication and connectivity, liaisons, processes, and the rules of interaction between all forces. The ultimate goal should be the optimal, phased employment of MP forces in support of a commander's operational plan. MP functions not only reflect and capture current capabilities, they define the MP Corps in the twenty-first century.

1-6. As the Army reshapes and focuses its resources on transformation, Force XXI, and other redesign efforts, the MP Corps stands proud and ready to support this progress and reiterate its commitment to assist, protect, and defend.

OPERATIONAL FRAMEWORK

1-7. The operational framework consists of the arrangement of friendly forces and resources in time, space, and purpose with respect to each other, the enemy, or the situation (see Figure 1-1). The operational framework for Army forces (ARFOR) rests within the combatant commander's theater organization. Each combatant commander has an assigned geographical area of responsibility (AOR), also called a theater, within which he has the authority to plan and conduct operations. Within the theater, joint-force commanders at all levels may establish subordinate operational areas such as areas of operation (AOs), joint-operations areas (JOAs) and joint rear areas

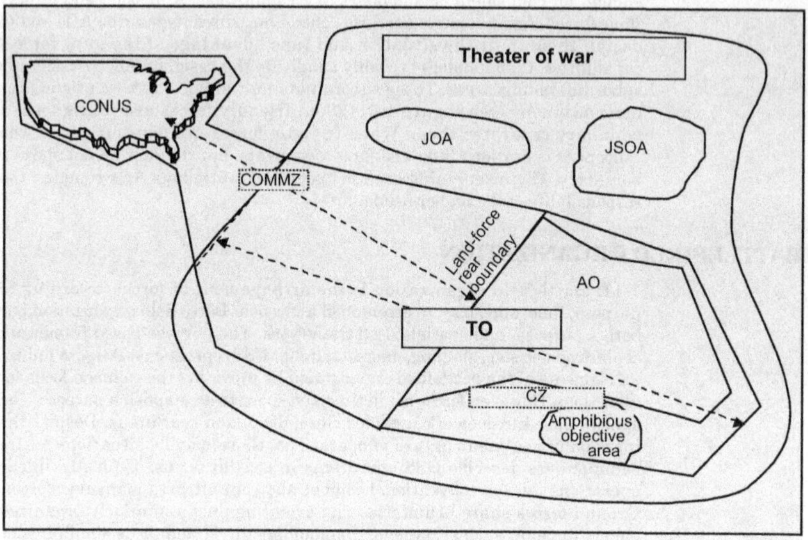

Figure 1-1. Operational Framework

(JRAs). The JRAs facilitate the protection and operation of bases, installations, and forces that support combat operations. When warranted, combatant commanders may designate theaters of war, theaters of operations (TOs), combat zones (CZs), and communications zones (COMMZs).

1-8. A theater of war is that area of air, land, or water that is, or may become, directly involved in the conduct of the war. A theater of war may contain more than one TO. It does not normally encompass the geographic combatant commander's entire AOR. A TO is a subarea (defined by a geographic combatant commander) within a theater of war in which specific combat operations are conducted or supported.

1-9. A CZ is the area required by combat forces for conducting operations. It normally extends forward from the land force's rear boundary. The COMMZ is the rear part of the TO (behind but contiguous to the CZ) that contains the lines of communication (LOC) and provides supply and evacuation support. Other agencies required for the immediate support and maintenance of field forces may also be located in the COMMZ. The COMMZ spans back to the continental US (CONUS) base, to a supporting combatant commander's AOR, or both.

1-10. An AO is an operational area defined by the joint-force commander for land and naval forces. An AO does not typically encompass the entire operational area of the joint-force command (JFC), but it should be large enough for component commanders to accomplish their mission and protect their forces. Army commanders use control measures to describe AOs and to design them to fit the situation and take advantage of the joint force's capabilities. Commanders typically subdivide the assigned AO by assigning subordinate-unit areas. These subordinate-unit areas may be contiguous or noncontiguous (see Figure 1-2). When friendly forces are contiguous, a boundary separates them. When friendly forces are noncontiguous, the concept of operations links the force's elements, but the AOs do not share a boundary. The intervening area between noncontiguous AOs remains the responsibility of the higher headquarters.

BATTLEFIELD ORGANIZATION

1-11. Battlefield organization is the arrangement of forces according to purpose, time, and space to accomplish a mission. Battlefield organization has both a purpose- and spatial-based framework. The purpose-based framework centers on decisive, shaping, and sustaining (DSS) operations. Purpose unifies all elements of the battlefield organization by providing the common focus for all actions. However, forces act in time and space to accomplish a purpose. The spatial-based framework includes close, deep, and rear areas. Despite the increasing nonlinear nature of operations, there may be situations where commanders describe DSS operations in spatial terms. Typically, linear operations involve conventional combat and concentrated maneuver forces. Ground forces share boundaries and orient against a similarly organized enemy force. In such situations, commanders direct and focus simultaneous DSS operations in deep, close, and rear areas, respectively (see FM 3-0).

1-12. The MP battlefield organization supports every Army echelon, from the Army service component command (ASCC) and the theater-support command

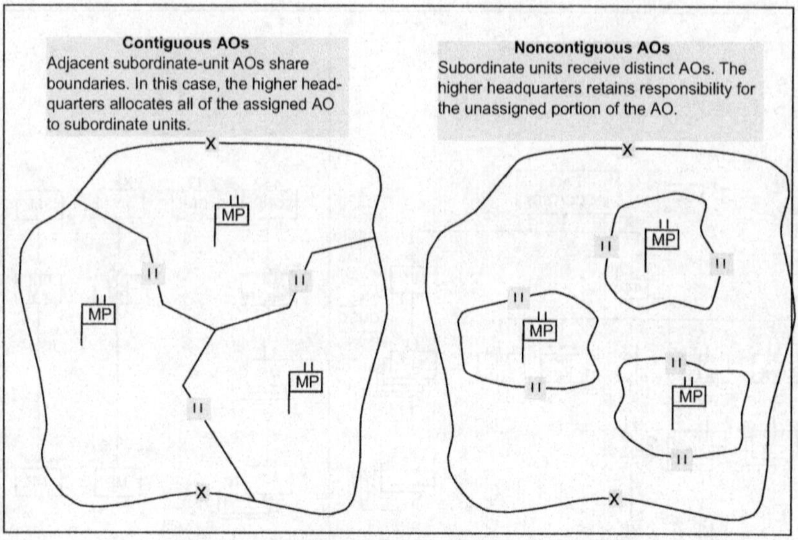

Figure 1-2. Contiguous and Noncontiguous AOs

(TSC) to the maneuver brigade. Regardless of the battlefield organization (purpose or spatial based), MP support to the Army commander is based on available resources and mission, enemy, terrain, troops, time available, and civilian considerations (METT-TC).

1-13. MP support throughout the theater of war may include MP units in the JOA and in the TO. If the combatant commander designates a COMMZ and a CZ within his TO, MP support will come from the established MP modified table of organization and equipment (MTOE) at the subordinate-command echelon. MP support to the JOA is also provided based on METT-TC and available MP assets. Typical MP support may include an I/R brigade liaison detachment (BLD), MP brigades and battalions, a division MP company, a military-working-dog (MWD) team, an L&O team, and a customs team. Figure 1-3, page 1-6, depicts a typical MP organization throughout the TO. In the COMMZ, Figure 1-3 depicts the different types of MP units that are assigned to echelons above corps (EAC) (the ASCC or the TSC). In the CZ, Figure 1-3 depicts the different types of MP units that are assigned to corps, division, and the separate brigades.

SUPPORT IN THE COMMZ

1-14. MP support in the COMMZ is provided by an array of multifunctional MP units. The following MP units provide MP support to EAC:

- The MP brigade (I/R). The MP brigade (I/R) may augment the ASCC or the TSC during wartime. Its mission is to provide command, staff

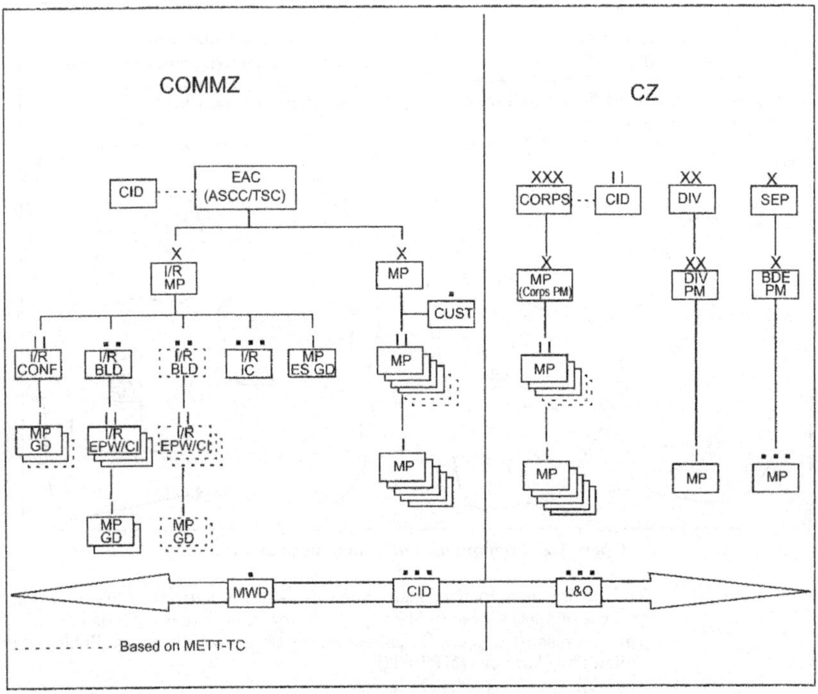

Figure 1-3. MP Structure in the TO

planning, and supervision of I/R operations. This includes coordination with joint and host-nation (HN) agencies, civilian police authority, nongovernmental organizations (NGOs), and US federal agencies.

- The MP brigade (CS). The MP brigade (CS) is assigned to the ASCC or the TSC during wartime (based on METT-TC). The MP brigade (CS) is capable of performing all five MP functions.
- The Criminal Investigation Division (CID) group. The CID group is a stovepipe organization that reports directly to the Commander, US Army Criminal Investigation Command (USACIDC). The CID group provides support to the ASCC and subordinate commands (TSC, corps, or division). See Chapter 9 for further discussion of CID support.

1-15. MP support to other EAC subordinate commands is performed only if MP resources are available. See Chapter 5 for further discussion of MP support to EAC.

SUPPORT IN THE CZ

1-16. MP support is provided in the CZ to each corps, division, and brigade (separate teams or initial/interim brigade combat teams [IBCTs]). An MP brigade (CS) is assigned to each corps, and the MP brigade commander is the corps's provost marshal (PM). A PM and his section, along with an organic division MP company, are assigned to each division. A PM cell and an MP platoon are organic to a separate brigade. A two-person PM cell is organic to the IBCT. The MP units assigned to corps, divisions, and separate brigades are capable of performing all five MP functions. They provide combat, CS, and combat-service-support (CSS) operations within their command's AO.

TYPES OF MILITARY POLICE UNITS

1-17. Most MP units supporting a TO and a JOA are capable of performing all five MP functions. However, the functions must be prioritized based on METT-TC and the availability of MP assets. Current MP structures are designed and tailored to better support the level of command deployed. For example, at the division level, division MP companies are organized as light, heavy, airborne, or air assault and are organic to their respective divisions. The EAC and corps MP brigades and battalions are equally designed to command and control a force mix of up to six battalions or companies. An MP escort-guard and guard company are designed to transport, guard, and provide security to EPWs, civilian internees (CIs), or dislocated civilians. The MP escort-guard company is assigned to the MP brigade (I/R), and the MP guard company is assigned to the MP battalion (I/R).

1-18. MP units can also be tailored and augmented to accomplish multiple, diverse, or specific missions. Customs, L&O, and MWD teams are examples of MP capabilities and flexible responses to a combatant commander's operational requirements. (See Table 1-1, pages 1-9 through 1-19, for a more complete description of MP units. See FM 19-10 and Army Regulation (AR) 190-12 for further information.) The battlefield workload analysis (BWA) is a tool used to determine the number of MP units required to perform some of these multiple missions (see Appendix C).

JOINT, MULTINATIONAL, AND INTERAGENCY OPERATIONS

1-19. In today's environment, the Army will rarely operate or fight alone. The high probability that the Army will operate in concert with its sister services, in an alliance with the forces of foreign nations, or in support of United Nations (UN) operations (when it is committed) is fully reflected in joint doctrine. In such operations, protecting LOC, key facilities, and command and control (C^2) centers will be a shared responsibility. Under this framework, MP units can expect to share the AO with joint, combined, multinational, or interagency resources. MP forces must be prepared to conduct a number of full-spectrum operations with a variety of government and nongovernmental agencies, other services, allied nations, and international agencies.

1-20. Corps and division commanders and staffs must plan (in advance) the transition from a single-service headquarters with joint representation to a joint headquarters capable of functioning as a joint task force (JTF) headquarters. When tasked to form a JTF headquarters, the corps or division must ensure that all of the staff sections and agencies have joint representation (see FMs 100-15 and 71-100). To this end, MP planners must ensure that the JTF is augmented with the appropriate MP forces and with the appropriate echeloned C^2.

1-21. Regardless of the force mix, the MP provide the force with unparalleled, multifunctional capabilities. Among these capabilities is the MP's ability to generate firepower or to handle populations such as EPWs/CIs, dislocated civilians, and refugees. Additionally, MP expertise in investigations and law enforcement enhances the capabilities of other joint, multinational, and interagency police and security forces.

1-22. MP security plans must reflect the joint synergy derived from combining the multiple and diverse capabilities of all participants. To capitalize upon that synergy, MP leaders must keep an open line of communication and coordination to offset the challenges presented by interoperability. Some of these interoperability challenges include—

- Differing political objectives.
- Differing capabilities.
- Cultural/language differences.
- Legal and policy constraints.
- Media impacts.
- Compromise of sensitive processes, procedures, and equipment.
- C^2.
- Communications (digital- and analog-equipment differences).

1-23. MP plans must also accommodate differences in planning capabilities, as well as differences in doctrine, training, and equipment. The intent is to match security missions with force capabilities. MP leaders must understand that operations will often involve multinational teams. While US forces routinely task-organize, this may be more difficult to accomplish with some multinational security forces. This kind of orchestration requires employing standardized procedures, communications, equipment, and liaison within the constraints of operations security (OPSEC).

1-24. Coordination is the key to mission accomplishment in multinational and interagency operations. A military coordination center or a civil-military operations center (CMOC) may meet this coordination requirement. The CMOC provides access for nonmilitary agencies desiring military (to include MP/CID) assistance and coordination. These nonmilitary agencies may include—

- Government organizations (GOs).
- NGOs.
- International organizations (IOs).
- International humanitarian organizations (IHOs).
- HN authorities and agencies.

1-25. The introduction of US Army MP forces in any joint, multinational, or interagency operation is based on METT-TC and the capabilities they bring to the operation. Effective integration of MP forces with other security forces reduces redundant functions, clarifies responsibilities, and conserves resources.

Table 1-1. Unit Descriptions

MP Unit	Mission	Action Elements	Support Role	Capabilities (Include But Are Not Limited to [See TOEs for Details])	Augmentation Needed for Mission Accomplishment
DIVISION MP UNITS					
MP Company (Airborne Division)	Perform the 5 MP functions.	A PM cell and 4 platoons. Each platoon has 3 squads and each squad has 2 three-man teams. Total: 24 three-man teams.	1 platoon deploys with each division brigade and provides DS during the assault phase. Then platoons revert to GS.	During the assault phase, 6 mounted, mobile or fixed-position teams deploy to provide DS for each brigade and 6 teams deploy to provide GS near the division main CP OR after the assault phase, 1 division EPW collection point (6 teams) and 18 mounted, mobile patrols or fixed-position teams provide GS OR 24 mounted, mobile or fixed-position teams are used in any combination.	The division band to provide close-in security for the division CP or to guard EPWs AND the corps CS MP to conduct combat operations within the division rear and provide sustained MMS, area security, special operations support, and escort for evacuating EPWs/CIs between division collection points and to the corps holding area AND the corps L&O MP and CID to conduct sustained L&O operations and criminal investigations AND division/corps transportation assets for evacuation of EPWs **NOTE: Both heavy division companies require corps CS MP augmentation for each forward support platoon (1 corps MP squad per forward support platoon).**
MP Company (Light Infantry Division)	Perform the 5 MP functions.	A PM cell and 3 platoons. Each platoon has 3 squads and each squad has 2 three-man teams. Total: 18 three-man teams	Provide GS for all units in the AO.	Provide screening security for the division main CP (6 teams), the division EPW collection point (6 teams), and 6 mounted, mobile or fixed-position teams.	
MP Company (Heavy Division)* *19333F000	Perform the 5 MP functions.	A PM cell and 2 division support platoons. Each platoon has 3 squads and each squad has a squad leader and 3 three-man teams AND 3 forward support platoons. Each platoon has 2 squads and each squad has a squad leader and 3 three-man teams. Total: 36 three-man teams	2 platoons provide GS and 3 platoons provide DS.	Provide screening security for the division main CP (6 teams), the division EPW collection point (6 teams), and 6 mounted, mobile or fixed-position teams AND EITHER 3 DS platoons providing their brigade with either 1 forward EPW collection point (3 teams) and 3 mounted, mobile or fixed-position teams or 6 teams in any combination OR 18 mounted, mobile or fixed-position teams are used in any combination.	
MP Company (Heavy Division)* *19333L000	Perform the 5 MP functions.	A PM cell and 5 platoons. Each platoon has 2 squads and each squad has 3 three-man teams. Total: 36 three-man teams	3 platoons provide GS and 3 platoons provide DS.	Provide screening security for the division main CP (6 teams), the division EPW collection point (6 teams), and 6 mounted, mobile or fixed-position teams AND EITHER 3 DS platoons providing their brigade with either 1 forward EPW collection point (3 teams) and 3 mounted, mobile or fixed-position teams or 6 teams in any combination OR 18 mounted, mobile or fixed-position teams are used in any combination.	

Table 1-1. Unit Descriptions (continued)

MP Unit	Mission	Action Elements	Support Role	Capabilities (Include But Are Not Limited to [See TOEs for Details])	Augmentation Needed for Mission Accomplishment
MP Company (Air Assault Division)	Perform the 5 MP functions.	A PM cell and 4 platoons. Each platoon has 3 squads and each squad has 2 three-man teams. Total: 24 three-man teams	Provide GS for all units in the AO.	24 mounted, mobile or fixed-position teams are used in any combination.	(See augmentation needs on previous page.)
MP COMBAT SUPPORT UNITS					
HHC MP Brigade	Provide C2 and coordinate the combat, CS, and CSS of all MP elements assigned or attached.	Command and staff elements and long-range planning section	Provide command, control, planning, and supervision for up to 6 MP battalions and other assigned or attached subordinate HQ.	Provide continuous command, control, coordination, planning, and supervision for subordinate units.	None
HHD MP Battalion (Combat Support)	Provide C2 and coordinate the combat, CS, and CSS of all MP elements assigned or attached.	Command and staff elements plus a support element	Provide command, control, planning, and supervision for up to 6 MP companies and other assigned or attached subordinate HQ.	Provide continuous command, control, coordination, planning, and supervision for subordinate units.	None
MP Company (Combat Support)	Perform the 5 MP functions.	4 platoons with 3 squads. Each squad has a squad leader and 3 three-man teams. Total: 36 three-man teams	Provide GS for all units in the AO.	Provide security for a unified or combined HQ or higher HQ or 1 main CP and 1 tactical CP OR operate 1 EPW holding area (9 teams) and 27 mounted, mobile or fixed-position teams for use in any combination.	Corps band assets to guard EPWs AND corps transportation for evacuation of EPWs AND external MP to escort EPWs
MP Company (Combat Support)	Perform the 5 MP functions.	4 platoons with 3 squads. Each squad has a squad leader and 3 three-man teams. Total: 36 three-man teams	Provide GS for all units in the AO.	Provide security for a unified or combined HQ or higher HQ or 1 main CP and 1 tactical CP OR operate 1 EPW holding area (9 teams) and 27 mounted, mobile or fixed-position teams for use in any combination.	Corps band assets to guard EPWs AND corps transportation for evacuation of EPWs AND external MP to escort EPWs
MP Company Arctic Support Augmentation Detachment	Augment MP units when operating in an arctic environment.	1 mechanic (63B10)	Provide mobility support to MP units.	Provide support as designated by the commander.	None
MP Detachment (C2)	Provide command and administrative personnel for MP law enforcement teams.	19503LA has a platoon HQS; 3 individuals. 19503LD has a commander and support personnel; 8 individuals.	Provide L&O. 19503LA provides platoon-level C2. 19503LD provides company-level command, control, and support.	Commands and controls law enforcement teams, investigations teams, MWD teams, and physical-security teams.	None

Table 1-1. Unit Descriptions (continued)

MP Unit	Mission	Action Elements	Support Role	Capabilities (Include But Are Not Limited to [See TOEs for Details])	Augmentation Needed for Mission Accomplishment
MP Detachment (Old Guard)	Perform the 5 MP functions.	4 squads. Each squad has 3 three-man teams. Total: 12 teams	Support the HHC old guard battalion.	Provide support as directed by the commander.	None
MP LAW AND ORDER (L&O) UNITS					
MP HQ Team	Provide C2, personnel administration, and logistical support to attached L&O augmentation teams.	Commander and support personnel	Provide L&O.	Provide C2, personnel administration, and logistical support to attached L&O augmentation teams.	None
MP Operations Team	Provide mission supervision, staff planning, and technical supervision required to support the L&O mission.	L&O operations officer and operations staff	Provide L&O.	Provide overall mission supervision, planning, employment, and coordination of support. Provide technical supervision and administrative support of MPI, TAI, and force protection MP. Provide an evidence custodian/police intelligence NCO to be responsible for confiscated property used in criminal offenses and to receive and collect the initial analysis of criminal/operational information/intelligence.	None
MP Desk or Desk and Record Team	Serve as the primary control point for L&O-related incidents and operations initiated by MP patrols.	2 three-man teams	Provide L&O.	Receive/record complaints, dispatch investigative persons, maintain control of offenders/detained persons, maintain the status of investigations, prepare reports, and provide criminal data to the police intelligence NCO for analysis.	None
MP Traffic Accident Investigations Team	Provide the technical capability to investigate traffic accidents.	1 two- or three-man team: 19517AD00 has 1 two-man team. 19523LE00 has 1 three-man team.	Provide L&O.	Augment MP capabilities to enforce MSR regulations; man checkpoints, roadblocks and dismount points; patrol traffic areas; and perform traffic escort duties.	The 19517AD00 accident investigation team is dependent on the supported unit for transportation.
MP Investigation Team	Provide the technical capability to investigate criminal incidents and conduct surveillance operations.	One, two, or three-man teams: 19517AE has 1 two-man team. 19533LA has 1 one-man team. 19533LB has 1 one-man team. 19533LC has 1 three-man team.	Provide L&O.	Investigate criminal incidents, conduct surveillance, work with HN military and civilian police, and collect criminal and operational intelligence. 19533LA00 provides supervision for the teams. 19533LB00 provides investigative support for populations of not less than 700 troops. 19533LC00 provides investigative support for populations of not less than 2,100 troops.	Dependent on the supported unit for transportation

Table 1-1. Unit Descriptions (continued)

MP Unit	Mission	Action Elements	Support Role	Capabilities (Include But Are Not Limited to [See TOEs for Details])	Augmentation Needed for Mission Accomplishment
MP Force Protection and Physical-Security Teams	Provide the expertise and technical capability to assist units in safeguarding personnel, equipment, and facilities.	One, two, or three-man teams: 19517AF has 1 two-man team. 19593LA has 1 three-man team. 19593LB has 1 one-man team.	Provide L&O and area security.	Conduct vulnerability assessments and identify and prioritize critical facilities and key terrain within the AO. Identify mission-essential activities that are vulnerable to criminal acts or disruptive activities. Ensure that activities are inspected to determine if safeguards are adequate. Assist base/base cluster commanders with the development of internal defense plans.	The 19517AF is dependent on the 19517AB operations team for transportation.
MP Detach-ment (Patrol Supervision Team)	Provide supervision for 3 MP teams.	1 MP NCO	Provide L&O.	Plan, direct, and supervise the employment of assigned or attached MP teams.	Dependent on the supported unit for transportation
MP Detach-ment (Motor and Dismounted Patrol Team)	Perform L&O operations.	1 three-man team	Provide L&O.	Protect designated personnel or facilities. Perform route and area reconnaissance, enforce MSR regulations, provide refugee and straggler control, and disseminate information. Perform peacekeeping operations to maintain surveillance over an area, observe activities, and report findings. Preserve or establish L&O. Apprehend absentees or deserters (US military personnel) in conjunction with civil law enforcement agencies.	The dismounted patrol team is dependent on the supported unit for transportation, when required.
MP Detach-ment (Registration Team)	Perform vehicle and firearm registration.	1 three-man team	Provide L&O.	Service up to 10,000 personnel for registration of individuals, vehicles, and firearms on a closed post or area. Issue credentials prescribing limits of circulation and privileges. Process up to 50 personnel daily for fingerprints and photography services.	Dependent on the supported unit for transportation
Senior Military Customs Inspector Supervisor	Perform technical supervision, staff planning, and coordination for customs inspection teams.	1 MP NCO	Provide L&O.	Supervise two senior military customs teams to ensure that personnel, equipment, and material meet customs, immigration, Department of Agricul-ture, and other federal agency requirements for units and personnel redeploying to the US.	None

Table 1-1. Unit Descriptions (continued)

MP Unit	Mission	Action Elements	Support Role	Capabilities (Include But Are Not Limited to [See TOEs for Details])	Augmentation Needed for Mission Accomplishment
Senior Military Customs Inspection Team	Conduct customs inspections and train and supervise redeploying unit personnel to augment US customs inspectors.	1 five-man team	Provide L&O.	Conduct inspections and advise redeploying units. Train selected individuals from the redeploying unit to augment US customs efforts. NOTE: One team supports one port of embarkation.	The redeploying unit undergoing customs inspection, to augment US Customs efforts. The number of personnel is proportional to the size of the unit.
MILITARY WORKING DOG (MWD) UNITS					
MWD Kennel Master Team	Supervise MWD teams.	One- or two-man team. 19537AA has 1 two-man team. 19583LG has 1 one-man team.	Provide L&O, area security, and I/R operations support.	Supervise, plan, and coordinate MWD operations and support requirements. Provide MWD handler/ canine proficiency certification.	None
MWD Explosive, Narcotics, and Patrol Team	Detect explosives and controlled substances and search for, detect, and control personnel in support of crime scene searches. Provide personal protection, MOUT, health and welfare inspections, and I/R and customs operations.	One-man/one-dog team. The 19537A000-series TOEs have 3 teams each. The 19583L000-series TOEs have 1 team each.	Provide L&O, area security, and I/R support.	The 19537A-series dog teams are capable of providing a 24-hour explosive, narcotics, and tracking handler/dog requirement or three concurrent short-duration missions each requiring 1 handler/dog team. The duration of the missions will vary based on the climate, the environment, and the individual dog's ability. The 19583L-series dog team's provide short- duration missions based on the dog's ability. Explosive and narcotics teams provide MWD patrol support when not employed with explosive and narcotics detection.	None
MILITARY POLICE INTERNMENT/RESETTLEMENT (I/R) UNITS					
MP Command (I/R)	Provide command, control, staff planning, and supervision of I/R operations performed by all assigned or attached elements.	Command and staff elements	Provide I/R operations support.	Provide command, control, and staff planning for 2 or more MP brigades.	MP IRIC AND MP EPW evacuation detachment

Table 1-1. Unit Descriptions (continued)

MP Unit	Mission	Action Elements	Support Role	Capabilities (Include But Are Not Limited to [See TOEs for Details])	Augmentation Needed for Mission Accomplishment
MP Brigade (I/R)	Provide command, control, and staff planning for I/R operations performed by all assigned or attached MP elements.	Command and staff elements	Provide I/R operations support.	Provide command, control, and staff planning for 2 to 7 MP I/R battalions or up to 21 MP I/R battalions when augmented by 7 brigade liaison detachments (BLDs). Plan and provide staff supervision of I/R collection and evacuation operations. Coordinate with HN military territorial organizations and civilian police authorities, nongovernment organizations, private volunteer organizations, and US federal agencies on I/R matters. Provide coordination and support for out-of-theater evacuation of EPWs, if required. Support posthostilities operations.	MP IRIC AND MP EPW evacuation detachment AND MP I/R BLD (TOE 19453LD00) **NOTE: The BLD expands the staff planning and coordination capabilities on a ratio of 1 BLD to 3 MP I/R battalions.**
MP Battalion (I/R)	Provide command, staff planning, administration, and logistical support for the operation of an I/R facility.	Command and staff elements	Provide I/R operations support.	Operate an internment facility for either EPWs/CIs or DCs or a confinement facility for US prisoners. (Never more that one category at the same time.)	MP detachment (I/R) (EPW/CI) for EPW/CI/DC internment missions AND MP detachment (I/R) (confinement) for US prisoner confinement missions AND MP company (EG) for security of the evacuation and/or movement of US prisoners or EPWs/CIs/DCs AND MP company (guard) for prisoner guard services AND Corps/EAC transportation assets for movement of EPWs/CIs/DCs

Table 1-1. Unit Descriptions (continued)

MP Unit	Mission	Action Elements	Support Role	Capabilities (Include But Are Not Limited to [See TOEs for Details])	Augmentation Needed for Mission Accomplishment
MP Detachment (I/R) (EPW/CI)	Augment the MP battalion (I/R) to provide supervision, administration, combat health support, and logistical support for operating an internment facility interning EPWs/CIs or housing DCs.	2 compound control sections (2 three-man teams), a work project section (1 two-man team), a personnel section, and a supply section.	Provide I/R operations support.	Provide command, control, and supervision of up to 1,000 EPWs/CIs or 2,000 DCs. Provide supply and subsistence support (to include supervision of food preparation) for up to 1,000 EPWs/CIs or 2,000 DCs. Direct the activities related to assigning and supervising work projects for EPWs/CIs. Augment the MP I/R battalion in processing and maintaining records for up to 1,000 EPWs/CIs or 2,000 DCs.	MP company (guard) for prisoner guard services
MP Company (Escort Guard)	Provide supervisory and security personnel for evacuating and/ or moving EPWs/CIs.	4 platoons with 3 squads each. Each squad has a squad leader and 3 three-man teams. Total: 36 teams	Provide I/R operations support.	Provide security for the movement of the following numbers of POWs or CIs by the methods indicated: Marching - 1,000 to 1,500. Vehicle - 1,500 to 2,000. Rail - 2,000 to 3,000. Air - 2,125 to 2,406.	Corps/EAC transportation assets for movement of EPWs/CIs
MP Company (Guard)	Provide guards for EPWs/CIs or US prisoners, installations, and facilities.	3 platoons with 3 squads each. Each squad has a squad leader and 2 five-man teams Total: 18 teams.	Provide I/R operations support.	Provide security for a confinement facility containing up to 500 US military prisoners. Provide guards for securing 1 EPW/CI compound containing up to 2,000 EPWs/CIs or 4,000 DCs. Provide security guards for 3 railway terminals, each having up to 8 tracks. (Includes EPW/CI rail movement operations and protection of sensitive material within the terminal.) Provide guards for 1 military installation or facility up to 240,000 square yards in size, containing sensitive material. (This unit can provide guards for material transit.)	None
MP I/R Processing Squad	Process EPWs/ CIs.	1 squad leader and 1 eight-man processing squad	Provide I/R operations support. Augment the I/R battalion processing capability to meet operational requirements.	Provide processing capabilities of about 8 EPWs/CIs per hour.	None

Table 1-1. Unit Descriptions (continued)

MP Unit	Mission	Action Elements	Support Role	Capabilities (Include But Are Not Limited to [See TOEs for Details])	Augmentation Needed for Mission Accomplishment
I/R Brigade Liaison	Expand the MP brigade (I/R) or ASCC C2 capabilities.	Liaison officer and support staff	Provide I/R operations support.	When assigned to the MP I/R brigade, provide staff augmentation that expands the brigade's staff planning, coordination, and C2 capabilities for 3 MP I/R battalions. When assigned to the ASCC, provide I/R staff augmentation and a liaison link with allied/ HN forces to ensure that the care and handling of US captured EPWs/CIs is in compliance with the Geneva Conventions.	None
I/R Camp Liaison Team	Provide continuous accountability of EPWs/CIs captured by US forces and transferred to an allied/HN for internment.	Liaison officer and support staff	Provide I/R operations support.	Provide advice, as requested, to commanders and staffs of allied/HN-operated internment facilities. Verify arrival, forward records, and provide continuous accountability for US captured EPWs/CIs interned in allied/HN facilities. Monitor prisoner treatment to ensure compliance with the Geneva Conventions. Receive/certify allied/HN requests for reimbursement of expenses associated with interning EPWs/CIs captured by US forces.	None
I/R Processing Liaison Team	Process and verify the trans-fer of EPWs/ CIs captured by US forces to a allied/HN facility.	Liaison officer and support staff	Provide I/R operations support.	Provide processing and transfer capability for US-captured prisoners to an allied/HN EPW/CI facility.	None
MP Detachment (EPW Evacuation)	Provide evacuation support of EPWs.	EPW coordination officer and staff plus 10 one-man POD/POE teams and 3 two-man POE teams	Provide I/R operations support.	Coordinate administrative, logistical, and transportation support for up to 7 PODs and 6 POEs used for out-of-theater EPW evacuation and the escort guard personnel that support the movements. Coordinate for theater security at PODs/POEs and intermediate refuel points.	Responsible POD/POE security forces

Table 1-1. Unit Descriptions (continued)

MP Unit	Mission	Action Elements	Support Role	Capabilities (Include But Are Not Limited to [See TOEs for Details])	Augmentation Needed for Mission Accomplishment
MP Detachment (Internment Resettlement Information Cener [IRIC])	Provide a US central agency in each theater of war to receive, process, maintain, disseminate, and transmit the required information and data relating to EPWs, American POWs, CIs, and DCs within the theater.	Command and staff personnel	Provide I/R operations support. Serve as the single source for collection and storage of EPW/CI information in theater. Forward information to the National Prisoner of War Information Center at DA.	Collect, process, and disseminate (to authorized agencies) information regarding interned personnel detained in theater and those released to the custody of allied/HN authorities within theater. Receive, document, and disseminate to the theater commander information received regarding American POWs, CIs, and foreign nationals who are captured, missing, or otherwise detained. Receive, store, and dispose of personal property belonging to interned personnel who have died, escaped, or been repatriated and any property belonging to enemy soldiers killed in action that is not disposed of through grave registration channels.	None
MP Detachment (I/R) (Confinement)	Augment an MP battalion (I/R) in operating a confinement facility. Provide supervision, administration, combat health support, and logistical support of US military prisoners.	Corrections officer and confinement facility staff	Provide I/R operations support.	Provides a control team for the command, control, and supervision of US military prisoners and a disciplinary guard team to supervise custodial personnel for a maximum custody of 500 prisoners. Provide administration, health service, supply, and food service personnel to augment the battalion in providing these services for internees. (The food service personnel supervise and train US military prisoners working in the internee dining facility.)	MP battalion (I/R) for command, staff, planning, and operational support AND MP company (guard) for prisoner guard services

Table 1-1. Unit Descriptions (continued)

MP Unit	Mission	Action Elements	Support Role	Capabilities (Include But Are Not Limited to [See TOEs for Details])	Augmentation Needed for Mission Accomplishment
MP Confinement Facility Detachment	Provide C2, staff planning, administration, and logistical support for a confinement facility for US military prisoners.	Commander and facility staff	Provide I/R operations support.	Provide command, control, and administrative support for a confinement facility.	MP guard company to provide exterior perimeter security and required guard support to the confinement facility AND MP correctional team supervisor (TOE19553LE) to provide supervision of the guard force AND MP correctional team work supervisor (TOE 19553LF) to provide supervision of prisoners
MP Correctional Team Supervisor	Provide correctional supervision of prisoner work guard forces within a confinement facility.	2 two-man teams	Provide I/R operations support.	Provide correctional supervision to guard force personnel and US military prisoners within a modular-configured confinement facility.	None
MP Correctional Team Work Supervisor	Provide direct control and supervision of military prisoners within a confinement facility.	1 two-man team	Provide I/R operation support.	Provide correctional control, supervisory functions, and escort duties for military prisoners within a modular-configured confinement facility.	None
MILITARY POLICE CRIMINAL INVESTIGATION (CID) UNITS					
MP Group (CID)	Provide command, control, staff planning, and supervision for all CID elements within the theater.	Command and staff elements	Provide L&O.	Provide command, control, staff planning, and supervision over all CID elements within the theater. **NOTE: The Criminal Investigation Command is a stovepipe organization.**	None

Table 1-1. Unit Descriptions (continued)

MP Unit	Mission	Action Elements	Support Role	Capabilities (Include But Are Not Limited to [See TOEs for Details])	Augmentation Needed for Mission Accomplishment
MP Battalion (CID)	Provide command, control, staff planning, and supervision for all assigned or attached CID elements.	Command and staff elements	Provide L&O.	Provides command, control, staff planning, and supervision over all assigned and attached CID elements.	None
MP Detachment (CID) HQ Cell	Provide C2, evidence custody control, and investigative administration support. Coordinate for personnel administration and logistic support.	Commander and support staff	Provide L&O.	Provide C2 of assigned CID SA sections and/or teams performing all CID operations in their area of responsibility.	None
MP Detach-ment (CID)	Provide criminal investigative support to Army commanders at all echelons.	2-man teams. The number of teams varies by organization and is as follows: DSE heavy - 4 DSE light - 3 Section A - 4 Section B - 4 Supervisory team - 1 Senior team - 1 SA team - 1	Provide L&O.	Supervise and conduct criminal investigations.	When directed, MP CS, L&O, and MWD support

Chapter 2

Battle Command

Command is the authority a commander in military service lawfully exercises over subordinates by virtue of rank and assignment. Leaders possessing command authority strive to use it with firmness, care, and skill.

FM 101-5-1

Battle command is the exercise of command in an operation against a hostile, thinking opponent. Battle command includes visualizing the current state and the desired end state, then formulating concepts of operations to get from one state to the other at the least cost. In addition to visualizing and formulating concepts, battle command encompasses assigning missions; prioritizing and allocating resources; selecting the critical time and place to act; and knowing how and when to make adjustments in the fight. Battle command enables MP commanders to lead, prioritize, and allocate assets required in support of the Army commander. MP commanders must observe, orient, decide, and act on their decisions quickly. Information is the key element in the battle-command process; therefore, the commander must have accurate and timely information upon which to base his decisions.

OVERVIEW

2-1. The battle command of MP units is typically decentralized due to the nature of their CS functions, METT-TC, and the needs of the Army commander. This places the burden of sound, timely decision making to the lowest levels. MP leaders must develop a keen sense of situational awareness and visualization, and they must constantly track the actions of supported units.

BATTLEFIELD VISUALIZATION

2-2. The ability to visualize the battlefield is a critical element of battle command. Battlefield visualization is an essential leadership attribute and is critical to accomplishing the mission. It is learned and attained through training, practice, experience, technical and tactical knowledge, and available battle-command technologies. It results when the MP commander understands the higher commander's intent, his assigned mission, the enemy, and the friendly force's capabilities and limitations. See Appendix D for further information on command technologies.

2-3. Battlefield visualization includes the MP commander's view of what his forces will do and the resources needed to do the mission. He envisions a sequence of actions that will cause his MP forces to perform at the desired end state. Ultimately, the MP commander's battlefield vision evolves into his intent and helps him develop his concept of operations.

COMMANDER'S INTENT

2-4. The commander's intent is a key part of Army orders. It is a clear, concise statement of what the force must do to succeed with respect to the enemy, the terrain, and the desired end state. It provides the link between the mission and the concept of operations by stating key tasks. These tasks, along with the mission, are the basis for subordinates to exercise initiative when unanticipated opportunities arise or when the original concept of operations no longer applies. MP leaders at all echelons must ensure that the mission and the commander's intent are understood two echelons down (see FM 101-5).

CONCEPT OF OPERATIONS

2-5. The commander's intent does not include the method by which the MP units will accomplish the mission. This method is called the commander's concept of operations. It must—

- Convey the commander's vision of how to accomplish the mission in a manner that allows his subordinates maximum initiative.
- Build around intelligence gathering and the precise employment of MP resources.
- Provide the basis for task organization, scheme of maneuver, terrain organization, tasks to subordinates, and synchronization.

COMMAND AND CONTROL RELATIONSHIPS

2-6. MP units are assigned to, attached to, or placed under the operational control (OPCON) of MP or other units they support. OPCON is the authority to perform command functions over subordinate forces. This includes organizing and employing commands and forces, assigning tasks, designing objectives, and giving authoritative direction necessary to accomplish the mission. MP C^2 relationships may be changed briefly to provide better support for a specific operation or to meet the needs of the supported commander. MP units may be placed under the OPCON of another unit commander for short-term operations. The MP unit remains in this relationship only as long as it is needed for that operation.

> MP support to the Bosnian municipal elections consisted of one division and two corps MP companies. These MP assets, attached to Task Force (TF) Eagle, were task-organized from different sources. The division MP company and the PM cell were organic to TF Eagle's mechanized infantry division headquarters, but the two corps MP units were from US Army Forces Command (FORSCOM) MP battalions in CONUS.

SUPPORT RELATIONSHIPS

2-7. MP units on the battlefield provide two types of support—general support (GS) and direct support (DS). Corps and EAC MP units provide GS to their respective corps/EAC subordinate commands. Light, airborne, and air-assault MP companies provide GS to their respective divisions. Heavy-division MP companies provide GS to the division rear and DS to the division's subordinate brigades.

STAFF RELATIONSHIPS

2-8. The PM for each level of command is that command's advisor on MP combat, CS, and CSS operations. The PM—

- Advises the commander and staff about MP abilities/capabilities.
- Supervises the preparation of plans and dictates policies.
- Coordinates MP operations.
- Assists and supervises the interaction of supporting and supported units.
- Reviews current MP operations.
- Coordinates with allied forces and HN military and civil police.
- Ensures that MP plans and operations supporting the commander's tactical plan are carried out.
- Recommends when and where to concentrate the command's MP assets.
- Supervises or monitors MP support in the command's AO.

2-9. The PM works daily with the commander and staff officers who employ MP resources and whose AORs influence MP support. The PM works closely with the coordinating staff at the appropriate command level to coordinate MP support. He ensures that MP planning is practical and flexible, that plans are coordinated with staff sections and subordinate commands, and that plans reflect manpower and resources needed by MP. (This includes the need for C^2, fire support, equipment, and supplies. It also includes construction, communication, transportation, and aviation support.) As new information is received, the PM reviews, updates, and modifies the plans. He ensures that the echelon commander gets the necessary MP support.

2-10. In the absence of specific directions or orders, the PM plans the use of MP assets. He evaluates the current operations and projects the future courses of action (COAs). He bases his plans on assumptions consistent with the commander's intent and a thorough knowledge of the situation and mission. The PM considers—

- METT-TC.
- Current estimates developed by the intelligence preparation of the battlefield (IPB) and the police information assessment process (PIAP).
- The environment within the AO. This includes the climate, the terrain, and obstacles. It also includes the legal authority and status of the force; the width, depth, size, and location of built-up areas; and the attitudes and abilities of the local populace.

- The types of units operating in the area (to include joint, combined, multinational, and interagency units) and the missions and capabilities of these units. This knowledge is imperative to understand their capability to counter threats in their area.
- The specific missions of MP units in the area and the impact that rear-area security operations will have on the ability of these units to perform other functions.
- Personnel, vehicles, and equipment in the MP units.

2-11. Coordination and communication between the PM and Army commanders is essential. Such actions ensure timely and efficient MP support to all levels of command during any operation. The informal, technical chain of coordination is an open line of communication between PMs at different echelons. The informal chain of coordination fosters cooperation and help among the MP elements at each echelon. For instance, when the division PM needs more assets to accomplish added missions, he initiates coordination with the corps PM. If the corps PM can provide support, the division PM formalizes his request for assistance through the division Assistant Chief of Staff, G3 (Operations and Plans) (G3).

Chapter 3
The Threat

In the 40-odd years of the Cold War, in many locations around the world, the Army performed a deterrent role as part of the containment strategy. In other places, at other times, the Army fulfilled the Nation's expectation in operations too small to be called "wars," although no less dangerous. To the soldier on the ground, Operations Urgent Fury in Grenada and Just Cause in Panama were indistinguishable from combat operations of their forefathers. Operations Provide Comfort in Iraq and Restore Hope in Somalia, although peace operations, also proved to be dangerous.

FM 100-1

The end of the Cold War has reduced, but not eliminated, the most immediate threat to the security of the US and other western nations. However, the absence of a dominant, identifiable threat has produced a far more complex and confusing strategic environment than the one that was present during the Cold War. Forward-deployed and CONUS-based ARFOR and civilians are and will continue to be engaged in a range of military actions. These actions stem from deterring conflicts to conducting peacetime engagement operations to providing support to civil agencies at home and abroad.

OVERVIEW

3-1. During the past decade, the US has deployed forces in multiple operations that have included crisis response in combat situations as well as participation in noncombat activities. The Army's presence in South Korea, Saudi Arabia, and Kuwait and its deployments to Somalia, Haiti, Bosnia, and Kosovo are clear indicators that the military must be prepared to face not only the traditional threat, but also a nontraditional, nonecheloned enemy. To support Army commanders successfully, MP leaders must understand the nature and complexity of these threats and how they can potentially affect the desired strategic, operational, and tactical end states.

REAR-AREA AND SUSTAINMENT OPERATIONS

3-2. The rear area for any particular command is the area extending forward from its rear boundary to the rear of the area assigned to the next lower level of command. This area is provided primarily for the performance of support functions. Operations in the rear area assure freedom of action and continuity of operations, sustainment, and C^2. Sustainment operations are those that enable shaping and decisive operations by assuring freedom of action and continuity of operations, CSS, and C^2 (see FM 3-0). Sustainment operations include the following elements:

- CSS.
- Rear-area and base security.
- Movement control.
- Terrain management.
- Infrastructure development.

3-3. During the Cold War, the danger to rear areas included forces that would be deployed in support of major soviet-style operations. The adversaries using the soviet model could be expected to engage in intense combat activity in their enemy's rear area. Their forces were prepared to penetrate into the enemy's rear and to attack and destroy its reserve forces and rear-area installations. To protect the rear areas, the MP were among the first mobile fighting forces available to the battlefield commander and thus, a source of combat power. Today, the Army commander uses the MP's flexibility and their modular-force training, adaptability, and mobility to serve as a combat multiplier throughout his entire AO. During sustainment operations, the MP perform all functions to ensure freedom of maneuver in support of the overall operational effort.

3-4. Failure to protect our forces during sustainment operations normally results in failure of the entire operation. Sustainment operations determine how fast ARFOR reconstitute and how far they can exploit success. The likelihood of MP units encountering the enemy and engaging in direct combat (not only in the rear area, but also during sustainment operations) cannot be underestimated.

3-5. Threats to rear-area and sustainment operations exist throughout the full spectrum of military operations. These threats may be related or independently engaged, but their effects are frequently cumulative. Threats to rear-area and sustainment operations are usually theater-dependent and are not limited to those outlined in this manual. Joint Publication (JP) 3-10 further discusses the threat in the rear area. Although JP 3-10 defines the threat in the context of a JRA, MP leaders can expect the same level of activity anywhere that US forces are deployed.

RECEPTION, STAGING, ONWARD MOVEMENT, AND INTEGRATION OPERATIONS

3-6. Reception, staging, onward movement, and integration (RSOI) operations consist of essential and interrelated processes in the AO that transform arriving personnel and materiel into forces capable of meeting operational requirements. During RSOI operations, the threat encountered will depend mostly on the type of entry, the nature of the operation, and the enemy. During major contingencies, forces deploy from power-projection platforms within the US or forward bases. The PM must plan MP support during the initial stages of the deployment to ensure the protection of follow-on forces and the detection of potential threats (see FM 100-17-3).

3-7. MP support to RSOI operations includes, but is not limited to—

- Conducting AS operations to counter or prevent enemy actions against marshalling and staging areas.
- Conducting convoy, airport, and rail security operations.
- Conducting populace- and resource-control operations.
- Conducting other physical-security and force-protection measures.
- Conducting other MP functions (as determined by the PM).

THREAT LEVELS

3-8. The threat is divided into three levels. These levels provide a general description and categorization of threat activities, identify the defense requirements to counter them, and establish a common reference for planning guidelines. MP leaders must understand that this does not imply that threat activities will occur in a specific sequence or that there is a necessary interrelationship between each level.

LEVEL I

3-9. Level I threats include the following types of individuals or activities:

- **Enemy-controlled agents.** Enemy-controlled agents are a potential threat throughout the rear area. Their primary missions include espionage, sabotage, subversion, and criminal activities. Their activities span the range of military operations and may increase during both war and military operations other than war (MOOTW). These activities may include assassinating or kidnapping key military or civilian personnel or guiding special-purpose individuals or teams to targets in the rear area.
- **Enemy sympathizers.** Civilians sympathetic to the enemy may become significant threats to US and multinational operations. They may be the most difficult to neutralize because they are normally not part of an established enemy-agent network, and their actions will be random and unpredictable. During war and MOOTW, indigenous groups sympathetic to the enemy or those simply opposed to the US can be expected to provide assistance, information, and shelter to guerrilla and enemy unconventional or special-purpose forces operating in the rear area.
- **Terrorism.** Terrorists are among the most difficult threats to neutralize and destroy. Their actions span the full spectrum of military operations.
- **Civil disturbances.** Civil disturbances, such as demonstrations and riots, may pose a direct or indirect threat to military operations. Although this threat may not be of great impact during war, it may significantly change and affect MOOTW.

LEVEL II

3-10. Level II threats include the following types of forces:

- **Guerilla forces.** Irregular and predominantly indigenous forces conducting guerrilla warfare can pose a serious threat to military forces and civilians. They can cause significant disruptions to the orderly conduct of the local government and services.

- **Unconventional forces.** Special-operations forces (SOF) are highly trained in unconventional-warfare techniques. They are normally inserted surreptitiously into the rear area before the onset of an armed conflict. They establish and activate espionage networks, collect intelligence, carry out specific sabotage missions, develop target lists, and conduct damage assessments of targets struck.
- **Small tactical units.** Specially organized reconnaissance elements are capable of conducting raids and ambushes in addition to their primary reconnaissance and intelligence-gathering missions. Small (size or capability), bypassed conventional units, as well as other potential threat forces, are also capable of conducting raids and ambushes to disrupt operations.

LEVEL III

3-11. Level III threats are made up of conventional forces. Potential threat forces are capable of projecting combat power rapidly by land, air, or sea deep into the rear area. Specific examples include airborne, heliborne, and amphibious operations; large, combined-arms, ground-forces operations; and bypassed units and infiltration operations involving large numbers of individuals or small groups infiltrated into the rear area, regrouped at predetermined times and locations, and committed against priority targets. Level III forces may use a combination of the following tactics as a precursor to a full-scale offensive operation:

- **Air or missile attack.** Threat forces may be capable of launching an air or missile attack throughout the rear area. It is often difficult to distinguish quickly between a limited or full-scale attack before impact; therefore, protective measures will normally be based on the maximum threat capability.
- **Nuclear, biological, and chemical (NBC) attack.** Commanders must be aware that NBC munitions may be used in conjunction with air, missile, or other conventional-force attacks. The NBC weapons could also be used at Level I or II by terrorists or unconventional forces in order to accomplish their political or military objectives.

THREAT-LEVEL MATRIX

3-12. Table 3-1 lists the threat levels and their likely appropriate responses. The threat levels listed are based on the type of threat. The table should not be construed as restricting the response options to any particular threat.

THREAT PRIORITIES

3-13. The threat will attempt to perform the following operations against targets in the rear area:

- Detect and identify targets.
- Destroy or neutralize operational weapons-system capabilities.
- Delay or disrupt the timely movement of forces and supplies.

Table 3-1. Threat Levels

Threat Level	Example	Response
I	Agents, saboteurs, sympathizers, and terrorists	Unit, base, and base-cluster self-defense measures
II	Small tactical units, unconventional-warfare forces, guerrillas, and bypassed enemy forces	Self-defense measures and response forces with supporting fires
III	Large tactical-force operations (including airborne, heliborne, amphibious, infiltration, and bypassed enemy forces)	Timely commitment of a TCF

- Weaken the friendly force's C^2 network.
- Disrupt support to combat forces.
- Set the stage for future enemy operations.
- Create panic and confusion throughout the rear area.

3-14. Typical examples of enemy priority targets include the following:

- NBC-weapons storage sites and delivery systems.
- Key command, control, communications, computers, and intelligence (C4I) facilities.
- Air-defense artillery (ADA) sites.
- Airfields and air bases.
- Port facilities.
- Main supply routes (MSRs) and MSR checkpoints.
- Key LOC.
- Reserve assembly areas (AAs).
- Troop barracks.
- Critical civilian and logistics facilities.

THREAT LOCATION

3-15. The fact that the Cold War has ended does not imply that our traditional threat has ended. North Korea and Iraq are constant reminders of this fact. For the near future, Army commanders will fight units with Cold-War-era equipment and tactics. The Army trains and is prepared to fight an enemy capable of interfering with our freedom of maneuver throughout the battlefield. On an extended battlefield with asymmetric threats, the danger to high-value assets (HVAs) (including CSS, C^2, communication nodes, and MSRs) only increases. The idea that the danger to the rear area decreases as you travel farther away from the forward edge of the battle area (FEBA) is not true. Threat intensity does not depend on geographical location; it depends on what operations the enemy believes must be initiated (and to what degree) to achieve its objective in the rear area. Military commanders depend on the MP to delay and defeat threats in their AO with a mobile reaction force.

COMMUNICATIONS ZONE

3-16. The nature of the COMMZ will encourage Level I and II threats to concentrate along the LOC and other areas of military significance. MP units will encounter an enemy that is capable of disrupting operations throughout the COMMZ while employing terrorist activities, enemy-controlled agent activities, enemy sympathizers, and saboteurs. If the enemy is Level III capable, MP leaders must expect infiltrations and air, missile, or NBC attacks as a precursor to a major Level III operation.

CORPS REAR AREA

3-17. The activities in Levels I and II will be similar in composition and density as in the COMMZ, but they will target key corps units, key facilities, and corps sustainment capabilities. The threat activities, especially at smaller unit levels, may even precede hostilities. MP leaders must be alert and prepared to encounter unconventional forces conducting diversionary or sabotage operations and small combat units conducting raids, ambushes, or reconnaissance operations or collecting special warfare intelligence. With the fast tempo of offensive operations, MP leaders must also be alert and prepared to encounter bypassed forces that can disrupt operations in the corps rear area.

DIVISION REAR AREA

3-18. The division rear area (DRA) contains many types of CS and CSS units and conducts many complex operations. As in the COMMZ and the corps rear area, the full spectrum of Level I, II, and III activities may occur in the DRA. The main target will be the division's HVA (including key C^2 facilities; airfields; artillery, aviation, and air-defense assets; LOC; and essential CSS units). The threat may conduct diversionary attacks, sabotages, raids, ambushes, and reconnaissance operations to affect the commander's freedom of maneuver and the continuity of operations. Unlike corps MP, the likelihood of division MP encountering bypassed enemy forces is expected. Failure to delay or defeat these forces will impact division operations.

OTHER TYPES OF THREATS

3-19. As US forces are deployed throughout the world, they will have to face nontraditional, asymmetric threats (other than those listed in Table 3-1, page 3-5) that may be geographically specific. As part of situational awareness, and in coordination with military intelligence (MI) and CID personnel, MP leaders must evaluate and assess the impact of these threats in their AO. A TO is vulnerable to any or a combination of the following threats:

- National or international organized crime.
- Narcotics traffickers.
- Narcotics terrorists.
- Extremist groups.
- Paramilitary groups.
- Ethnic or religious disputes.
- Trade in illegal weapons or strategic materials.

3-20. MP leaders must be aware that other threats exist and that they have the same potential as the Level I and II threats to disrupt operations in rear-area or sustainment operations. In some instances, the above threats' capabilities or the massing of personnel may have the same potential threat as a Level III threat.

COUNTERING THE THREAT

3-21. The disruption of rear-area and sustainment operations directly affects military efforts. Three types of forces may be used to counter the threat in these areas—a base/base-cluster self-defense force, a response force, or a tactical combat force (TCF).

BASE/BASE-CLUSTER SELF-DEFENSE FORCE

3-22. A base cluster is established when the appropriate echelon rear-operations cell or command post (CP) places geographically contiguous or noncontiguous bases under the control of a headquarters. The base cluster becomes the next higher tactical C^2 headquarters of those bases. The rear-operations cell or the rear CP may also establish a base cluster for a corps support group (CSG), an area support group (ASG), or other CSS units operating in the corps or division rear areas.

3-23. US ARFOR have the inherent responsibility to contribute as many forces as possible for base defense and local security for themselves and their facilities, installations, and activities. Each base and base-cluster commander must develop a defense plan to detect, defeat, and minimize the effects of Level I and limited Level II threat attacks on his base or base cluster (including NBC attacks). To maximize the unit's mission accomplishment, defense plans must be flexible and allow for differing degrees of security based on the probability of threat activity. Defense plans are given to MP units operating near the base or base cluster. The base commander most often employs a series of defense measures providing internal and perimeter security. His internal reaction forces use organic weapons to neutralize and defeat most low-level threat activity. Although not fully equipped to engage major conventional or unconventional enemy forces that may confront him, a base commander must deploy his personnel to defend themselves until MP, HN, local police, or combat forces (if available) can respond.

RESPONSE FORCE

3-24. A response force is summoned when the base or base cluster is faced with threat forces that are beyond their self-defense capability. If the MP are the designated response force, they must—

- Coordinate with the supported bases or base-cluster commanders to conduct a joint IPB.
- Review base and base-cluster self-defense plans.
- Exchange signal-operating-instructions (SOI) information.
- Identify MP contingency plans to counter likely enemy activities.
- Integrate ADA, engineer, chemical, field-artillery (FA), Army-aviation, and close-air-support (CAS) fire support into their plans (if available).

3-25. MP units help the base or base cluster return to its primary mission by defeating Level II threats. MP units closely watch likely avenues of enemy approach, possible landing zones (LZs), drop zones (DZs), C^2 facilities, and other key installations. They accomplish this through the MMS, AS, and PIO functions.

3-26. If MP units are not the designated response force, they may become the initial response force for units within their AO. When this occurs, they block, delay or, if possible, destroy enemy elements within their capability. If the attack is by a larger or more capable force, they will maintain contact and continue to develop the situation or delay until the appropriate response force appears or the battlefield commander commits the TCF.

TACTICAL COMBAT FORCE

3-27. When the MP response force encounters or engages threats beyond its ability to defeat, it immediately notifies the higher headquarters. The battlefield commander will then evaluate the situation and commit the TCF to defeat the Level III threat. The TCF is normally a combined-arms organization tailored by the corps or division G3, based on METT-TC. The TCF normally receives fire, aviation, or other support needed to fight and defeat the threat. Once the TCF is identified and before it is committed to battle, it will conduct direct coordination with the MP or other response forces regarding the exchange of reconnaissance information, battle-handoff procedures, and contingency plans for TCF operations. Once the TCF is committed, the MP unit performing as the response force becomes OPCON to the TCF commander.

MP AS THE TCF

3-28. The MP brigade or battalion is capable of conducting TCF missions only when properly augmented. MP augmentation may be in the form of fires, small combat units, aviation assets, or CAS. The specific type of augmentation is METT-TC dependent. Additionally, the MP C^2 headquarters must receive the respective liaison officers to ensure that augmentation forces are synchronized and employed according to their capabilities. The MP commander's situational awareness and battlefield visualization are key elements to TCF operations. Once designated as the TCF, the MP unit commander establishes liaison with the appropriate rear CP to obtain—

- The current rear-area IPB.
- The friendly unit disposition.
- Defense plans.
- Priorities for protection.
- The fire-support plan.

3-29. Based on the above information, the MP commander conducts his own IPB and develops a concept of operations. He then forwards it to the appropriate higher echelon for coordination and approval.

3-30. The MP's ability to employ organic MP assets as part of the TCF is limited by the following factors:

- **MP availability.** Normally, all MP assets available are committed at all times. The specific function and scope that MP units perform during the operation are determined by the Army commander's needs, the intensity of the conflict, and the availability of MP resources. The commander, with advice from the MP leader, must decide which MP operations must be scaled back, delayed, or shifted before the MP unit can be designated as part of the TCF.
- **MP dispersion.** MP units are normally displaced over a large geographical area. Technological capabilities and mobility allow them to operate over great distances. In today's battlefield, a typical MP company employment covers between 1,000 and 1,200 square kilometers and performs numerous missions in support of all five functions. The distance between elements, the reprioritization and movement of other MP units, the difficult terrain, poor roads, and bad weather may slow down the MP's commitment as a TCF.

Chapter 4

Military Police Functions

Military police support the Army commander's mission to win the battle. They help the commander shape the battlefield so that he can conduct decisive operations to destroy enemy forces, large or small, wherever and whenever the Army is sent to war.

MAJ(P) Anthony Cruz

The MP Corps supports shaping and sustainment operations while performing its five basic functions as a flexible, economy-of-force organization. Through these functions, MP units are able to provide the commander with an array of CS operations across the full spectrum of military operations (see Table 4-1).

Table 4-1. MP Functions

	Functions				
	MMS	**AS**	**PIO**	**L&O**	**I/R**
Subtasks	Support to river-crossing and breaching operations and passage of lines Straggler and dislocated-civilian control Route R&S MSR regulation enforcement	Recon operations ADC Base/air-base defense Response-force/ TCF operations Critical site, asset, and HRP security Force protection/ physical security Antiterrorism	IPB support PIAP Active and passive roles Information collection and dissemination Joint, interagency, and multinational coordination	Law enforcement Criminal investigations US Customs operations Related L&O training	EPW/CI handling US military prisoner handling Populace and resource control Dislocated civilians
NOTE: Subtasks not all-inclusive.					

OVERVIEW

4-1. MP assets are limited. Specific functions are performed at any given time and are determined by the supported commander's need, the intensity of the conflict, and the availability of MP resources. The supported commander, through the command's PM, sets the priorities for MP operations.

4-2. The PM (based on METT-TC and the available assets) continuously evaluates the trade-off between the MP support that the commander requires and the MP support that can be provided. To meet the priorities set by the commander's tactical plan, the PM recommends the allocation and employment of MP assets for MP combat, CS, and CSS operations.

MANEUVER AND MOBILITY SUPPORT

4-3. The MMS function involves numerous measures and actions necessary to support the commander's freedom of movement in his AOR. The MP expedite the forward and lateral movement of combat resources and ensure that commanders get forces, supplies, and equipment when and where they are needed. This is particularly important in the modern battlefield where there is a greater geographical dispersal of forces and lengthened LOC.

4-4. The MP maintain the security and viability of the strategic and tactical LOC to ensure that the commander can deploy and employ his forces. The MP support the commander and help expedite military traffic by operating traffic-control posts (TCPs), defilades, or mobile patrols; erecting route signs on MSRs or alternate supply routes (ASRs); or conducting a reconnaissance for bypassed or additional routes. The MP move all units quickly and smoothly with the least amount of interference possible.

4-5. As part of the MMS function, the MP support river-crossing operations, breaching operations, and a passage of lines. They also provide straggler control, dislocated-civilian control, route reconnaissance and surveillance (R&S), and MSR regulation enforcement.

RIVER-CROSSING OPERATIONS

4-6. US forces conduct river-crossing operations to move a large force across a river obstacle with a minimum loss of momentum. The MP play a vital role by assisting the force commander in crossing the river as quickly and efficiently as possible. The crossing is usually planned and conducted by the headquarters directing the crossing. As such, a division crossing operation is conducted by a corps. Whether a brigade or division is crossing, the division MP company may also cross to provide uninterrupted support to the division. In these instances, there is a total reliance on the corps MP to support the crossing. The same is true for breaching operations and a passage of lines.

4-7. MP support for river-crossing operations reduces the crossing time and promotes the efficient movement of vehicles. It reduces congestion, speeds the crossing, and enables the maneuver commander to continue his momentum toward his primary objective. The MP establishes staging and holding areas and TCPs to control movement to and from these areas (according to the traffic-control plan). The MP may be called on to provide security for crossing forces at the crossing sites. In most cases, the MP TCPs and engineer regulation points (ERPs) are located on both sides of the river to improve communications and coordination between the units.

4-8. MP employment for river crossing is influenced by METT-TC. The number and placement of MP assets supporting a river-crossing operation varies with the size of the crossing force, the direction of the crossing (forward or retrograde), and the degree of enemy resistance expected or encountered.

4-9. The MP operating inside the crossing areas are OPCON to the crossing-area commander for the duration of the operation. The MP operating outside of the crossing area are under the command of their appropriate echelon commander.

4-10. The main thrust of MP support to river-crossing operations is within the immediate river-crossing site. The MP direct units to their proper locations (such as holding areas and staging areas) and ensure that units move through the area within the time listed on the movement schedule. This is a highly critical aspect of river crossing because the number of crossing sites is limited. MP assets are placed where they can stress MMS operations on MSRs leading into the crossing area.

4-11. The MP also provide AS to allow crossing forces to cross the river without losing momentum or forces. On both near and far sides, the MP are used to recon the crossing unit's flanks and rear to enhance security (see FM 19-4).

BREACHING OPERATIONS

4-12. Breaching operations are conducted to allow forces to maneuver through obstacles. Obstacle breaching is the employment of a combination of tactics and techniques to advance an attacking force to the farside of an obstacle that may be covered by fire. It is perhaps the most difficult combat task a force can encounter. Breaching operations begin when friendly forces detect an obstacle, and they end when the battle handover has occurred between the follow-on forces and a unit conducting the breaching operation (see FM 90-13-1).

4-13. The MP support breaching operations in numerous ways. MP assets are employed based on METT-TC, the available resources, and the commander's priorities. As a minimum, MP support may include, but is not limited to—

- Establishing TCPs along routes leading to or departing from the breaching site.
- Establishing holding areas.
- Establishing TCPs at the breaching site.
- Assisting engineers with temporary route signs.
- Establishing straggler-control operations.
- Conducting AS operations.

4-14. The most critical MP support is provided at the breaching site. The MP provide the commander with a means to control traffic flow to appropriate lanes. When multiple lanes branch off of a single far-recognition marker, the MP assist in directing the formation through various lanes. They also assist in modifying the traffic flow when lanes have been closed for maintenance or expansion. The MP conduct close coordination with the crossing-force commander and the TF commander executing the breaching operation. The MP enable the commander to make last-minute changes in traffic flow, thereby giving him increased flexibility to react to the enemy situation.

PASSAGE OF LINES

4-15. A passage of lines is an operation in which a force moves forward or rearward through another force's combat positions with the intention of moving into or out of contact with the enemy. The passage of lines is a high-risk military operation that requires close coordination between the passing unit, the stationary unit, and the MP providing the support.

4-16. The MP help reduce confusion and congestion during a passage of lines. They provide security in areas surrounding passage points and passage lanes to ensure that the passing unit has priority for using routes to and through the areas. The headquarters directing the operations sets the route's priority. The MP can support a forward, rearward, or lateral passage of lines. Before the actual operation, the MP in the AO conduct an area or zone reconnaissance to become familiar with the routes to, through, and beyond the area of passage. This enables the MP to extend the commander's C^2 by providing directions at passage points and by guiding the units through the passing lanes. Maintaining unit integrity and reducing incidents of stragglers is vital to maintaining the passing unit's momentum in a forward passage of lines. The MP perform aggressive straggler- and dislocated-civilian-control operations to prevent possible infiltration of the enemy.

4-17. A passage of lines is usually planned and coordinated by the headquarters directing the passage. A division's passage of lines is planned and coordinated by the corps headquarters. The detailed plans are made and coordinated between the divisions involved. Close coordination between the division and corps PMs is essential. An MP unit may be the unit involved in passing through the lines of another unit. When conducting a delay of a Level II threat, the MP are likely to conduct a passage of lines with the TCF. To avoid fratricide, close coordination between the MP response-force commander and the TCF is imperative (see FM 19-4).

STRAGGLER CONTROL

4-18. Mobile patrols, TCPs, and checkpoint teams return stragglers to military control as part of their operations. Most stragglers are simply persons who become separated from their command by events in the CZ or while moving through the COMMZ. If a straggler is ill, wounded, or in shock, an MP must give him first aid and, if needed, call for medical evacuation (MEDEVAC). If a straggler is uninjured, an MP directs him to his parent unit or to a replacement unit (as command policies dictate). The MP ensure that stragglers attempting to avoid return to their units are escorted back to their command (as a minimum).

4-19. The MP set up special posts for straggler control following NBC attacks or major enemy breakthroughs that result in large numbers of lost, dazed, and confused military personnel. Mobile MP teams operate between posts, and they also direct or collect stragglers. Straggler collection points may be needed if many stragglers are present in a combat theater. If allied forces are present in the theater, each nation establishes a collection point for its own personnel. MP teams are aware of each allied location and are prepared to assist allied soldiers in returning to their respective command. The MP use available transportation assets to transfer stragglers from TCPs and checkpoints to a straggler collection point. At the collection point, they are screened and sorted for removal to a medical facility or returned to their units to reconstitute the tactical commander's combat force.

4-20. The MP report information about stragglers with whom they come in contact. This information is compiled by the MP headquarters and forwarded through appropriate channels to the higher command. Information given by stragglers that is of immediate tactical value is reported without delay.

DISLOCATED-CIVILIAN CONTROL

4-21. The MP expediting traffic on MSRs may encounter dislocated civilians that could hinder military traffic. The MP assist and divert dislocated civilians from MSRs and other areas to I/R facilities. They may also deny the movement of civilians whose location, direction of movement, or actions may be a threat to themselves or to the combat operation. The HN government is responsible for identifying routes for the safe movement of dislocated civilians out of an AO. If needed, the MP assist the civil-affairs unit and the HN in redirecting dislocated civilians to alternate routes established by the HN government.

4-22. The US forces do not assume control of dislocated civilians unless requested to do so by the HN or unless operating in an environment with a hostile government. When the senior US commander assumes responsibility, the MP coordinate with civil affairs to set up TCPs at critical points along the route to direct dislocated civilians to secondary roadways and areas not used by military forces. As required, MWD teams may be used as a show of force or as a deterrent to assist with uncooperative personnel.

ROUTE R&S

4-23. The MP conduct hasty and deliberate route reconnaissances to obtain information on a route and nearby terrain from which the enemy can influence troop movement. A route reconnaissance focuses on continually monitoring the condition of MSRs, ASRs, and other areas. MP patrols look for restricting terrain, effects of weather on the route, damage to the route, NBC contamination, and enemy presence or absence. When enemy activity is spotted, the MP report it, maintain surveillance, and develop the situation. To gather information for proposed traffic plans, they look at the type and number of available routes; and they check load classifications, route widths, obstructions, and restrictions.

MSR REGULATION ENFORCEMENT

4-24. The MP undertake MSR regulation enforcement to keep the routes free for DSS operations. MP units support the command's MSR regulation measures as stated in the traffic-regulation plan (TRP). The TRP contains specific measures to ensure the smooth and efficient use of the road network. It assigns military route numbers and names, the direction of travel, highway regulation points, and preplanned MP TCPs. Most importantly to the MP, it gives the route's control classification. The MP ensure that classified routes are used only by authorized traffic. Vehicles traveling on roads too narrow for their passage or on roads unable to support their weight can obstruct the route.

4-25. To expedite traffic on MSRs, the MP operate special circulation control measures such as—

- Temporary route signing.
- Static posts such as TCPs, roadblocks, checkpoints, holding areas, and defilades at critical points.
- Mobile teams patrolling between static posts and monitoring traffic and road conditions.

4-26. They also gather information on friendly and enemy activities and help stranded vehicles and crews. The MP also place temporary route signs to warn of hazards or to guide drivers unfamiliar with the route. Using these measures, the MP exercise jurisdiction over the road network in the AO and coordinate with the HN (whenever possible) to expedite movement on MSRs.

AREA SECURITY

4-27. The MP perform the AS function to protect the force and to enhance the freedom of units to conduct their assigned missions. The MP who provide AS play a key role in supporting forces in rear-area and sustainment operations. The MP are a response force that delays and defeats enemy attempts to disrupt or demoralize military operations in the AO. The MP's mobility makes it possible for them to detect the threat as they aggressively patrol the AO, MSRs, key terrain, and critical assets. The MP's organic communications enable them to advise the appropriate headquarters, bases, base clusters, and moving units of impending enemy activities. With organic firepower, the MP are capable of engaging in decisive operations against a Level II threat and delaying (shaping) a Level III threat until commitment of the TCF.

4-28. Throughout all aspects of the AS function, the MP perform counteractions to protect the force and to prevent or defeat a Level II threat operating within the MP's AO. MP countermeasures may include implementing vulnerability assessments, developing procedures to detect terrorist actions before they occur, hardening likely targets, and conducting offensive operations to destroy the enemy. The MP use checkpoints and roadblocks to control the movement of vehicles, personnel, and materiel and to prevent illegal actions that may aid the enemy. The use of these control measures serves as a deterrence to terrorist activities, saboteurs, and other threats. However, at the same time, checkpoints and roadblocks expose the MP to these potential threats. To counter this fact, the MP may upgrade or harden vehicles and defensive positions.

4-29. The MP provide combat power to protect the C^2 headquarters, equipment, and services essential for mission success. The MP provide the battlefield commander with a light, mobile fighting force that can move, shoot, and communicate against any threat. Major subtasks associated with the AS function include reconnaissance operations; area damage control (ADC); base/ air-base defense; response-force operations; and critical site, asset, and high-risk personnel (HRP) security.

RECONNAISSANCE OPERATIONS

4-30. As part of their AS mission, the MP serve as the eyes and ears of the battlefield commander by seeking out the enemy and reporting information obtained by recon patrols. The MP conduct area and zone reconnaissances, screening, surveillance, and countersurveillance to gain information to help guard against unexpected enemy attacks in the AO. The MP monitor likely avenues of approach and potential LZs and DZs. They become familiar with towns and other populated areas, ridgelines, woods, and other terrain features from which the enemy can influence movements along road networks. The MP pay close attention to areas near facilities designated critical by the commander. These areas include key MSR bridges and tunnels, depots, terminals, logistics-support bases, ammunition supply points (ASPs), communications centers/nodes, and C^2 headquarters. The MWD teams provide explosive detection and personnel detection/tracking capabilities that enhance reconnaissance operations (especially in urban terrain).

AREA DAMAGE CONTROL

4-31. MP units take measures to support ADC before, during, and after hostile actions or natural and man-made disasters. The ADC actions integrate CS and CSS functions for many units. Engineers, medical personnel, and Army aviators work closely to ensure quick relief operations. The MP provide MSR regulation enforcement, refugee control, and some local security when required. As with reconnaissance operations, the MP may use MWD explosive-and personnel-detection capabilities to augment all MP missions in rear-area and sustainment operations.

BASE/AIR-BASE DEFENSE

4-32. The MP are the base and base-cluster commanders' links for detection, early warning, and employment against enemy attacks. The information gathered is dispersed throughout the rear area to help apprise the commander of enemy activities near bases. Base defense is the cornerstone of rear-area security. When the threat exceeds the base/base-cluster capability, the base/ base-cluster commander requests MP assistance through the appropriate C^2 element.

4-33. Air-base defense requires special MP coordination with the US Air Force (USAF). The MP treat air bases like any other base or base cluster. A USAF air base may house the base-cluster commander, or it may be a cluster by itself. The MP are responsible for the air base's external defense. Its internal defense is primarily the responsibility of the Air Force's security forces. The security force provides in-depth defense for weapons, weapons systems, command centers, personnel, and other priority resources established by the base commander.

4-34. The security force is trained and equipped to detect, delay, and deny Level I and II threats. If a Level III threat is present, the security force is tasked with delaying actions; however, the HN, a sister service, or other support must be employed to defeat this threat. If the security force requires assistance to defeat a Level II threat, it may rely on MP response forces or another response force to assist in the defense. If available, the MP response force will react to the air-base defense, just as it would for any other base or base cluster within the MP's AO. However, the key to successful MP employment depends on the critical exchange of information before and during the MP employment. Good communications, an understanding of the defense plan, and liaison operations are vital in preventing responding forces from entering a situation that could result in fratricide.

RESPONSE-FORCE OPERATIONS

4-35. The MP are the base and base-cluster commanders' response force against enemy attacks in rear-area or sustainment operations. The MP gather information about the enemy while performing missions throughout the AO. This information provides commanders with enemy activity near bases. When needed, the MP provide a mobile response force to respond to bases under attack and to destroy the enemy. A base commander's defense plan is the cornerstone for protecting rear-area and sustainment operations. The base commander is responsible for defeating all Level I threats. When this threat exceeds his capabilities, he requests MP support. The MP located near bases or patrolling or conducting AS operations consolidate their forces, respond as quickly as possible, and conduct combat operations to destroy the enemy. If needed, the MP conduct a battle handover to the TCF.

4-36. MP forces performing as a response force are capable of conducting the following offensive operations:

- A movement to contact.
- A hasty ambush.
- A hasty attack.
- A delay.
- A call for fire.
- A repel attack against critical sites.
- A defense of critical sites.

4-37. To conduct these missions, the MP consolidate into squads or platoons to delay, defeat, or defend against the threat. See FMs 71-3 and 71-100 for more information on battle-handover operations.

CRITICAL SITE, ASSET, AND HRP SECURITY

4-38. The MP perform their AS function across the entire designated AO. When the MP provide security around a critical site or asset, they usually provide a mobile security screen, taking advantage of its weapons and communications platforms. This standoff protection detects and defeats the threat before it can move within direct-fire range of the facilities. The MP may be tasked to provide detail security to key facilities, assets, and personnel.

4-39. The MP provide security to major CPs and other facilities within the AO. Their employment maximizes mobility, lethality, and communications capabilities as a security screen. They may be required to establish local AS measures (such as checkpoints and listening/observation posts) to further protect these facilities. The MP provide internal access-control points to critical facilities, and they act as a response force. When the critical CP relocates, the MP provide in-transit security. Other types of critical site security include ASPs; deep-water ports; petroleum, oil, and lubricants (POL) terminals and pipelines; trains and railways; and air bases.

4-40. The MP may provide convoy security for top-priority units transporting especially critical supplies to combat forces. MP assets should be employed primarily on aggressive patrolling, route, area, and zone reconnaissance measures that would create a safe and secure environment for all types of vehicular and unit movement.

4-41. The MP and the CID provide protective services to designated key personnel by providing access control to restricted areas within CPs, providing in-transit security, or providing static security measures around the clock. The MP coordinate with the CID when close-in protection of key personnel is needed. The MP and the CID also provide training for personal-protection countermeasures. The MWD teams may be employed to enhance MP and CID detection and protection capabilities.

INTERNMENT AND RESETTLEMENT

4-42. The Army is the Department of Defense's (DOD's) executive agent for all EPW/CI operations. Additionally, the Army is DOD's executive agent for long-term confinement of US military prisoners. Within the Army and through the combatant commander, the MP are tasked with coordinating shelter, protection, accountability, and sustainment for EPWs/CIs. The I/R function addresses MP roles when dealing with EPWs/CIs, dislocated civilians, and US military prisoners.

4-43. The I/R function is of humane as well as tactical importance. In any conflict involving US forces, safe and humane treatment of EPWs/CIs is required by international law. Military actions on the modern battlefield will result in many EPWs/CIs. Entire units of enemy forces, separated and disorganized by the shock of intensive combat, may be captured. This can place a tremendous challenge on tactical forces and can significantly reduce the capturing unit's combat effectiveness. The MP support the battlefield commander by relieving him of the problem of handling EPWs/CIs with combat forces. The MP perform their I/R function of collecting, evacuating, and securing EPWs throughout the AO. In this process, the MP coordinate with MI to collect information that may be used in current or future operations.

4-44. Although the CS MP unit initially handles EPWs/CIs, modular MP (I/R) battalions with assigned MP guard companies and supporting MWD teams are equipped and trained to handle this mission for the long term. A properly configured modular MP (I/R) battalion can support, safeguard, account for, guard, and provide humane treatment for up to 4,000 EPWs/CIs; 8,000 dislocated civilians; or 1,500 US military prisoners.

EPW/CI HANDLING

4-45. The MP are tasked with collecting EPWs/CIs from combat units as far forward as possible. The MP operate collection points and holding areas to temporarily secure EPWs/CIs until they can be evacuated to the next higher echelon's holding area. The MP escort-guard company assigned to the MP brigade (I/R) evacuate the EPWs/CIs from the corps's holding area to the COMMZ's internment facilities. The MP safeguard and maintain accountability, protect, and provide humane treatment for all personnel under their care.

4-46. In a mature TO, MP (I/R) units process EPWs/CIs collected by MP teams and other units in the CZ. MP guard companies assigned to the MP (I/R) units guard EPWs/CIs at designated camps (see FM 19-40).

POPULACE AND RESOURCE CONTROL

4-47. Populace and resource control (PRC) denies adversaries or insurgents access to the general population and resources. The MP supports civil-affairs personnel and the tactical commander in planning and conducting PRC programs employed during all military operations. These programs may consist of curfews, movement restrictions, resettling dislocated civilians, licensing, ration control, regulation enforcement, amnesty programs, inspecting facilities, and guarding humanitarian-assistance distributions. The MP also direct dislocated civilians to resettlement camps where they are cared for while NGOs work to coordinate their relocation.

4-48. The MP's security capability, acceptability, and interface with the populace make them suitable as the primary forces of choice in these operations. The MP I/R units are specifically designed to fill this need (see FM 41-10).

US MILITARY PRISONERS CONFINEMENT

4-49. The MP detain, sustain, protect, and evacuate US military prisoners. Whenever possible, soldiers awaiting trial remain in their units, unless reasonable grounds exist to believe that they will not appear at the trial, the pretrial hearing, or the investigation or that they will engage in serious criminal misconduct. Under either of these two pretrial confinement instances, the commander must also reasonably believe that a less severe form of restraint (such as conditions of liberty, restriction in lieu of apprehension, or apprehension) are inadequate. When these circumstances exist and other legal requirements are met, US military personnel may be placed in pretrial confinement under the MP's direct control. Convicted military prisoners are moved as soon as possible to confinement facilities outside of the AO.

4-50. MP confinement operations parallel (but are separate from) the MP's EPW/CI operations. No member of the US armed forces may be placed in confinement in immediate association with an EPW or other foreign nationals who are not members of the US armed forces. A confinement facility is maintained within the TO only if distance or the lack of transportation requires such a facility. When military prisoners are retained in theater, temporary field detention facilities may be established in the CZ and a field confinement facility in the COMMZ (see FM 19-40 and AR 190-47).

LAW AND ORDER

4-51. The L&O function consists of those measures necessary to enforce laws, directives, and punitive regulations. The MP's L&O function extends the battlefield commander's C^2. The MP, in close coordination with the CID, work to suppress the chance for criminal behavior throughout the AO. By coordinating and maintaining liaison with other DOD, HN, joint, and multinational agencies, the MP at all levels coordinate actions to remove conditions that may promote crime or that have the potential to affect the combat force. Crime-prevention measures and selective enforcement measures are also performed as part of other functions. For example, the MP investigate traffic accidents and regulate traffic as part of the MMS function. The L&O function includes major areas such as law enforcement, criminal investigations, and support to US Customs operations. The primary units conducting L&O are the L&O detachments, customs teams, and CID units. Both the MWD team and the MP company (CS) also support the L&O function.

LAW ENFORCEMENT

4-52. Law-enforcement operations assist the battlefield commander in preserving his force. The MP dedicate assets to conduct law-enforcement operations based on the commander's needs. Since MP L&O and CID assets may be limited during the initial stages of any operation, the PM depends on the supported commander's development of an effective crime-prevention program and uses established investigative tools (such as inquiries and AR 15-6 investigations) to enforce rules. This will allow the PM to employ limited MP assets to perform other functions. Law-enforcement operations include responding to civil disturbances, conducting raids, investigating traffic accidents, conducting vehicle searches, supporting the commander's force-protection program, and providing support to HN and civil-enforcement agencies. Law enforcement also includes employing special-reaction teams (SRTs), marksman/observer (M/O) teams, and MWD teams.

4-53. The Army conducts counterdrug-support operations that generally fall within several DOD counterdrug-mission categories. The MP support the Army's role rather than directly participating in civil law-enforcement activities (such as searches, seizures, and arrests). When tasked, the MP provide training to law-enforcement agencies in common soldier skills, physical security, and tactical planning and operations. US Code (USC) Title 18, Posse Comitatus Act, Section 1385 prohibits the use of DOD assets to enforce civilian law (federal and state) except in cases and under circumstances expressly authorized by the Constitution or by an act of Congress.

4-54. In multinational operations, the MP may assist with the creation of multinational police units. Circumstances that may support the establishment of these police forces include existing or negotiated terms of international agreements or security-assistance programs, a multinational operational agreement, or appropriate military directives. The MP provide the capability to train foreign MP and/or reconstitute indigenous constabulary forces as part of stability and support operations. The MP can provide the initial mentoring to these forces and provide temporary, emergency law-enforcement capabilities until the foreign military or civilian police forces are functional (see FM 100-8).

CRIMINAL INVESTIGATIONS

4-55. The MP investigate offenses against US forces or property committed by persons subject to military law. Investigations against minor crimes (such as low-value, personal-property thefts or simple assaults) are normally investigated by the MP's L&O detachment. Investigations against major incidents involving wrongful death, serious bodily injury, and war crimes are referred to special agents of the USACIDC. The USACIDC conducts death investigations in the absence of HN agreements or in conjunction with the HN. The USACIDC special agents are authorized to investigate any alleged criminal conduct in which there is an Army interest unless prohibited by law or higher authority.

4-56. The USACIDC's investigative authority and investigative responsibility outside of the US are determined by international treaty or agreement (including status of forces agreements [SOFAs]), the policies of the HN government, the US ambassador, and AR 195-2. In the absence of such provisions, the following guidelines apply:

- On Army-controlled installations, the USACIDC has the authority to investigate alleged crimes.
- Outside of an Army-controlled installation, the USACIDC may investigate after coordinating with HN authorities.

4-57. In all environments, the USACIDC has the responsibility to investigate all felony crimes involving Army personnel, DA civilians and agencies, and companies working for the Army. The USACIDC investigates war crimes and crimes involving personal and government property affecting the Army's mission (logistics security [LOGSEC]). Other investigations (such as those based on international treaties, SOFAs, and joint investigations with the HN) may be undertaken if requested by the supported commander in support of the overall Army mission. See Chapter 9 for a complete discussion of the USACIDC.

US CUSTOMS OPERATIONS

4-58. The MP support the US Customs Service (USCS), the US Department of Agriculture (USDA), other federal agencies, joint staffs, and commanders who enforce the laws and regulations of the US concerning customs, agriculture, and immigration border clearances. Support to the USCS also includes assistance to federal agencies to eliminate the illegal traffic of controlled substances and other contraband through Army channels. MP support to customs operations are normally performed by specially trained MP customs teams. Although other MP units are not trained in all facets of customs operations, they may assist MP customs teams, the USCS or the USDA, and other federal agencies in the enforcement of applicable laws and regulations. When tasked, the MP/CID supports the investigation of violations of US Codes, DOD or DA regulations, and applicable provisions of SOFAs.

4-59. The MP report violations of customs laws, regulations, inspections, and investigative results to the installation's PM, the supported commander, and affected units. During redeployment from outside the continental US (OCONUS) to CONUS installations, the MP support the USCS or USDA efforts to ensure that personnel, equipment, and materiel meet customs, immigration, and agriculture requirements as stated by all applicable laws and regulations. As with other functions, MWD teams may be employed in support of customs operations for the detection of explosives or narcotics.

> During operation Just Cause, an MP platoon temporarily assumed the customs mission at the main terminal of the Torrijos International Airport located just outside of Panama City. Their mission supported the air evacuation of foreign-national civilians and the redeployment of some of the initial-entry US forces.

POLICE INTELLIGENCE OPERATIONS

4-60. The PIO function supports, enhances, and contributes to the commander's protection program, situational awareness, and battlefield visualization by portraying relevant threat information that may affect his operational and tactical environments. This threat information—whether it is police, criminal, or combat information—is gathered while conducting MP functions. The PIO function—

- Demonstrates the MP's/CID's capability to collect relevant threat information actively or passively.
- Ensures that all information collected while conducting MMS, AS, I/R, and L&O functions continues to be reported through the proper channels so that it can be analyzed by the Intelligence Officer (US Army) (S2) or the Assistant Chief of Staff, G2 (Intelligence) (G2) with support from the appropriate MP echelon.
- Coordinates with USACIDC elements to employ data developed by the USACIDC's programs. These programs include—
  - The Combating Terrorism Program as outlined in AR 525-13 and CIDR 195-1.
  - The Criminal Intelligence Program (CIP).
  - Personal-security vulnerability assessments (PSVAs).
  - A crime threat analysis.
  - Logistics-security threat assessments (LSTAs).

NOTE: The MP/CID must ensure that criminal information is released according to existing controls and restraints.

- Maintains constant liaison and communication with the higher echelon S2/G2; psychological-operations (PSYOP) units; HN police and other law-enforcement agencies; joint, combined, interagency, and multinational forces; the staff judge advocate (SJA); the CMOC; civil-affairs teams; and the force-protection officer.

4-61. The MP brigade commander, the battalion commander, and the PM are responsible for the PIO function. As such, each one must determine the best way to employ the available staff resources to monitor the execution of the PIO function within his command.

IPB PROCESS AND PIO FUNCTION

4-62. The IPB is a systematic, continuous process for analyzing the threat and environment in a specific geographic area. It is designed to support staff estimates and military decision making. Applying the IPB process helps the commander selectively apply and maximize his combat power at critical points in time and space on the battlefield by—

- Determining the threat's likely COA.
- Describing the environment the unit is operating within and the environmental effects on the unit.

4-63. The IPB process consists of the following four steps:

- **Define the battlefield environment.** The S2/G2 identifies the battlefield characteristics that will influence friendly and threat operations, establishes the limits of the area of interest (AOI), and identifies gaps in current intelligence holdings.
- **Describe the battlefield effects.** The S2/G2 evaluates the environmental effects with which both sides must contend. The S2/G2 identifies the limitations and opportunities that the environment offers on the potential operations of friendly and threat forces. This evaluation process focuses on the general capabilities of each force until COAs are developed later in the IPB process. This environmental assessment always includes an examination of terrain and weather, but it may also include discussions of characteristics of geography and infrastructure and their effects on friendly and threat operations.
- **Evaluate the threat.** The S2/G2 and his staff analyze the command's intelligence holdings to determine how the threat normally organizes for combat and conducts operations under similar circumstances. When facing a well-known threat, the S2/G2 can rely on historical databases and threat models. When operating against a new or less known threat, he may need to develop his intelligence databases and threat models concurrently.
- **Determine the threat's COA.** Given what the threat normally prefers to do and the effects of the specific environment in which he is operating, his likely objectives and the COAs available to him are determined. The S2/G2 develops enemy COA models that depict the threat's available COAs. He also prepares event templates and matrices that focus intelligence and identify which COA the enemy will execute (see FM 34-130).

4-64. Although the S2/G2 has the staff responsibility for the command's IPB, he is not the only one who conducts or needs to understand and use the IPB. Every Army commander and staff member must understand and apply the IPB process during the staff planning process. The MP on the battlefield are no exception. The MP employ the IPB process as their first step in developing and implementing the PIO function within their commands.

4-65. During the IPB process, the S2/G2 uses all available databases, intelligence sources/products, and related MI disciplines to analyze the threat and the environment. The PIO function supports this process by providing the S2/G2 with collected police, criminal, and combat information that can directly and significantly contribute to the success of the MI effort. In addition to the combat information, the PIO function provides additional information on possible criminal threats and COAs that may support the S2's/G2's IPB process and that can be used by the commander to upgrade the force-protection posture.

POLICE INFORMATION ASSESSMENT PROCESS

4-66. Like the S2/G2 uses the IPB process to continuously analyze the threat and the environment in a specific geographical area, MP leaders use the PIAP as a tool to continuously collect, organize, interpret, and gain access to police/criminal information in support of the IPB process. Criminal trends may have an impact on the tactical scenario, and the PIAP is a method used to consider this threat and its impact on friendly forces. MP leaders cannot use the PIAP as a substitute for the IPB process—the PIAP complements the IPB (see Appendix E).

PASSIVE MODE PIO

4-67. Every MP conducts the PIO function in a passive mode during their normal day-to-day operations and across the full spectrum of military operations. In the passive mode, PIO are not a stand-alone function; as such, they cannot be separated from other MP functions.

4-68. During the performance of MMS, AS, I/R, and L&O functions, the MP develop and exchange information with other organizations in the AO. The MP obtain information through contact with civilians, NGOs, IHOs, local and HN police, multinational police, and other security forces. If the MP receive, observe, or encounter information (police, criminal, or combat) while performing these functions, they will immediately submit a report to relay information up the chain of command. This report may be in the form of a size, activity, location, unit, time, and equipment (SALUTE) report; a spot report (SPOTREP); or another appropriate report. When the higher echelon (brigade, battalion, or PM) receives this information, it is simultaneously integrated into the ongoing IPB/PIAP and forwarded to the higher echelon S2/G2 (see Figure 4-1, page 4-16).

4-69. If police/criminal information is obtained, the MP—

- Update the previous police/criminal estimates provided to the S2/G2.
- Identify new or potential criminal threats or trends in the AO.
- Consider recommending that the supported commander upgrade the force-protection level.
- Notify adjacent units of the potential criminal threat that may affect their forces.
- Consider reprioritizing MP support to the identified threat area.
- Share information with HN/local police and other agencies.

4-70. If combat information is obtained, the MP—

- Forward the information to the higher headquarters S2/G2.
- Forward the information to the MP chain of command, integrate it into the MP's IPB process and, if necessary, take appropriate action.
- Notify the adjacent unit of the potential threat that may affect their forces.

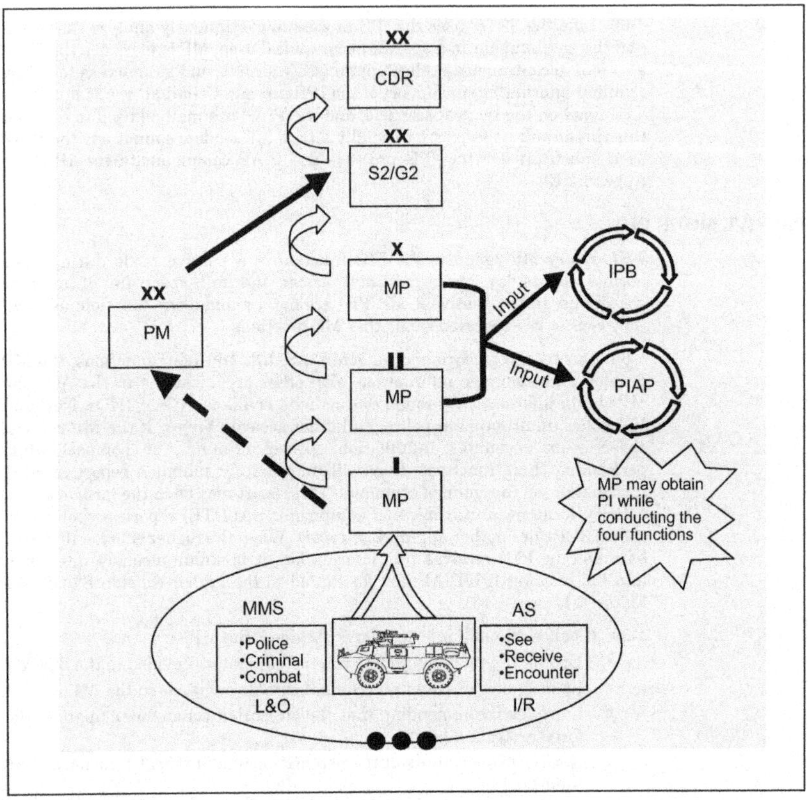

Figure 4-1. PIO—Passive Mode

> During Operation Uphold Democracy, an MP team was conducting a TCP as part of a cordon and search operation in Port-au-Prince, Haiti. While performing the task, two civilians approached the MP team informing them of criminal activity in the neighborhood. During the interview, the MP team prepared a sketch of a house and surrounding areas. The team also obtained information describing the criminals and their weapons. Recognizing that the criminal activity was in fact the actions of a political/mercenary group named FRAP, the MP team radioed the platoon leader and forwarded the field interview to higher headquarters. Two days later, a unit from the 10th Mountain Division raided the house, capturing weapons, ammunition, and equipment.

4-71. The preceding vignette demonstrates the MP performing the PIO function in the passive mode. The MP team received the information while conducting a TCP and submitted it through the appropriate chain of command, which resulted in an action taken. This example stresses the importance of submitting information up the chain of command regardless of whether it may be police, criminal, or combat information.

ACTIVE MODE PIO

4-72. The MP perform the PIO function in the active mode and across the full spectrum of military operations when directed by higher headquarters. In this mode, the MP conduct specific MMS, AS, I/R, and L&O missions with the intent to collect information actively in support of the S2's/G2's IPB process or the PIAP.

4-73. When the S2/G2 identifies a gap in the command's knowledge of the threat and the current threat situation, it may be recommended to be included as priority intelligence requirements (PIR). The S2/G2 will then develop a collection plan to assist him in filling this gap. Part of his collection strategy is to select the best collectors available to cover each intelligence requirement. After a thorough analysis (which includes availability, capability, and performance history), the collection manager identifies which collection asset can best be used in response to a given requirement, and the Operations and Training Officer (US Army) (S3)/G3 tasks the asset. If the S2/G2 determines that the MP is the right force to serve as collectors, the S3/G3 will go through the appropriate request channels and task the PM. The S2/G2 will then provide the PM with a specific guideline and a prioritized collection requirement (see FM 34-2).

4-74. On receipt of the mission, the PM will conduct a mission analysis to decide which specific MP function (MMS, AS, I/R, or L&O) is needed to satisfy the S2's/G2's requirement. Once the analysis is completed and the appropriate function selected, the PM will then task subordinate units with the collection mission. Once the mission is completed, the PM may receive another collection tasking or continue with previous MP tasks.

4-75. When the MP commander or the PM conducts the PIAP, he may also encounter a police/criminal information gap. This gap may become the MP commander's police/criminal information requirements (PCIR). If the gap cannot be filled with available data from the S2/G2, the CID, the HN, and other agencies, the MP commander/PM may task subordinate MP units or request CID assistance to support the collection effort. On receipt of the mission, the MP will then conduct a mission analysis and decide which specific MP function (MMS, AS, I/R, or L&O) is needed to satisfy the PM requirement. Once the mission is completed, the MP may receive another collection tasking or continue with previous MP tasks (see Figure 4-2, page 4-18).

4-76. Since any soldier on the battlefield can report police or criminal information, the MP commander or the PM must constantly coordinate and communicate with the S2/G2, PSYOP, and other agencies to obtain information that could be of MP/CID interest. This constant coordination is a key factor for ensuring that the MP/CID has visibility over the police/criminal information that is reported through non-MP channels.

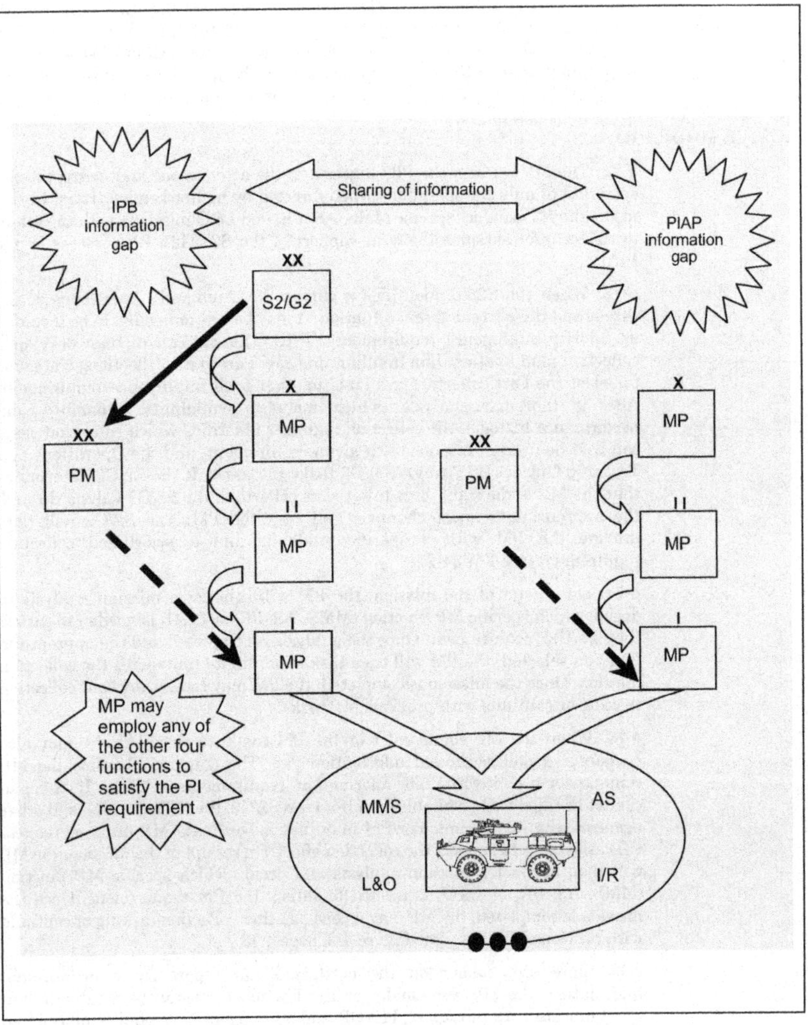

Figure 4-2. PIO—Active Mode

Chapter 5

Military Police Support to Echelons Above Corps

MP units supporting EAC perform combat, CS, and CSS operations. Like the MP supporting corps and divisions, MP units supporting EAC units perform the five MP functions based on available assets and the supported commander's needs.

USAMPS

MP support to EAC includes support to the ASCC and the TSC. The ASCC is responsible for Army Title 10 requirements in support of a combatant commander. This support includes recruitment, organization, supply, equipment, training, servicing, mobilizing, demobilizing, administration, and maintenance functions.

OVERVIEW

5-1. The ASCC may also be responsible for significant DOD- and combatant-commander-designated Army support to other services. As the senior Army commander in the AOR, the ASCC commander tailors and designates ARFOR to accomplish operational-level tasks while conducting major land operations. The ASCC's operational responsibilities include—

- Recommending the proper employment of Army-component forces to the joint-force commander or to the subunified commander.
- Accomplishing operational missions as assigned.
- Selecting and nominating specific Army units for assignment to subordinate theater forces.
- Informing the combatant commander of the Army's CSS effects on operational capabilities.
- Providing data to the supporting operations plans (OPLANs) as requested.
- Ensuring signal interoperability.

5-2. The ASCC provides administrative and logistics (A/L) services to assigned ARFOR and to those of subordinate JFCs. When appropriate, the ASCC delegates the authority for support tasks to a single subordinate Army headquarters. In major operations, the TSC (along with other EAC support commands) would be the ASCC's lead organization for planning, coordinating, executing, or providing required support functions (see FM 100-10).

5-3. The TSC is the senior Army support organization in a theater. Its commander reports to the ASCC or ARFOR commander. The TSC normally operates at the operational level of CSS with links to the strategic and tactical levels. Unity of command is the critical element that the TSC brings to the fight. The TSC is a multifunctional organization that centralizes the command, control, and supervision of support functions at EAC as directed by the ASCC or ARFOR commander. The TSC's mission is to maximize throughput and follow-on sustainment, including all CSS functions, of ARFOR and other designated supported elements. The TSC is capable of synchronizing logistics and other support operations for the ASCC. It provides area support to EAC units in the COMMZ and sustainment support to tactical forces. This support may include supply, procurement, property disposal, maintenance, transportation, field services, health services, civil-military affairs, MP support, engineer support, religious support, finance support, and personnel and administrative services.

5-4. Units and commands requiring support coordinate with the TSC support-operations staff to secure their initial support, to reestablish support, or to resolve support problems. In a fully developed theater, the TSC coordinates with a corps support command (COSCOM) for support of combat forces, although direct coordination with a division support command (DISCOM) is sometimes necessary. The TSC, augmented by a rear operations center, is also responsible for security operations as directed by the ASCC/ARFOR commander (see FM 63-4).

MILITARY POLICE SUPPORT

5-5. MP support to EAC units is provided through an array of multifunctional MP units. The nature of the operation, METT-TC, and the requirements of the supported commander will determine which type of MP unit is appropriate to augment, assign, attach, or place under OPCON to an EAC unit. The types of MP units that support EAC include CS, I/R, CID, and L&O teams (such as MWD or customs teams).

THE MP BRIGADE (CS)

5-6. The MP brigade (CS) provides MMS and AS to extended LOC within the COMMZ. These supply corridors include ports, inland waterways, railways, pipelines, airfields, MSRs, and MSR critical points. The MP support the users of the COMMZ's LOC by aggressively patrolling the area along the LOC. They play an important role in securing rear areas by performing combat operations against the threat. When properly augmented, the MP brigade headquarters may serve as the TSC's/ASCC's TCF headquarters. The MP provide MMS on the COMMZ MSRs leading into the corps's rear area. The MP implement the plans of HN and US staff elements to control the forward movement of combat resources along the LOC.

5-7. If resources are available, the MP brigade (CS) provides escorts to move US noncombatants (if present) from AA points to theater embarkation terminals. Until the MP brigade (I/R) arrives in theater, the CS MP units also perform EPW, confinement, and other operations normally performed by the MP I/R units.

5-8. The organization of an MP brigade (CS) supporting EAC includes the following:

- A brigade headquarters and headquarters company (HHC).
- Up to six MP battalions (each with up to six companies).

- Numerous L&O detachments and MWD teams.

5-9. Additionally, the ASCC's PM or commander may attach or direct OPCON of customs teams to the MP brigade (CS). Battalion and company organization in the MP brigade is the same as that in the corps MP brigade; however, METT-TC determines the number of battalions and companies. The MP brigade (CS) has additional MP companies to provide security for EAC-specific units/activities such as—

- A unified command.
- An ASCC and TSC headquarters.
- LOC seaports, airfields, and railways.
- EAC ammunition storage areas.
- EAC petroleum terminals and pipelines.

5-10. While the corps MP brigade (CS) MWD teams are employed in a GS role, the MP brigade (CS) MWD teams are employed to augment seaport security and to conduct inspections of postal items to detect explosive materiel and narcotics.

5-11. The C^2 within the MP brigade (CS) is consistent with that in any Army brigade. The MP brigade commander works directly for the EAC commander, the battalion commander works for the brigade commander, and the company commander works for the battalion commander. However, this usual C^2 relationship may be altered briefly (based on METT-TC) to enhance the overall EAC combat capability for responding to a Level II threat. For example, MP units operating within an ASG's AO may be under the OPCON of the ASG's rear-area operations center (RAOC), which directly tasks MP units responding to Level II threats. The same is true for placing MP units under the OPCON of the EAC's TCF headquarters for responding to Level III threats. Any conflict in mission priorities is resolved through MP C^2 channels.

5-12. The MP brigade commander is both the MP brigade commander and the EAC's PM. He employs his assets according to METT-TC and the commander's concept of operations. Factors affecting his employment of MP assets include the—

- Nature of the operation (joint, combined, or multinational).
- HN's ability to provide MP-related support (such as port security).
- Custody and location of EPWs/CIs during internment operations until I/R units arrive in theater.
- Number of kilometers of the MSR in relationship to movement-control requirements.
- Number and kinds of critical facilities.
- Number of HRP requiring close-in security.
- HN's ability to control the civilian populace, refugees, and dislocated civilians.
- Supply distribution strategy.
- Risk acceptance and threat in the AO.
- Communications requirements (such as using teams as relays).

5-13. Whenever possible, the MP brigade's AO coincides with the territorial responsibility of the supported command. The MP brigade commander assigns the MP battalion's AO by the above factors as well as by METT-TC. For example, the AOR for an MP battalion may be a large population center of larger geographical areas in which CSS complexes and MSRs are located. But as employment factors and the commander's needs change, so will the MP's AOR. The MP brigade commander must move and tailor his forces to meet the current and projected mission requirements. Unlike many other EAC assets, MP units require 100 percent mobility to shift AOs frequently and rapidly. The following vignette depicts the required MP flexibility to support EAC operations:

> During Operation Desert Shield/Storm, MP units were supporting and moving units throughout their AOs at a such an extraordinary rate that many of them had to relocate their headquarters multiple times just to keep pace with their changing AO.

5-14. Most EAC MP assets are employed along LOC and around areas of high troop concentration. Few EAC MP are dedicated to support fixed commitments (such as ports, air bases, and headquarters [discussed in paragraph 5-8]). When supporting fixed commitments, MP units provide a mobile security screen, and they man static positions when directed or when based on METT-TC. The MP brigade (CS) may have to plan for and actually perform the evacuation and internment of EPWs/CIs and the confinement of US military prisoners until the MP brigade (I/R) arrives.

THE MP BRIGADE (I/R)

5-15. US policy requires that all persons held in Army custody be accorded humane care and treatment from the moment of custody to their final release. The policy applies to detained or interned civilians as well as to EPWs and confined US military personnel. This policy is equally binding on all US troops (see FM 19-40).

5-16. The ASCC or the TSC supports US laws, regulations, policies, and international agreements by providing personnel, administrative, morale, internment, resettlement, and confinement services for the TO. The TSC's MP brigade (I/R) in the TO provides this support. However, since most I/R units are in the reserve components, the initial I/R operations (as mentioned above) may have to be conducted by the MP brigade (CS). Once the I/R unit arrives in the AO, it is responsible for—

- Providing firm but humane control of EPWs/CIs and dislocated civilians.
- Coordinating with HN personnel, military territorial organizations, civilian police authorities, NGOs, private volunteer organizations, and US federal agencies on matters pertaining to I/R operations.
- Performing C^2 operations for all I/R units.
- Controlling, employing, and releasing EPWs/CIs as set forth by the Geneva convention and other international laws and by the UN and other governmental bodies.
- Handling US military prisoners.

5-17. In a mature theater in which large numbers of EPWs are captured, the EPWs' requirements may exceed the capacity of the MP brigade (I/R). In this instance, an I/R command is established. An MP command (I/R) has two or more MP brigades (I/R) and will normally be assigned to the ASCC. When the MP command (I/R) assumes OPCON of the MP brigades (I/R) from the lower echelon, it assumes that echelon's I/R mission.

5-18. If the US decides to transfer captured EPWs/CIs to the HN or to another nation, the US must ensure that the nation is a party to the Geneva convention and is willing and able to comply with the convention. In this case, the number and type of I/R MP units required for processing and retaining EPWs/CIs before the transfer is based on agreements and on METT-TC. Additionally, the MP brigade (I/R) is assigned I/R teams that are located at the processing and transfer points and at the HN or third-country EPW camps. The MP brigade (I/R) liaison team will supervise these dispersed teams to ensure that the HN or the third country provides adequate care and security of US-captured EPWs/CIs and that accountability is maintained according to the Geneva convention.

5-19. The MP brigade (I/R) HHC is the C^2 element for the brigade's assets. It consists of the following elements:

- A brigade command section.
- A company headquarters.
- An Adjutant (US Army) (S1).
- An S2.
- An S3.
- A Supply Officer (US Army) (S4).
- A Civil-Affairs Officer (US Army) (S5).
- A Communications Officer (US Army) (S6).
- Finance and accounting.
- Medical operations.
- Public affairs.
- A unit ministry team.
- An SJA.
- An inspector-general (IG) section.

5-20. Other brigade assets may include the following:

- An MP I/R BLD.
- An I/R information center.
- An MP escort-guard company.
- An MP I/R battalion headquarters.
- MP I/R (EPW/CI) detachments.
- MP I/R (confinement) detachments.
- MP guard companies.
- MWD teams.
- Processing squads, processing liaison teams, camp liaison teams, and evacuation teams (all as required to support EPW transfer or to conduct an out-of-theater evacuation).

5-21. The MP brigade (I/R) subordinate units are employed most often in the COMMZ near CSS facilities and are placed near sea, air, and rail terminals. They receive, process, and intern EPWs/CIs; confine US military prisoners; or assist in the resettlement of refugees or dislocated civilians.

5-22. The MP brigade (I/R) escort-guard company supports the evacuation of EPWs/CIs from the CZ. The company has a company headquarters and four platoons. The escort-guard company requires sufficient mobility to transport MP personnel to the CZ and to operate mobile teams while escorting the EPWs/CIs. The MP go forward to the corps's holding area to take custody of the EPWs/CIs. They may go forward to division collecting points, if distances and resources permit. Using any means of available transportation, the MP ensure that the EPWs/CIs are quickly evacuated to MP battalions (I/R) in the COMMZ. Close coordination with the EAC and corps movement-control centers and the corps MP brigade is required to ensure that transportation assets returning to the COMMZ are employed to evacuate EPWs/CIs from the corps's holding area. Walking wounded EPWs/CIs are evacuated by the same means as other EPWs/CIs, while litter patients are evacuated through medical channels. Guarding EPWs/CIs while in the MEDEVAC channels and during their hospitalization is not an MP mission; therefore, there is not an MP force structure to support this mission. In most instances, the impact of having the MP perform this mission causes trade-offs in missions for which they are responsible.

5-23. The theater MP brigade (I/R) and out-of-theater MP brigade (I/R) subordinate units will evacuate EPWs to internment sites within CONUS (if directed). The theater brigade structure is based on the projected capture rate over time and available out-of-theater transportation assets (frequency and capacity). The out-of-theater brigade structure is based on the total EPW/CI population, the number of internment sites, transportation nodes, and escort requirements. Theater escort-guard MP move the EPWs/CIs to the seaport and aerial port of embarkation (SPOE/APOE). The escort-guard MP assigned to the out-of-theater brigade escort the EPWs/CIs from the theater ports of embarkation (POEs) to the out-of-theater internment sites. The out-of-theater brigade is assigned an I/R evacuation detachment, which is employed at and coordinates the evacuation from the theater POEs, through the out-of-theater ports of debarkation (PODs), to the out-of-theater facilities.

THE MP BATTALION (I/R)

5-24. The MP battalion (I/R) is a modular organization and can be configured to operate internment facilities for EPWs/CIs, confine US military prisoners, or resettle dislocated civilians. When performing EPW/CI internment operations, the MP brigade (I/R) has up to 7 MP battalions (I/R); when augmented with the appropriate number of BLDs, it has up to 21 MP battalions (I/R). The ASCC, the TSC, and the MP brigade (I/R) must consider that the requirement to establish an MP battalion (I/R) internment facility is resource intensive. Therefore, MP I/R units, other supporting units, supplies, and equipment for the EPWs/CIs should arrive in theater ahead of the projected EPW/CI arrival at the internment facilities. Early arrival should be based on the time required to establish fully operational facilities (construct and man) and resupply operations before the EPWs/CIs arrive.

5-25. The MP battalion (I/R) has a command section, a company headquarters, and various staff sections. The staff sections provide the core battalion-level capabilities to conduct internment operations. The modulated design expands as the EPW/CI population increases. The battalion is assigned up to four detachments, two guard companies and, if needed, a processing squad. When fully operational, an MP battalion (I/R) operates an enclosure capable of interning 4,000 EPWs/CIs. The battalion mission centers on eight 500-man compounds. The battalion operates the compounds in close proximity to maximize its resources for the security and internment of the EPWs/CIs. This includes the resources needed to employ EPWs/CIs as a labor force according to the provisions of the Geneva convention.

5-26. Each MP detachment (I/R) (EPW/CI) operates two 500-man compounds and provides augmentation to the battalion staff sections to support 1,000 EPWs/CIs. Each guard company is capable of providing security for 2,000 EPWs/CIs. The guard company has a company headquarters and three platoons. The guard company requires sufficient mobility and communications to support routine battalion missions. While minimum mobility and communications is required to support EPW/CI internment operations, on-site guard personnel must often move considerable distances guarding labor groups performing work projects throughout the COMMZ.

5-27. The MP (EPW/CI) processing squad is capable of processing eight EPWs/CIs per hour and includes interpreters to support the processing. If processing squads are required to augment MP battalions (I/R), the operational requirements will be based on METT-TC.

MP-Battalion Resettlement Operations

5-28. The basic organization used for EPW/CI internment is used for resettlement operations. The primary mission-focus change is from guarding EPWs/CIs to protecting and controlling dislocated civilians. As such, an MP battalion (I/R) with four detachments and two guard companies is capable of supporting 8,000 dislocated civilians. However, the MP battalion (I/R) may require augmentation to conduct L&O operations associated with the resettlement. Augmentation may include the full scope of PM functions (operations, investigations, physical security, MP-station operations, and patrols) and civil affairs.

MP-Battalion Confinement Operations

5-29. When configured with the MP detachment (I/R) (confinement), the MP battalion (I/R) is capable of confining US military prisoners. The MP detachment (I/R) (confinement) provides trained corrections and support personnel required for confinement operations. As with the EPW/CI configuration, the modular confinement structure expands as the US prisoner population increases. With three confinement detachments and three guard companies, the MP battalion (I/R) has a maximum capacity of handling 1,500 US prisoners. Generally, only one MP brigade battalion (I/R) is configured for confinement, but the actual number will depend on the number of US prisoners requiring confinement within a theater. While theater policy for confinement operations remains with the ASCC commander, it is the MP brigade (I/R) that executes the mission.

5-30. When possible, soldiers awaiting trial remain in their units unless reasonable grounds exist to believe that they will not appear at the trial, the pretrial hearing, or the investigation or that they will engage in serious criminal misconduct. Under either of these two pretrial confinement instances, the commander must also reasonably believe that a less severe form of restraint (such as conditions of liberty, restriction in lieu of apprehension, or apprehension) is inadequate. When these circumstances exist and other legal requirements are met, US military personnel may be placed in pretrial confinement under the MP's direct control. Commanders may choose to establish field confinement facilities within their AO. However, corps and division MP companies have the expertise to operate only a field detention facility for a limited period of time. These units cannot operate a confinement facility and have neither the resources nor the capability to operate such a facility on an extended basis. Therefore, all confinement is consolidated in the COMMZ whenever possible.

5-31. All assets of the MP battalion (I/R) (confinement) are employed to detain, confine, sustain, and protect US prisoners. As with the battalions conducting EPW/CI and resettlement operations, the battalion conducting confinement operations is generally located in the rear of the COMMZ, near logistics and transportation support. This allows US prisoners to be moved as quickly as possible from the corps's area to the COMMZ's confinement facility. Movements of US prisoners from the COMMZ to CONUS will be according to DA policy.

Chapter 6

Military Police Support to Corps

Corps MP support their command by performing the MP functions critical to the success of their tactical commander's concept of operations.

USAMPS

Corps are the largest tactical units in the US Army. They are the instruments by which higher echelons conduct operations at the operational level. Higher headquarters tailor corps for the theater and the mission for which they are deployed. They contain organic combat, CS, and CSS capabilities to sustain operations for a considerable period of time. Corps are capable of operating in a joint and multinational environment, providing C^2 for up to five divisions and covering up to 35,000 square kilometers.

OVERVIEW

6-1. MP support to a corps is provided by an MP brigade (CS) assigned to each corps. The MP brigade provides combat, CS, and CSS throughout the corps's AO. However, subordinate MP units are not assigned to subordinate corps units. Instead, the MP brigade commander gives them an AO based on the corps commander's concept of operations. When possible, MP battalion AOs coincide with those of the CSG RAOCs.

6-2. The corps MP provide combat power within the command's rear area. They perform combat operations to counter Level II forces and to support the defeat of Level III forces. When properly augmented, the MP brigade may serve as the corps's TCF. The corps MP also provide a critical link between MP operations in the division and in the COMMZ. The corps MP support division commanders by helping the division MP conduct sustainment operations. The corps MP coordinate with the division MP for mutual support.

THE MILITARY POLICE BRIGADE (COMBAT SUPPORT)

6-3. The MP brigade (CS) supporting a corps contains a brigade headquarters, up to six MP battalions (CS), numerous L&O detachments, and MWD teams. Each MP battalion (CS) has up to six MP companies (CS). As with the EAC's MP brigade (CS), the number of battalions and companies is determined by METT-TC. The corps MP brigade (CS) has additional companies to support each division and to provide security for the corps and COSCOM headquarters and corps ammunition storage areas (see Figure 6-1, page 6-2).

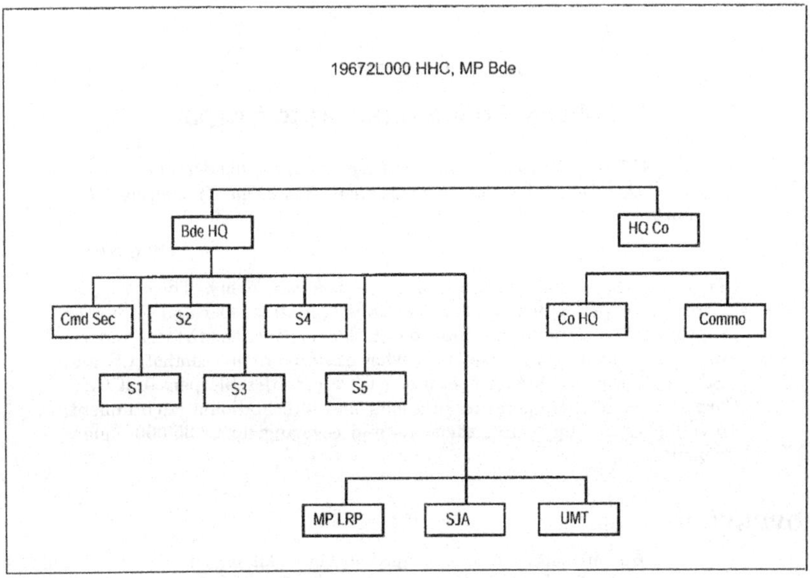

Figure 6-1. Military Police and USACIDC Organizations

6-4. The MP brigade HHC provides C^2 and A/L support to the brigade. The brigade HHC consists of a company headquarters and a brigade headquarters that contains the commander's immediate staff. The staff officers supervise the brigade's major organizational elements, including the—

- S1, S2, S3, S4, and S5.
- MP long-range plans (LRP) section.
- Communications section.
- SJA.
- IG.
- Public affairs.
- Unit ministry team (UMT).

6-5. The MP brigade (CS) command sergeant major (CSM) requires mobility and communications capabilities to execute his duties and responsibilities. The MP LRP section works with the corps G3 plans element, operating out of the corps main CP.

6-6. The MP battalion headquarters and headquarters detachment (HHD) provides C^2 for MP CS companies and any other assigned or attached MP elements. The battalion HHD consists of a detachment headquarters and a battalion headquarters that contains the battalion commander's staff. The staff officers supervise the major organizational elements, including the—

- S1, S2, S3, and S4.
- Communications and support sections.
- UMT.

6-7. As with the MP brigade (CS), the battalion CSM requires mobility and communications. The support section is vital to an MP battalion commander's ability to sustain his widely dispersed assigned or attached units during the performance of all five MP functions, primarily the MMS and AS functions.

6-8. The MP company (CS) provides support to an assigned AO. The company has a company headquarters, an MP operations center, a combat-medic section, and four platoons. The company headquarters provides maintenance, supply, communications, mess, and medical support to the unit. An MP operations center supports the unit's operation, conducting and planning for all five MP functions. The MP operations center includes three MP teams as the company-level response force.

6-9. The MP L&O detachments provide support to an assigned AO. The headquarters provides A/L support. The operation team plans and supervises desk operations, the traffic-accident and MP-investigation teams, and the force-protection teams. As with the EAC MP brigade (CS) detachments, requirements are based on the population supported and on METT-TC.

6-10. The number of assigned MWD teams is significantly less than those assigned to the EAC MP (CS) or (I/R) brigades. A kennel master, five explosive/patrol teams, and four narcotics/patrol teams are normally assigned to the corps MP brigade (CS). They are employed based on METT-TC.

COMMAND AND CONTROL

6-11. The C^2 in a corps MP brigade (CS) is consistent with that of any Army brigade. The MP brigade commander commands the brigade and all attached personnel. Battalion commanders work for the brigade commander. The MP company commanders receive their orders and work for their respective battalion commanders and direct their platoon leaders according to mission requirements.

6-12. When corps MP assets are sent to augment the division PM or the MP company, they are placed under the OPCON of the division PM for as long as the unit is needed in the division area. This command relationship is applicable to company-size organizations and smaller. If an MP battalion (or larger MP unit) augments the division, then the division PM will not exercise OPCON of that MP unit. The MP battalion will be placed OPCON as designated by the division commander but not under the division PM. One L&O detachment is normally attached to the supported division and placed OPCON to the division PM. However, MP leaders at each level establish an MP C^2 relationship based on METT-TC and the supported commander's needs.

EMPLOYMENT

6-13. Each corps MP brigade commander employs his assets according to METT-TC, the needs of the forces operating in his AO, and the priorities of the corps commander. Few MP assets in the corps area are employed to support fixed commitments. Instead, based on a broad consideration of the enemy and friendly situations, the corps MP are employed to support friendly forces engaged in combat, CS, and CSS operations. Since MP forces are dispersed throughout the corps area, the concentration of US forces, the location and vulnerability of critical sites, and the number of kilometers of the MSR to be controlled influence the designation of MP assets in the AO.

6-14. The MP brigade commander establishes his MP concept of operations based on the corps commander's concept of operations. His successful employment of MP assets depends on his foreseeing where the battle will be rather than where the battle is. Based on the rear-area IPB and PIAP, the MP brigade commander allocates and shifts resources to ensure the accomplishment of priority missions. This ensures the continuous support and forward sustainment of combat units and the safety of CSS units operating in the corps rear and sustainment areas.

6-15. To support the MP brigade commander's planning, the MP at brigade headquarters operate from several locations. Most of the staff locates in the vicinity of the corps rear CP where they can interface with the corps staff responsible for planning and executing rear operations. The control element of the brigade headquarters must be located where it can command and control its subordinate units. The brigade commander and his staff decide the best place to locate this element. The planning element of the headquarters locates near the corps main CP where it can interface with the corps commander's coordinating, special, and personal staffs. From there they monitor MP operations, integrate MP support with the corps plans cell for future operations, and learn the enemy situation through the G2's threat analysis almost immediately. The detailed information on rear-area activities and operations provided by the corps staff enhances the accuracy of the MP LRP.

6-16. The brigade S3 section provides the day-to-day planning and execution of an MP mission. The section provides a responsive CP that can relocate frequently throughout the AO. The S3 will normally provide liaison personnel to the corps rear CP, selected COSCOM units, interagencies, or other headquarters (based on mission requirements). When possible, battalion AOs coincide with the AOs of the CSG RAOCs. The MP brigade commander usually tailors battalions' AO boundaries to ensure responsive and flexible support across the corps's AO. He pays particular attention to the LOC behind the most heavily committed division and the critical bases and facilities in that area. He also ensures that the MP are available to respond quickly to combat operations throughout the entire corps rear area or during sustainment operations.

6-17. The MP brigade commander, coordinating with his battalion commanders, locates the MP companies where they can provide combat and CS power throughout the corps's AO. He bases his decision on the—

- Number and composition of urban areas.

- Location of CS and CSS units.
- Location of critical facilities (such as the headquarters, ammunition storage areas, and airfields).
- Expected threat.
- Level and frequency of support needed by the commander.
- Current and projected tactical situation.
- MSR network, including choke points and critical bridges and tunnels.
- Number of supported divisions and requirements.

6-18. The prioritization of MP missions is especially crucial during the early stages of the deployment when it is unlikely that an MP brigade commander will have a full complement of CS companies. Augmenting the division MP company with corps MP assets may not be possible at that time. Until it is, the MP brigade commander must locate corps MP assets to help meet the needs of the division while fulfilling the needs of the corps.

6-19. Like the brigade commander, battalion commanders plan the employment of their companies and platoons using METT-TC. Certain corps needs are constant. One MP company is allocated to provide security for the COSCOM, and one MP company is allocated to provide security for the corps's main CP. One platoon from that company may be used to secure the corps tactical CP or the jump CP. One or more platoons will also help secure the corps's rear CP. The number of MP assets allocated for a corps-level EPW/CI holding area and escort is based on METT-TC. However, a minimum of one platoon is dedicated to operate the corps's EPW/CI holding area and a minimum of one platoon per division is allocated for evacuating EPWs/CIs from division collection points. Additional MP assets may be allocated to provide security for the corps's ammunition storage area and ASPs supporting the divisions. The MP battalion commander places his companies where—

- MP assets support the brigade commander's concept of operations.
- The MP can support troop concentration, bases and base clusters, road networks, and critical areas.
- The MP can aggressively patrol critical terrain and monitor LZs and DZs to detect or deny enemy interference.
- The MP can respond to Level II threats.
- The MP can support the movement of combat resources throughout the AO.
- The MP can remove EPWs/CIs from division collection points.
- The MP can influence stragglers, refugees, and dislocated civilians.

6-20. Battalion commanders may choose to place a company behind the division rear boundary. This can help to ease the coordination between the corps MP and the division MP.

Chapter 7

Military Police Support to Division

Division MP provide essential support to the forward tactical commander. The MP provide a flexible, mobile, and lethal force ready to be employed in combat or CS operations.

USAMPS

The division is a large Army organization that trains and fights as a tactical team. Largely self-sustaining, it is capable of independent operations. The division is a unit of maneuver organized with varying numbers and types of combat, CS, and CSS units. Divisions fall mainly into two categories—heavy or light. The first category includes heavy divisions, their variants, and the Division XXI digitized heavy division. The second category includes the light infantry, airborne, and air-assault divisions. Each division can conduct operations over a wide range of environments. The success of Army operations depends on the success of its divisions.

OVERVIEW

7-1. In corps operations, divisions are normally comprised of three maneuver brigades, each with up to nine maneuver battalions, artillery battalions, aviation battalions, and supporting CS and CSS units. Divisions perform a wide range of tactical missions and are self-sustaining for limited periods of time. The corps augments divisions as the mission requires. All divisions must be able to deploy and conduct offensive, defensive, stability, and support operations. Airborne and air-assault divisions must be able to conduct forced-entry operations. Divisions may be part of a JTF or serve as the ARFOR headquarters.

7-2. MP support to a division is provided by an organic MP company that includes the PM and his staff (assigned to each division) and is augmented by the corps MP brigade. The division MP company provides support in the DRA and in the maneuver brigade rear area. It is fully mobile in order to relocate frequently under short notice. While division MP companies are capable of performing all five MP functions, they must be augmented by corps MP/CID to fully perform all five MP functions throughout the division's battle space. The division PM must receive at least one corps MP company for GS missions in the DRA, one L&O detachment for L&O missions, and one CID detachment for criminal investigations and LOGSEC. In a mature theater (or based on METT-TC), the division PM can expect two corps MP companies. In addition, the division PM may request a corps MP battalion headquarters to provide C^2 to corps MP units operating in the DRA or to perform missions as a TCF. Because the need for MP support exceeds division organic assets (and many times exceeds augmenting corps MP assets), careful planning of MP employment is essential.

DIVISION MILITARY POLICE COMPANY

7-3. The organization of the division MP company is set by specific tables of organization and equipment (TOE) and is designed to support a specific type of division. Regardless of the type of division, the assets in a division MP company enable a PM and his supporting staff to supervise the performance of all five MP functions. The PM coordinates the employment of the MP assets in the division's AO through his PM section. The division MP company contains a headquarters, a PM section, a medic section, and three to six platoons.

COMMAND AND CONTROL

7-4. The C^2 of the division MP company extends downward from the tactical commander to the division PM. The division PM is the division commander's special staff officer. Under normal circumstances, he works for the chief of staff. The division PM has OPCON of the division MP company and MP assets that have been provided by the corps. The division PM plans and employs all of the organic, assigned, or attached MP according to METT-TC. He anticipates the support needs of the division commander and plans the employment to meet his needs. After learning the division commander's concept of operations and estimating the situation, the division PM decides which MP functions are critical to accomplish the commander's mission. He then recommends the allocation of resources, tasks his assets, and establishes the priority of support.

7-5. When the division MP company is augmented by corps MP platoons, the platoons are placed under the direction of the division MP company commander, who assigns an AO to each platoon. When corps MP assets augmenting the division make up a company, the company is attached to the division. The company then comes under the OPCON of the PM, who assigns it an AO. The augmenting corps MP company is usually given an AO from the division rear boundary forward. However, this assignment is flexible and based on the division commander's needs as set forth by the division PM. The four platoons of that corps MP company perform all five MP functions in that AO. Additionally, the corps MP brigade (CS) L&O detachment supporting a division is attached to the division and OPCON to the division PM.

7-6. The division MP company commander directs the employment of company assets. Through his company headquarters, he provides administrative, maintenance, and logistical support to the PM section and to the company's platoons that are dispersed throughout the division's AO. These sections are mobile to support the platoons without delay despite the extended distances that may be separating them. The company commander is also mobile. He travels the extended distances from one platoon area to another when checking on the status of his soldiers and resolving the problems affecting mission accomplishment. The division HHC provides mess support. Personnel administration is handled by the company headquarters and the DISCOM. The number and kind of assets in a division MP company and the configuration of its platoons and squads are determined by the type of division to which the company is assigned.

NOTE: See the consolidated TOE update that is published yearly for complete details (base and objective) for all MP TOEs.

7-7. At the direction of the division commander, the division band may be available to perform its secondary mission of supporting MP operations. When the intensity of the conflict prevents the band from performing its musical mission, the division commander may direct it to augment the MP company. The band, normally employed as a unit under the direction of the bandmaster, is under the OPCON of the MP company commander while augmenting the MP. When tasked to augment MP assets, band members may be employed to augment MP security at the division main CP or to augment perimeter security at the division EPW/CI collection point. The band is released to perform its primary mission as soon as the tactical situation permits.

MILITARY POLICE EMPLOYMENT

7-8. The employment of division MP companies differs somewhat with the type of division to which they are assigned. In heavy divisions, the division MP companies provide GS to the DRA and DS to the division's maneuver brigades. When the platoon is providing DS to a maneuver brigade, the MP platoon leader coordinates all logistical support with the supported brigade. A platoon's DS relationship is seldom interrupted. However, METT-TC may require weighing MP support to the maneuver brigade tasked as the main effort or employing all platoons in a division MP company to function as a unit (such as during division-size river-crossing operations or deliberate breaching operations). In light divisions, the MP provide only GS to the DRA or during sustainment operations. Support exceptions are addressed under MP support to airborne and air-assault divisions, discussed later in this chapter.

7-9. In the division (where flexible support of an austere force is crucial), the division PM must have a clear understanding of situational awareness. To obtain current information for projecting MP needs in the division area, he must be mobile and be able to conduct split-cell operations. The assets available to the PM include the division MP company and at least one corps MP company. Corps augmentation is required for sustained operations and for special operations such as river crossings, dealing with dislocated civilians, and refugee internment or resettlement. The division PM coordinates with the corps PM and the MP brigade or CID commanders for—

- Evacuating and guarding EPWs/CIs from division to corps.
- Providing law-enforcement assistance to HN forces in the division's AO.
- Providing corps augmentation for the division's AO, convoy security, LOC security, AS, R&S, L&O, and other missions. These other missions may include augmentation for security of the division main CP, ASPs, and other critical facilities.
- Integrating, sharing, and exchanging police intelligence between corps and division MP elements.
- Providing CID support.

SUPPORT TO THE HEAVY DIVISION

7-10. The Army's armored and mechanized infantry divisions (normally referred to as heavy divisions) provide mobile, armor-protected firepower that is normally employed for their mobility, survivability, lethality, and psychological effect (shock) on the enemy. These divisions destroy enemy armored and mechanized forces, and they can seize land areas and secure key terrain. Because of their strategic lift requirements, heavy divisions are slow to deploy from home staging bases into an AO. They have high consumption rates of supplies and have limited use in restrictive terrain. These capabilities and limitations are key factors in planning effective and efficient MP support throughout the division's AO.

7-11. In the heavy divisions, the PM section is organized to support split-cell operations at the main and rear CPs. The PM must be mobile to ensure that he is fully aware of the current status of critical MP operations. Therefore, the deputy PM (DPM) locates in one of the PM cells and—

- Handles the section's routine operations.
- Monitors ongoing division operations and MP support.
- Helps provide long-range planning and interface with the primary division staff.
- Forwards PM taskings to the division MP company commander.

7-12. The DPM and the operations sergeant normally set up operations at the division main CP where they can coordinate requirements with the division staff personnel. The company headquarters is initially located near the division rear CP in the division support area (DSA). Once augmentation arrives from the corps, the headquarters relocates with one of its GS platoons operating behind the brigade rear boundary. However, this location could be changed based on METT-TC, the supported commander's needs, and the company commander's idea of where he can exercise better C^2 for his unit.

7-13. The Army of Excellence (AOE) heavy division MP company has six platoons. Three platoons provide support to each maneuver brigade and are designated as DS. The other three platoons are designated as GS platoons. One MP platoon provides security for the division main CP; one provides security for the division's EPW central collection point; and one performs other MP operations within the division rear.

7-14. The GS MP platoons' AOs are configured based on METT-TC and the availability of MP augmentation from the corps. The DS MP platoons' AOs coincide with the supported maneuver brigade's boundary. Each platoon headquarters locates within its brigade's support area or any other area where it can best provide and receive support. To accomplish its mission, each DS platoon requires a minimum of two squads, each with three teams. One squad operates the EPW/CI collection point. The other squads perform MMS and AS operations. All MP platoons are capable of performing all five MP functions. However, performance of these functions is prioritized based on METT-TC and the division commander's concept of operations. The division PM, the company commander, and METT-TC dictate how these platoons should be tasked-organized to accomplish the mission.

SUPPORT TO THE LIGHT INFANTRY DIVISION

7-15. The AOE light infantry division (LID) is one of the most rapidly and strategically deployable divisions. It fights as part of a larger force in war or conducts missions as part of a joint force in MOOTW. Its C^2 structure readily accepts any augmentation forces, permitting task-organizing for any situation. The augmentation required for the division is largely determined by METT-TC. The division's capabilities allow it to exploit the advantages of restricted terrain and limited visibility. It achieves mass through the combined effects of synchronized small-unit operations and fires rather than through the physical concentration of forces on the battlefield. These characteristics are key factors in planning and employing MP assets in support of the LID.

7-16. The LID MP companies are capable of performing all five MP functions. However, their performance of these functions is prioritized based on METT-TC and the division commander's concept of operations. Contrary to the heavy division MP company, the LID MP companies are much smaller. The constrained size of the LID MP companies makes corps MP augmentation crucial to the sustainment of MP operations. Additionally, the LID MP company is the only MP unit with the capability of antiaircraft support through the use of shoulder-fired air-defense weapons.

7-17. The company has three GS platoons to support the division. No platoons are provided to the maneuver brigade. One platoon is normally located in the vicinity of the division main CP so that its resources can help support CP security. Another platoon locates in the DSA and operates the division EPW/CI collection point. The last platoon has an AO configured according to METT-TC and the commander's priority of MP missions. Each GS MP platoon has a headquarters and three squads, each with two teams. The PM section is located in the vicinity of the division main CP. The exact location is based on the current operational status and on METT-TC.

SUPPORT TO THE AIRBORNE DIVISION

7-18. The AOE airborne division can rapidly deploy anywhere in the world to seize and secure vital objectives. It conducts parachute assaults to capture initial lodgments, executes large-scale tactical raids, secures intermediate staging bases (ISBs) or forward operating bases (FOBs) for ground and air operations, and rescues US nationals besieged overseas. It can also serve as a strategic or theater reserve as well as a reinforcement for forward-presence forces. The airborne division can assault deep into the enemy's rear area to secure terrain or to interdict enemy supply and withdrawal routes. It can seize and repair airfields to provide a FOB and airheads for follow-on air-landed forces. It is capable of all other missions assigned to LIDs. The airborne division does not have sufficient armored protection to defeat heavier armored formations at close range. Therefore, engagements with enemy armored formations require special considerations. Antitank weapons in the division compensate for, but do not completely offset, this deficit.

7-19. Airborne divisions conduct operations in two phases—assault and defense. The division accomplishes the assault phase in three stages. First, they deploy; second, they establish the airhead; and third, they extend the logistics base and build their force.

7-20. Like other division MP companies, those assigned to the airborne division are employed to support their division commander's concept of operations. The airborne MP company has four platoons, each having three squads with two teams. Normally, the company headquarters and one of the platoon headquarters colocates with the PM section. The division PM's exact location depends on METT-TC. During the assault phase, the airborne division MP company provides DS to the assault brigade. The remaining platoons provide MMS and AS in the vicinity of the division main CP.

7-21. The nature of airborne operations makes the capture of EPWs likely. Therefore, during the first stage of the assault phase, the priority of MP support is given to EPW operations. After assembling the DZ or LZ, the MP collect EPWs captured during the assault. Combat elements are relieved of EPWs as far forward as possible. In airborne operations, EPWs are held for later movement to a central collection point. During the first stage of the assault, the MP perform limited straggler and refugee control and undertake AS operations, when possible.

7-22. MP support is reevaluated after the airhead is established in the second stage of the assault. The PM takes in consideration personnel and equipment flow, roadways, and security requirements to shift MP support priorities. As the entire operation matures, MP support may expand to include all five MP functions.

SUPPORT TO THE AIR-ASSAULT DIVISION

7-23. The air-assault division executes tactical missions at operational depths to achieve strategic results. It is capable of launching brigade-size air assaults of nearly 4,000 soldiers from either an ISB or a tactical assembly area (TAA). Within 6 hours, this air-assault task force (AATF) can attack an opposing force, occupy and defend key terrain, or establish a FOB (out to 150 kilometers) from which even deeper operations can be executed. Air-assault operations are terrain independent, but they rely on suitable weather and a detailed attrition/assessment of enemy capability—particularly air-defense assets along air corridors and in the objective area. The air assault is preceded by detailed, lethal and nonlethal condition settings, culminating in a comprehensive condition check before execution.

7-24. As the corps AATF unit, the brigade combat team (BCT) task-organizes and habitually trains with both aviation lift and attack (Apache) battalions. An air-assault division MP company has four platoons, each having three squads with two teams. To facilitate operations, air-assault division MP platoons are habitually aligned with each of the three AOE BCTs; however, as with aviation assets, the MP are task-organized to support the main effort's BCT/AATF. When conducting these operations, two MP platoons may be tasked to provide support; one is available to posture and participate in the air assault, and the other conducts AS and MMS operations at the pickup zone (PZ). This second MP platoon may further provide MMS and accompany a ground-assault convoy (GAC) to the objective, or it may remain at the PZ to receive EPWs returning on aircraft from the objective area. The mission of conducting MMS along this vulnerable ground LOC is particularly critical if the AATF objective is a FOB from which further division operations will expand.

7-25. MP participation in the actual air assault competes for heavy lift with FA and air-defense systems, as well as with vital Class V resupply. However, the lethality and versatility of the MP bode well for their employment, and two MP platoons are available to support the brigade air assault as the division's center of gravity.

7-26. When possible, habitually aligned platoons remain with their brigades, and corps assets perform GS missions. However, when no corps assets are available and two division platoons are employed as stated above, the two remaining platoons conduct division EPW collection-point operations and other MP functions based on METT-TC. Normally, the EPW platoon and the MP company headquarters colocate in the DSA. As required (and based on METT-TC), airflow planning includes EPW/CI evacuation from the AATF/ FOB collection point back to the DSA. The PM section operates from the division rear CP to facilitate I/R operations and to coordinate MMS and AS with key logistical staff. Due to potentially extreme distances on the air-assault battlefield, the DPM normally locates with the division main CP to serve as a key G3 battle-staff member and to coordinate PIO with the G2.

SUPPORT TO DIVISION XXI (HEAVY DIVISION)

7-27. Division XXI represents a significant change in the manner in which division operations are conducted. These changes are brought on by information-age capabilities; an increased integration of service components into an effective battle team; more lethal, survivable, and agile systems; and more capable soldiers and leaders. The Division XXI operates in a larger battle space and at a higher tempo than the AOE division. The division is evolutionary in design, but revolutionary in its use of information technology. It improves the Army's deployability while enhancing its ability to dominate in decisive fights. The following are characteristics of the Division XXI operational environment:

- **Multidimensional.** The division will operate in an extended battle space that goes beyond the traditional physical dimension of width, depth, and height. It includes portions of the electromagnetic spectrum; and it extends beyond the physical boundaries of the division through its communication and digital connectivity to other Army, joint, and coalition elements, even reaching back to CONUS.
- **Precise.** Precision operations go beyond a precision strike to include every aspect of military operations from deployment through combat and through redeployment or transition to other operations. Precision in decisive operations is enabled by three emerging capabilities. First, digitization provides soldiers and leaders at each echelon the information required for making decisions. Second, a full suite of strategic, operational, and tactical sensors linked to analytical teams fuses combat information into situational awareness across the battle space. Third, simulations enable Army elements to be tailored and operations to be planned, war-gamed, and rehearsed—yielding precision execution.

- **Nonlinear.** Nonlinear operations do not seek a battle-space grid of close, deep, and rear operations. Instead, the battle space is fluid, changing as METT-TC changes through the duration of mission preparation and execution. Another dimension of this characteristic is the synchronization of near-simultaneous operations to achieve nonlinear effects across the battle space.
- **Distributed.** Division operations are distributed or executed where and when required to achieve decisive effects concentrated at a decisive point. Dispersion empowers subordinates to operate independently within the commander's intent, leading to synergistic effects that exceed the effects of a centralized headquarters.
- **Simultaneous.** The concept of decentralization operations that are multidimensional, precise, distributed, and nonlinear yields the capability to conduct simultaneous operations across the battle space. Simultaneous operations seize the initiative and present the enemy leadership with multiple crises and no effective responses. Rather than a single, concentrated attack, the division executes a series of attacks (lethal and nonlethal) as simultaneously as possible.
- **Integrated.** Division operations are fully integrated with joint, multinational, and nongovernmental partners. Integrated operations enable the Army to leverage the full suite of capabilities that the services bring to the battle space.

7-28. The Division XXI heavy division is one of total integration—not rounding up or down. Reserve-component soldiers assigned to the division are an integral part of the division mission and task accomplishment. The division cannot operate without them. The division is focused on massing effects on the enemy—enabled by a new foundation of information technology and distributed logistics. The division is built around the pattern of "conduct decisive operations," accomplished by either fires or maneuver. The design performs complementary battle-space shaping in concert with the overall corps plan. Its long-range assets are not typically held in reserve nor just employed in the close fight; they are used to best support the corps and division commanders' plans of operations.

7-29. Although the basic structure of the division has not changed—three maneuver brigades, a division artillery, and a DISCOM—the internal structure and operational construction have changed in a near revolutionary way. The C^2 structure and the basic organizational structure are very similar to the divisions discussed above. However, by integrating the capabilities of digital technology and by using enhanced situational awareness, it is better able to execute combat operations.

7-30. The maneuver brigades have their own scouts—a brigade reconnaissance troop (BRT); therefore, they no longer rely exclusively on the battalion scouts. The maneuver battalions are reduced to three companies with a total number of 45 combat platforms. The division artillery has three Multiple-Launch Rocket System (MLRS) batteries per battalion. The cavalry squadron is under the aviation brigade's C^2 but retains and performs the division's traditional cavalry mission. Figure 7-1 depicts the organizational structure of Division XXI.

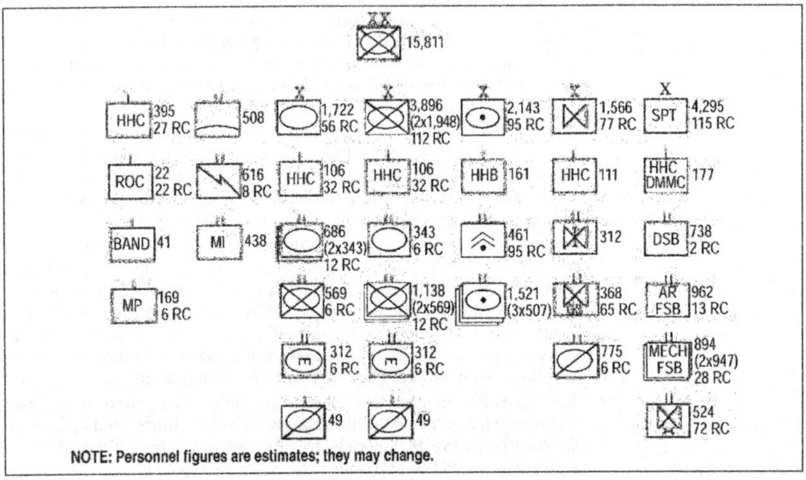

Figure 7-1. Division XXI Structure

7-31. The division's CSS is centralized. Instead of each unit having its own organic CSS personnel, they are now centralized in CSS units. The CSS assets are reorganized and reassigned to the DISCOM. Forward-support companies (FSCs) are habitually associated with each maneuver battalion and forward-support battalions (FSBs) will support each maneuver brigade. This operational structure enables distribution-based rather than supply-based logistics concepts to support the division. The overall logistics footprint for the division is now reduced and the division has organic self-sufficiency to operate unsupported for up to 30 hours.

7-32. A major difference between the AOE fix-forward doctrine and the Division XXI maintenance support is the change in the actual repair site. Division XXI maintenance units directly supporting a given customer will diagnose faults and will replace components forward in the battlefield and will repair rearward. Units will be manned with maintenance personnel capable of performing DS and limited GS maintenance.

7-33. The division operates in a battle space two and a half times larger than an AOE division, and it will usually fight as part of a corps or a JTF. The division conducts offensive, defensive, stability, and support operations in cooperation with other elements of the corps/JTF to gain tactical or operational results significant to the joint-force commander's campaign goals.

7-34. Although the division is a heavy division, MP support is somewhat different from that of an AOE heavy division. The PM (based on METT-TC) establishes operations where MP assets can best be commanded. At the same time, overall MP operations can be synchronized to support operations within the division rear and the division commander's intent can be effectively supported. To this end, the PM conducts split-cell operations within the division main CP. Normally, the PM section is distributed between the security-operations cell (SOC) and the mobility/survivability (M/S) cell. The DPM (who works out of the M/S cell) is also part of the division planning team. He ensures that MP operations are integrated into division plans according to the PM's intent. The DPM attends coordination meetings and assumes coordination responsibility with the rest of the division staff when the PM relocates away from the division main CP.

7-35. The PM, as a staff officer, is involved with every aspect of planning and with the military decision-making process (MDMP). He uses the Force XXI Battle Command—Brigade and Below (FBCB2) C^2 system to understand the battlefield. He understands the mission and the commander's intent and priorities. The PM has complete situational awareness, establishes an effective staff relationship with the G3, and recommends the best way to employ MP resources.

7-36. The Division XXI heavy division MP company has five platoons. Although tasked-organized based on METT-TC and the commander's needs, three MP platoons provide DS for each maneuver brigade. The DS platoon requires a minimum of three squads (each having three teams) to accomplish the mission. The other MP platoons are normally employed forward in sector, providing AS and force protection for the division main and tactical (TAC) CPs and for other HVAs in the vicinity of the maneuver brigade rear boundary. As with the AOE heavy division, the Division XXI heavy division requires augmentation with at least one corps MP company for CS missions in the DRA and can expect two companies in a mature theater (based on METT-TC) as the main effort. In addition, the division may request an MP battalion headquarters to command and control CS MP operations in the rear or to conduct TCF operations.

7-37. While the division MP company is capable of performing all five MP functions, it must be augmented by the corps MP to fully perform all of these functions throughout a mature division's battle space. The PM prioritizes MP functions (based on the supported commander's needs), conducts a mission analysis, and requests needed corps assets through the division G3. Corps MP companies, L&O detachments, and CID assets are needed for sustainment operations. When corps MP companies and L&O detachments are attached to the division, they are placed under OPCON of the PM. The PM will then assign an AO and missions based on METT-TC.

7-38. The PM supports the division's distribution-based logistics concept by conducting aggressive R&S operations that secure the LOC. Since the main effort is to secure the LOC, traditional MMS missions (such as convoy security and escorts) are performed by exception. Through the allocation of increased MP to conduct LOC R&S operations aggressively, the division is able to detect and destroy threat forces before they interrupt operations in rear-area or sustainment operations.

7-39. Digitization gives the division the situational awareness that allows it to cover much larger distances and to move much faster. The systems that provide this digitization are generally unprotected HVAs on the battlefield, which require aggressive AS and force-protection operations.

7-40. The EPW operations are much more fluid, sometimes requiring the division GS MP to move forward and assume brigade collection points, thus allowing DS platoons to move forward with their brigades. Similarly, corps MP may be required to move forward and assume the division's central collection points as the division's rear boundary is pushed forward to shorten the LOC.

7-41. As with all AOE divisions, until corps L&O and CID detachments augment the division, L&O operations are on an exception basis. The MP collect intelligence during every facet of mission execution. The PM develops police intelligence in the division main CP through coordination with the SOC, the G2, the Assistant Chief of Staff, G5 (Civil Affairs) (G5), and the Assistant Chief of Staff, G6 (Communications) (G6).

Chapter 8

Separate Brigades and Initial/Interim Brigade Combat Teams

The MP support separate brigades and IBCTs designed to provide the brigade commander with MP assets that can serve as a force multiplier and that can be employed as an economy of force. The brigade MP platoon is as lethal, flexible, and capable as any other platoon in the MP Corps.

USAMPS

The Army uses separate brigades to inject a small but powerful force where it is needed. It must be able to fight and win while operating on its own for a sustained period of time. It must be able to defend itself on a 360-degree front in war or MOOTW. While there are some variations of separate brigades, it is the heavy separate brigade that is most commonly found within the force.

OVERVIEW

8-1. The Army's IBCT is a full-spectrum, wheeled combat force. It is employed in all operational environments against all projected future threats. However, it is designed and optimized primarily for employment in small-scale contingency operations in complex and urban terrain, confronting low-end and midrange threats that may employ both conventional and asymmetric capabilities. The IBCT deploys very rapidly, executes early entry, and conducts effective combat operations immediately on arrival to prevent, contain, stabilize, or resolve a conflict through shaping and decisive operations. The IBCT participates in war (with augmentation) as a subordinate maneuver component within a division or a corps and in a variety of possible roles. The IBCT also participates (with appropriate augmentation) in stability and support operations as an initial-entry force or as a guarantor to provide security for stability forces by means of its extensive capabilities.

SUPPORT TO SEPARATE BRIGADES

8-2. MP support to separate brigades is provided by a four-squad MP platoon organic to the brigade HHC. A separate PM cell within the brigade HHC serves as the C^2 element for the platoon (see Figure 8-1, page 8-2). Support to the platoon and the PM section for maintenance, supply, mess, and communications is provided by the brigade HHC. Since the platoon and PM section have no organic support, the MP leadership must perform close coordination for this support. However, the MP platoon must compete with other brigade HHC assets for priority of repair for weapons, vehicles, and communications equipment.

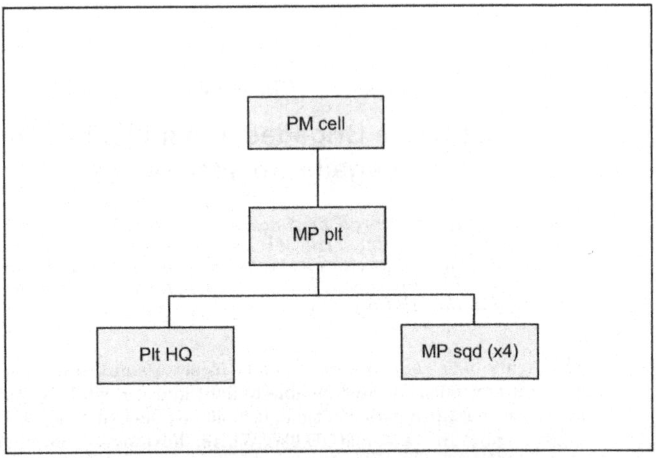

Figure 8-1. MP Support to a Separate Brigade

8-3. The C^2 of an MP unit supporting separate brigades also extends down from the tactical commander. The separate brigade PM has OPCON of brigade MP assets the same way the division PM has OPCON of division MP assets. This includes any MP assets that may have been provided from the corps. The PM advises the commander of a separate brigade on matters pertaining to MP operations. The platoon leader directs the execution of his platoon's mission based on the priorities set forth by the PM and the supported commander. Since the separate brigade's organic MP platoon is more robust than an MP platoon supporting a division maneuver brigade, METT-TC will determine the requirement for augmentation. However, corps L&O and CID augmentation is required.

8-4. The separate brigade's MP platoon is capable of performing all five MP functions. However, its resources are quite limited. Although the MP squads are employed according to METT-TC, the platoon supporting the separate brigade may have—

- One squad operating in the EPW collection point.
- One squad providing a mobile security screen and providing AS around the brigade's main CP.
- Two squads conducting MMS and AS throughout the brigade's rear area.

SUPPORT TO THE INITIAL/INTERIM BRIGADE COMBAT TEAM

8-5. The IBCT is a divisional brigade. It is designed to optimize its organizational effectiveness and to balance the traditional domains of lethality, mobility, and survivability with the domains required for responsiveness, deployability, sustainability, and a reduced in-theater footprint. Its two core qualities are its high mobility (strategic, operational, and tactical) and its ability to achieve decisive action through a dismounted infantry assault. The major fighting components are its motorized infantry battalions. The IBCT has a unique reconnaissance, surveillance, and target acquisition (RSTA) squadron to enhance situational understanding.

8-6. To achieve a rapid deployment threshold, the brigade's design capitalizes on the widespread use of common vehicular platforms—particularly a highly mobile, medium-weight, combat/CS platform coupled with the minimization of the personnel and logistical footprint on theater. Encompassing a personnel strength of about 3,500 and preconfigured in ready-to-fight combined-arms packages, the entire IBCT can deploy within 96 hours of "first aircraft wheels up" and begin operations immediately upon arrival at the APOD. In essence, the APOD is the TAA. The IBCT cannot conduct forced-entry operations, but it provides the JFC with an improved capability to arrive immediately behind forced-entry forces and begin operations to shape the battle space and to execute decisive action to expedite decisions. Once committed, the IBCT can sustain operations for up to 180 days without relief.

8-7. The IBCT's organization is expandable either through augmentation or scalability (according to METT-TC) in any given contingency. The IBCT is scalable in terms of its ability to accept like-type additional forces to expand core tasks and functions already resident within the IBCT (for example, adding additional infantry or RSTA organizations). The IBCT is also capable of accepting augmentation consisting of units or elements that execute tasks or functions not resident within the IBCT (for example, adding armor, air-defense, MP, or aviation assets). In both cases, added units execute their normal mission-essential task list (METL) tasks and, therefore, will not require extensive training to deploy and operate with the IBCT. In both cases, the IBCT includes the command, control, and communications (C^3) capabilities necessary to permit the rapid integration of additional enabling capabilities, particularly for operations outside the scope of small-scale contingency operations (such as stability operations, support operations, and war).

8-8. The IBCT organization excludes other unit-based capabilities often provided in a division slice. However, for each missing capability, the brigade headquarters includes staff cells capable of conducting rudimentary planning and analyses to ensure that all functional-area considerations are incorporated in route planning and preparation for operations. The first MP elements in support of the IBCT are at the brigade headquarters level.

8-9. The MP planning cell is composed of a two-person (a major [MAJ] and a sergeant first class [SFC]) planning cell located within the HHC's maneuver-support cell and under the direct staff supervision of the brigade S3. The role of the MP planning cell is significantly different from that of a traditional division PM or a separate brigade PM. The main difference is in the IBCT's lack of organic or habitual MP assets. However, the absence of organic MP assets makes the job of the MP planning cell that much more critical. The MP planning cell must—

- Understand the organization, capabilities, and limitations of the IBCT.

- Conduct effective liaison with higher headquarters PM elements.
- Become an effective planner and anticipator of MP requirements.
- Prepare MP staff estimates and employment recommendations.
- Plan for MP deployment via air, sea, rail, or land.
- Task-organize MP units effectively and efficiently.
- Assume C^2 of incoming MP forces or, if operating under a division or corps, relinquish C^2 to the division/corps PM or battalion commander (if appropriate for effective employment of MP forces).

8-10. Depending on METT-TC, the brigade may be augmented by MP elements ranging from a platoon to a battalion. In any case, the MP planning cell then becomes a staff planner and coordinator of functional matters pertaining to—

- Force protection, physical security, and vulnerability countermeasures.
- Operations for collecting, processing, and evacuating (possibly) EPWs, CIs, and detainees.
- US military prisoner operations.
- Operations for processing and controlling dislocated civilians and refugees.
- Customs and counterdrug operations.
- Criminal investigations and CID-/MP-investigator support.
- PIO and establishing links with HN police agencies and other international/interagency law-enforcement agencies.
- MMS, AS, and L&O operations.
- Coordination of MP or CID division/echelons above division (EAD) augmentation forces, MWD teams (explosives, narcotics, and patrol), L&O detachments, and MP I/R units through the division, corps, and major Army command (MACOM) PMs.
- Training guidance to other US forces required to perform MP functions in the absence of MP forces.

8-11. Since the IBCT is a divisional brigade, the division PM and the IBCT MP planning cell play an important role in developing an optimum MP force package to support the brigade commander's concept of operations. Despite the brigade's early-entry timelines, the MP planning cell must consider and plan for MP augmentation forces as early as possible to free valuable combat resources. Small-scale contingency operations that result in numerous EPWs, civilian detainees, and refugees will hamper momentum and freedom of maneuver.

8-12. Once the initial brigade receives MP augmentation (see Figure 8-2), the MP priority of effort during the offense will focus on ensuring that routes remain unencumbered and secure for movement of ground combat, CS, and CSS forces. The MP's priority of effort during the defense will focus on conducting AS and counterreconnaissance along the LOC, C^2 centers, and CSS bases. The MP may conduct response-force operations or become part of the TCF.

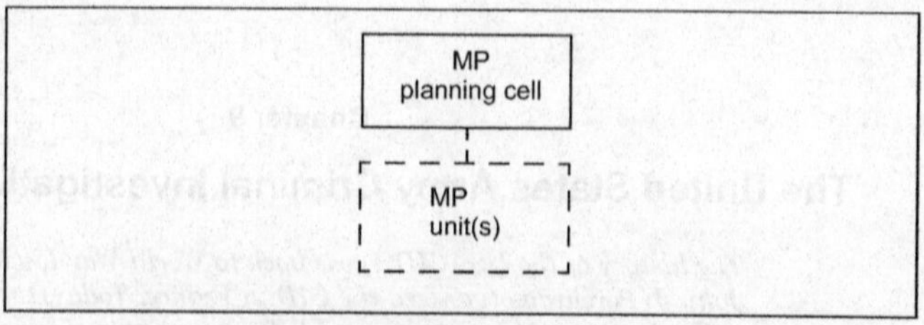

Figure 8-2. MP Support to the IBCT

8-13. Stability and support operations present some unique challenges. The IBCT may be deployed to a geographical area that is politically unstable, that lacks civil control, or that is in complete turmoil. The MP planning cell plans for and coordinates MP support according to available resources and the supported commander's needs. In this scenario (and based on METT-TC), an MP battalion TF could be required to deal with the challenges of—

- A significant number of refugees or dislocated civilians.
- AS or force-protection issues.
- MMS operations.
- Black-market and criminal investigations.
- Restoration of order.
- Civil unrest.
- Intervention of private and nongovernmental organizations.

Chapter 9

The United States Army Criminal Investigation Command

The history of the USACIDC goes back to World War I, when General John J. Pershing organized the CID in France. Today, the USACIDC is the Army's sole agent responsible for investigating felony crimes on and off the battlefield. The USACIDC provides investigative support to commanders at all echelons.

USACIDC

The USACIDC investigates offenses committed against US forces or property, or those committed by military personnel or civilians serving with US forces or where there is a military interest. The USACIDC agents investigate violations of international agreements on land warfare. They conduct special investigations at the direction of the USACIDC's commanding general (CG) or a higher authority. In general, the USACIDC's missions include—

- Investigating and deterring serious crimes.
- Conducting sensitive/serious investigations.
- Collecting, analyzing, processing, and disseminating criminal intelligence (CRIMINTEL).
- Conducting protective-service operations for designated personnel.
- Providing forensic-laboratory support.
- Maintaining Army criminal records.
- Enhancing the commander's crime-prevention and force-protection programs.
- Performing LOGSEC operations.

OVERVIEW

9-1. The USACIDC's operations help the commander maintain discipline and order by preventing or investigating crimes that reduce a unit's ability to fight. During the investigation of serious crimes, the USACIDC concentrates its efforts on investigating serious crimes such as wrongful deaths, controlled-substance offenses, theft, fraud, sex crimes, and assaults. The USACIDC also conducts sensitive and special investigations involving matters pertaining to senior Army officials and those associated with classified programs.

9-2. The USACIDC agents collect, analyze, process, and disseminate criminal intelligence/information relating to crime within or directed toward the Army. Specific information relating to modus operandi, crime techniques, investigative leads, gang violence, and terrorism is shared with the appropriate intelligence and law-enforcement agencies. Conversely, USACIDC agents solicit and receive crime-related information from the MP and from local, national, and foreign law-enforcement and intelligence agencies. Special agents identify and evaluate crime-conducive conditions and indicators of potential attacks against Army property, facilities, or personnel. They then provide reports to the appropriate commander.

COMMAND AND CONTROL

9-3. The USACIDC is a centralized (stovepipe) MACOM whose special agents in the field report through the USACIDC's chain of command (detachment to battalion to group) to the CG, who in turn reports directly to the Army Chief of Staff and the Secretary of the Army (see Figure 9-1).

9-4. As an independent MACOM, the USACIDC's tactical units are not under the C^2 of supported organizations. The USACIDC's elements may be temporarily attached to a supported commander when required to accomplish a nonstandard mission. Reassignment, promotion, accreditation, and disciplinary actions are retained by the USACIDC. Attachments will be coordinated with the appropriate USACIDC headquarters planning the specific mission and approved by the CG, USACIDC. Additionally, although there is no formal staff relationship, USACIDC commanders advise their supported commanders on criminal-investigation matters. This enhances the quality, reliability of information, support, and trust between USACIDC elements and their supported commanders.

WARTIME SUPPORT

9-5. The USACIDC supports each echelon of command from the division to the ASCC. The theater USACIDC structure is comprised of a C^2 headquarters

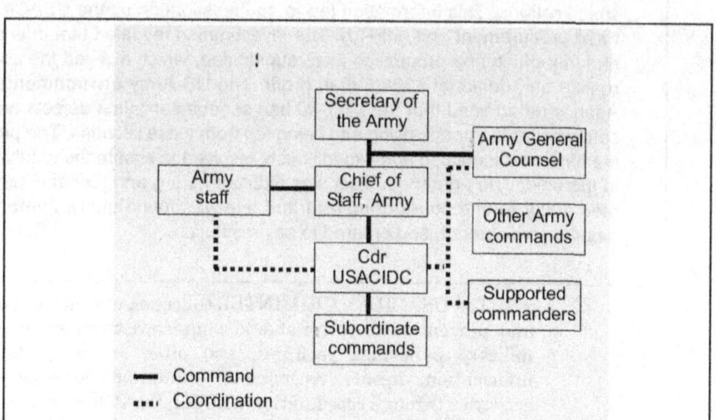

Figure 9-1. USACIDC Chain of Command

and mobile, modular, and tailorable investigative detachments. The USACIDC supports combatant commanders with the following functions:

- **LOGSEC.** Tracking and protecting materials and equipment from the manufacturer to the soldier on the battlefield.
- **CRIMINTEL.** Collecting, consolidating, analyzing, and disseminating intelligence associated with criminal and terrorists activities targeted at Army interests.
- **Criminal investigations.** Investigating war crimes and, in some cases, crimes against coalition forces and HN personnel.
- **Protective-service operations.** As in peacetime, protecting key personnel anywhere on the battlefield.

9-6. The USACIDC's LOGSEC function protects the Army's supply pipeline against criminal activities from the manufacturer, through logistics channels, all the way to the frontline soldier. It involves preventing, detecting, and investigating criminal and terrorist activities such as supply diversion, destruction, and sabotage or product substitution. The USACIDC's LOGSEC operations assess LOGSEC, identify weaknesses, and provide a prioritization of threats so that commanders can implement preventive measures to reduce the vulnerability of the logistics pipeline. Whenever possible, the USACIDC will initiate actions to recover logistical losses and return them to Army control.

The Army let a $22 million contract for the manufacture of its mobile kitchen trailer (MKT)-90 mobile field kitchens in 1990. These kitchens were distributed to Army units worldwide, including those units serving in Saudi Arabia during Operation Desert Shield/Desert Storm. After distribution of the MKT was well underway, a US government quality-assurance representative received information from the manufacturer's employees indicating that the kitchens were not built to contract specifications. This information led to an investigation by the USACIDC's major fraud procurement unit (MFPU). The investigation revealed that inferior material and manufacturing processes were substituted, which allowed the contractor to realize an additional $228,000 in profit. The US Army Environmental Hygiene Agency determined that the MKT-90 had serious sanitation defects and posed a safety hazard to troops using and being fed from these kitchens. The problem with the MKT was identified and expeditiously resolved to ensure the safety of soldiers in the field. The primary subject was indicted by a grand jury and subsequently pled guilty to one count of major fraud. He was found guilty, sentenced to 10 months confinement, and ordered to pay restitution.

9-7. The USACIDC's CRIMINTEL collection efforts focus on the identification and prevention of terrorist and nonterrorist crimes against US and allied military personnel, facilities, and other interests. Upon collecting this information, agents recommend countermeasures to combat subversive activities through coordination with MI, the PM, rear-area operations officers, and HN military and civil intelligence agencies. Additionally, when directed, the USACIDC becomes the lead US military investigative agency at theater level. When this happens, it is tasked with leading the prevention-of-terrorism effort from all services, not just the Army component.

> The MI analysts, the MP, and US Army criminal investigators deployed in Kosovo formed a combined all-source information center. This center provided threat assessments to units assigned to TF Falcon. Fifty-five percent of these assessments referenced some form of criminal activity. These intelligence products provided commanders with reliable information concerning potential incidents or criminal activity. With nearly 7,000 US soldiers deployed within the region, these assessments were absolutely crucial for the force-protection planning effort.
>
> The fusion cell within the information center developed intelligence products from national, theater, and operational sources. Due to the significant threat posed by criminal elements in the region, two CID military agents and two CID civilian analysts were attached to the fusion cell to facilitate the police-intelligence function. The CID personnel, in cooperation with MP soldiers, played a key role by linking CRIMINTEL to specific groups and events. The CRIMINTEL collection effort was specifically targeted on weapons, drugs, organized crime, and the identification of smuggling routes. The identification of smuggling routes resulted in a significant increase in the number of weapons being confiscated. The timely transfer of CRIMINTEL products to tactical units enabled a rapid response to serious confrontations, increased confiscation of arms and ammunition, and improved stability in TF Falcon's AO.

9-8. During war and MOOTW, the criminal-investigation effort includes—

- War crimes and, when directed, crimes against coalition forces and HN personnel.
- Major felony crimes committed by US personnel or EPWs/CIs and, when directed, crimes committed by foreign nationals in which there is an Army interest.
- Criminal acts by indigenous personnel, factions, and ad hoc groups.
- Special/sensitive investigations, hostage negotiations, polygraph operations, and force-protection operations.
- CRIMINTEL operations.

9-9. The USACIDC detachments provide enhanced coverage and protection for designated key and essential leaders during war and MOOTW. The protection requirement for senior JTFs and Army commanders may be significantly greater during MOOTW than during peacetime or war as the propensity of asymmetrical threats (such as criminal and terrorist groups) operating in the AO also increases.

ORGANIZATION

9-10. The USACIDC is organized with an array of multifunctional units. As with many other Army organizations, its force is in a transition process. During peacetime, the USACIDC groups provide brigade-level area support for worldwide Army operations. The group provides C^2, staff planning, and administrative support for USACIDC elements assigned to an AOR or deployed to a contingency operation. During war, a group provides the same capabilities in support of an ASCC and provides C^2 for all USACIDC units in the theater. Regardless of the operational environment, the group ensures the connectivity between all USACIDC units. It establishes and maintains links with supported units and interagency, joint, multinational, allied, and HN authorities on matters pertaining to Army and USACIDC operations. The group headquarters has a command section, a detachment headquarters, an S1, an S2, an S3, an S4, an SJA, and a communications section. It provides C^2 for up to six subordinate battalions. Major differences in the AOE and Force XXI groups are in an increase of support and mission personnel and the movement of all polygraph support down to battalion level.

9-11. During peacetime, the USACIDC battalions provide area support for worldwide Army operations. They provide C^2, staff planning, forensic-science support, CRIMINTEL, polygraph support, technical guidance and supervision of USACIDC operations, and administrative support for USACIDC elements assigned to an AOR or deployed to a contingency operation. During war, the USACIDC battalion provides the same capabilities in support of each corps and TSC. Under AOE designs, one battalion supports each corps and TSC. Under the Force XXI design, up to two battalions support each TSC. Like the groups, each USACIDC battalion ensures connectivity and establishes and maintains links with other units/agencies within its AOR. The battalion headquarters consists of a command section, a detachment headquarters, an S1, an S2, an S3, an S4, and a communications section. It provides C^2 to five USACIDC detachments on a sustained basis. Its span of control can accommodate up to seven detachments for a short period of time. Major differences in the new battalion are the addition of a CSM, an S6 officer, a detachment commander and support personnel, and CRIMINTEL and forensic personnel.

9-12. The USACIDC investigative team is the smallest operational investigative element. Each team consists of two special agents (a warrant officer and a noncommissioned officer [NCO]). METT-TC requirements dictate that these teams have the capability to operate independently from the detachment headquarters. The investigative teams are the building block for both AOE and Force XXI USACIDC detachments. Both designs provide levels of flexibility to task-organize without splitting units apart. The detachment (with its complement of teams) performs the full range of criminal-investigative functions in all operational environments, but it has limited CRIMINTEL and LOGSEC management capabilities and no forensic or polygraph capability. Each detachment includes a headquarters section and up to eight investigative teams. METT-TC requirements dictate that these teams have the capability to operate independently from the detachment headquarters. The major differences between the AOE and Force XXI detachments' design begin with replacing the heavy and light division support elements (DSEs), the corps-area support element (CASE), the TSC-area support element, (TASE), and the port-area support element (PASE) designs with a modular design consisting of a standardized headquarters element and two special-agent sections. Each section consists of four investigative teams. Personnel administration, supply, and the bulk of administrative support found in the AOE organizations have been consolidated at battalion.

Additionally, the Force XXI detachment is commanded by a special-agent warrant officer.

9-13. Both the AOE and the Force XXI units retain the Army's standard dependencies upon other units for support requirements (religious, health services, finance, photographic processing, and so forth). The new USACIDC TOEs have enhanced A/L support capabilities. However, the new units still retain the AOE support requirement for food services and emergency Class III/V resupply from CS MP brigades, battalions, and companies.

THE FIELD INVESTIGATIVE UNIT

9-14. Sensitive investigations are normally conducted by the field investigative unit (FIU). The FIU is a one-of-a-kind organization within the DOD that enables the Secretary of the Army to conduct sensitive investigations requiring access to special information or programs that are highly classified. The FIU works closely with the Army IG, the Judge Advocate General of the Army, and the Army General Counsel to support commanders in the special-operations and intelligence communities and in the area of classified acquisition programs. The FIU may also be assigned investigations involving senior Army personnel or those of special interest to the Army leadership.

THE COMPUTER-CRIME INVESTIGATION UNIT

9-15. The computer-crime investigative unit (CCIU) deals with intrusions involving classified networks or multijurisdictional offenses. The CCIU works closely with MI and federal law-enforcement agencies to coordinate military actions, nonmilitary-affiliated offenders, and foreign-intelligence services. In addition, it provides technical assistance to CID elements that are conducting computer-related investigations.

THE PROTECTIVE-SERVICE UNIT

9-16. The protective-service unit (PSU) provides worldwide protective services to designated personnel to protect them from assassination, kidnapping, injury, or embarrassment. The PSU plans, coordinates, and executes executive protection for—

- The Secretary of Defense.
- The Deputy Secretary of Defense.
- The Chairman of the Joint Chiefs of Staff.
- The Vice Chairman of the Joint Chiefs of Staff.
- The Secretary of the Army.
- The Army Chief of Staff.
- The Army Vice Chief of Staff.
- Others (as directed).

9-17. During war and MOOTW, the USACIDC routinely provides protective-service support to senior Army ground commanders and, in some cases, to JTF commanders. The USACIDC agents also provide training to the MP details providing protective services to senior field commanders.

The President of the US hosted the North Atlantic Treaty Organization's (NATO's) 50th summit anniversary celebration in Washington, DC during the period of 22 to 26 April 1999. The President invited 28 heads of state or prime ministers, 48 foreign ministers, 88 ministers and chiefs of defense, and the chairman of the military committee to the summit. During the conference, the principal guests resided in 23 hotels in Washington, DC. The responsibility for providing protection for the ministers, chiefs of defense, and the chairman for the military committee was given to the USACIDC's PSU. In preparation for this mission, additional training on protective-service operations and evasive driving techniques was given to over 200 reserve agents. These individual mobilization augmentees, troop program units, and other USACIDC agents were critical to the successful execution of this mission.

US ARMY CRIMINAL-INVESTIGATIONS LABORATORY

9-18. Forensic-laboratory support is provided by the US Army Criminal-Investigations Laboratory (USACIL) that is currently located at Fort Gillem, Georgia. The USACIL is a nationally accredited forensic laboratory that provides all three services with dedicated, highly responsive, deployable, state-of-the-art scientific and forensic investigative support. The six forensic divisions of the USACIL include the following:

- Chemistry (serology, drugs, trace, and deoxyribonucleic-acid [DNA] testing).
- Questioned documents (alterations and comparisons of handwriting, typing, and printing).
- Firearms and tool marks (weapons and obliterated numbers).
- Photography (evidence, charts and photos for court).
- Latent prints.
- Computer forensics.

9-19. The USACIL collects, receives, and examines evidence and prepares reports of findings. It also provides expert-witness testimony at court-martials.

US CRIME RECORDS CENTER

9-20. The US Crime Records Center (CRC) maintains the Army's criminal records. It maintains more than two million MP reports and USACIDC reports of investigations. These records are retained for at least 40 years before being destroyed. During their retention, the records are actively maintained and frequently researched.

During the course of an investigative search, civilian police authorities seized 19 hand grenades. A routine search of stolen property and criminal index files failed to identify the grenades as having been stolen from a nearby military installation. However, a search of the CRC files identified a USACIDC investigation relating to the theft of a case of grenades that occurred 20 years earlier. Further investigation revealed that the lot number of the recovered grenades was identical to that of the stolen grenades. The investigative leads resulting from the CRC file search contributed to the resolution of both cases.

9-21. The CRC conducts more than 10,000 criminal-history name checks each month to identify victims and perpetrators of criminal offenses. The checks are requested not only by USACIDC agents, but also by other military and civilian law-enforcement officials. The CRC is also the Army's agent for Freedom of Information Act (FOIA) and Privacy Act requests relating to criminal investigations and MP reports. The CRC also manages the Army's polygraph programs and their support to Army installations around the world.

EMPLOYMENT

9-22. Whenever possible, USACIDC units are colocated with the MP. This provides unified MP support across the battlefield, and it facilitates logistical support (primarily mess) provided by MP units. However, the USACIDC maintains its vertical C^2. Based on METT-TC, the MP group (CID) will colocate with the supported ASCC and the MP battalions (CID) supporting the corps and EAC will colocate with the MP brigade (CS). When feasible, MP detachments (CID) will colocate with MP battalions/companies and division MP companies. Figure 9-2 illustrates a typical battlefield array for AOE organizations, and Figure 9-3 illustrates a Force XXI array.

9-23. Since USACIDC detachments are austere organizations with little organic A/L capability, they rely on the battalion for unit-level maintenance, supply, and personnel support. However, detachments must often operate independently when performing criminal-investigation functions throughout the AOR. During these times of long-distance separation, detachments must seek A/L support from the supported unit.

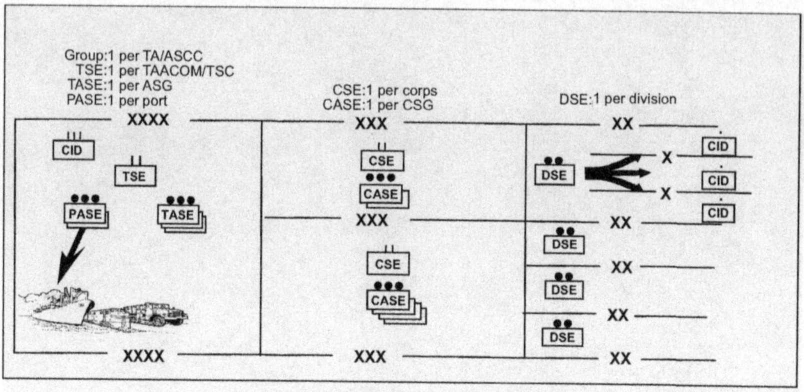

Figure 9-2. AOE Battlefield Array

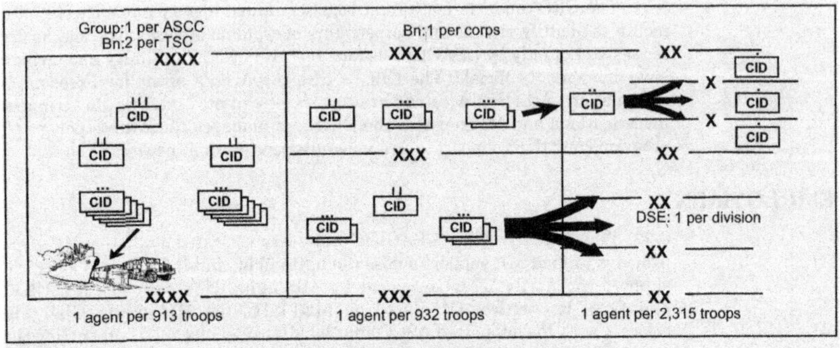

Figure 9-3. Force XXI Battlefield Array

Chapter 10

Offensive and Defensive Operations

The whole of military activity must relate directly or indirectly to the engagement. The end for which a soldier is recruited, clothed, armed, and trained; the whole object of his sleeping, eating, drinking, and marching is simply that he should fight at the right place and the right time.

Clausewitz

Army doctrine considers the full range of operations from war to MOOTW. An operation is a military action or the carrying out of a military mission needed to gain the objectives of any battle or campaign. Offensive and defensive operations normally dominate military operations in war and in some smaller-scale contingencies. An offensive operation is aimed at destroying or defeating the enemy and imposing our will on him for a decisive victory. Defensive operations defeat an enemy attack, buy time, economize forces, or develop conditions favorable for a counteroffensive that regains the initiative and defeats the enemy.

OVERVIEW

10-1. The offense is the decisive form of battle. Offensive campaigns and major operations are designed to achieve operational and strategic objectives quickly and decisively at the least cost. Offensive operations combine the terrain and the force to achieve the objective. The main purpose of the offense is to defeat, destroy, or neutralize the enemy force. Offensive operations are characterized by a rapid shift in the main effort to take advantage of opportunities by momentum and simultaneous destruction of enemy defenses.

10-2. The immediate purpose of defensive operations is to defeat an attack. Military forces defend only until they can gain sufficient strength to attack. Commanders choose to defend when they need to buy time, to hold a piece of terrain, to facilitate other operations, or to erode enemy resources at a rapid rate while reinforcing friendly operations. The ARFOR conduct defensive operations as part of major operations and campaigns and in combination with offensive operations (see FM 3-0).

SUPPORT TO OFFENSIVE OPERATIONS

10-3. MP support to offensive operations varies according to the type of operation being conducted. The MP functions support each echelon commander based on METT-TC. In the offense, MP priorities are often placed on MMS, I/R, and AS operations. Regardless of the offensive effort, MP leaders supporting an offensive operation must—

- Understand the commander's intent and his concept of operations.
- Anticipate that selective elements of the offensive force may need to pause, defend, resupply, or reconstitute while other forces attack.
- Anticipate changes in the operational tempo and prepare to weight the MP effort to meet the maneuver commander's intent.
- Provide MP support within the context of linear and nonlinear operations. They must understand how both types of operations affect the security of rear areas and LOC, the protection of C4I, the means of sustainment, and the land force's mobility.
- Support the maneuver commander's intelligence efforts by conducting the PIO function. The MP must observe, identify, and report information.
- Assist maneuver forces to mass. The MP protect C^2 nodes such as the CP that orchestrates the mass efforts and resources that sustain the offensive move. The MP must deny incursions by enemy reconnaissance forces into the AO. They also maintain surveillance, provide early warning, and impede and harass the enemy with supporting and organic fires.
- Speed the commitment of follow-on forces and reserves. They must also conduct MMS operations to assist the commander in a smooth and quick shift of forces to support the main effort.
- Anticipate the pursuit and exploitation by positioning MP forces that can support continuous operations. The MP support as far forward as possible while protecting resources needed to seize the opportunity. The EPWs/CIs are collected as far forward as possible. The MP coordinate the treatment of dislocated civilians and refugees with the HN or foreign forces.
- Tailor the MP support based on the unit being supported (corps, divisions, and brigades).
- Consider the location and composition of response forces and the TCF and the strength of rear-area threats.
- Integrate a force into the offensive plan that is capable of conducting aggressive combat operations against designated targets.

> During Operation Just Cause, an MP platoon received the mission to disable an armed Panamanian Defense Force (PDF) patrol boat docked near a key objective. After receiving the mission, the platoon leader maneuvered his force into position and attacked after receiving enemy fire. The platoon successfully disabled the patrol boat and captured the objective while suffering minimum casualties.

10-4. During offensive operations, MP support provided by teams, squads, and platoons provide the commander with an agile, flexible, versatile, and capable force ready to contribute to the overall mission success.

SUPPORT TO DEFENSIVE OPERATIONS

10-5. The MP support defensive operations consistent with the commander's intent and priorities. They support a defending force by assisting subordinate elements in the execution of different, yet complementary actions. For example, the MP may support the maneuver of an attacking unit by conducting MMS operations while simultaneously providing AS for a defending unit in a base cluster. The MP support reflects the maneuver commander's focus—destroying an attacking force, retaining or denying key terrain, moving away from an enemy force, or a combination of the aforementioned. However, despite the intent of the defensive operation being conducted, MP leaders must—

- Conduct MMS operations to help a force maneuver and mass. The MP must anticipate the shift from the defense to the offense and assist the forward movement of reserves.
- Conduct AS operations to deny information to enemy reconnaissance elements seeking out the exact location of the defending force. The MP are positioned where they can control key terrain or improve the defensive capability of bases and base clusters. The MP conduct aggressive R&S to locate and deny enemy access to critical logistical facilities.
- Protect sustainment resources while supporting the lateral, forward, and rearward movement of forces.
- Consider the type and size of the AO, the LOC security, the rear threat, and the number of EPWs/CIs and dislocated civilians to determine how they may affect the movement of forces.

> During Operation Desert Shield, EAC, corps, and division MP units conducted a myriad of missions to support both the XVIII Corps's (Airborne) and the VII Corps's movement to their TAA. The MP's main focus was on MMS and AS throughout the AO to ensure freedom of movement and to protect US and coalition forces against terrorists and other threats in the rear area.

10-6. The success of MP operations in the defense depends on the MP leader's situational awareness, the commander's intent, and the precise employment of MP resources when and where they are needed. In the defense, the MP perform their five functions to assist the defending commander by providing a lethal mobile force that allows him to concentrate his efforts and resources in fighting the next battle.

Chapter 11

Stability and Support Operations

ARFOR conduct stability operations in a complex, dynamic, and often asymmetric environment. Stability operations are usually nonlinear and noncontiguous. They are often time and manpower intensive. The purpose of support operations is to meet the immediate needs of designated groups for a limited time until civil authorities can accomplish these tasks without Army assistance.

FM 3-0

The Army's mission is to prepare for war and, when deterrence fails, to achieve a quick, decisive victory. The DOD projects that many of its future missions will be stability and support operations. However, these operations are not new to the Army. The Army has participated in stability and support operations in support of national interests throughout history. The Army has protected its citizens, assisted nations abroad, and served America in a variety of other missions. The pace, frequency, and variety of stability and support operations have quickened in the last three decades. As demonstrated in Operations Provide Comfort, Restore Democracy, and JTF Los Angeles, these types of operations present a challenge to all services.

OVERVIEW

11-1. Stability operations are conducted outside of the US and its territories to promote and protect US national interests by influencing political, civil, and military environments and by disrupting specific illegal activities. Stability operations may include both developmental and coercive actions. Developmental actions enhance a HN government's willingness and ability to care for its people. Coercive actions apply carefully prescribed limited force or the threat of forces to change the AO's environment.

11-2. Support operations provide essential supplies and services to relieve suffering and to help civil authorities prepare or respond to crises. In most cases, the ARFOR's efforts are focused on overcoming conditions created by man-made or natural disasters. The ultimate goal of support operations is to meet the immediate needs of designated groups (for a limited time) until civil authorities can accomplish these tasks without Army assistance.

11-3. Stability and support operations are distinctly different from offensive and defensive operations and must be analyzed differently. For instance, the application of the rules of engagement (ROE) will be unique to the operation. The restrictions on combat operations and the use of force must be understood and obeyed at all levels. MP leaders must clearly take the initiative and ensure that the applicable guidance is implemented and followed by all subordinates.

11-4. Stability and support operations have more diverse political considerations than are usually encountered in war. There are also complexities that must be addressed in joint, combined, and interagency stability and support operations. The Army's challenge is to execute these operations successfully in a joint or combined arena without degrading its capability to wage war successfully.

11-5. If the US is to fight and win future wars, it must develop, train, and employ its force structure in a manner that ensures success in every operation across the spectrum of military operations. Under these auspices, MP forces can provide commanders with especially valuable assets for stability and support operations. The MP have the unique capability of serving as a combat, CS, and CSS force. The domestic and international acceptability of the MP operational image frequently makes the MP the most appropriate force for stability and support operations. Additionally, the MP are flexible and capable of rapidly transitioning from one end of the spectrum to the other if the environment changes.

FORCE SUITABILITY

11-6. The MP's capabilities and their operational and supporting tasks are as integral to stability and support operations as they are to offensive and defensive operations. MP branch-/core-related skills are highly compatible with the capabilities required for stability and support operations. The MP provide a highly capable, politically acceptable force that is suitable for a variety of missions. They possess robust moving, shooting, and communicating capabilities; and they project an assist, protect, and defend image that is particularly important when tailoring a force that requires significant capabilities but a low political profile. The MP provide a flexible, but limited, economy-of-force organization. Mission requirements will always exceed available MP resources, especially in stability and support operations. For this reason, MP assets must be prioritized to deliver the greatest mission capability. The specific operations that MP units perform at a given time are determined by the echelon commander's needs, the mission requirements, and the availability of MP resources. As discussed in previous chapters, the echelon commander, through the command's PM, sets the priorities for MP operations. As in offensive and defensive operations, the PM knows that while the MP force can perform all MP functions, they may not be robust enough to perform all assigned MP missions all of the time. The PM must continuously make trade-offs between the numbers and kinds of MP missions that can be supported.

11-7. The MP's capabilities are further enhanced by their training, mind-set, and experience in dealing with people in highly stressful and confusing situations. Although capable of conducting combat operations when needed, the MP are highly practiced in de-escalation and in employing the minimum-essential force to contain potentially violent situations. This mind-set serves as the framework for MP law-enforcement training and is especially applicable in MP support for stability and support operations. Furthermore, this mind-set is exercised and reinforced daily in peacetime law-enforcement operations. This is a significant capability. MP soldiers learn and receive constant reinforcement training in controlling a situation on the spot before it escalates and cannot be controlled. The MP have a unique understanding of human nature and are adept at handling emotion-laden situations.

> During operations in Kosovo, as both Serbs and Albanians attempted to create unfavorable conditions and responses by US forces, the MP kept their cool under fire in the streets of Gnjilane and Kamenica. The MP's reaction led to a quick understanding by all sides that the US Army MP were fair and just and understood the tense situations.

11-8. The MP stress four essential training competencies—skill proficiency, human dimension and attitude, camaraderie and teamwork, and leadership. These competencies are the foundation for all MP operational performances. They equip the MP to exercise discretion in dealing with others, to protect and assist those in need, and to accomplish the varying demands of MP missions in stability and support operations and, when needed, in war.

STABILITY AND SUPPORT

11-9. Stability and support operations are performed in support of US authorities or in support of international bodies and foreign governments when so required by US authorities. The MP have long provided the essential support for stability and support operations; however, the frequency of these operations has increased significantly in recent years. Consequently, the MP have been deployed in greater numbers with greater frequency. MP units have a substantial history of successful, simultaneous, and continuous deployments in both a lead and a support role.

SUPPORT TO DOMESTIC CIVIL AUTHORITY

11-10. The purpose of support to domestic civil authorities is to—

- Render humanitarian assistance.
- Provide disaster relief.
- Restore order.
- Combat terrorism.
- Support counterdrug operations.

11-11. Military operations supporting domestic civil authorities are governed by the provisions of the Stafford Act, the Posse Comitatus Act, and other laws and regulations. Peacetime experience in civil-military affairs equips the MP to oversee, assist, and train other units and agencies for law enforcement, sensitive security, and operations to control dislocated civilians (within the legal limits of a particular stability or support operation). For this reason, the MP are often needed after other forces have redeployed, and they stay until government order is restored. Since active-component military missions are constrained by the Posse Comitatus Act, nonfederalized national guard (NG) units are often employed to conduct these operations (particularly counterdrug operations). The NG units can operate under gubernatorial authority within state jurisdiction without Posse Comitatus restrictions. The NG MP units are the only NG units trained, skilled, and experienced in providing law enforcement. They offer civilian authorities a unique, specialized capability not available from other NG units.

SUPPORT TO HUMANITARIAN ASSISTANCE AND DISASTER RELIEF

11-12. Humanitarian assistance and disaster-relief operations provide emergency assistance to victims of natural and man-made disasters and other endemic conditions. Natural disasters include earthquakes, floods, and hurricanes. Man-made disasters include riots and civil conflicts. Humanitarian assistance and disaster relief conducted by US forces alleviate urgent nonmilitary needs of a populace until the appropriate civilian agencies can provide the needed services without military support. These types of operations include, but are not limited to—

- Conducting searches and rescues.
- Evacuating, sheltering, sustaining, and protecting dislocated civilians.
- Making food, medical care and treatment, and other civilian-aid programs available.
- Reducing opportunities for criminal activity and restoring L&O.
- Maintaining other operations needed to ensure the well being of the affected population.
- Providing relief operations for man-made disasters. These operations are characterized by civil conflict, and they entail all of the requirements for other disasters but focus on the restoration of L&O.
- Providing force protection and security.

11-13. MP support can be an important asset in all disaster-relief operations. The MP are trained and equipped for decentralized operations. MP teams operate in highly mobile vehicles equipped with radios, which are invaluable in disaster-relief operations. Each MP platoon, company, and battalion headquarters is equipped with high-frequency radios. These capabilities allow the MP (even when widely dispersed) to maintain centralized communications; to provide L&O support; and to assist in search, rescue, and evacuation efforts. Because of their decentralized operations and communications equipment, the MP can also play a vital role in notification, ADC, and access and egress control. They can deploy with disaster-assessment teams to assist in advising and developing a disaster-assistance plan. The MP can also deploy to an affected area to—

- Assist in restoring order.
- Search for, rescue, and evacuate victims.
- Provide personnel and vehicular circulation controls.
- Provide ADC.
- Conduct detention operations.
- Provide physical and area security to protect life and property immediately following a disaster.
- Protect fire and emergency-services personnel.
- Gather, record, and report information.
- Provide force protection and security.

11-14. MP (I/R) battalions can support civil-affairs units by establishing I/R facilities to shelter, sustain, protect, and account for disaster-relief victims. These units can also train and assist government agencies (US and foreign) or police in the operation of facilities for dislocated civilians and detainees.

11-15. The MP possess capabilities critical to relief in man-made disasters characterized by civil conflict. The MP have the essential training and skills needed for relief operations that focus on restoring civil order. Restoration of civil order demands—

- **Objectivity and neutrality.** The objective of civil-disturbance control is to restore L&O, not to impose punishment or suppress peaceful dissent. Government forces quelling civil disorders must be emotionally objective and politically neutral. MP law-enforcement training stresses objectivity and neutrality, an important attribute for forces employed in stability or support operations.
- **Minimal use of force.** The application of more than minimum force may be detrimental to mission success. The MP are particularly suited for these situations, having been trained on the technique for the use and measured application of minimum forces to accomplish mission goals.
- **Public acceptance.** Government forces must avoid actions that will alienate the populace. Any action offensive to community sensitivities must be avoided. The MP receive training in crowd-control operations, interpersonal communications skills, and minimum use-of-force techniques—skills that are critical in gaining public acceptance.
- **Positive image.** In many disturbance operations, demonstrators and government forces compete for the approval and sympathy of the general population. The MP's assist-and-protect image provides the government with a distinct advantage in such situations.
- **Threat awareness.** Government forces must pay attention to and try to learn the organization, intentions, and activities of demonstrators. Equal attention must be paid to how the community will respond to the demonstrator's actions and the government's response. The MP interact daily with the populace, placing them in a position to develop criminal information and to determine the collective attitude of the populace.
- **Effective coordination.** Disturbance operations often involve many agencies from various levels of government. Thorough, timely, and coordinated preparation, planning, training, and execution are essential for mission success. MP civil-disturbance training (including planning and coordination measures) and their communications assets enhance coordination, particularly during mission execution.
- **Operational unity.** Multiple control forces representing different agencies with varying jurisdictional authority can pose C^2 problems. When unity of command cannot be achieved, operational unity must be achieved by colocating operational centers, integrating communications, and delineating organizational responsibilities. The MP are well trained in police and military operations, possessing a sound understanding of both civilian police and military measures, legal constraints, and effective coordination.

11-16. The MP operational tasks supporting man-made disasters vary according to the nature of the disaster and the operational environment. For example, in CONUS where the commitment of military forces to civil disturbances is considered a last resort and military involvement is limited by law, nonfederalized NG MP units have long been employed by civil authorities for their unique capabilities. MP forces can—

- Isolate and contain an area with barricades, roadblocks, and perimeter patrols. The MP can also assist civil-affairs units by enforcing pass and identification systems and public-utility controls.
- Provide mobile patrols to enhance the security of high-priority targets (such as buildings, utilities, and services that are critical to the community's economic and physical well being).
- Monitor, disperse, contain, or limit crowd movement through—
  - Observation points and patrols.
  - The communication of interest and intent to the crowd.
  - The channelization or diversion of the crowd.
  - Gaining the cooperation of the dissident leadership.
  - The proclamation of the illegal nature of the crowd's actions.
  - Show of force.
  - The use of crowd-control formations.
  - The apprehension of crowd members and leaders.
  - The application of minimum-force measures.
- Establish area control to prevent looting, to protect businesses and other likely targets, and to prevent arson. The MP accomplish this mission with saturation patrolling, including vehicle and foot patrols. With augmentation, the MP can perform air and water patrols. They also enforce populace control measures (such as ordinances to prevent gathering, permits to gather, restrictions on circulation, restrictions on the interference with government and public functions and personnel, restrictions on possessing weapons, and other measures instituted by civil authorities).
- Neutralize special threats (such as snipers and bombings) that are highly dangerous to both government forces and the community. Snipers pose a particularly grave danger in disaster-relief operations. The MP can take immediate protective actions, secure an area, isolate the threat and, when authorized, use SRTs to assist civilian or HN authorities in apprehending or neutralizing the sniper. The MP are trained to perform the necessary actions when an explosive device is discovered. In such an event, they can secure and evacuate an area, organize search teams, and isolate the site.

11-17. The MP can also support in rescue and recovery operations, which may include rescuing US or foreign nationals. Such operations are also employed to locate, identify, and recover sensitive materials deemed critical to national security. The operations may be conducted in benign or hostile environments. The MP provide OPSEC, and they process and account for civilians affected by rescue and recovery operations. The MWD teams can also help search for lost or hidden personnel. In stability and support operations, the MP can perform their police functions for rescue and recovery operations while keeping the perceived military signature low.

SUPPORT TO COUNTERDRUG OPERATIONS

11-18. Counterdrug operations are measures taken to disrupt, interdict, and destroy illicit drug activities. In a counterdrug role, the DOD may offer certain forms of support to HN counterdrug personnel, to US civilian law-enforcement agencies operating in a HN, to Department of State (DOS) counterdrug personnel, and to the Bureau of International Narcotics Matters. Without direction from the National Command Authority acting under constitutional or statutory authority, US forces engaged in counterdrug activities may not engage in direct law-enforcement activities (such as apprehension, search, seizure, and other similar activities). As discussed earlier, the Posse Comitatus Act authorizes specific DOD assistance in counterdrug activities, but US military forces usually may not participate directly in arrests, searches, seizures, and similar activities. Conversely, as discussed earlier, USC Title 32 authorizes state governors to use NG MP units with full authority to engage directly in law-enforcement activities. When authorized, US advisors may accompany counterdrug forces on operations.

11-19. MP operational contributions can include the use of force, drug identification, crime-scene protection, customs operations, surveillance of criminal activity, serious-incident reporting, and other related activities. Employing the MP for these operations ensures force suitability and avoids a mismatch in terms of unit capability and mission effectiveness.

11-20. Border-screening operations consist of three separate but related measures—mobile patrolling, the use of listening/observation posts, and airfield surveillance. These measures are part of the MP's AS and MMS functions used in tactical rear-area operations. The MP's operational effectiveness may be further increased through augmentation by a civilian law-enforcement officer who would be responsible for search, seizure, and arrest actions. Active-component units are permitted to assist in marijuana eradication in support of domestic law-enforcement agencies (DLEAs). Marijuana-eradication operations are an application of the MP AS, mobility, and L&O functions, thus providing realistic training while supporting an operation.

11-21. As discussed in previous chapters, the MP are well suited to augment customs operations. They routinely perform customs inspections for US forces in Korea and Germany. The NG MP regularly conduct vehicle, cargo, and container searches under the supervision of US Customs agents and at US ports of entry.

11-22. The USACIDC agents can provide support to counterdrug operations. The CID units identify the sources of illegal-drug distribution systems. When directed, the MP and the CID complement interagency counterdrug efforts to stop the production, flow, and distribution of drugs and to provide planning, training, equipment, and facilities to support domestic counterdrug operations. The CID also supports CRIMINTEL programs, internal security, and site surveys at borders where Army units will deploy. In addition to the aforementioned support, the DOD General Counsel has ruled that MWD teams can be loaned to DLEAs to assist in detecting smuggled contraband.

SUPPORT IN COMBATING TERRORISM

11-23. The Department of Justice (DOJ) is the lead federal agency for dealing with acts of terrorism within the US, its territories, and its possessions. Within the DOJ, the Federal Bureau of Investigation (FBI) has the lead. The FBI can train the police of friendly nations in antiterrorism and counterterrorism operations. Usually, US military forces act in a technical-advisory but not a law-enforcement capacity in combating acts of terrorism. It is often difficult to distinguish the acts of politically motivated terrorists from violent acts of criminals or individuals in society at large. In fact, all terrorist acts are criminal; however, they also may be part of an insurgency.

11-24. The MP are key players in the defensive planning, implementation, and control of combating-terrorism operations, whether in garrison or deployed to war or MOOTW. In whatever countries US forces operate, MP units conduct antiterrorism operations daily as part of their L&O and AS functions. The MP also respond to terrorist incidents. They help secure the area and search for evidence, and they help the lead agency. The MP provide advice and assistance in developing antiterrorism and counterterrorism programs for deployed US forces and DOS agencies and for police, paramilitary, and military forces of assisted nations (when so authorized). MP training, liaison, and joint patrols help strengthen HN-police organizations, improving their effectiveness and efficiency in combating terrorism. The MP also assist HN police in performing PRC operations.

11-25. The training of police (domestic and foreign) requires DA approval, and restrictions exist on some forms of training. Legal advice and coordination are a must before training can occur. The MP and the CID provide personnel and equipment for planning, advising, equipping, and training agencies to prevent, deter, and respond to terrorism. The training focuses on deterring terrorist acts such as bombings, kidnappings, assassinations, taking hostages, and hijackings. Related assistance may include training and equipment for explosive detection, management of hostage situations, physical security, protection of key personnel, and defensive/evasive driving methods. For counterterrorism operations on US military installations, the MP provide highly trained, specially equipped SRTs for situations requiring the selective use of force. Additionally, CID agents are highly trained in hostage negotiation, which increases the commander's options before resorting to force.

11-26. The MP and the CID assist in counterterrorism information-gathering activities through liaison and joint patrols with HN police. These activities are part of the PIO function. The MP have extensive contacts with civilian police, who in turn have contacts with the populace. These contacts can help surface criminal information concerning terrorist plans, recruitment, and support structures sooner than military sources. The CID can provide commanders with CRIMINTEL on a daily basis. At the same time, MP antiterrorism support includes measures to reduce the vulnerability of people and property. Such measures include—

- Personal awareness.
- Personal-protection operations and techniques.
- Crime prevention.
- Physical-security programs.

11-27. The MP and CID crime-prevention programs employ proactive measures to protect people and their property and US and HN property. Crime prevention is a continuous process of planning, implementing measures, evaluating effectiveness, modifying measures, and developing further information. The primary crime-prevention tool is the crime-prevention survey. It is used to examine the physical features of various critical sites. The MP physical-security programs help identify, reduce, eliminate, or neutralize conditions favorable to criminal, terrorist, and insurgent activities. The key measures used to accomplish this are physical-security inspections and surveys.

SUPPORT TO NONCOMBATANT EVACUATION OPERATIONS

11-28. Noncombatant evacuation operations (NEOs) relocate civilians who are endangered in a foreign nation. The NEOs are classified as benign (unopposed) or hostile (opposed). MP roles in support to NEOs differ accordingly. In a benign environment, NEOs are usually conducted with the help and full cooperation of the affected nation, with little or no opposition to evacuation. The use of force is limited to self-defense (as outlined by the ROE) and protection of the evacuees. In a hostile environment, NEOs are opposed by forces with specific intentions of preventing or destroying the operation. In this scenario, the MP will follow the ROE as established by higher headquarters.

11-29. MP operations in NEOs closely parallel EPW/CI evacuation operations. The objective of both is to safely collect, secure, protect, transport, process, and account for people, whether they are EPWs or noncombatants. However, due to the dissimilarities of these groups, the techniques employed differ.

11-30. The MP can secure selected facilities, reception and transfer points, or other locations required to collect, process, and evacuate people. They can provide route reconnaissance and mobile security teams to escort HRP, DOS personnel, evacuees, and sensitive material. If needed, MP units assist civil-affairs personnel with the movement and control of dislocated civilians. The MP may support civil affairs by setting up and operating evacuation sites, holding areas for foreign nationals denied evacuation, and reception and processing stations. This support can be provided by MP CS units for a short duration and by MP I/R units for a prolonged duration. Additionally, the MP will establish and maintain liaison with HN police to ensure a mutual understanding of jurisdiction, to coordinate efforts, and to provide criminal and tactical intelligence. In hostile environments requiring the employment of combat forces, the MP would be employed in their traditional CS role.

SUPPORT TO SECURITY-ASSISTANCE OPERATIONS

11-31. Security-assistance operations are one of the main tools of US foreign policy. A critical concern in these operations is the HN's ability to plan and manage its own defense resources. In security-assistance operations, the US provides defense materiel, military training, and defense-related services through grants, loans, credit, or cash sales. When authorized, the MP support security assistance with military and police training through the International Military Education and Training Program (IMETP). This program provides instruction and training to foreign military and civilian personnel on a grant/aid basis. Many foreign civil law-enforcement agencies request training in the following areas:

- Basic police procedures (such as patrolling and crime-scene protection).
- Physical security.
- Corrections.
- Civil-disturbance operations.
- Customs operations.
- Traffic control.
- Use of force.

11-32. The MP possess the expertise and experience to advise, train, and assist a HN's military and police forces. The MP's participation in the IMETP includes—

- Formal and informal instruction of foreign students.
- Technical education and tactical training with applicable training aids.
- Assistance to foreign police and military elements by MP mobile training teams (MTTs).
- Providing MP units or small, modular, functionally specific teams to support security-assistance needs. These teams can train HN police in both field and institutional facilities.
- Support to nation-assistance operations.

11-33. Nation-assistance operations support a nation's effort to promote development, ideally by using its national resources. This is achieved through education and the transfer of essential skills to the HN. The goals of such operations are to—

- Promote long-term regional stability.
- Develop sound and responsive democratic institutions.
- Develop supportive infrastructures.
- Promote an environment that allows for orderly political change and economic progress.

11-34. The MP support nation-assistance operations with training and joint MP and HN-police operations. The MP know and understand police operations, military operations, and C^2 procedures. Consequently, the MP are ideal for training the law-enforcement personnel in both field and institutional settings. The MP's nation-assistance goals are to—

- Enhance L&O in democratic societies.
- Improve efficiency and effectiveness.
- Promote the proper usage and maintenance of equipment.
- Establish a sound training base for police.

- Standardize procedures that enhance combined police operations.
- Promote friendship and goodwill toward the US.
- Preclude the need for US military interventions to counter acts of violence.

11-35. Combined police operations reinforce training and provide HN police with mobility, security, and communications to operate more effectively while conserving critical personnel resources. The MP provide support through training and the following:

- Law-enforcement, security, and criminal-information support to the HN police force (to include L&O operations and administration in a democratic society).
- Patrol and desk operations.
- Circulation-control operations.
- PRC and civil-disturbance operations.
- MWDs.
- Physical-security operations.
- Personnel-security operations.
- Area- and route-security operations.
- Counterdrug operations.
- Antiterrorism operations.
- Mass immigration operations.
- Customs and border operations.
- Confinement operations.
- Crime prevention.
- Community and police relations.
- C^2 of police operations.

11-36. When conducting nation-assistance operations, as well as other operations, the MP will often be tasked to support civil-affairs operations. Close coordination with the G5/S5 and liaison with the appropriate HN representatives have added emphasis to these types of operations. The scope of civil-affairs operations is influenced by the economic, social, psychological, and political background of the country and its people. Civil-affairs personnel will coordinate the MP's effort to restore stability, contribute to national development, and promote support for the host government. Civil-affairs personnel assist in preventing civilian interference with military operations and coordinate all other military and civil affairs (such as community relations, PRC, civil defense, and community- and area-security programs).

11-37. The MP can conduct ADC measures to assist the HN in the aftermath of a natural or man-made disaster. They help rescue and evacuate the injured for medical care, and they issue food, water, and essential supplies (See FM 41-10).

SUPPORT TO PEACEKEEPING OPERATIONS

11-38. Peacekeeping operations (PKOs) are military operations conducted with the consent of the belligerents in a conflict. The PKOs are intended to maintain negotiated truces and to facilitate diplomatic solutions. The US may participate in a PKO under the auspices of an international organization, in cooperation with other countries, or unilaterally. A PKO usually occurs in an ambiguous situation in which there is extreme tension and the possibility of violence. This violence may range from terrorism and sabotage to minor military conflicts involving known and unknown belligerents. MP forces may serve as an element of a UN peacekeeping force or as part of a multinational observer group. The overall operational control of a peacekeeping force is exercised by a multinationally staffed military peacekeeping command. However, the commander of each assigned national force retains command integrity of his unit.

11-39. The basic MP force structure and appropriate augmentation are situation dependent. The size and composition of the MP forces are determined by diplomatic negotiations. Personnel spaces are allocated based on the US contingents' respective missions. The principle of consent affects the composition and function of the peacekeeping force. Consent applies to the degree of interest the disputing countries have when contributing their forces to participate in the peacekeeping effort.

11-40. Each PKO is unique but is generally characterized by constraints in the use of force. In a PKO, force is normally limited to self-defense, the defense of others, and the protection of national security. Therefore, the soldiers in a PKO must possess the skills required for warfare, but they must also be trained so that they are equally able to act as intermediaries. In this aspect, the MP are a distinct asset. They are experienced in a case-by-case approach to the use of force. They are experienced in using discretion and in the de-escalation of potential violence. The MP training in EPW operations are especially useful in handling EPWs if the peacekeeping force supervises the exchange of prisoners of war.

11-41. The peacekeeping force employs a combination of the following techniques to control the potential for violence inherent to PKOs:

- Observation is the peacekeeping force's primary responsibility and basic requirement. It provides timely and accurate reports of any suspicious situation, incident, or occurrence.
- Information gathering is a continuous requirement. Peacekeepers must be constantly alert to what takes place around them and to any changes or inconsistencies in the behavior, attitude, and activities of military and civilian personnel.
- Surveillance and supervision operations help oversee the implementation of agreements. The operations require restraint, tact, and patience.
- Complaint investigations must be thorough and objective, and they must result in a fair assessment.
- Negotiation and mediation skills are often required of peacekeepers. If peacekeepers can resolve minor problems at the lowest level, they can prevent major issues from rising, which may prevent increased tensions or the resumption of fighting.
- Patrolling (mounted and dismounted) among the population publicizes presence. It also serves as a means of gathering information, informally enhancing supervision, and investigating complaints.

11-42. The above techniques are compatible with MP battlefield functions. Operationally, the MP conduct route and area R&S, operate mounted or dismounted patrols, operate listening/observation posts, provide humanitarian assistance, investigate possible terrorist and criminal acts and cease-fire and sanction violations, and gather information. The MP provide a flexible, wide-ranging means of information dissemination. Additionally, they provide crucial support for force protection. The MP and the CID employ active and passive measures to deter and defeat threats directed at service members, their families, DOD civilians, facilities, equipment, and very important persons (VIPs). The MP use emerging technologies in force protection to conduct security, detection, and identification operations and to warn of intruders.

11-43. As part of a PKO, the MP conduct crowd- and riot-control operations, including the extraction of mob leaders. The MP are an ideal force for controlling antagonistic masses engaged in rioting, looting, and demonstrating. The MP understand how to make the transition from the lower end of the use-of-force spectrum to the use of lethal force, if so required or directed. Finally, the MP deploying with peacekeeping forces can provide early-on force protection, including headquarters and initial aerial port/seaport of debarkation (APOD/SPOD) security. The MP's mobility, firepower, and communications provide critical reconnaissance, information-collection, and response-force capabilities and acceptability not found in combat forces.

11-44. When conducting crowd- and riot-control operations in a PKO or peace-enforcement operation (PEO), nonlethal weapons (NLWs) are an additional means of employing force for the purpose of limiting the probability of death or serious injury to noncombatants or belligerents. However, the use of lethal force must always remain an inherent right of individuals in instances when they, their fellow soldiers, or personnel in their charge are threatened with death or serious bodily harm. The NLWs add flexibility to the control of disturbances by providing an environment where military forces can permissively engage threatening targets with limited risk of noncombatant casualties and collateral damage (see FM 90-40).

11-45. The use of lethal force (employed under the standing ROE) will never be denied. At no time will forces be deployed without the ability to defend themselves against a lethal threat nor will they forego normal training, arming, and equipping for combat. Nonlethal options are a complement to, not a replacement for lethal force. They seek to expand a proactive response across the range of military operations.

11-46. The decision to use NLWs against an adversary during a confrontation should be delegated to the lowest possible level, preferably to the platoon or squad. However, this requires that all personnel (not just leaders) have a clear understanding of the ROE and the commander's intent.

11-47. Commanders and public-affairs officers must be prepared to address media questions and concerns regarding the use and role of NLWs. They must be prepared to address the role of NLWs, and they must make it clear that the presence of NLWs in no way indicates abandoning the option to employ deadly force in appropriate circumstances.

11-48. The PEOs are in the high end of the PKO's spectrum. These operations are just short of combat and require a different operational approach. MP operations to support PEOs correspond to MP battlefield functions supporting combat forces. The PEOs often pose special operational considerations for dealing with dislocated civilians. The MP provide PRC operations and I/R, when required.

ORGANIZATIONS AND CAPABILITIES

11-49. MP units are well suited for stability and support operations given their unique capabilities and low force signature. The MP brigade, battalion, company, and I/R units may be tailored for stability and support operations based on their unique METT-TC.

MP BRIGADE TF

11-50. In essence, the organization and capabilities of the MP organization in stability and support operations is the same as those for other operations. However, its augmentation is tailored based on METT-TC unique to stability and support operations and the commander's needs. The MP brigade has been the centerpiece of successful MP deployment during previous stability and support operations. The MP brigade provides comprehensive planning, C^2, and A/L support to assigned and attached subordinate units. The brigade also has the command and staff experience to coordinate and supervise HN security and law-enforcement activities.

11-51. The MP brigade headquarters provides senior MP leadership and the communications capabilities required to direct a variety of functional elements—CSS, PSYOP, public affairs, and combat-arms augmentation—in support of joint or unilateral TF operations. At authorized level of organization (ALO) 1, the MP brigade headquarters has an SJA section to coordinate the legal and jurisdictional issues inherent to stability and support operations. The brigade headquarters also has an S5 section and a liaison capability to coordinate support and assistance to HN agencies, friendly forces, other US forces, DOS personnel, and nongovernmental or private volunteer organizations.

11-52. The MP brigade TF (MPBTF), with its robust capabilities and low force signature, is well suited for the unique mission requirements and sensitivities of stability and support operations. The MPBTF can be employed as part of a combat TF in high-threat, opposed-entry operations or as the lead TF for a low-threat, unopposed-entry operation. The following are examples of previous MPBTFs and the operations in which they participated:

- **Operation Provide Comfort.** The 18th MP Brigade deployed to northern Iraq to provide humanitarian assistance and protection to the Kurds.
- **JTF Los Angeles.** The 49th MP Brigade (NG) provided civil-disturbance assistance to Los Angeles in the aftermath of the Rodney King trial.
- **Operation Hawkeye.** The 16th MP Brigade and units of the 89th MP Brigade provided security, protection, and humanitarian

assistance to the population of Saint Croix in the aftermath of Hurricane Hugo.

- **JTF Andrew.** The 16th MP Brigade provided humanitarian assistance in the aftermath of Hurricane Andrew.
- **Operation Sea Signal.** The 89th MP Brigade deployed to Guantanamo Bay, Cuba, to provide humanitarian assistance to Haitian migrants.
- **Operation Restore Democracy.** The 16th MP Brigade deployed to Haiti to provide humanitarian assistance.

11-53. Early deployments of the MPBTF headquarters or a mission-tailored portion of the headquarters positions a senior, effective MP C^2 headquarters to respond to all MP requirements. It can assist in the coordination with the media, governmental and nongovernmental agencies, and other forces, and it can coordinate civil-military aspects of the operation. The MP brigade headquarters can assimilate and employ additional forces without degrading operations. The augmentation of brigade headquarters is driven by METT-TC. With the brigade headquarters in place, subordinate or supporting units can be significantly expanded without the loss of C^2. In essence, little or no loss in operational momentum occurs when the C^2 element deploys first.

11-54. The MP LRP section of the MP brigade headquarters colocates and works with the higher headquarters' G3/Operations Directorate (J3) plans element. This allows the MP LRP to incorporate MP operations effectively into higher-level planning, and it enables an efficient transition to field operations. Also, because in peacetime this section functions as a PM cell, it provides the PM with efficient and effective technical support during stability and support operations. As the operations approach termination or enter sustainability and fewer forces are required, the brigade headquarters can redeploy, leaving a battalion TF in charge of MP forces.

MP BATTALION TF

11-55. The MP battalion (CS) provides command, control, and support for assigned MP companies (CS). MP battalions (CS) are best employed as part of a TF—either as part of a combat TF (high threat, opposed) or a support TF (low threat, unopposed). With sufficient augmentation, the battalion can provide the logistical, administrative, personnel, and operational support needed to lead a diverse TF. During stability and support operations, an MP battalion TF may be placed under the OPCON of a brigade-size TF (MP or non-MP) or a division TF. If placed under the OPCON of a division TF, both the division PM and the MP battalion commander will synchronize their efforts to maximize the employment of available divisional and nondivisional MP assets in support of the division commander's concept of operations.

11-56. As with the MPBTF, the MP battalion's contributions to stability and support operations are not a new concept. The following are examples of previous battalion TFs and the operations in which they were involved:

- **Cuban Refugee Movement.** The 519th, 716th, 720th, and 759th MP Battalions deployed for humanitarian-assistance operations.
- **Operations Restore Hope and Continued Hope.** The 720th MP Battalion provided humanitarian assistance in Somalia.

- **Operation Sea Signal.** The 720th MP Battalion deployed to Guantanamo Bay, Cuba, to conduct humanitarian-assistance operations for Haitian migrants. Later, the 716th and 759th MP Battalions also deployed to Guantanamo Bay to perform like functions for the Cubans.

MP COMPANY IN STABILITY AND SUPPORT OPERATIONS

11-57. The MP company is the MP Corps's primary mission performer. It executes a wide variety of missions. In these operations, MP companies—

- Provide mobility support for both vehicles and personnel.
- Provide AS, including security for critical personnel, sites, cargoes, and railways.
- Secure, safeguard, shelter, and control detainees and dislocated civilians.
- Restore order in civil disturbances and conduct PRC operations.
- Carry out limited L&O operations.
- Provide liaison, coordination, and training for all aspects of law enforcement to other agencies.
- Conduct police-intelligence collections and disseminate information to the populace, friendly forces, and other HN agencies.
- Support force protection and security.

11-58. The MP company possesses robust mobility, lethality, and communications. Each of the MP companies' three-man teams are equipped with either a high-mobility, multipurpose, wheeled vehicle (HMMWV), an up-armored HMMWV, or an armored security vehicle (ASV); an AN/VRC 91 radio; a Light-Vehicle Obscuration Smoke System (LVOSS); an MK-19 grenade machine gun; a squad automatic weapon (SAW); two M-16 or M-4 rifles; an M203 grenade launcher; and a 9-millimeter pistol as a side arm for each team member. In addition to conducting combat operations, the MP company commander has the capability to conduct limited L&O operations (desk and investigative operations). With a tremendous operational reach, the MP company will displace its platoons as far as possible to accomplish all of the MP functions.

MP I/R UNITS IN STABILITY AND SUPPORT OPERATIONS

11-59. The MP I/R units are specifically equipped and trained to support, safeguard, account for, and provide proper and humane treatment for EPWs/CIs, US military prisoners, or dislocated civilians. When conducting dislocated-civilian operations, the MP I/R units support and interact with civil-affairs and PSYOP units who are responsible for monitoring and handling dislocated civilians. Should more than one category of personnel have to be supported, the MP I/R units will establish separate compounds for each to preclude contact with other populations. When operation of the facility is going to be more than a short-term endeavor, the intent is to involve (as appropriate) NGOs, IHOs, HN authorities, and dislocated civilians in the administration of the facility. The tasks performed by MP I/R units for stability and support operations are basically the same as those for war, but the operations and procedures are modified according to the operational environment and the categories of people to be housed in the facility (see FM 19-40).

11-60. Criminal detainees (who are detained against their will) are subject to controls and procedures that closely parallel those used for EPWs/CIs. The primary emphasis is on guarding and preventing escape until their disposition is determined by recognized civilian judicial/law-enforcement authorities. When the operation is conducted outside of the US and its territories, criminal detainees are treated according to established legal procedures.

11-61. Dislocated civilians (who are being assisted rather than detained) are provided aid, shelter, and protection. The emphasis is on protecting them from harm by natural forces or hostile personnel. A special category of personnel arises when the MP I/R unit is required to house personnel that are dislocated civilians but who must be detained against their will. Such is the case of mass migrants who flee their countries and find themselves under US custody while formal proceeding guidance is being developed. In this case, MP I/R units must be sensitive to the situation and attempt to strike a balance between security, shelter, protection, and detention procedures.

11-62. Military detainees (who are detained against their will) are subject to controls and procedures that closely parallel those used for EPWs/CIs. In stability and support operations, US forces detain opposing-force military personnel. Without a formal declaration of war, opposing military personnel who are captured cannot be categorized as EPWs. However, they are afforded many of the EPW's rights and privileges under the Geneva convention. Such a situation occurred during Operation Just Cause when PDF personnel were detained by US forces.

11-63. In an operational environment in which hostile groups are engaged against one another, an I/R facility may be set up to protect one group from another. In this case, the facility's purpose is to shelter, sustain, account for, and protect its dislocated civilians from the hostile group. The MP I/R units and other supporting units concentrate on providing AS to protect the facility from direct fire. Other MP or combat forces provide protections beyond the direct-fire zone. Accountability for the dislocated civilians is coordinated with the SJA and civil affairs. They focus on maintaining a record of the people in the facility and their physical conditions. In operations in which no hostile groups are engaged (such as natural disasters), the I/R facility may be set up to provide shelter, sustain, and account for personnel. There may not be a need for external security personnel.

11-64. The MP I/R C^2 structure for stability and support operations is based on METT-TC. The nature and complexity of the mission, the number and types of detainees, and the operational duration should be considered. At one end of the operations spectrum, an MP brigade (I/R) may be required; while at the other end, an MP battalion (I/R) may be assigned to an MP brigade (CS).

Chapter 12

Force Protection

There is no more important responsibility than force protection. It is an inherent command responsibility and must be fully integrated into every unit's mission. A commander must continually review his unit's force-protection posture and crosswalk it with current and changing policy and threat levels. Force protection demands the personal involvement of commanders to ensure the best possible security consistent with the threat to personnel and mission-essential resources.

AR 525-13

Force protection consists of those actions that prevent or mitigate hostile actions against DOD personnel (to include family members), resources, facilities, and critical information. It coordinates and synchronizes offensive and defensive measures to enable the joint force to perform while degrading opportunities for the enemy. It includes air, space, and missile defense; NBC defense; antiterrorism; defensive-information operations; and security to operational forces and means.

OVERVIEW

12-1. The MP contribute significantly to the Army's overall force-protection effort. They are directly responsible for executing certain security-related measures (law enforcement and physical security) and for providing support to others (HRP security and antiterrorism), all of which contribute to force protection. Commanders use the Army's risk-management methodology and established security programs and capabilities within US forces to help prioritize and implement force-protection measures.

SUPPORT TO FORCE PROTECTION

12-2. MP support to force protection is accomplished under the leadership of the respective command's PM. It is accomplished mainly through an active role in physical security, HRP security, law enforcement, and antiterrorism.

PHYSICAL SECURITY

12-3. Physical security is an integral part of security of forces and means. Physical-security measures deter, detect, and defend against threats from terrorists, criminals, and unconventional forces. These measures include—

- Fencing and perimeter standoff space.
- Lighting and sensors.

- Vehicle barriers.
- Blast protection.
- Intrusion-detection systems and electronic surveillance.
- Procedural measures such as security checks, training and awareness, property accountability/inventory requirements, physical-security inspections of mission-essential or vulnerable areas (MEVAs), and physical-security surveys of installations.

12-4. These physical-security measures are not exclusively peacetime measures. They are designed to protect US personnel and equipment regardless of the mission or the geographical location.

12-5. The MP support the commander's security of forces and means programs by—

- Controlling or monitoring installation, base, or base-cluster access or entrance points.
- Monitoring intrusion-detection systems and providing a response force.
- Conducting physical-security inspections.
- Conducting perimeter security or site surveillance.
- Recommending the placement of walls, berms, gates, or barriers around designated MEVAs, high-value areas, or perimeters.
- Supporting the commander's risk-analysis effort.
- Conducting roving patrols, checkpoints, or roadblocks.
- Performing other physical-security measures as required by the commander.

HRP SECURITY

12-6. The HRP security program provides additional security to designated individuals (and their families) whom by virtue of their rank, assignment, symbolic value, vulnerabilities, or location or a specific threat are at a greater risk than the general population. This security program is applicable across the full spectrum of military operations.

12-7. The MP support the commander's HRP security measures by—

- Conducting protective-service details for HRP (normally conducted by USACIDC special agents).
- Responding to duress alarms and intrusion-detection systems.
- Conducting a PSVA (performed by the USACIDC).
- Employing MWDs.
- Coordinating with local, state, and federal agencies as well as joint, multinational, and HN law-enforcement agencies.
- Performing other HRP security measures as required by the commander.

LAW ENFORCEMENT

12-8. The Army's law-enforcement program supports the security of forces and means through the prevention, detection, response, and investigation of crimes. The MP, the USACIDC, and DOD police play a crucial role in the law-enforcement program. They provide a visible deterrent against a broad portion of the threat spectrum, investigate crimes, and provide an initial response force to threat incidents. As with the previous programs, the law-enforcement program is applicable anywhere across the spectrum of military operations.

12-9. The MP support law-enforcement operations by—

- Providing liaison teams with local, state, and federal agencies; HN police; and joint and multinational agencies.
- Employing SRTs and hostage-negotiation teams (normally provided by the USACIDC).
- Providing traffic enforcement, MSR regulation enforcement, and other route-control measures.
- Employing MWDs.
- Conducting MP investigations (criminal and noncriminal).
- Conducting patrolling, AS, and surveillance measures.
- Implementing applicable threat-condition (THREATCON) measures.
- Conducting and implementing other law-enforcement measures as required by the commander.

ANTITERRORISM

12-10. Terrorism is characterized as the unlawful use of violence or the threat of violence to coerce or intimidate a government or a society. Protection against a terrorist threat requires both an offensive counterterrorism capability and a defensive antiterrorism program. The primary focus of antiterrorism is on training and awareness and thorough planning. Antiterrorism is not a stand-alone program. It leverages other force-protection-related programs (such as physical security and law enforcement) to provide much of the physical defense against terrorism.

12-11. The MP support the commander's antiterrorist program by—

- Collecting CRIMINTEL (within the limits of AR 190-45).
- Responding to terrorist threats and incidents.
- Performing aggressive patrols and R&S operations.
- Detecting suspicious activities.
- Enforcing joint and cooperative agreements with the FBI, the USACIDC, MI, and other pertinent agencies (home or abroad).
- Conducting information collection and dissemination.
- Protecting critical assets, facilities, and personnel.
- Performing other antiterrorism measures as required by the commander.

Appendix A

Metric Conversion Chart

This appendix complies with current Army directives which state that the metric system will be incorporated into all new publications. Table A-1 is a conversion chart.

Table A-1. Metric Conversion Chart

Metric to English			English to Metric		
Multiply	**By**	**To Obtain**	**Multiply**	**By**	**To Obtain**
Length					
Centimeters	0.0394	Inches	Inches	2.54	Centimeters
Meters	3.28	Feet	Feet	0.0305	Meters
Meters	1.094	Yards	Yards	0.9144	Meters
Kilometers	0.621	Miles (stat)	Miles (stat)	1.5609	Kilometers
Kilometers	0.540	Miles (naut)	Miles (naut)	1.853	Kilometers
Millimeters	0.039	Inches	Inches	25.40	Millimeters
Area					
Square centimeters	0.1550	Square inches	Square inches	6.45	Square centimeters
Square meters	10.76	Square feet	Square feet	0.0929	Square meters
Square meters	1.196	Square yards	Square yards	0.836	Square meters
Volume					
Cubic centimeters	0.610	Cubic inches	Cubic inches	16.39	Cubic centimeters
Cubic meters	35.3	Cubic feet	Cubic feet	0.0283	Cubic meters
Cubic meters	1.308	Cubic yards	Cubic yards	0.765	Cubic meters
Milliliters	0.0338	US liq ounces	US liq ounces	29.6	Milliliters
Liters	1.057	US liq quarts	US liq quarts	0.946	Liters
Liters	0.264	US liq gallons	US liq gallons	3.79	Liters
Weight					
Grams	0.0353	Ounces	Ounces	28.4	Grams
Kilograms	2.20	Pounds	Pounds	0.454	Kilograms
Metric tons	1.102	Short tons	Short tons	0.907	Metric tons
Metric tons	0.984	Long tons	Long tons	1.016	Metric tons

Appendix B

Doctrine and Training Literature

Users of this manual must understand the elements of doctrinal literature and their relationship to each other. The commonly used terms doctrine and tactics, techniques, and procedures (TTP) are interrelated and mutually supportive; yet, each term has its own usage, level of detail, and place in the hierarchy of doctrinal publications. Because there is a close interrelationship, publications overlap in what they provide to the user. Figure B-1 illustrates the relationship of these doctrinal terms and publications.

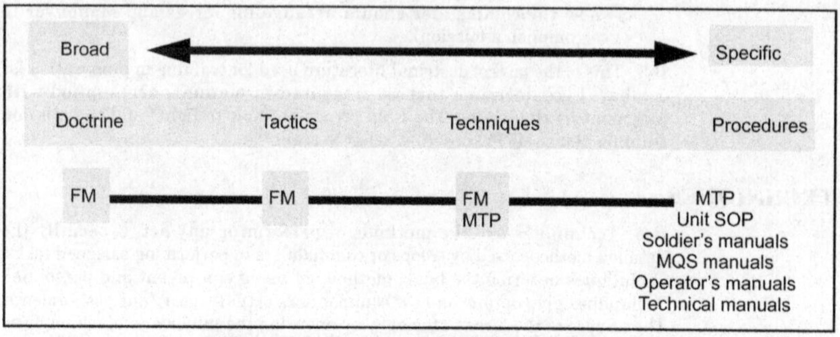

Figure B-1. Doctrinal and Training Literature

DOCTRINE

B-1. Doctrine is the body of fundamental principles by which the military guide their actions in support of national objectives. It is authoritative, but requires judgment in application. Doctrine provides the Army's fighting philosophy, establishes a common language, and structures the manner within which units plan and conduct combat operations.

B-2. Doctrine is not absolute; it is applied by the leader to meet the circumstances of the situation being faced. Doctrine provides for this flexibility with broad fundamentals and principles for conducting operations. Leaders well grounded in doctrine understand how to synchronize the elements of combat power and how to apply TTP using the assets available to accomplish the mission.

B-3. Doctrine is described in FMs. FM 3-0 defines the systems of war fighting practiced by the US Army. FMs 100-15 and 71-100 expand on this foundation. This manual describes the CS doctrine of the MP; it is the capstone manual for MP operations. It defines the role, operational requirements, functions, and C^2 relationships of MP units. This manual is not oriented on a specific organizational structure. Leaders at all levels apply this doctrine to their unique organizational and operational situations and use various tactics and techniques to accomplish the doctrine.

TACTICS

B-4. Tactics describe how the leader carries out doctrine (relating to the arrangement of forces for battle). Tactics include the—

- Employment of units in combat (to accomplish a mission assigned by the commander).
- Ordered arrangement and maneuver of units in relation to each other and to the enemy in order to use their full potential (a description of how the commander should arrange his forces and maneuver to accomplish a mission).

B-5. This is the part of doctrinal literature used for training in preparation for combat. Like doctrine, tactics presented in manuals are applied with judgment by the leader. The FMs prescribe "how to fight" and the mission training plans (MTPs) describe "what to train."

TECHNIQUES

B-6. Techniques are the methods of performing any act, especially the detailed methods used by troops or commanders in performing assigned tasks. Techniques describe the basic methods of using equipment and personnel. Techniques give details on how commanders actually carry out assignments. They improve the force's efficiency by ensuring the uniformity of action or by ensuring that the actions of various individuals and elements complement those of other individuals or elements.

B-7. The FMs and MTPs provide a description of techniques using the personnel and equipment available in actual organizations. These techniques show (at a more detailed level) how to use available assets to carry out tactics.

PROCEDURES

B-8. Procedures are the lowest level of detail. They address "how to" at the task level. Procedures are a particular course or mode of action that describe how to perform a certain task. Procedures include the—

- Standing methods used by units to accomplish tasks.
- Operating procedures for weapons and equipment.
- Crew drills.
- Staff actions and coordination requirements.
- Methods of target engagement by direct- and indirect-fire weapons.

B-9. Procedures are building blocks of individual- and collective-task accomplishment that serve as the foundation of tactics and techniques. Procedures are explained in unit standing operating procedures (SOPs), MTPs, soldier's manuals, operator's manuals, technical manuals (TMs), and similar publications.

Appendix C

Battlefield Workload Analysis

With minor adjustments in the capability block, the BWA can be used to determine MP requirements for the division. A BWA work sheet is shown in Figure C-1. A separate matrix is prepared for each EAC and corps MP brigade. Figure C-2, page C-3, is a sample of a BWA matrix for a 3-division corps MP CS company.

Mission/Function[1]	Capability[1]	Workload[2]	Rqmt	HN[18]	US Rqmt	Risk[19]	Mission Rqmt
Divisions supported[3]	1						
MMS	360 km MSR and ASR						
Refugees on control routes[4]	150,000 per day						
AS—corps rural terrain[5]	2,000 sq km						
AS—EAC rural terrain[5]	3,000 sq km						
AS—urban terrain[5]	800 sq km (urban)						
AS—USAF main opns base[5]	1,000 sq km/USAF MOB						
AS—convoy escort[6]	# tms per convoy	# convoys					
AS—ammo security[7]	1.3 modular ammo co						
	2 modular ammo co						
AS—pipeline security[8]	400 km						
AS—EAC pipeline co[8]	3						
AS—trains[9]	7						
AS—small critical sites[10]	12						
AS—major HQ security[11]	1						
AS—EAC port security[12]	0.66 of port						
AS—combat operations	When required, assets are diverted from other missions.						
L&O—incident response[13]	75,000 nondivision troops						
I/R—US prisoner detention[14]	700						
I/R—EPW escort/corps[15]	1,900 walking						
	2,500 by vehicle	Min: 1 plt					
	3,800 by train						
I/R—EPW holding/corps[16]	2,000	Min: 1 plt					
Other assigned missions[17]	Mission	Force size					
		Totals[20]					
Battalion HHD requirement = 0.199 x number of companies[21]							

Figure C-1. Sample BWA Work Sheet

NOTE: Workload ÷ capability = doctrinal requirement - HN offset = US requirement - acceptable risk = mission requirement

[1]Mission/capabilities are as stated in section I of TOEs 19477L000 and 19677L000. The unit capability for each mission is what the unit can accomplish when performing no other mission (for example, if it is only doing MSRs, the unit can cover 360 kilometers in a 24-hour period). Unless otherwise specified, the capabilities are the same at corps and EAC.

[2]Workload is the amount (MSR or area) or number of supported units/facilities.

[3]Corps support to each division is covered by allocating one MP company per supported division. Any additional support requirements should be based on a detailed division PM/MP mission analysis and included under other assigned missions.

[4]Unless specifically tasked, the Army normally relies on HN support to manage refugees. Redirecting refugees from MSRs and keeping them off of MSRs is part of (as is straggler control and TCP) the mission and MSR workload. When assigned the mission, the capability is for control of refugees on established refugee control routes that are in addition to the MSRs.

[5]Compute each square kilometer mission requirement separately. Do not compute the same terrain in more than one computation (for example, the area around the Air Force's main operating base (MOB) is not included with either rural or urban square kilometers).

[6]Convoy-escort capability is not stated in the TOE and is an AS function rather than an MMS function. Determine the average number of teams per mission and the average number of simultaneous missions. Multiply the number of teams required per mission by 0.028, then multiple the result by the number of simultaneous missions to determine the total requirement. State the number of teams per convoy and the number of convoys, then convert the force size to a 3-decimal-place company equivalent. Each additional team, squad, and platoon equivalent equals 0.028, 0.083, and 0.25 respectively.

[7]Security for modular ammo companies (ASPs) supporting divisions is three platoons and is normally conducted directly behind the division (corps forward). However, METT-TC may require the ammo units to move into the DRA. In this case, corps units should be attached to the division for the duration. Security for corps storage areas (CSAs) and theater storage areas (TSAs) is one MP company per two ammo companies. Security is provided on an area-support basis in the form of screening, with limited fixed posts.

[8]The EAC petroleum terminal/pipeline operating company manages the EAC pipeline that is normally laid adjacent to MSRs. Pipeline security is inherently provided by the MP operating on the MSRs. Unless the pipeline is not near the MSR, MP support is only allocated against the EAC petroleum terminal/pipeline operating company's bulk storage facilities. Security is provided on an area-support basis in the form of screening, with limited fixed posts.

[9]Train security is normally an EAC mission. This capability equates to simultaneously providing five MP teams to each of seven trains. Actual workloads and requirements are computed in the same manner as a convoy escort.

[10]Small critical facilities are normally defined as key MSR bridges and tunnels and communication nodes that may require additional security.

[11]The major headquarters are corps, COSCOM, TSC, unified command and, if directed, the JTF. The ASCC depends on the MP to provide security for two CPs. The ASCC requirement is an additional three platoons to support a second CP.

[12]The requirement for one seaport is 1.5 companies.

[13]Teams provide incident response, crime-scene security, and traffic control. They also assist L&O detachment teams and CID investigators.

[14]For deliberate planning in support of an OPLAN, personnel to be confined will normally be held by either the unit or the EAC I/R MP battalion performing the confinement mission. This does not preclude a requirement for the echelon PM to operate a field detention facility until prisoners are released back to their units or evacuated to the EAC confinement facility.

[15]The primary mode is vehicle (back-haul doctrine). Regardless of the mode, a minimum of one platoon (0.25 requirement) per supported division is allocated for EPW escort from the division central collection point (DCCP) to the corps holding area (CHA). Because the actual number of EPWs is unknown, coupled with no time/distance study for each supported division, one platoon per division provides a minimum capability for continuous evacuation within the stated doctrinal timeline (evacuate to the CHA within 12 hours of arrival at the DCCP).

[16]At a minimum, one platoon (0.25 requirement) is allocated to operate the CHA (provides the capability to hold 500 EPWs at any given time).

[17]Add any additional continuous missions assigned by the echelon commander not covered above. Examples include VIP escort/security for other than the CG and the deputy commanding general (DCG) of major headquarters, screening exposed corps flanks, and increasing teams per mission/function which decreases doctrinal capability (for example, permanent TCPs or checkpoints). Each additional team, squad, or platoon equivalent equals 0.028, 0.083, and 0.25 respectively.

Figure C-1. Sample BWA Work Sheet (continued)

[18]For each mission/function, any workload either fully or partially accomplished by the HN is subtracted from the doctrinal mission requirement. Enter the MP company equivalent offset and subtract from the doctrinal requirements to determine the US requirement.

[19]For each mission/function, if the operational/MP commander will accept mission risk, subtract the MP company equivalent to determine the mission requirement.

[20]Add columns to determine the total doctrinal company requirement, HN offsets, US requirement, acceptable risk, and mission requirement to support the operation/AO. Round the requirement to the nearest whole number (0.4 or fewer rounds down, 0.5 or greater round up).

[21]Determine battalion HHD requirements by multiplying the company requirements by 0.199 and rounding.

Figure C-1. Sample BWA Work Sheet (continued)

Mission/Function	Capability	Workload	Rqmt	HN	US Rqmt	Risk	Mission Rqmt
Divisions supported	1	3	3.000		3.000		3.000
MMS	360 km MSR and ASR	1,050	2.917		2.917		2.917
Refugees on control routes	150,000 per day	25,000	0.167	0.167	0.000		0.000
AS—corps rural terrain	2,000 sq km	6,500	3.250		3.250	0.500	2.750
AS—EAC rural terrain	3,000 sq km						
AS—urban terrain	800 sq km (urban)	1,500	1.875		1.875	0.500	1.625
AS—USAF main opns base	1,000 sq km/USAF MOB	1,000	1.000		1.000		1.000
AS—convoy escort	5 tms per convoy	4	0.560		0.560		0.560
AS—ammo security	1.3 modular ammo co	3	2.308		2.308		2.308
	2 modular ammo co	1	0.500		0.500		0.500
AS—pipeline security	400 km						
AS—EAC pipeline co	3						
AS—trains	7						
AS—small critical sites	12	9	0.750		0.750	0.250	0.500
AS—major HQ security	1	2	2.000		2.000		2.000
AS—EAC port security	0.66 of port						
AS—combat operations	When required, assets are diverted from other missions.						
L&O—incident response	75,000 nondivision troops	45,000	0.600		0.600	0.300	0.300
I/R—US prisoner detention	700						
I/R—EPW escort/corps	1,900 walking						
	2,500 by vehicle	Min: 1 plt/div	0.750		0.750		0.750
	3,800 by train						
I/R—EPW holding/corps	2,000	Min: 1 plt/	0.250		0.250		0.250
Other assigned missions	PSD for Corps CofS	1 sqd	0.083		0.083		0.083
	Flank screen	1 plt	0.250		0.250		0.250
		Totals	20	0.177	20	1.3	19
Battalion HHD requirement = 0.199 x number of companies			4		4		4

SAMPLE

NOTE: Workload ÷ capability = doctrinal requirement - HN offset = US requirement - acceptable risk = mission requirement

Figure C-2. Sample BWA Matrix

Appendix D

Automated Information Systems in Support of Battle Command

As the MP Corps continues to advance into the twenty-first century, it must continue to pursue and stay in tune with the Army's initiative to acquire advanced technologies and operational concepts. This will give MP soldiers on the battlefield an information advantage that is critical for successful performance of their functions. The following information opens a window into the near future and explains some of the automation initiatives that will enhance the Army's and the MP Corps's soldiers.

ARMY INFORMATION SYSTEMS

D-1. The Global Command and Control System (GCCS) is the midterm solution and the bridge to the concepts outlined in the concept of command, control, communications, computers, and intelligence for the warrior (C4IFTW). The GCCS's core consists of the basic functions required by the war fighter to plan, execute, and manage military operations. The Army's selected C^2 system is the Army Battle Command System (ABCS).

D-2. The ABCS integrates the C^2 systems found at each echelon—from the ground-force component commander at the theater or JTF level to the individual soldier or the weapons platform. Whether ARFOR are deployed for land combat or are conducting peace operations, providing humanitarian assistance, or giving aid to civil authorities, the ABCS supports the mission by integrating the battle-space automation systems and communications that functionally link strategic and tactical headquarters. The ABCS is interoperable with joint and multinational C^2 systems at upper echelons across the full range of C^2 functionality, and it is vertically and horizontally integrated at the tactical and operational levels. The ABCS has three major components—

- The Global Command and Control System—Army (GCCS-A).
- The Army Tactical Command and Control System (ATCCS).
- The FBCB2.

GLOBAL COMMAND AND CONTROL SYSTEM—ARMY

D-3. The GCCS-A is the Army's component of the joint GCCS. It was built from application programs developed by the Army Worldwide Military Command and Control System [WMCCS] Information System (AWIS), namely the—

- Strategic Theater Command and Control System (STCCS).

- Theater Automated Command and Control Information Management System (TACCIMS).
- EAC portion of the Combat Service Support Control System (CSSCS) (CSSCS/EAC).

D-4. The GCCS-A is required in the MP LRP, S1, S3/NBC, and S4 sections within the MP brigade at EAC; in the Assistant Chief of Staff, G1 (Personnel) (G1), the G3, and the Assistant Chief of Staff, G4 (Logistics) (G4) sections within the MP command (I/R); and in the S1, S3, and S4 sections within the MP brigade (I/R).

ARMY TACTICAL COMMAND AND CONTROL SYSTEM

D-5. The ATCCS embodies the commander and his staff and the doctrine, procedures, and tools used to command and control forces on the tactical battlefield. The ATCCS's functional subsystems provide tools to support the C^2 process across the spectrum of low- to high-intensity conflicts worldwide; for MOOTW; and for combat, CS, and CSS functional areas at corps-and-below units. The ATCCS provides a common set of C^2 tools and functions for commanders that extends the ATCCS to the battalion level. The component's ATCCS's are the—

- Maneuver Control System (MCS).
- All-Source Analysis System (ASAS).
- Advanced Field-Artillery Tactical Data System (AFATDS).
- Air- and Missile-Defense Planning and Control System (AMDPCS).
- CSSCS.

D-6. The FBCB2 provides situational awareness and C^2 to the lowest tactical echelons. It facilitates a seamless flow of battle-command information and situation-awareness information across the battle space at brigade and below on Army platforms.

FORCE XXI BATTLE COMMAND—BRIGADE AND BELOW

D-7. The FBCB2 system, a battle-command information system, provides the technology to complete the ABCS information flow process from brigade to platform and across all platforms within the brigade TF. The FBCB2 enables commanders to remotely operate and maintain ABCS database connectivity (regardless of the command vehicle) and to digitally control and monitor their subordinate units' statuses and positions. Interoperable FBCB2 systems will enhance the battle commander's ability to have a detailed picture of the battlefield (situational awareness), to maximize force survivability, and to facilitate battlefield synchronization to reduce or destroy the enemy's capability to fight. The FBCB2 relies on integrating, processing, and moving information in a digital form. To support projections of future conflicts with an accompanying increase in the tempo of battle, the FBCB2 must be able to obtain information contained in databases from higher, lower, and adjacent elements without formally requesting that information through command channels. The war fighter requires the ability to pull data from all battlefield-system databases to keep current with the rapid changes caused by the dynamics of battle. The FBCB2 is required for divisional MP company commanders, PMs, platoon leaders, teams, and squads.

MANEUVER CONTROL SYSTEM

D-8. The MCS is the ABCS's horizontal and vertical integrator. The MCS will enable commanders and their staffs to collect, coordinate, and act swiftly on near-real-time battlefield information. It will quickly, accurately, and automatically move tactical information on the battlefield. This allows a commander to readily access information and display current situation reports (SITREPs) that assess enemy strength and movement as well as the status of friendly forces. The MCS will then aid the battle staff in rapidly disseminating the commander's orders. It is intended to provide two major functional capabilities—maneuver functional-area control and force-level information. The MCS is the ATCCS's functional subsystem that—

- Supports the combined-arms force commander directly by providing automated support for planning, coordinating, controlling, and using maneuver functional-area assets and tasks.
- Coordinates and synchronizes supporting arms in conducting operational planning, field operations, and training.

D-9. The MCS is required for division PM operations sections, company CPs at echelons corps and below (ECB), MP battalion commanders, S3 sections at corps, MP brigade commanders, and corps MP LRP and S3 sections.

ALL-SOURCE ANALYSIS SYSTEM—REMOTE WORK STATION

D-10. The ASAS—remote workstation (RWS) subsystem will operate within the existing intelligence and electronic warfare (IEW) battlefield operating system (BOS) and ABCS architectures to provide automation support to G2/S2 intelligence staff elements and other designated intelligence organizations from the maneuver battalion through EAC. It will operate at the security level appropriate to its echelon, mission, and functions. It assists the G2/S2 in providing IEW support to the commander and other staff elements in planning, coordinating, and conducting current and future tactical operations. It interfaces with the sensitive compartmented information (SCI) analysis and control element (ACE) subsystem to provide operational guidance and to exchange IEW information and products. It also operates as the technical control portion of the IEW C^2 node of the ABCS to provide current IEW and enemy-situation information to the force-level information database for access and use by ABCS users. Two RWSs are also provided to the brigade analysis control team (ACT) of the DS MI company to process data input from DS collection systems, to filter intelligence data from broadcast systems, and to pass tactically relevant enemy-situation information or situation-development data to the brigade S2's RWS. The ASAS-RWS is required in the S2 section of the CID group, the S2/S3 section of the CID battalion, the G2 section of the MP command (I/R), the S2 section of the MP battalion (I/R), and the S2 section of the MP battalion and brigade at all echelons.

COMBAT SERVICE SUPPORT CONTROL SYSTEM

D-11. The CSSCS will function as an integral part of the ATTCS. It is designated as an information system with the capability to rapidly collect, analyze, and disseminate logistical, medical, and personnel information to accelerate and improve decision making by CSS and force commanders. It also enables the rapid exchange of C^2 information with other battlefield functional areas (BFAs) through automated interfaces. The system will process unclassified through secret information. The CSSCS will be employed in all divisions, corps, EAC, separate brigades, and armored cavalry regiments. The MP require CSSCS functionality in all S1 and S4 sections in battalion or higher organizations except where the GCCS-A is required as stated above.

TRANSPORTATION COORDINATOR AUTOMATED MOVEMENT SERVER

D-12. The Transportation Coordinator Automated Movement Server (TC-AIMS II) is used by transportation agents and deploying units of each service to automate the processes of planning, organizing, coordinating, and controlling deployment and redeployment operations worldwide during peace as well as contingencies. The TC AIMS II will link all DOD-component unit movements with the installation transportation officer/traffic-management officer (ITO/TMO). This will provide the capability to plan and execute deployments rapidly to meet CINC-required arrival and departure dates. The TC AIMS II will also provide the ITO's/TMO's day-to-day traffic management functions and the theater transportation managers with the capabilities to use theater common lift assets effectively. The TC AIMS II will also allow CSS organizations and supporting commands to sustain the force and perform effective movement control, coordination, distribution, and sustainment once in theater. This is also defined as transportation movement and support of DOD personnel and cargo during all phases of military operations in all environments (including RSOI and battlefield operations). The TC AIMS II's functionality is required at all division PM operations and S3 sections within MP battalions and brigades to facilitate MSR regulation enforcement. Additionally, MP companies, battalions, brigades, and commands require the TC AIMS II to plan, organize, coordinate, and control deployments and redeployments.

GLOBAL COMBAT SUPPORT SYSTEM—ARMY

D-13. The GCSS-A is an integrated logistical support system with accompanying hardware that will connect the service's supply, finance, transportation, maintenance, and personnel databases. Under this system, the Army will integrate 13 Legacy tactical logistics systems and will eventually interface with the service's wholesale logistics systems as well as the consolidated logistics systems of the other services and the DOD. The system is critical for modernizing automation, for reducing logistical footprints in theaters, and for expanding war fighters' capabilities. The system will also replace the Standard Army Management Information System (STAMIS). The GCSS-A is required at all MP division company CPs and MP battalion and brigade S4 sections. Additionally, it is required at all supply functions within MP detachments that support I/R missions.

MILITARY POLICE AUTOMATED SYSTEMS

D-14. In an effort to maintain an information advantage and to acquire advance technologies and operational concepts, the MP Corps has acquired the following systems:

- The Military Police Management Information System (MPMIS) and its subsystems
- The CID's automated systems and their subsystems.

MILITARY POLICE MANAGEMENT INFORMATION SYSTEM

D-15. The MPMIS is a DA-mandated reporting system that does not interface with the ABCS. The MPMIS provides automated tools for MP records management and reporting. It reduces the administrative burden on the PM's staff. It increases the awareness of problem areas by managing crime information. The MPMIS consists of the following elements:

- **Offense Reporting Systems Windows (ORS-2W).** The ORS-2W provides automation support to track and report complaints, criminal incidents, and traffic incidents. It incorporates the CRC, the federally mandated National Incident-Based Reporting System (NIBRS), the DOD-mandated Defense Incident-Based Reporting System (DIBRS), the military community (MILCOM), and the installation's data requirements. The ORS-2W's functionality is required at PM activities at all echelons of command within the Army.
- **Correctional Reporting System (CRS-3).** The CRS-3 provides automation support to track and report inmates and corrections staff at all Army-controlled correctional facilities.
- **Security Management System (SMS).** The SMS provides automated support to assist physical-security inspectors and force-protection officers in reporting inspections and surveys. It tracks intrusion-detection devices and MWD training and health requirements.
- **Vehicle Registration System (VRS).** The VRS provides installations with an automated means to track vehicle, weapon, and bicycle registrations.
- **Prisoner-of-War Information System (PWIS-3).** The PWIS-3 provides the automated means of tracking an EPW's status from capture through release. This system is required at all EPW management activities.

CID AUTOMATED SYSTEMS

D-16. The Headquarters (HQ) USACIDC mandates CID automated systems. Similar to the MPMIS, it does not interface with the ABCS. The system facilitates data-collection efforts within the command. The CID's automated systems consist of the—

- **Automated Criminal Investigative Reporting System (ACIRS).** The ACIRS supports management and data-collection functions at the CRC and HQ USACIDC. It supports the following:
  - Maintaining investigative data.
  - Conducting agent investigative activities.
  - Reporting investigative data.
  - Managing the investigative processes.
  - Conducting criminal-activity analyses.

- **CRIMINTEL.** Data sources for CRIMINTEL include MP reports, CRIMINTEL reports and bulletins, reports of investigation, crime-prevention surveys, case files, collection folders, raw-data folders, significant data files, and target-analysis files. The CRIMINTEL interfaces with the ACIRS, but it does not share CRIMINTEL products.

Appendix E

Police Information Assessment Process

The PIAP differs somewhat from the tactical IPB. While the IPB allows the maneuver commander to see the effects that a variety of factors may have on his forces, the PIAP looks beyond this approach to determine how other relevant information may impact his forces. This information may be obtained from police or criminal actions or incidents encountered during the performance of MP functions.

IMPLEMENTATION

E-1. The PIAP is not a substitute for the IPB. MP leaders must continue to rely on the IPB and to use its estimates as a starting point for the PIAP. However, the PIAP is a dynamic and continuous cycle that complements the IPB (Figure E-1). The following paragraphs represent the doctrinal approach to implementing the PIAP. These steps provide the basic foundation to the process. They are not exclusive; the successful development of the PIAP will depend on the MP leader's ability to apply the process to his specific environment, METT-TC, and the commander's priorities.

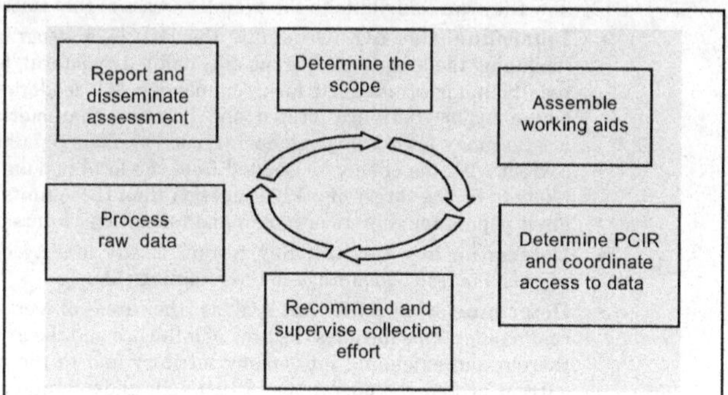

Figure E-1. The PIAP

DETERMINE THE SCOPE OF THE PIAP

E-2. While determining the scope of the PIAP might sound simple, it is one of the most crucial elements of the entire process. It entails identifying what you want to be the end result of your PIAP. To help determine the scope of the PIAP, MP leaders must perform the following subtasks:

- **Conduct a mission analysis.** A detailed mission analysis will ensure that all available resources are focused toward one goal. The more specific the mission is, the more focused and productive the effort. This involves more than just looking at the mission statement, the commander's intent, and the concept of operations. It involves translating the mission and situation into easily understood, manageable goals. The following are examples of these missions:
  - Identifying criminal threats from local national to US Army logistics operations within the port of Mogadishu, Somalia.
  - Determining the impact of gang violence on US forces operating in the vicinity of Camp Bondsteel, Kosovo.
  - Determining how rioters' violence will affect force protection before deploying forces to Los Angeles, California.
  - Determining the status of public service agencies (such as the police, the fire department, and the emergency medical services [EMS]) in Homestead, Florida, after the city was struck by Hurricane Andrew.

NOTE: As you can see from the examples, these missions are quite specific. In all likelihood, it would be necessary to establish several different PIAP missions to adequately protect Army elements operating in the MP's AOR.

- **Prioritize missions.** MP leaders must prioritize each PIAP mission since MP resources will probably be limited. The commander's intent is the key factor for determining priorities. When the commander's most important mission has been determined, MP leaders must decide how they can contribute to the overall success of that mission.
- **Determine the AO.** Generally, the AO is a geographical area (including the airspace above) usually defined by lateral, forward, and rear boundaries assigned to a commander. MP leaders must know where major units are located and how the characteristics of the location may impact ongoing operations. For example, units in direct contact with the enemy or isolated from the local population are less likely to be the target of a PIAP mission than those units adjacent to great population centers or in rest and relaxation centers.
- **Determine key terrain.** Key terrain is any locality or area that affords a marked advantage to the retaining side.
- **Determine the AOI.** An AOI is the area of concern to the commander. This includes the area of influence and the areas adjacent thereto and extending into enemy territory and to the objectives of current or planned operations. A thorough understanding of the AOI may help predict potential threat actions. For example, how far are local agitators and ringleaders willing to travel to interfere with US operations in Pristina, Kosovo? Would they go to other US compounds?
- **Determine required information products/sources.** The desired end-state product must be decided. Should it be a report? An updated SITREP? An upgrade in force-protection measures? Or a combination of products? It should also be determined who can provide the best

information (the CID; MI; the local police chief; or joint, interagency, or multinational agencies).

ASSEMBLE WORKING AIDS

E-3. During this step, MP leaders assemble the necessary graphic and nongraphic aids and information that will help them accomplish their mission. The CID and the MP may form an analysis team to track and analyze police information that might assist tactical operations or population-control operations. Working aids include, but are not limited to—

- **Maps.** Besides the standard military maps that depict the military AOs, the MP must attempt to procure indigenous maps of AOs. Detailed city and street maps will prove to be invaluable even if they are in a foreign language.
- **Language aids.** Phrase books and translation dictionaries cannot take the place of intensive language training, but they are better than nothing. These are essential even if the MP have assigned interpreters.
- **Open-source information.** This can cover an extensive range of invaluable information, but it should be available from a variety of sources. Examples include—
  - Demographic information on the HN population.
  - Societal information on the HN.
  - The locale and disposition of HN police forces.
  - Historical crime data of the area.
  - Environmental information (terrain and weather factors).
  - Local newspaper articles.
  - Internet sources on all of the above.

NOTE: Sources for this type of information can come from intelligence estimates provided by the higher headquarters' S2/G2, other military units, or the HN.

- **OPSEC information.** This information includes the disposition of US forces (to include MP, MI, CID, and other services' investigating units) deploying or being used within the AO.
- **Other working aids.** These may include items such as computers, printers, boards, and acetate that will be necessary depending on the missions the MP are undertaking.

DETERMINE INFORMATION REQUIREMENTS AND COORDINATE ACCESS TO DATA

E-4. Once the mission analysis is completed, the next step is to determine if there are information gaps. An information gap is a missing piece of information that is critical to the analysis. MP leaders must identify and prioritize the gaps using the time, the available resources, and the commander's intent. Once an information gap is identified, it becomes PCIR. Examples of PCIR are as follows:

- Where is the nearest police station, and how did they handle a certain problem?

- What is the composition of the two gangs operating in our AO, and how can they influence US operations?
- Why is river-crossing point Bravo a danger to US crossing forces? Why are local rioters disrupting only US operations on the east bank?

E-5. The next step is to determine if PCIR can be answered with data already available. To do this, the available military and nonmilitary sources must be identified. Some sources may include—

- **USACIDC.** The USACIDC is responsible for programs such as combating terrorism, CRIMINTEL, personal-security assessment, crime analysis, and LOGSEC threat assessment. Some of the information gathered by the USACIDC (although releasable under existing controls and restraints) may help with the PIAP mission. In fact, USACIDC special agents may already be working in the AO and may be available to help the MP with the collection effort.
- **MI element.** Like the USACIDC, the MI element operating in the AO may have come across a piece of information (police or criminal) that may help accomplish the PIAP mission. The MP must make daily contact with the S2/G2 and coordinate access to invaluable information. As with the USACIDC, available intelligence information may only be releasable under strict controls and restrictions imposed by the commander.
- **Other MP units.** Coordination with an adjacent MP unit may prove to be beneficial, especially if the other MP unit has conducted a similar PIAP mission or has conducted operations in the AOI.
- **HN law-enforcement agencies.** Local law-enforcement agencies will provide the essential populace knowledge otherwise not available to US forces. Chiefs of police, sheriffs, and other key community leaders may help MP leaders fill the information need of the PIAP mission.
- **Joint, interagency, and multinational forces.** Close and effective liaison with these elements will result in quick access to information that can produce the same benefits as the agencies listed above.

NOTE: If, after consulting with all available sources, the PCIR is not resolved, the MP leader must decide the best collection effort to solve the PCIR.

RECOMMEND AND SUPERVISE THE POLICE/CRIMINAL COLLECTION EFFORT

E-6. The collection effort is the means by which specific PCIR will be met. This implies developing a collection strategy, tasking specific collectors, and supervising the collection effort.

- **Develop a collection strategy.** After a through study of the availability, capability, and disposition of the potential collecting resources, MP leaders select which asset is better suitable to perform the mission. Are organic MP the best collectors for this mission? Or do we need to request support from USACIDC units? What unique or organic capabilities do each bring to the effort? Which MP function is the best? Do we conduct MMS, AS, or L&O operations? Part of the collection strategy includes coordinating with the S2/G2, the SJA, the

CID, and other agencies before launching the collection effort. This coordination will eliminate duplication of effort, interference with an ongoing effort, or stepping out of legal limits.

- **Task or request specific collectors.** The missions that will be tasked to respective collectors must be determined. Information collectors can be tasked with more than one mission at a time. However, it is imperative that their tasks be prioritized based on mission requirements and time available. The appropriate tasking or request chain must be used to request an MP team, a platoon, or USACIDC special agents.
- **Supervise the collection effort.** The collectors must be provided with reporting guidelines. How often should they report? Should they report "no information?" Specific reporting instructions should be provided, including how, when, and where they report. A collection tasking chart should also be used (Figure E-2). As more and more collectors are being used, it becomes important to track their missions, capabilities, and success. A brief note on the reporting instructions of each collector should be included.

PIAP Missions	Unit Tasked	Echelon Employed	POC	Collection Strategy Employed	Date Collection Started	Location	Date Collection Ended
PCIR 1							
PCIR 2							
PCIR 3							
PCIR 4							

Figure E-2. Collection Tasking Chart

PROCESS POLICE/CRIMINAL RAW DATA

E-7. This processing step converts raw data into police intelligence through analysis. The key to processing the data is to understand it. For example, as collectors report raw data it may often appear to be meaningless information. Rarely will one collector receive all of the necessary information to answer the PCIR at one time. Instead, a variety of information must be fitted together to form a coherent, even if incomplete, picture.

E-8. Every piece of data is important. This is the cardinal rule of the entire process. Until the data is analyzed for reliability and compared with other data, all reports must be treated equally. This is important to remember, for the trend will be to use data that either fits your expectations or gives the most complete picture.

E-9. A report of "no activity" can be as important as reports with information. If the collector is capable of collecting the information and is actively trying to collect, reports of "no activity" can be critical. First, if there is no indication of a change, they can be used to reprioritize PIAP missions. Second, they can be used to retask the collectors to other areas. It is important to check the mission's original intent periodically to ensure that it is still valid and merits further efforts.

E-10. To determine the reliability of the data, the MP leader must evaluate the source. Does the information come from the police chief, local youth, or other military forces? One of the best ways to determine the reliability of the information is to have more than one collector reporting on the same PCIR. If more than one collector reports the same information, it can normally be presumed to be accurate.

E-11. Information can be analyzed in a variety of ways—chronologically, geographically, by the impact on the force, by the type of activity or the modus operandi, or by the association of participants. The MP leader must evaluate the information after it is analyzed to determine whether the information collected answers the PCIR or whether it helps accomplish the PIAP mission. If the information does not answer these questions, he must decide whether it is necessary to task other collectors, retask the same collectors, or wait until the situation develops.

REPORT AND/OR DISSEMINATE POLICE AND CRIMINAL ASSESSMENT

E-12. Reporting and disseminating is the most important aspect of the PIAP. Collection and analysis is valueless unless the data is communicated to commanders who can best use the information.

E-13. Reporting requirements must be established based on the unit's SOP and other established processes. MP leaders must decide if the information collected must be sent using an established priority format or if it can wait to be sent using SITREPs. Does the information collected affect the force-protection measures or does it immediately impact the tactical scenario? Do we need to alert units operating in the vicinity of AA Fox to avoid sector 2? Additionally, MP leaders must ensure that any information released follows existing constraints and regulations.

E-14. It is also advantageous to provide feedback to the collectors. Collectors often receive no feedback on their performance. Informing them as to which part of their information was vital and beneficial can often increase their productivity in future operations.

SAMPLE SCENARIO

E-15. The following example helps illustrate the implementation of all six steps of the PIAP:

E-16. While deployed in support of Operation Control Chaos, US forces assigned to TF Blue Thunder are tasked with stability and support operations in the city of Corbina. This city, although currently implementing a peace treaty, was the site of extreme civil unrest. Even though the overall situation is not as chaotic as it was before US intervention, there are some sectors that are still in turmoil. Common crimes (such as robberies, assaults, looting, and vandalism) still occur daily and there is fear that this condition may spread out of control throughout the city. The TF commander knows that in order to bring peace to the entire city, he must attempt to stabilize the situation in some of these bad sectors. He decides that the best way to deal with this situation is to assign a sector (sector 51) to an MP battalion in an attempt to control the situation. If the MP battalion succeeds, he may use the same concept throughout the city. He is aware of the five MP functions and how they can help bring this sector under some control. Under this framework, the TF commander issues the following mission: *On order, the MP battalion will conduct stability and support operations in the vicinity of sector 51, the city of Corbina, in support of TF Blue Thunder.* Upon receipt of this mission, the battalion commander gathered his staff and initiated the planning process. Before the MP staff could initiate the PIAP, they made contact with the S2/G2 and received an updated IPB estimate. Now that the IPB estimates are on-hand, the staff initiates the PIAP.

DETERMINE THE SCOPE OF THE PIAP

E-17. The staff performs the following steps to determine the scope of the PIAP:

- **Conduct a mission analysis.** The staff looks at the given mission, the higher echelon commander's intent, and the concept of operations. After considering all of these factors, they produce a restated mission—*On order, the battalion will conduct MP operations in sector 51 to determine the type of threat and to bring this sector under control in support of TF Blue Thunder.*
- **Prioritize PIAP missions.** Since there is only one mission, this subtask is not applicable.
- **Determine the AO and the AOI.** The staff determines that the AO is the entire sector 51 and that the AOIs are the sectors adjacent to 51.
- **Determine key terrain.** The staff determines that the key terrains are the market place, the plaza, and the city park. They select these areas because most of the incidents occurred in these populated areas. Stabilization in these key areas may be advantageous to mission success.
- **Determine required information products/sources.** The end result is the stabilization of sector 51. Minimize the crime incidents and restore L&O. Information should come from the local police chief, local community leaders, civil affairs, PSYOP, and the CID.

ASSEMBLE WORKING AIDS

E-18. The staff assembles the following working aids in support of the PIAP:

- **Maps.** The staff acquires local maps that detail the market place, the plaza, and the city park. Although the maps are in a foreign language, the staff is able to identify these three prominent features and surrounding areas.
- **Language aids.** In addition to dictionaries, the staff is able to get two translators from civil affairs.

- **Open-source information.** Open-source information includes a variety of encyclopedias and country books. Newspaper reports and historical crime data is available via the internet or through MI sources. The emphasis is in obtaining information on the history and composition of the local police force.
- **Classified intelligence.** The staff receives a briefing from the S2/G2 describing the operations of SOF units in the area.
- **Other working aids.** Computers, charts, templates, and acetate are available.

DETERMINE INFORMATION REQUIREMENTS AND COORDINATE ACCESS TO DATA

E-19. After a careful analysis of all available information, the following information gaps were turned into PCIR:

- Why is the police force not actively engaged in L&O?
- What is the major threat to civilians in each area?
- Are local gangs involved?
- Is their local judicial system in place?
- Where are prisoners housed?

E-20. The staff consults with the USACIDC, MI, other MP units, and other sources but determines that the information provided is not clear or sufficient. Since the PCIR are not resolved, the staff decides to recommend that the commander initiate a collection effort.

RECOMMEND AND SUPERVISE THE POLICE/CRIMINAL COLLECTION EFFORT

E-21. The staff performs the following steps to recommend and supervise the police/criminal collection effort:

- **Develop a collection strategy.** Upon coordinating with the S2/G2, the SJA, and the CID, the MP commander decides that the best collectors will come from within organic assets. He decides that he will employ the MP to go out and seek the needed information. The battalion commander will employ a company commander to interview the local police chief and an MP/interpreter will patrol around the designated AOI. The MP commander will need additional interpreters to help him with his collection effort.
- **Task or request specific collectors.** The MP staff requests an interpreter through the normal chain of command. The TF G3 assigns six interpreters for a period of 90 days.
- **Supervise the collection effort.** All collectors are instructed to report not later than (NLT) 1800 every day. The report will be in the form of an "end of day" outbrief to the S2 and the battalion commander. A collection tasking chart (Figure E-3) will also be completed.

PROCESS POLICE/CRIMINAL RAW DATA

E-22. After receiving all of the above information, the S2 concludes that the inability of the police force to do their job has had a domino effect on all areas of peace and order. This information is confirmed through other reports from information collected by the CID (through their own efforts) and the CA. The information also reveals that the gangs have intentions of attacking US forces if they interfere with their "business." The S2 determines that although the situation may seem to be chaotic, once the locals are able to trust the police forces, everything should return to normal.

REPORT AND/OR DISSEMINATE POLICE AND CRIMINAL ASSESSMENT

E-23. Since the collection effort identifies a potential threat to US forces, the commander decides to notify higher headquarters and increase his own force-protection measures. He requires MP elements patrolling in designated areas to be no less than a squad. If patrolling dismounted, a HMMWV or an ASV will trail the squad. The MP commander uses the information provided during the PIAP to develop his implementation plan. This plan includes joint patrols, training, and construction of a joint MP/local national police station.

PIAP Mission: Bring Sector 51 Under Control								
PCIR	Unit Tasked	Echelon Employ-ed	POC	Collection Strategy Employed	Date Collection Started	Location	Date Collection Ended	PCIR Result
Why is the police force not actively engaged in L&O operations?	A Co	2 teams	CPT Smith	Direct interview	31 Jan 00	Sector 51, Police HQ	2 Feb 00	The police force does not know how to interact with the population.
What is the major threat to civilians in each area?	A Co	6 teams	1LT Thomas	Patrol and interview	31 Jan 00	Market place, plaza, and park	7 Feb 00	Muggings, robberies, and stabbings are the major threats to civilians.
Are local gangs involved?	A Co	2 teams	CPT Smith	Direct interview	31 Jan 00	Sector 51, Police HQ	2 Feb 00	Two rival gangs exist.
Is a local judicial system in place?	A Co	2 teams	1LT Jones	Direct interview	31 Jan 00	Courthouse	3 Feb 00	A judicial system exists, but the judge is afraid to sentence anyone.
Where are the prisoners housed?	A Co	4 teams	1LT Lucas	AS and MMS	31 Jan 00	Local prison	2 Feb 00	The local prison is abandoned.

SAMPLE

Figure E-3. Sample Collection Tasking Chart

Glossary

1LT first lieutenant

A/L administrative and logistics

AA assembly area

AATF air-assault task force

ABCS Army Battle Command Systems

ACE analysis and control element

ACIRS Automated Criminal Investigative Reporting System

ACT analysis control team

ADA air-defense artillery

ADC area damage control

AFATDS Advanced Field-Artillery Tactical Data System

ALO authorized level of organization

AMDPCS Air- and Missile-Defense Planning and Control System

AO area of operation

AOE Army of Excellence

AOI area of interest

AOR area of responsibility

APOD aerial port of debarkation

APOE aerial port of embarkation

AR armored

AR Army regulation

ARFOR Army forces

AS area security

ASAS All-Source Analysis System

ASCC Army service component command

ASG area support group

ASP ammunition supply point

ASR alternate supply route

ASV armored security vehicle

ATCCS Army Tactical Command and Control System

attn attention

AWIS Army WMCCS Information System

BCT brigade combat team

bde brigade

BFA battlefield functional area

BLD brigade liaison detachment

bn battalion

BOS battlefield operating system

BRT brigade reconnaissance troop

BWA battlefield workload analysis

C^2 command and control

C^3 command, control, and communications

C4I command, control, communications, computers, and intelligence

C4IFTW command, control, communications, computers, and intelligence for the warrior

CAS close air support

CASE corps-area support element

CCIU computer-crime investigative unit

cdr commander

CG commanding general

CHA corps holding area

CI civilian internee

CID Criminal Investigation Division

CIDR Criminal Investigation Division Regulation

CINC commander in chief

CIP Criminal Intelligence Program

cmd command

CMOC civil-military operations center

co company

COA course of action

CofS Chief of Staff

commo communications

COMMZ	communications zone
conf	confinement
CONUS	continental US
COSCOM	corps support command
CP	command post
CPT	captain
CRC	Crime Records Center
CRIMINTEL	criminal intelligence
CRS-3	Correctional Reporting System
CS	combat support
CSA	corps storage area
CSE	corps support element
CSG	corps support group
CSM	command sergeant major
CSS	combat service support
CSSCS	Combat Service Support Control System
cust	customs
CZ	combat zone
DA	Department of the Army
DC	District of Columbia
DCCP	division central collection point

DCG deputy commanding general

DIBRS Defense Incident-Based Reporting System

DISCOM division support command

div division

DLEA domestic law-enforcement agency

DMMC division materiel management center

DNA deoxyribonucleic acid

DOD Department of Defense

DOJ Department of Justice

DOS Department of State

DPM deputy provost marshal

DRA division rear area

DS direct support

DSA division support area

DSB division support battalion

DSE division support element

DSS decisive, shaping, and sustaining

DZ drop zone

EAC echelons above corps

EAD echelons above division

ECB echelons corps and below

EG escort guard

EMS emergency medical services

EPW enemy prisoner of war

ERP engineer regulation point

ES escort

FA field artillery

FBCB2 Force XXI Battle Command—Brigade and Below

FBI Federal Bureau of Investigation

Feb February

FEBA forward edge of the battle area

FIU field investigative unit

FM field manual

FOB forward operating base

FOIA Freedom of Information Act

FORSCOM US Army Forces Command

FSB forward-support battalion

FSC forward-support company

G1 Assistant Chief of Staff, G1 (Personnel)

G2 Assistant Chief of Staff, G2 (Intelligence)

G3 Assistant Chief of Staff, G3 (Operations and Plans)

G4 Assistant Chief of Staff, G4 (Logistics)

G5 Assistant Chief of Staff, G5 (Civil Affairs)

G6 Assistant Chief of Staff, G6 (Communications)

GAC ground-assault convoy

GCCS Global Command and Control System

GCCS-A Global Command and Control System—Army

GD guard

GO government organization

GS general support

HHB headquarters and headquarters battery

HHC headquarters and headquarters company

HHD headquarters and headquarters detachment

HMMWV high-mobility, multipurpose, wheeled vehicle

HN host nation

HQ headquarters

HRP high-risk personnel

HVA high-value asset

I/R internment and resettlement

IBCT initial/interim brigade combat team

IC information center

IEW intelligence and electronic warfare

IG inspector general

IHO international humanitarian organization

IMETP International Military Education and Training Program

IO international organization

IPB intelligence preparation of the battlefield

IRIC internment/resettlement information center

ISB intermediate staging base

ITO installation transportation officer

J3 Operations Directorate

Jan January

JFC joint-force command

JOA joint-operations area

JP joint publication

JRA joint rear area

JSOA joint special operations area

JTF joint task force

km kilometer(s)

L&O law and order

LID light infantry division

liq liquid

LOC lines of communication

LOGSEC logistics security

LRP long-range plan

LSTA logistics-security threat assessment

LVOSS Light-Vehicle Obscuration Smoke System

LZ landing zone

M/O marksman/observer

M/S mobility/survivability

MACOM major Army command

MAJ major

MCRP Marine Corps reference publication

MCS Maneuver Control System

MDMP military decision-making process

mech mechanized

MEDEVAC medical evacuation

METL mission-essential task list

METT-TC mission, enemy, terrain, troops, time available, and civilian considerations

MEVA mission-essential or vulnerable area

MFPU major fraud procurement unit

MI military intelligence

MILCOM military community

min minimum

MKT mobile kitchen trailer

MLRS	Multiple-Launch Rocket System
MMS	maneuver and mobility support
MOB	main operating base
MOOTW	military operations other than war
MP	military police
MPBTF	military police brigade task force
MPMIS	Military Police Management Information System
MQS	military qualification standards
MSR	main supply route
MTOE	modified table of organization and equipment
MTP	mission training plan
MTT	mobile training team
MWD	military working dog
NATO	North Atlantic Treaty Organization
naut	nautical
NBC	nuclear, biological, and chemical
NCO	noncommissioned officer
NEO	noncombatant evacuation operation
NG	national guard
NGO	nongovernmental organization
NIBRS	National Incident-Based Reporting System

NLT	not later than
NLT	nonlethal weapon
NWP	Naval warfare publication
OCONUS	outside the continental US
OPCON	operational control
OPLAN	operation plan
opns	operations
OPSEC	operations security
ORS-2W	Offense Reporting Systems Windows
PASE	port-area support element
PCIR	police/criminal information requirements
PDF	Panamanian Defense Force
PEO	peace-enforcement operation
PI	police intelligence
PIAP	police information assessment process
PIO	police intelligence operations
PIR	priority intelligence requirements
PKO	peacekeeping operation
plt	platoon
PM	provost marshal
POC	point of contact

POD	port of debarkation
POE	port of embarkation
POL	petroleum, oil, and lubricants
PRC	populace and resource control
PSD	protective security detail
PSU	protective-service unit
PSVA	personal-security vulnerability assessment
PSYOP	psychological operations
pub	publication
PWIS-3	Prisoner-of-War Information System
PZ	pickup zone
R&S	reconnaissance and surveillance
RAOC	rear-area operations center
RC	reserve component
ROC	rear operations center
ROE	rules of engagement
rqmt	requirement
RSOI	reception, staging, onward movement, and integration
RSTA	reconnaissance, surveillance, and target acquisition
RWS	remote workstation
S1	Adjutant (US Army)

S2 Intelligence Officer (US Army)

S3 Operations and Training Officer (US Army)

S4 Supply Officer (US Army)

S5 Civil-Affairs Officer (US Army)

S6 Communications Officer (US Army)

SALUTE size, activity, location, unit, time, and equipment

SAW squad automatic weapon

SCI sensitive compartmented information

sec section

SEP separate

SFC sergeant first class

SITREP situation report

SJA staff judge advocate

SMS Security Management System

SOC security-operations cell

SOF special-operations forces

SOFA status of forces agreement

SOI signal operating instructions

SOP standing operating procedures

SPOD seaport of debarkation

SPOE seaport of embarkation

SPOTREP	spot report
spt	support
sq	square
sqd	squad
SRT	special-reaction team
STAMIS	Standard Army Management Information System
stat	statute
STCCS	Strategic Theater Command and Control System
TA	theater Army
TAA	tactical assembly area
TAACOM	Theater Army Area Command
TAC	tactical
TACCIMS	Theater Automated Command and Control Information Management System
TAI	traffic accident investigation
TASE	TSC-area support element
TC-AIMS	Transportation Coordinator Automated Movement Server
TCF	tactical combat force
TCP	traffic-control post
TF	task force
THREATCON	threat condition
TM	technical manual

TMO traffic-management officer

tms teams

TO theater of operations

TOE tables of organization and equipment

TRADOC US Army Training and Doctrine Command

TRP traffic-regulation plan

TSA theater storage area

TSC theater support command

TSE tactical support element

TTP tactics, techniques, and procedures

UMT unit ministry team

UN United Nations

US United States

USACIDC US Army Criminal Investigation Command

USACIL US Army Criminal Investigations Laboratory

USAF US Air Force

USAMPS US Army Military Police School

USC US Code

USCG United States Coast Guard

USCS US Customs Service

USDA US Department of Agriculture

VIP very important person

VRS Vehicle Registration System

WMCCS Worldwide Military Command and Control System

Bibliography

AR 15-6. *Procedures for Investigating Officers and Boards of Officers.* 11 May 1988.

AR 190-12. *Military Police Working Dogs.* 30 September 1993.

AR 190-45. *Law Enforcement Reporting.* 30 September 1988.

AR 190-47. *The Army Corrections System.* 15 August 1996.

AR 195-2. *Criminal Investigation Activities.* 30 October 1985.

AR 525-13. *Antiterrorism / Force Protection (AT / FP): Security of Personnel, Information, and Critical Resources.* 10 September 1998.

CIDR 195-1. *Special Agent Operations Manual.* October 1994.

DA Form 2028. *Recommended Changes to Publications and Blank Forms.* 1 February 1974.

FM 3-0. *Operations.* To be published within the next 6 months.

FM 19-1. *Military Police Support for the AirLand Battle.* 23 May 1988.

FM 19-4. *Military Police Battlefield Circulation Control, Area Security, and Enemy Prisoner of War Operations.* 7 May 1993.

FM 19-10. *The Military Police Law and Order Operations.* 30 September 1987.

FM 19-40. *Enemy Prisoners of War, Civilian Internees and Detained Persons.* 27 February 1976.

FM 34-2. *Collection Management and Synchronization Planning.* 8 March 1994.

FM 34-130. *Intelligence Preparation of the Battlefield.* 8 July 1994.

FM 41-10. *Civil Affairs Operations.* 14 February 2000.

FM 63-4. *Combat Service Support Operations—Theater Army Area Command.* 24 September 1984.

FM 71-3. *The Armored and Mechanized Infantry Brigade.* 8 January 1996.

FM 71-100. *Division Operations.* 28 August 1996.

FM 90-13-1. *Combined Arms Breaching Operations.* 28 February 1991.

FM 90-40. *Multiservice Procedures for the Tactical Employment of Nonlethal Weapons (NLW) {MCRP 3-15.8; NWP 3-07-31; USCG Pub 3-07.31}.* 6 October 1998.

FM 100-1. *The Army.* 14 June 1994.

FM 100-5. *Operations.* 14 June 1993.

FM 100-8. *The Army in Multinational Operations.* 24 November 1997.

FM 100-10. *Combat Service Support.* 3 October 1995.

FM 100-15. *Corps Operations.* 29 October 1996.

FM 100-17-3. *Reception, Staging, Onward Movement, and Integration.* 17 March 1999.

FM 101 5. *Staff Organization and Operations.* 31 May 1997.

FM 101-5-1. *Operational Terms and Graphics (MCRP 5-2A).* 30 September 1997.

JP 3-10. *Joint Doctrine for Rear Area Operations.* 28 May 1996.

USC Title 18. *Crimes and Criminal Procedure.* 23 January 2000.

USC Title 32. *National Guard.* 5 January 1999.

Index

A

AATF. *See* air-assault task force (AATF).
ADC. *See* area damage control (ADC).
air-assault division, 7-7
air-assault task force (AATF), 7-7
air-base defense, 4-7
airborne division, 7-6
alternate supply route (ASR), 4-2, 4-5
ammunition supply point (ASP), 4-7
antiterrorism, 12-3
AOE. *See* Army of Excellence (AOE).
AOR. *See* area of responsibility (AOR).
area damage control (ADC), 4-6
area of responsibility (AOR), 1-3
area security (AS), 1-1, 3-3, 3-8, 4-3, 4-6—4-9, 5-2, 7-7, 10-1
area support group (ASG), 3-7
Army information systems, D-1
Army of Excellence (AOE), 7-5, 7-6, 7-8, 7-10, 9-5
Army service component command (ASCC), 1-4, 1-5, 1-6, 5-1, 5-5
AS. *See* area security (AS).
ASCC. *See* Army service component command (ASCC).
ASG. *See* area support group (ASG).
ASP. *See* ammunition supply point (ASP).
ASR. *See* alternate supply route (ASR).
asymmetric threat, 3-5
augmentation, 7-4, 7-5, 8-1, 8-2, 8-4, 11-6
automated systems, D-5

B

band, 7-3
base cluster, 3-7, 4-7
base defense, 4-7
battle command, 1-2, 2-1
battlefield,
    circulation and control, 1-1
    effects, 4-14
    environment, 4-14
    missions, 1-2
    organization, 1-4—1-7
    visualization, 2-1—2-2
battlefield workload analysis (BWA), 1-7, C-1
BCT. *See* brigade combat team (BCT).
breaching operations, 4-2, 4-3, 7-4
brigade combat team (BCT), 7-7
BWA. *See* battlefield workload analysis (BWA).

C

C^2. *See* command and control (C^2).
capabilities, 1-2
CCIU. *See* computer-crime investigative unit (CCIU).
centralized MACOM, 9-2
CI. *See* civilian internee (CI).
CID. *See* Criminal Investigation Division (CID).
CIP. *See* Criminal Intelligence Program (CIP).
civil affairs, 4-5, 4-13
civil disturbances, 3-3, 4-11
civil-military operations center (CMOC), 1-8, 4-13
civilian internee (CI), 1-7, 1-8, 4-9, 4-10, 5-4, 5-5
CMOC. *See* civil-military operations center (CMOC).
COA. *See* course of action (COA).
collection point, 4-4, 6-5, 7-5
collection strategy, E-4
combat zone (CZ), 1-4, 1-7, 4-4
combating terrorism, 11-8—11-9
Combating Terrorism Program, 4-13
command and control (C^2), 2-2, 7-2, 9-2
commander's intent, 2-2
communications zone (COMMZ), 1-4, 1-5, 3-6, 4-4, 5-2
COMMZ. *See* communications zone (COMMZ).
computer-crime investigative unit (CCIU), 9-6
concept of operations, 2-2
confinement, 4-10
contiguous, 1-4, 3-7
control element, 6-3
convoy security, 4-8
corps rear area, 3-6
corps support command (COSCOM), 5-2, 6-4
corps support group (CSG), 3-7
COSCOM. *See* corps support command (COSCOM).
counterdrug operations, 11-7
countering the threat, 3-7
course of action (COA), 2-3, 4-14
CRC. *See* Crime Records Center (CRC).
Crime Records Center (CRC), 9-7—9-8
crime threat analysis, 4-13
criminal activities, 3-3
criminal intelligence (CRIMINTEL), 9-1, 9-2, 9-3, D-6
Criminal Intelligence Program (CIP), 4-13
criminal investigation, 4-11—4-12, 9-2
Criminal Investigation Division (CID), 1-6, 3-6
CRIMINTEL. *See* criminal intelligence (CRIMINTEL).
CSG. *See* corps support group (CSG).
customs team, 1-5, 1-7, 4-11, 5-2
CZ. *See* combat zone (CZ).

D

defeat, 4-6
defense plans, 3-7
delay, 4-6
detainees, 11-17
digitization, 7-11
direct support (DS), 2-3

disaster,
man-made, 1-2
natural, 1-2
relief, 11-4—11-6
DISCOM. *See* division support command (DISCOM).
dislocated civilian, 1-7, 1-8, 4-4, 4-5, 6-5, 11-17
distributed, 7-8
diversionary attacks, 3-6
division MP company, 7-2—7-3
division rear area (DRA), 3-6
division support area (DSA), 7-5
division support command (DISCOM), 5-2, 7-2
Division XXI, 7-1, 7-8—7-11
divisional brigade, 8-2
doctrinal publications, B-1
doctrine, B-1
DRA. *See* division rear area (DRA).
DS. *See* direct support (DS).
DSA. *See* division support area (DSA).

E

EAC. *See* echelons above corps (EAC).
echelons above corps (EAC), 5-1
effective coordination, 11-5
enemy-controlled agents, 3-3
enemy priority target, 3-5
enemy prisoner of war (EPW), 1-1, 1-7, 1-8, 4-9, 4-10, 5-4, 5-5, 7-11
enemy sympathizers, 3-3
EPW. *See* enemy prisoner of war (EPW).
escort-guard company, 1-7, 5-6
espionage, 3-3

F

FBCB2. *See* Force XXI battle command—brigade and below (FBCB2).
FEBA. *See* forward edge of the battle area (FEBA).
field investigative unit (FIU), 9-6
FIU. *See* field investigative unit (FIU).
FOB. *See* forward operating base (FOB).
force-projection doctrine, 1-2
force-protection, 1-2, 3-3, 4-11, 7-11, 12-1
Force XXI, 1-3, 9-5
Force XXI battle command—brigade and below (FBCB2), 7-10
forced-entry operations, 7-1
forensic laboratory, 9-7
forward edge of the battle area (FEBA), 3-5
forward operating base (FOB), 7-6, 7-7
forward support battlation (FSB), 7-10
forward support company (FSC), 7-10
FSB. *See* forward support battalion (FSB).
FSC. *See* forward support company (FSC).
functional-area considerations, 8-3

G

general support (GS), 2-3
Geneva convention, 5-5
GO. *See* government organization (GO).
government organization (GO), 1-8
GS. *See* general support (GS).
guard company, 1-7
guerilla forces, 3-3

H

heavy division, 7-4—7-5
high-risk personnel (HRP), 4-6, 4-8, 12-2
high-value asset (HVA), 3-5, 3-6, 7-10
holding area, 4-2, 6-4
HRP. *See* high-risk personnel (HRP).
humanitarian assistance, 11-4—11-6
HVA. *See* high-value asset (HVA).

I

I/R. *See* internment and resettlement (I/R).
IBCT. *See* initial/interim brigade combat team (IBCT).
IHO. *See* international humanitarian organization (IHO).
initial-entry force, 8-1
initial/interim brigade combat team (IBCT), 1-7, 8-1, 8-2—8-5
integrated, 7-8
intelligence preparation of the battlefield (IPB), 2-3, 4-13, 4-14, 6-3, E-1
interagency operations, 1-7
intermediate staging base (ISB), 7-6
international humanitarian organization (IHO), 1-8
international organization (IO), 1-8
internment and resettlement (I/R), 1-2, 4-9—4-10, 10-1
investigative team, 9-5
IO. *See* international organization (IO).
IPB. *See* intelligence preparation of the battlefield (IPB).
ISB. *See* intermediate staging base (ISB).

L

L&O. *See* law and order (L&O).
language aids, E-3
law and order (L&O), 1-1, 1-5, 4-10—4-12
law enforcement, 4-11, 12-2—12-3, E-4
LID. *See* light infantry division (LID).
light infantry division (LID), 7-5—7-6
lines of communications (LOC), 1-4, 4-2, 5-2, 6-4
LOC. *See* lines of communication (LOC).
logistics security (LOGSEC), 4-12, 7-1, 9-1, 9-2
logistics security threat assessment, 4-13
LOGSEC. See logistics security (LOGSEC).
long-range plans (LRP), 6-2
LRP. *See* long-range plans (LRP).

M

M/O. *See* marksman/observer (M/O).
main supply route (MSR), 4-2, 4-3, 4-5, 5-3
maneuver and mobility support (MMS), 1-2, 3-8, 4-2—4-6, 4-11, 5-2, 7-7, 10-1
marksman/observer (M/O), 4-11
MDMP. *See* military decision-making process (MDMP).
METL. *See* mission-essential task list (METL).
METT-TC. *See* mission, enemy, terrain, troops, time available, and civilian considerations (METT-TC).

MEVA. *See* mission-essential or vulnerable area (MEVA).
MI. *See* military intelligence (MI).
military decision-making process (MDMP), 7-10
military intelligence (MI), 3-6, E-4
military operations other than war (MOOTW), 3-3, 9-4, 9-6
military police
  automated systems, D-4
  availability, 3-9
  battalion (CS), 11-15
  battalion (I/R), 5-6—5-8
  brigade, 11-13
  brigade (CS), 5-2—5-4, 6-1—6-2
  brigade (I/R), 5-4—5-6
  company, 11-15
  dispersion, 3-9
  functions, 1-2, 4-1
  I/R units, 11-16—11-17
Military Police Management Information System (MPMIS), D-5
military prisoners, 4-10, 5-4
military working dog (MWD), 4-5, 4-7, 4-11, 4-12, 5-2, 5-3, 6-2
mission analysis, E-2
mission, enemy, terrain, troops, time available, and civilian considerations (METT-TC), 1-5, 1-7, 1-8, 2-3, 4-3, 5-2, 5-3, 6-4, 7-2, 9-5
mission-essential or vulnerable area (MEVA), 12-2
mission-essential task list (METL), 8-3
MMS. *See* maneuver and mobility support (MMS).
modified table of organization and equipment (MTOE), 1-5
MOOTW. *See* military operations other than war (MOOTW).
MPMIS. *See* Military Police Management Information System (MPMIS).
MSR. *See* main supply route (MSR).
MTOE. *See* modified table of organization and equipment (MTOE).
multidimensional, 7-8
multifunctional capabilities, 1-8
multinational operations, 1-7
MWD. *See* military working dog (MWD).

N

NBC. *See* nuclear, biological, and chemical (NBC).
NEO. *See* noncombatant evacuation operation (NEO).
neutrality, 11-5
NGO. *See* nongovernmental organization (NGO).
NLW. *See* nonlethal weapon (NLW).
noncombatant evacuation operation (NEO), 11-9
noncontiguous, 1-4, 3-7
nongovernmental organization (NGO), 1-6, 1-8
nonlethal weapon (NLW), 11-13
nonlinear, 7-8
nuclear, biological, and chemical (NBC), 3-4, 4-4

O

objectivity, 11-5
OPCON. *See* operational control (OPCON).
open-source information, E-3
operational control (OPCON), 2-2, 4-2, 5-2
operational framework, 1-3—1-4
operational unity, 11-5
operations,
  combat, 1-7, 2-3, 4-2, 6-3
  combat service support (CSS), 1-7, 2-3, 4-2, 6-3
  combat support (CS), 1-7, 2-3, 4-2, 6-3
  confinement, 5-7—5-8
  decisive, shaping, and sustaining (DSS), 1-4
  defensive, 10-1, 10-3
  offensive, 10-1—10-2
  protective service, 9-3
  rear area, 3-1—3-2, 3-7, 4-7, 4-8
  reconnaissance, 3-6, 4-6—4-7
  resettlement, 5-7
  river crcossing, 4-2
  security, 5-2
  security-assistance, 11-9—11-11
  stability, 1-2, 8-4, 11-1, 11-3, 11-15
  support, 1-2, 8-4, 11-1, 11-3, 11-15
  sustainment, 3-1—3-2, 3-7, 4-7, 4-8
operations security (OPSEC), 1-8, E-3
OPSEC. *See* operations security (OPSEC).

P

passage of lines, 4-2, 4-3—4-4
PCIR. *See* police/criminal information requirements (PCIR).
peacekeeping operation (PKO), 11-11—11-13
peacetime engagement operations, 3-1
personal-security vulnerability assessment (PSVA), 4-13
physical security, 3-3, 12-1—12-2
PIAP. *See* police information assessment process (PIAP).
PIO. *See* police intelligence operations (PIO).
PIR. *See* priority intelligence requirements (PIR).
PKO. *See* peacekeeping operation (PKO).
planning element, 6-3
POD. *See* port of debarkation (POD).
POE. *See* port of embarkation (POE).
police/criminal information requirements (PCIR), 4-17
police information assessment process (PIAP), 2-3, 4-15, 4-17, 6-3, E-1
police intelligence operations (PIO), 1-2, 3-8, 4-13—4-18
  active mode, 4-17
  passive mode, 4-15
populace and resource control (PRC), 4-10
port of debarkation (POD), 5-6
port of embarkation (POE), 5-6
positive image, 11-5
Posse Comitatus Act, 11-3
PRC. *See* populace and resource control (PRC).
precise, 7-8
priority intelligence requirements (PIR), 4-17
protective-service unit (PSU), 9-6
provost marshal (PM) cell, 2-3
PSU. *See* protective-service unit (PSU).
PSVA. *See* personal-security vulnerability assessment (PSVA).
psychological operations (PSYOP), 4-13, 4-17

PSYOP. *See* psychological operations (PSYOP).
public acceptance, 11-5

R

R&S. *See* reconnaissance and surveillance (R&S).
raid, 4-11
RAOC. *See* rear-area operations center (RAOC).
rear area, 3-1
rear-area operations center (RAOC), 5-3, 5-1, 6-5
rear-area security, 4-7
reception, staging, onward movement, and integration (RSOI), 3-2—3-3
reconnaissance, 4-2, 4-5
  area, 4-6
  zone, 4-6
reconnaissance and surveillance (R&S), 4-5, 7-11, 10-3
reconnaissance, surveillance, and target acquisition (RSTA), 8-3
refugees, 1-8, 6-5
replacement unit, 4-4
response force, 3-7—3-8, 4-6, 4-8
river-crossing operations, 4-2, 7-4
ROE. *See* rules of engagement (ROE).
RSOI. *See* reception, staging, onward movement, and integration (RSOI).
RSTA. *See* reconnaissance, surveillance, and target acquisition (RSTA).
rules of engagement (ROE), 11-1, 11-9

S

sabotage, 3-3
security, 4-8
security force, 4-7
separate brigades, 8-1—8-2
simultaneous, 7-8
situational awareness, 2-1, 3-6
small tactical units, 3-4
special-reaction team (SRT), 4-11
SRT. *See* special-reaction team (SRT).
staff relationships, 2-3
Stafford Act, 11-3
staging area, 4-2
straggler control, 4-4, 6-5
subversion, 3-3
support functions, 3-1
support relationships, 2-3

T

TAA. *See* tactical assembly area (TAA).
tables of organization and equipment (TOE), 7-2, 7-3
tactical assembly area (TAA), 7-7
tactical combat force (TCF), 3-8
tactics, techniques, and procedures (TTP), B-1—B-2
TCF. *See* tactical combat force (TCF).
TCP. *See* traffic control post (TCP).
terrorism, 3-3, 12-3
theater of war, 1-4, 1-5
theater support command (TSC), 1-4, 1-5, 1-6, 5-1
threat,
  awareness, 11-5
  level, 3-3—3-4
  location, 3-5—3-7
  priorities, 3-4—3-5
threat-level matrix, 3-4
TOE. *See* tables of organization and equipment (TOE).
traffic-control post (TCP), 4-2, 4-3, 4-5
traffic-regulation plan (TRP), 4-5
TRP. *See* traffic-regulation plan (TRP).
TSC. *See* theater support command (TSC).
TTP. *See* tactics, techniques, and procedures (TTP).

U

unconventional forces, 3-4
United States Army Criminal Investigation Command (USACIDC), 4-12, 9-1, E-4
  agents, 11-7
  organization, 9-4—9-6
  wartime support, 9-2—9-4
United States Army Criminal-Investigations Laboratory (USACIL), 9-7
United States Customs Service (USCS), 4-12
USACIDC. *See* United States Army Criminal Investigation Command (USACIDC).
USACIL. *See* United States Army Criminal-Investigations Laboratory (USACIL).
USCS. *See* United States Customs Service (USCS).
use of force, 11-5